CONTENTS

Appendixes

Index

PREFACE

The urge to make geology more quantitative has led to the widespread inclusion of chemistry, physics, and mathematics—the so-called "basic" sciences—among the required courses in undergraduate curricula. One can only applaud this practice, for certainly many of the new and exciting ideas in earth science have come from application of quantitative reasoning to geologic problems. It is reasonable to ask, however, if the mere requiring of courses in basic science is an efficient way to give most students facility in handling ideas quantitatively. For the few who take easily to mathematical symbolism, the answer would be "yes"; they need only be shown the tools that the basic sciences can supply, and thereafter they will instinctively put the tools to good use. But many who are attracted to geology need special incentives to keep them using the quantitative ideas learned in basic science, and too often such incentives are missing from advanced geology courses. This is not a criticism of the geology courses, for their chief purpose is to develop geological judgment rather than to teach more physics and chemistry. But students who find few immediate applications of basic science in their geological work tend to forget what they have learned in the elementary courses. They take the courses in much the way they would take doses of bad medicine, because the doctor prescribes them, but they find them largely irrelevant to their principal interests. Thus methods and concepts painfully acquired become rusty, and later in their careers, when they need them, they must turn for help to someone else. For it is an unhappy fact that the language of fundamental science slips from the memory as fast as any other acquired language if it is not frequently used.

There is need, then, for courses in the undergraduate geology curriculum specifically designed to pick up ideas from elementary physics, chemistry, and mathematics and put them to work immediately in the handling of geologic problems. This book is intended as a text for the chemical part of such a program.

It is addressed to juniors and seniors, perhaps to an occasional sophomore—to students fresh from a year of chemistry and equipped with the geological background normally provided by courses in physical and historical geology and mineralogy. Anything beyond this minimum background, in either of the sciences, would help to make the subject easier. Especially to be recommended, as a means of giving more life to some of the discussions, is at least a little contact with actual geologic problems in the field.

The plan of the book is to review, at a moderately advanced level, those topics from elementary geology to which chemical ideas may be usefully applied. Chemical concepts are developed along with their geological applications; as far as possible an effort is made to avoid both long geological descriptions and long chemical arguments devoid of geological references. Knowledge of physical chemistry is not assumed, beyond the physicochemical ideas incorporated nowadays in most elementary chemistry courses. Concepts from physical chemistry are presented as the need for them arises out of geological arguments; by the time students have finished the book, they should have a fair grasp of the kind of physical chemistry most useful in geology. No mathematics is employed more difficult than the basic calculus required to justify a few physicochemical formulas.

The part of physical chemistry called "thermodynamics" poses a problem. It is by all odds the most important branch of the subject for geochemical purposes, but on first encounter the development of its basic rules seems to many students formidably abstract. To avoid an overlong theoretical presentation, and yet to give students some convenient working tools, the book attempts a compromise. In the body of the text only a bare minimum of thermodynamics is employed: emphasis is placed on free energy as a measure of how far reactions will go; entropy is mentioned briefly and is used only as a means of calculating free energies at temperatures higher than normal; activities and fugacities are treated simply as modified concentrations and vapor pressures. Then for students who find the subject of real interest, a supplementary chapter giving somewhat greater depth is included as Appendix XI.

In keeping with the pedagogical purpose of the book, most chapters end with a list of questions. These, it should be emphasized, are an essential part of the book. In a subject like geochemistry no real understanding comes from mere reading. Ideas become part of a student's useful mental equipment only when they are reflected on, digested, and manipulated, and problems are a necessary goad to such activity. The questions include not only those giving practice in obtaining numerical results from equations but also many requiring students to think about specific geologic problems, or to evaluate geologic hypotheses, in terms of the quantitative ideas just presented. Attention is continually directed, both in problems and in the text, to the degree of relevance of chemical ideas to geological questions—to the relations between theoretical formulas and results of controlled laboratory experiments, on the one hand, and geologic field observations on the other. Only through the incentive that problems give for meditating on such relations can a student develop geochemical judgment, the feel for the kinds of geological situations in which chemistry can be usefully employed and for the kinds in which the precise answers of chemistry are misleading or meaningless.

At the end of each chapter is also a list of books and articles, including those to which reference is made in the chapter and a few others intended as an introduction to current geochemical literature. The articles were chosen more with an eye to intrinsic interest and suitability for undergraduate readers than for completeness of coverage. With only a few exceptions the papers are published in standard geological publications that should be widely available.

Students should be cautioned that the book is no more than its title implies, an *introduction* to an enormously complex subject. Material in the book should provide enough background to enable students to read contemporary geochemical literature with understanding and pleasure and to form judgments about the quality of geochemical work. But if they wish to go beyond this and undertake geochemical work on their own, they will find themselves poorly prepared. For anyone seriously interested in geochemistry the book is by no means a substitute for a thorough course in physical chemistry or for extensive reading in some of the excellent volumes now available on more specialized aspects of the field.

The dozen years that have passed since the first edition was prepared have seen a prodigious growth of geochemistry on many fronts, and a major updating is needed. More than ever it is difficult to squeeze all branches of the subject into a single volume of reasonable size. One must pick and choose among the topics to be emphasized, knowing that the choices cannot satisfy the tastes of all readers. In general, the subject matter and organization are little changed in this new edition; many discussions have been modernized, but the focus remains on applications of chemistry to classical geologic problems, rather than on specialized subjects at the forefront of the advancing science. The new edition, like the old, is intended primarily for students at an intermediate level, those who have had a good introduction to chemistry and geology, and need to see how the two subjects interact.

Major additions are a new chapter on isotope geochemistry, included at the behest of many readers; an expansion of the chapter on element distribution; and, inevitably, a recasting of many discussions to incorporate the ideas of plate-tectonic theory. Speculative parts of the book, particularly sections dealing with magmatic differentiation, the nature and origin of ore-forming solutions, metamorphic petrology, and the earth's geochemical history, have been extensively revised in the light of recent work. Technical terms and symbols have been brought up to date throughout, and several tables in the appendix have been completely redone to include more modern data. In writing about a subject that changes as rapidly as geochemistry, one cannot expect to win the race against obsolescence, but at least one can hope that a new edition will markedly narrow the gap between textbook and current research.

To students, Stanford colleagues, and many elsewhere who have used the book, I am most grateful for pointing out errors and making suggestions for improvement in the subject matter and in the writing.

Konrad B. Krauskopf

INTRODUCTION TO
GEOCHEMISTRY

CHEMICAL EQUILIBRIUM

If two substances are mixed, under particular conditions of temperature and pressure, will a reaction take place? If so, how fast and how far will it go? These are some of the basic questions of chemistry. Applied to naturally occurring materials, they are central themes of geochemistry. We shall be seeking an answer to them, in one sense or another, throughout all the discussions in this book.

What kinds of answers is it reasonable to expect? Clearly no general answer is possible; if it were, the sciences of chemistry and geochemistry would long since have completed their mission. For particular questions, the simplest answers are the empirical ones that come from ordinary experience: we know that acids react with carbonates, that organic materials burn in oxygen, that active metals liberate hydrogen from water. A great deal of our ability to make chemical predictions comes from just such qualitative background knowledge. It is often useful to refine these predictions by attaching numbers to them, numbers like equilibrium constants, solubility products, adsorption coefficients, rate constants. Or we can look behind these descriptive numbers at energy relations, and try to express the ability of substances to react in terms of the energy they possess. At still another level of abstraction, we can seek an interpretation of the energy relations in the spacing and movements of electrons within atoms. For complex systems, to which these more sophisticated and quantitative treatments cannot be applied, we may have to rest content simply with knowing which ones of several possible products or sequences of products are most likely. At the present state of knowledge the answers we seek must be tailored to particular situations, and the limitations are often frustrating. The great fascination of geochemistry lies in the effort to make the answers ever more general, more quantitative, more applicable to a wider range of natural processes.

We begin by looking at the results obtainable by applying the simplest sort of numbers to chemical reactions, the numbers called equilibrium constants.

1-1 THE "LAW OF MASS ACTION"

Two Norwegian investigators of the last century, Guldberg and Waage, stumbled almost by accident on the key to the numerical handling of chemical equilibrium. They were concerned, as we are today, with the problem of what makes a reaction go, and they were thinking in terms of then-prevalent ideas about vaguely defined "chemical forces" and "chemical affinities." Working with slow reactions, they were impressed with the effect on these reactions of changing the amounts of reactants and concluded that the "driving force" of a reaction depends on the mass of each substance. The meaning they attached to "driving force" is not very clear; they speculated that it was somehow related to reaction rate but did not make the connection explicit. The "mass" in their statement, from the context, would be better labeled "concentration." This rather obscure formulation has come down to us as "the law of mass action," which in more recent times has generally been stated, "The rate of a reaction is directly proportional to the concentration of each reacting substance."

Even this formulation is beset with difficulties. If a reaction could be depended on to take place as a result of simple molecular collisions, with some fraction of the total number of collisions leading to reaction in a given time, the statement would be reasonable and accurate. Most reactions, unfortunately, are not this simple. If reaction rates are measured experimentally, they are found in general to depend on various powers of the concentration, even sometimes on negative powers, and only rarely on the first power as the statement requires. Presumably this means that reactions take place in steps, some steps being slower than others; each step may follow the "law," but the sum of steps going at different rates does not. Hence the law of mass action probably does not merit the importance often attached to it. As a rule that holds for ideally simple reactions, and for postulated simple steps in reaction mechanisms, it has some usefulness, but as a general law for reaction rates it is misleading.

A more important result of Guldberg and Waage's work was the recognition that many chemical reactions are incomplete because they are reversible and that the same final mixture of substances is obtained when a reversible reaction is started from either end. In symbols, a reversible reaction might be represented:

$$A + B \rightleftharpoons Y + Z$$

We can mix equal amounts of A and B in one container, and equal amounts of Y and Z in another; when reaction has stopped in each, the same mixture of all four substances is present. In Guldberg and Waage's language, this meant that both reactions go until their "driving forces" become equal. If we show each driving force as proportional to concentrations,

$$\text{Driving force of forward reaction} = k_1[A][B]$$
$$\text{Driving force of reverse reaction} = k_2[Y][Z]$$

then equality of the driving forces requires that

$$k_1[A][B] = k_2[Y][Z]$$

or
$$\frac{[Y][Z]}{[A][B]} = \frac{k_1}{k_2} = K$$

Brackets indicate concentrations, k_1 and k_2 are proportionality constants, and K is another constant obtained by dividing k_1 by k_2. In other words, when a reversible reaction has gone as far as it can, the quotient obtained by multiplying the concentrations of the products and dividing by the concentrations of reactants is a constant quantity. This relation turns out to be experimentally verifiable, quite apart from any meaning we attach to "driving force" or any complexities of reaction mechanism.

In modern language we would say that a reversible reaction stops when it has reached a state of *chemical equilibrium*, that the stopping is only apparent, because the two reactions are still going on, and that the state of equilibrium is merely a balance between opposing processes going at equal rates. The K is the *equilibrium constant*, a number which defines the equilibrium condition for a particular reaction.

Since the emphasis in our dynamic picture of equilibrium is on reaction rates, it is tempting to use the law of mass action to express the rates, and so to "explain" the constancy of K. We need only substitute "rate" for "driving force" in the two equations above, and then K becomes the quotient of the two "rate constants," k_1 and k_2. Qualitatively this argument makes good sense. It is often convenient, because the effect on an equilibrium mixture of changing the concentration of one of the reacting substances can be visualized as resulting from the corresponding change in rate; thus if more A is added, the rate of the forward reaction is temporarily increased, and the final equilibrium mixture will have more Y and Z and less B than before. The reasoning is justified because an increase in the concentration of A will practically always make the reaction A + B go faster, no matter what the specific relation may be. Quantitatively, however, the reasoning is fallacious, as explained above, because only in rare cases will k_1 and k_2 correspond to physically measurable quantities. Generally, measured rates are *not* proportional to simple first powers of concentrations, or to any other predictable simple powers. In subsequent discussions equilibrium will often be visualized as a state of balance between opposite reactions going at equal rates, the rates being qualitatively dependent on concentrations, but no attempt will be made to express the rates in numbers.

For a theoretical explanation of the constancy of K as a quotient of concentrations, it is more satisfactory to start with energy relations than with reaction rates. We shall explore these energy relations in later chapters.

The history of the law of mass action is an illuminating episode in the development of scientific ideas. The law as originally stated was so vague as to have little meaning, and attempts to reframe the law in terms of reaction rates have caused much confusion. Yet it had the germ of an idea that led to the correct formulation of equilibrium constants, a discovery that marks a milestone in the development of physical chemistry. No better refutation could be found of the common thesis that science progresses by orderly logical steps from clearly defined concepts.

1-2 AN EXAMPLE OF EQUILIBRIUM: HYDROGEN CHLORIDE

A more general equation for a chemical reaction is:

$$aA + bB + \cdots \rightleftharpoons yY + zZ + \cdots \qquad (1\text{-}1)$$

where A, B, Y, Z represent chemical formulas, a, b, y, z represent coefficients, the dots stand for additional reactants or products, and the double arrow indicates that the reaction is reversible. The corresponding equilibrium constant is†

$$K = \frac{[Y]^y[Z]^z \cdots}{[A]^a[B]^b \cdots} \qquad (1\text{-}2)$$

Note that the concentration of each substance is raised to a power given by its coefficient. For the present we may take this as a generalization from experimental results, without trying to justify it on the basis of reaction mechanism.

We apply Eq. (1-2) first to mixtures of hydrogen, chlorine, and hydrogen chloride, three gases often detectable at volcanic vents. From laboratory study it is well known that mixtures of hydrogen and chlorine can be made to react explosively to form hydrogen chloride, and likewise that hydrogen chloride raised to a high temperature decomposes into its elements. In other words, the reaction represented by the equation

$$H_2 + Cl_2 \rightleftharpoons 2HCl \qquad (1\text{-}3)$$

if → direction is dominant, the K is large

---

† A K defined in this manner, as a quotient of concentrations, is only approximately constant. When activities are used in place of concentrations, the K is a true constant. Because concentrations are easier to visualize than activities, the approximate "concentration constants" will be used at first, and activities will be introduced in Chapter 3.

is known to go both forward and backward, so that equilibrium among the three gases would be shown as

$$K = \frac{[HCl]^2}{[H_2][Cl_2]} \qquad (1\text{-}4)$$

The brackets in this equation indicate concentrations of the three gases.

In what units should concentrations of gases be expressed? Obviously there is a wide choice of possible units: grams per liter, moles per liter, moles per kilogram, pounds per cubic foot, and many others. By convention, the unit ordinarily used is no one of these, but a quantity proportional to them, the *partial pressure* in atmospheres. (In a mixture of several gases, each gas behaves approximately as if it were alone in the space, and exerts a pressure independently of the others; this is its partial pressure, and the total pressure of the mixture is the sum of all the partial pressures.) The constancy of the quotient in Eq. (1-4) can be expressed equally well with any units, but to compare different constants and to standardize numerical values in tables, some arbitrary choice must be made—and for most purposes the partial pressure is the most convenient. Except where otherwise stated, the brackets around gas molecules in equilibrium-constant formulas will always indicate atmospheres of pressure (abbreviated atm).

The measured value of K in Eq. (1-4) at 1000°C is 2.6×10^8, or $10^{8.4}$. In an equilibrium mixture at this temperature, suppose that H_2 and Cl_2 are present each at 1 atm pressure. What would be the partial pressure of HCl? Substitution gives

$$K = \frac{[HCl]^2}{1 \times 1_{atm}} = 10^{8.4} \qquad \text{or} \qquad [HCl] = 10^{4.2} \text{ atm}$$

Or suppose that the concentration of HCl is 0.1 atm and that of Cl_2 $10^{-5.0}$ atm. What is the equilibrium pressure of H_2?

$$\frac{(0.1)^2}{[H_2] \times 10^{-5}} = 2.6 \times 10^8 \qquad \text{and} \qquad [H_2] = 3.8 \times 10^{-6} \text{ atm}$$

Thus, whatever the pressure of two of the gases may be, the third must have a value such that K will remain constant.

Once equilibrium is established, a change in concentration of one constituent must lead to changes in the others. For example, if equilibrium exists among H_2 and Cl_2, each at 1 atm, and HCl at $10^{4.2}$ atm, and if enough H_2 is added to make its total pressure momentarily 10 atm, then the three gases must react to bring K back to its original value. This means that more HCl must be formed by reaction between Cl_2 and some of the added H_2. If x is the amount of H_2 that reacts, then x atm of Cl_2 must also react and $2x$ atm of new HCl must be produced. Hence

$H_2 + Cl_2 \rightleftharpoons 2HCl$

$$\frac{(10^{4.2} + 2x)^2}{(10 - x)(1 - x)} = 10^{8.4}$$

$\left(10^{4.2} + 2x \right)^2$

$\left(10 - x \right) \left(1 - x \right)$

H_2 changed to 10

$(10 - x)(1 - x) = 1$

x must be less than 1

$10 - 10x = 1$

$10x = 9 \qquad x = .9$

As often happens with equilibrium constants, some simple assumptions have led to a complicated algebraic equation. This one could be solved straightforwardly as a quadratic, but time is generally saved by hunting for a shortcut. Note here that x must be less than 1, so that $2x$ is negligible in comparison with $10^{4.2}$. Omitting $2x$ from the numerator simplifies the equation a great deal:

$$(10 - x)(1 - x) = 1$$

$10 - 10x = 1$ $10x = 9$ $x = .9$

Now x is fairly small in comparison with 10; if it is neglected in the first parenthesis, then $1 - x$ is approximately equal to 0.1, and x must be about 0.9. This is sufficiently accurate for most geologic purposes. If x is not neglected in the $10 - x$ term, a slightly better value is obtained: $x = 0.89$. Hence a new equilibrium mixture is established among 9.11 atm of H_2, 0.11 of Cl_2, and $10^{4.2}$ of HCl.

It is important to sense how this sort of equilibrium reacts to any disturbance, like the addition of H_2 in the last example. Whatever changes of concentration are imposed, the equilibrium automatically adjusts itself so that K remains constant. One way to see why this happens is to recall that a chemical equilibrium is *dynamic*, maintained by a balance between opposing reactions that go on continuously. If more H_2 is added to an equilibrium mixture, the forward reaction

$$H_2 + Cl_2 \rightarrow 2HCl$$

is favored, while the reverse reaction is unaffected:

$$2HCl \rightarrow H_2 + Cl_2$$

Hence for a brief time there is imbalance, until enough new HCl is built up to make its rate of decomposition equal to its rate of formation. Alternatively, one may recall that a general property of equilibrium is to respond to a disturbance in such a way as to counteract the disturbance as far as possible. When H_2 is added, the equilibrium responds by trying to cut down its concentration, hence by favoring for a time the forward reaction over the reverse reaction. From either point of view the equilibrium is pictured as consisting of active opposing processes, always ready to compensate for changes imposed from without.

The formula for the equilibrium constant depends on how the chemical equation is stated. The HCl reaction, for example, could equally well be written backward:

$$2HCl \rightleftharpoons H_2 + Cl_2 \tag{1-5}$$

for which the equilibrium constant would be

$$\frac{[H_2][Cl_2]}{[HCl]^2} = K' \tag{1-6}$$

The new constant K' is the reciprocal of K in Eq. (1-4). Or we could write

$$\tfrac{1}{2}H_2 + \tfrac{1}{2}Cl_2 \rightleftharpoons HCl \qquad \text{and} \qquad \frac{[HCl]}{[H_2]^{1/2}[Cl_2]^{1/2}} = K'' \tag{1-7}$$

where this time K'' is the square root of the original K. Either K' or K'' could be used just as easily as K in calculating how concentrations change when one of the substances is added or removed. *An equilibrium constant has no meaning except in terms of a specific chemical equation.* One cannot speak of "the equilibrium constant of HCl," but only of the equilibrium constant for Eq. (1-3), for Eq. (1-5), for Eq. (1-7), or for some other way of writing the reaction.

1-3 THE EFFECT OF TEMPERATURE

At low temperatures hydrogen burns readily in chlorine, and if the chlorine is in excess, no detectable H_2 is left over after the burning stops. Hydrogen chloride is stable at low temperatures, showing no tendency to decompose into its elements. These observations mean that the equilibrium constant for Eq. (1-3) under such conditions is a very large number; the forward reaction is practically complete, the backward reaction goes scarcely at all. On the other hand, at 2000°C hydrogen and chlorine react only slightly and HCl is rapidly decomposed, so that the equilibrium constant for Eq. (1-3) is a small number. Other equilibria show similar changes with temperature. An equilibrium constant, therefore, must be defined for a particular temperature and has no meaning unless the temperature is specified. By convention, equilibrium constants are most commonly given for 25°C, and this number is understood unless another temperature is stated.

At room temperature, then, an equilibrium mixture of the three gases we have been discussing might consist almost entirely of HCl alone, or it might contain HCl plus an excess of either element, with the second element present only in undetectable traces. A mixture of H_2 and Cl_2 at room temperature should be far from equilibrium and should react until at least one of the elements is used up. Now laboratory experience seems to contradict this prediction, for a mixture of the two elements at room temperature appears to be inert as long as it is not touched with a flame or exposed to strong illumination. What is the difficulty?

We can say only that the reaction between H_2 and Cl_2 at ordinary temperatures is a very slow one. In terms of molecular structure, a partial explanation is that the elements can react only when the diatomic molecules H_2 and Cl_2 are broken up, and molecular collisions at room temperatures are, for the most part, not sufficiently violent to accomplish the splitting. Thus equilibrium is indeed displaced far toward HCl, but attainment of equilibrium may be indefinitely delayed because of sluggishness of the reactions. There is nothing in either the chemical equation or the equilibrium constant to warn us that slow reactions may make true equilibrium difficult to reach: the concept of equilibrium refers only to the final result, not to the means of attaining it.

This poses a real problem. If nonequilibrium mixtures can be seemingly stable, how do we recognize true equilibrium mixtures? How would we know *experimentally*, for example, that a mixture of hydrogen and chlorine with only a trace of HCl at 25°C is not actually in equilibrium? In this particular case the answer is easy, because all we need do is introduce a fairly minor disturbance—heat the mixture, touch it with a flame, expose it to light—and the resulting loud explosion will tell us emphatically that the mixture was not at equilibrium. A less spectacular but more rigorous test for equilibrium is to see whether the same mixture can be obtained by starting from the other end of the reaction, in this case by starting with pure HCl. We would find, of course, that the HCl showed no tendency to decompose into our original mixture, which tells us immediately that one or the other cannot be in equilibrium. In general, *to see if a mixture of compounds is an equilibrium assemblage, we set up experiments to see whether the same mixture can be obtained by starting with pure components at each end of the reaction.* If no reaction takes place in either direction, we are probably dealing with slow processes, and we must find some way to speed them up. Until we do get reactions to take place, we have no way of being certain whether our original mixture is at equilibrium or is apparently inert because possible reactions are very slow.

Slow reactions are always speeded up by a rise in temperature. The factors determining reaction rates are so complex that no general quantitative expression for the speeding up can be formulated. A useful qualitative rule is that most reactions have their rates doubled or trebled by a 10° (Celsius) rise in temperature. This means that equilibrium is always more easily attained at high temperatures. Even at the temperatures of molten rocks, however, all chemical reactions cannot be depended on to reach equilibrium. In future discussions we shall often be faced with the question whether a given assemblage of minerals, liquids, and gases may be considered an equilibrium mixture.

To go back to the HCl example, hot gases from volcanic vents and fumaroles can sometimes be shown to contain the three gases H_2, Cl_2, and HCl. Furthermore, the relative amounts of the three are reported to be widely different even from fumaroles of approximately the same temperature. This is good evidence that not all of the mixtures can be equilibrium assemblages—in other words, that during the violent escape of gas from lava beneath the surface, not enough time has elapsed for the gas mixture to adjust itself to changing conditions of temperature and pressure. This observation, together with the fact that the reaction of hydrogen with chlorine is exothermic (heat-producing), has suggested a possible explanation for temperature measurements showing that the surface of a lava pool may be hotter than its interior.

We shall find that most reactions involving polyatomic molecules are similar to the H_2–Cl_2 reaction in that they are sluggish at room temperature and reach equilibrium readily only at temperatures of a few hundred degrees. Reactions involving ions in water solution, however, are generally almost instantaneous even at low temperatures, because the reaction process requires no preliminary breaking up of stable molecules. Reactions of ions in silicate melts, on the other hand, may be slow, because movements of the ions are impeded by the extreme viscosity of the liquid. These generalizations have many exceptions, and we shall have to examine geologic processes individually to see whether or not equilibrium conditions can be expected in any particular circumstance.

1-4 A SECOND EXAMPLE: CARBON DIOXIDE IN WATER

The hydrogen chloride reaction has only minor interest in geology but serves as a beautifully simple illustration of the principles of equilibrium because all the substances involved are gases. More technically, we say that all mixtures of H_2, Cl_2, and HCl constitute a single *phase* (unless, of course, the temperature is low enough for chlorine or hydrogen chloride to liquefy), the word phase meaning any part of a system which is homogeneous and which is separated from other parts of the system by sharp boundaries. The equilibria of greatest concern to geologists are those involving more than one phase—liquid and gas, liquid and solid, or more complicated systems containing both liquid and gas together with several solid phases. Such equilibria are described as *heterogeneous*, in contrast to *homogeneous* equilibria like the HCl reaction.

A simple two-phase system of everyday experience and of great importance geologically is the equilibrium between carbon dioxide and water. For a preliminary study of this system, one need only put a stopper in a bottle half full of water. Gaseous CO_2 from the air in the top of the bottle dissolves in the water, and dissolved CO_2 in the water escapes into the air. Soon a balance is established between these two processes:

$$CO_2 \text{ in air} \rightleftharpoons CO_2 \text{ dissolved in water} \qquad (1\text{-}8)$$

The importance of this reaction is due to the fact that the dissolved CO_2 reacts with water to form an acid, H_2CO_3, carbonic acid. Thus we can rewrite the equation for the equilibrium,

$$CO_2 \text{ (in air)} + H_2O \rightleftharpoons H_2CO_3 \qquad (1\text{-}9)$$

(We assume here that *all* the dissolved CO_2 forms H_2CO_3. This is not strictly correct; to be more accurate, we should consider also the equilibrium CO_2 (dissolved) $+ H_2O \rightleftharpoons H_2CO_3$, which actually is displaced far to the left. To simplify this and future discussions, we omit this step. As long as the assumption is used consistently, the omission does not affect geologic arguments.)

The equilibrium responds predictably to changes in conditions: if we increase the amount of CO_2 by adding some from a tank, we speed up the forward reaction and more CO_2 dissolves; if we decrease the pressure of CO_2 by attaching a vacuum pump, the forward reaction is slowed and the equilibrium shifts to the left; heating the bottle decreases the solubility of the gas (or the stability of H_2CO_3), and again the equilibrium shifts to the left. Or we could disturb the reaction chemically by adding a base to neutralize the acid: if a little NaOH solution is poured into the bottle, some of the H_2CO_3 is destroyed, thereby slowing the reverse reaction and permitting more CO_2 to dissolve.

A numerical value for the equilibrium constant,

$$K = \frac{[H_2CO_3]}{[CO_2][H_2O]} \qquad (1\text{-}10)$$

can be found by looking up the solubility of CO_2. This is given in tables as 0.76 liter/liter of water at 25°C when the pressure of CO_2 is maintained at 1 atm. To express concentrations in solution, a common unit is moles per liter; the 0.76 liter of CO_2 under these conditions would represent 0.76/24.5 or 0.031 mole, so that the concentration of H_2CO_3 in Eq. (1-10) may be written 0.031M. (The symbol M means moles of solute per liter of solution, and the number 24.5 is the volume in liters occupied by 1 mole of any gas at 25°C and 1 atm pressure.) For the concentration of CO_2 gas we use atmospheres of pressure, as we did for the gases in the HCl equilibrium; in this case its concentration will be 1 atm. For the concentration of H_2O we could find moles per liter (1,000/18.016 = 55.5), but in all dilute solutions the concentration of H_2O is so nearly the same that we can treat it as constant and include it in the value of K. Hence we write

$$K = \frac{[H_2CO_3]}{[CO_2]} = \frac{0.031}{1} = 0.031 = 10^{-1.5} \tag{1-11}$$

Note that the K in Eq. (1-11) is 55.5 times larger than the K in Eq. (1-10), since the unchanging concentration of water is included in the former.

Having found the constant for the CO_2 reaction, we can now use it to calculate how much H_2CO_3 is present in water exposed to ordinary air. Air contains 0.03% CO_2 by volume; this means a volume fraction of 0.0003 and therefore a partial pressure of 0.0003 atm (since partial pressure is approximately proportional to mole fraction, and this in turn to volume fraction). Hence we substitute in Eq. (1-11):

$$K = 0.031 = \frac{[H_2CO_3]}{0.0003}$$

from which

$$[H_2CO_3] = 0.031 \times 0.0003 = 10^{-1.5} \times 10^{-3.5} = 10^{-5}M$$

This seems a very small concentration of acid, but it is sufficient to make natural waters much better weathering agents than they could be without it.

1-5 A THIRD EXAMPLE: CALCIUM SULFATE

Calcium sulfate stirred in water dissolves to a slight extent, producing the free ions Ca^{2+} and SO_4^{2-}:

$$CaSO_4 \rightarrow Ca^{2+} + SO_4^{2-}$$

Soon the solution becomes *saturated*, after which no further $CaSO_4$ will dissolve, however long the stirring is continued or however much solid $CaSO_4$ is added. This behavior suggests that an equilibrium has been reached. To prove that equilibrium exists, we try to make the reaction go backward, by mixing (in another container) separate solutions containing equal amounts of Ca^{2+} and

SO_4^{2-}. A precipitate of $CaSO_4$ forms immediately, and we find that the concentration of Ca^{2+} and SO_4^{2-} left in solution when the precipitate settles is the same as their concentration in the original saturated solution. Hence we have shown that the reverse reaction

$$Ca^{2+} + SO_4^{2-} \rightarrow CaSO_4$$

leads to the same result as the forward reaction, so that the existence of equilibrium is proved. We combine the equations into

$$CaSO_4 \rightleftharpoons Ca^{2+} + SO_4^{2-} \tag{1-12}$$

(Actually the situation is not quite so simple. The precipitate formed when Ca^{2+} and SO_4^{2-} are mixed at room temperature is not anhydrous $CaSO_4$ but gypsum, $CaSO_4 \cdot 2H_2O$. This complication will be discussed later, but for the present discussion it is not important.)

The equilibrium constant for Eq. (1-12) is

$$K = \frac{[Ca^{2+}][SO_4^{2-}]}{[CaSO_4]} \tag{1-13}$$

The concentrations of Ca^{2+} and SO_4^{2-} may be expressed as moles per liter of solution. But what meaning can we give to the expression $[CaSO_4]$, referring to the "concentration" of solid calcium sulfate in equilibrium with its saturated solution? One might guess at first that the effect of $CaSO_4$ on the equilibrium, as determined by its rate of ionization, would depend on the amount of solid surface exposed to the solution. Actually, provided the mixture is kept well stirred, the amount of surface makes no difference; both reverse and forward reactions take place only at the solid surface, and one is no more affected than the other by the amount of surface exposed. (This argument disregards the effect of very minute crystals of the solid, which will be discussed later.) Hence the "concentration" of the solid is effectively constant, whether the amount present is small or large. We assign a value of 1 to this constant concentration, so that it does not appear in the expression for the equilibrium constant:

$$K = [Ca^{2+}][SO_4^{2-}] \tag{1-14}$$

Strictly speaking, the K here would be related to the K in Eq. (1-13) by the factor $[CaSO_4]$.

To generalize: *the concentration of any pure solid (or pure liquid) taking part in an equilibrium is assumed equal to 1*, so that it need not appear explicitly in the expression for the constant. For example, if solid $CaSO_4$ is dropped into a solution of barium chloride, the following equilibrium is quickly established:

$$Ba^{2+} + CaSO_4 \rightleftharpoons Ca^{2+} + BaSO_4 \tag{1-15}$$

Both the $CaSO_4$ and $BaSO_4$ in this equation are solid precipitates, so that the equilibrium constant is simply

$$K = \frac{[Ca^{2+}]}{[Ba^{2+}]} \tag{1-16}$$

Another example is the formation of hematite from magnetite by heating in oxygen:

$$2Fe_3O_4 + \tfrac{1}{2}O_2 \rightleftharpoons 3Fe_2O_3 \qquad (1\text{-}17)$$

Both Fe_3O_4 and Fe_2O_3 are solids, so that the constant must be

$$K = \frac{1}{[O_2]^{1/2}} \qquad (1\text{-}18)$$

meaning that the two oxides can exist together at equilibrium only at a single fixed pressure of oxygen (for a given temperature), and that this pressure is not at all dependent on the relative amounts of the solid oxides present.

A number like the K in Eq. (1-14) is an important kind of equilibrium constant called a *solubility product*. For any slightly soluble salt, a similar constant product of ionic concentrations can be set up. For example,

$$AgCl \rightleftharpoons Ag^+ + Cl^- \qquad \text{Solubility product} = K = [Ag^+][Cl^-]$$

$$CaF_2 \rightleftharpoons Ca^{2+} + 2F^- \qquad \text{Solubility product} = K = [Ca^{2+}][F^-]^2$$

$$As_2S_3 \rightleftharpoons 2As^{3+} + 3S^{2-} \qquad \text{Solubility product} = K = [As^{3+}]^2[S^{2-}]^3$$

Numerical values of solubility products are given in the appendix, Table VII-1.

The usefulness of solubility products can be illustrated by a few simple examples. For $CaSO_4$, the experimentally determined value of K is 3.4×10^{-5} at $25°$. From this we can readily calculate the solubility of $CaSO_4$ in pure water. The solubility in moles per liter is equal to the concentration of Ca^{2+} (or of SO_4^{2-}) in the saturated solution, since every mole of $CaSO_4$ that dissolves gives 1 mole of each ion in solution. Hence

$$\text{Solubility} = [Ca^{2+}] = [SO_4^{2-}]$$

and

$$(\text{Solubility})^2 = [Ca^{2+}][SO_4^{2-}] = K = 3.4 \times 10^{-5}$$

whence

$$\text{Solubility} = 5.8 \times 10^{-3} M$$

For this ideally simple case, solubility is just the square root of the solubility product. *In general, this is not true for geological environments*, since most natural waters contain Ca^{2+} and SO_4^{2-} from other sources.

Solubilities as small as this are often expressed in a different kind of unit, *parts per million* (ppm). This means simply weight of solute in a million parts of solution, in any units—grams per million grams, tons per million tons, etc. To convert

$5.8 \times 10^{-3}M$ $CaSO_4$ to ppm, for example, we first multiply by the molecular weight:

$$5.8 \times 10^{-3} \times 136 = 0.79 \text{ gram/liter}$$

$$= \text{approximately } 0.79 \text{ gram of } CaSO_4 \text{ per } 1,000 \text{ grams of solution}$$

and this is equivalent to 0.79×1000 or 790 ppm. An alternative expression often used in geochemistry is milligrams per kilogram (mg/kg).

The ideal relation between solubility and solubility product is less simple for a salt like CaF_2, because the ions are formed in unequal concentrations. Every mole of the solid that dissolves gives 1 mole of Ca^{2+} and 2 moles of F^-. Hence the solubility may be set equal to the concentration of Ca^{2+}, and the concentration of F^- may be expressed as $2[Ca^{2+}]$. The measured value of K (from Table VII-1) is $10^{-10.4}$ at 25°C. Then

$$K = 10^{-10.4} = [Ca^{2+}][F^-]^2 = 4[Ca^{2+}]^3$$

$$[Ca^{2+}] = (0.25 \times 10^{-10.4})^{1/3} = 10^{-3.7}M = \text{solubility of } CaF_2.$$

Note that this result holds *only* for the case of solid CaF_2 in contact with water containing no Ca^{2+} or F^- from other sources.

Returning to $CaSO_4$, let us see how the solubility is affected by other ions in solution, particularly by an excess of Ca^{2+} or SO_4^{2-}. What is the solubility, for example, of $CaSO_4$ in a solution of $0.1M$ $CaCl_2$? Let x be the solubility; then the amount of Ca^{2+} in solution will be $(x + 0.1)$, since the $CaSO_4$ contributes x moles/liter and the $CaCl_2$ contributes 0.1. The concentration of SO_4^{2-} will be x, and the equilibrium constant is

$$K = [Ca^{2+}][SO_4^{2-}] = (0.1 + x)(x) = 3.4 \times 10^{-5}$$

Multiplying,

$$0.1x + x^2 = 3.4 \times 10^{-5}$$

This equation could be solved by the familiar quadratic formula, but again we look for a possible shortcut. We might guess that x will not be larger than the solubility we calculated for pure water, $5.8 \times 10^{-3}M$; hence the term x^2 should be small in comparison with $0.1x$. If x^2 is neglected, the equation gives at once

$$x = 3.4 \times 10^{-4}M$$

Then we check the validity of our assumption that x^2 is small by back-substitution:

$$0.1x + x^2 = 3.4 \times 10^{-5} + 11.6 \times 10^{-8} \cong 3.4 \times 10^{-5}$$

This statement is true to an accuracy of about 0.3%, which justifies the assumption.

Thus the solubility of $CaSO_4$ in $0.1M$ $CaCl_2$ solution is $3.4 \times 10^{-4}M$, compared with $5.8 \times 10^{-3}M$ in pure water. The decrease in solubility is a result we might have anticipated qualitatively, by noting that the excess Ca^{2+} would speed up the reverse reaction of the equilibrium

$$CaSO_4 \rightleftharpoons Ca^{2+} + SO_4^{2-} \qquad (1\text{-}12)$$

but would not affect the forward reaction. Clearly, excess of SO_4^{2-} would also lower the solubility; and in general, for any salt in equilibrium with its saturated solution, the solubility is less if an excess of one of its ions is present. This decrease in the solubility of a salt due to the presence of one of its own ions in solution is called the *common-ion effect*. The presence of ions *different* from those furnished by the salt itself generally makes the salt *more* soluble, but this fact we could not predict from simple equilibrium reasoning.

As a final example, let us calculate the value of K for the reaction of a barium salt with $CaSO_4$ [Eq. (1-15)]. Here two cations are in equilibrium with two salts, and the concentrations must adjust themselves so that both solubility products are maintained:

$$K' = [Ca^{2+}][SO_4^{2-}] = 3.4 \times 10^{-5}$$
$$K'' = [Ba^{2+}][SO_4^{2-}] = 1.0 \times 10^{-10}$$

The $[SO_4^{2-}]$ can be eliminated by dividing one equation by the other:

$$\frac{[Ca^{2+}]}{[Ba^{2+}]} = K = 3.4 \times 10^5$$

In other words, $BaSO_4$ is so much more insoluble than $CaSO_4$ that the equilibrium concentration of Ca^{2+} is 340,000 times that of Ba^{2+}.

In all this discussion the temperature has been assumed fixed at 25°C. Solubilities change with a change in temperature, and the solubility products must increase or decrease accordingly. For most salts the solubility increases as the temperature rises, but the rate of increase is different for different salts and cannot be predicted from elementary rules.

The salt we have used as our principal example, calcium sulfate, is a well-known compound geologically, occurring as the two minerals gypsum and anhydrite (and also the very rare bassanite, $CaSO_4 \cdot \frac{1}{2}H_2O$). Gypsum differs from anhydrite chemically in that it contains water (gypsum is $CaSO_4 \cdot 2H_2O$; anhydrite, $CaSO_4$), and this difference leads to a slight difference in solubility. At ordinary temperatures gypsum is slightly less soluble (solubility product 2.0×10^{-5} instead of 3.4×10^{-5}), but at higher temperatures anhydrite is less soluble. The temperature at which their solubilities become equal is approximately 50°C in fresh water and 20°C in seawater, but the figures are not entirely certain. This makes it apparent that solubility products are by no means an adequate description of all the complications that may arise in the behavior of slightly soluble salts, but they are often useful in making approximate predictions for reasonably simple systems.

1-6 LE CHATELIER'S RULE

In a previous discussion (Sec. 1-2) a general property of systems in equilibrium was mentioned: *A chemical equilibrium responds to any disturbance by trying to undo the effects of the disturbance.* The example cited was the effect on the HCl equilibrium of the addition of H_2: the equilibrium shifts so as to use up part of the added H_2 by reaction with Cl_2. The common-ion effect provides another simple illustration: if equilibrium between solid $CaSO_4$ and its saturated solution is disturbed by the addition of SO_4^{2-}, the equilibrium responds by trying to use up the added SO_4^{2-} by increased precipitation of $CaSO_4$. This property of equilibria goes by the name of *Le Chatelier's rule*, after the French physical chemist who first gave it explicit statement.

The rule contains nothing very profound. In a sense it is hardly more than a restatement of the definition of dynamic equilibrium, for a system cannot be in equilibrium unless it seeks to maintain the *status quo*, in other words to counteract the effect of an outside disturbance. Still, Le Chatelier's statement is a handy device for predicting qualitatively which way an equilibrium will shift in various circumstances.

Let us apply it, for example, to the behavior of equilibria when temperature changes. We note first that the energy change in the two reactions of an equilibrium must be equal and opposite. If one reaction gives out heat (or is *exothermic*), the reverse reaction must absorb heat (or is *endothermic*), and according to the law of conservation of energy the amounts of heat must be the same. Now if equilibrium has been established, and the temperature is raised, Le Chatelier's rule tells us that the equilibrium must respond by trying to counteract the temperature increase. This it can do if the endothermic reaction, the one that absorbs heat, is favored over the exothermic reaction. Thus in the HCl example [Eq. (1-3)], the forward reaction between H_2 and Cl_2 is strongly exothermic and the reverse reaction is endothermic; hence the decomposition of HCl is favored by high temperatures, and the equilibrium will shift toward H_2 and Cl_2 as the temperature rises. For another example, the ionic dissociation of most salts on dissolving in water is an endothermic reaction (as is shown convincingly by the cooling effect when a very soluble salt like sodium thiosulfate is stirred in water); hence when solubility equilibrium is established at one temperature, a rise in temperature will cause more of the salt to go into solution.

In general, therefore, the effect of temperature on an equilibrium can be qualitatively predicted by the simple rule that endothermic reactions are favored by a rise in temperature, exothermic reactions by a fall in temperature. It would also seem reasonable to suppose that the magnitude of the temperature effect would depend on the amount of the energy change, but the quantitative relationship we had best defer to a later chapter.

Another simple prediction from Le Chatelier's rule relates to the effect of total pressure on an equilibrium. Hitherto we have spoken of increasing or decreasing the partial pressure of a single gas, and this pressure has been treated simply as a measure of concentration. But now suppose that an equilibrium is established and

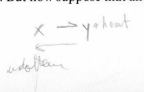

that we raise the pressure on the entire system. How will the equilibrium respond to this disturbance? According to the rule, that reaction must be favored which will lead to a diminution of pressure. This will evidently be the reaction that gives the smaller total volume. Thus, if equilibrium is established between water and ice at the melting point,

$$\underset{\text{Density 1.00}}{H_2O(\text{liquid})} \quad \rightleftharpoons \quad \underset{\text{Density 0.92 g/cm}^3}{H_2O(\text{solid})}$$

and if pressure is increased, the reverse reaction is favored because liquid water has a lower specific volume (higher density) than ice. This is a long-winded way of stating the familiar fact that ice may be melted by the application of pressure.

Equilibria involving gases are especially sensitive to pressure changes. Limestone at many granite contacts, for example, is partly converted to wollastonite according to the reaction

$$CaCO_3 + SiO_2 \rightleftharpoons CaSiO_3 + CO_2 \tag{1-19}$$

 The reaction involves three solids and the single gas CO_2. Hence the forward reaction produces a large increase in volume, so that high pressures would tend to prevent the formation of wollastonite.

When both reactions of an equilibrium involve gases, increase in pressure favors the reaction that leads to the gas or gas mixture having the smaller volume. This can be readily predicted from the number of gas molecules shown in the equation (as a consequence of Avogadro's law that equal numbers of gas molecules of any kind occupy the same volume, provided that the volumes are measured at similar temperatures and pressures). For example, one hypothesis regarding the formation of cassiterite deposits from magmatic gases is represented by the equation

$$SnCl_4 + 2H_2O \rightleftharpoons SnO_2 + 4HCl \tag{1-20}$$

At the high temperatures of cassiterite deposition, both $SnCl_4$ and H_2O would be gases, so that the equation shows three gas molecules on one side and four on the other. Hence the forward reaction would lead to an increase in volume and would be favored by low pressure.

These examples should make it clear that Le Chatelier's rule is a useful means for qualitatively predicting the effect on equilibria of changes in temperature and total pressure as well as changes of concentration. We have seen how concentration changes can be handled quantitatively also, by means of the equation for equilibrium constants. We shall return later (Chap. 8) to the problem of formulating quantitative rules for changes in temperature and total pressure.

1-7 STABILITY

One other concept needs qualitative mention now and quantitative refinement later. This is the idea of chemical stability, widely but often ambiguously used in geological discussions.

A stable substance is one that does not react readily in a particular environment. Gold under ordinary conditions is stable, because it is not noticeably attacked by air or most acids; metallic iron is unstable because it rusts when exposed to air. Water at usual temperatures is a stable compound, TNT an unstable compound. Sanidine is a form of potassium feldspar stable at high temperatures, adularia a form stable at low temperatures. Liquid water supercooled to $-5°C$ is unstable because it freezes quickly when disturbed. A supersaturated solution of potassium chloride is unstable, because addition of a grain of the salt causes rapid crystallization throughout the solution. These are all familiar examples of the stability concept.

Two precautions are needed to make usage of the term precise. First, stability has no meaning except when referred to particular external conditions. To say "iron is unstable" makes no sense; it is only mixtures of iron and air, or iron and acid, that are unstable. Water is stable enough at ordinary temperatures, but at 4000°C decomposes spontaneously into its elements. We often speak of olivine as a relatively unstable mineral, on the grounds that it weathers readily when exposed to air and that it reacts readily with silica at high temperatures; but there is nothing inherently unstable about olivine itself. Metamorphic rocks represent assemblages of minerals stable under certain temperature-pressure conditions but unstable when these conditions change. In using the term stability, one must keep in mind that a qualifying phrase is always implied—stable *with respect to* particular conditions of temperature and pressure, and particular kinds of associated substances.

The second precaution is to note that stability may be defined *by reference to either equilibrium or reaction rate.* In discussing mixtures of H_2, Cl_2, and HCl, one can say that a mixture rich in HCl is stable at ordinary temperatures, whereas a combination of H_2 and Cl_2 is stable above 2000°C, on the grounds that equilibrium is known to shift in these directions as the temperature changes. On the other hand, a mixture of H_2 and Cl_2 is apparently stable in the dark at room temperature, in the sense that no visible reaction occurs, but here the stability is due to a very slow rate of reaction. A similar example is galena exposed to air. The mineral is apparently perfectly stable, for museum specimens in contact with air persist indefinitely without visible change. Yet outcrops of veins containing galena commonly have anglesite entirely replacing the sulfide or surrounding remnants of the sulfide, suggesting that galena is slowly oxidized; experimental work confirms this deduction by showing that equilibrium in the reaction

$$PbS + 2O_2 \rightleftharpoons PbSO_4$$
$$\text{Galena} \qquad\qquad \text{Anglesite}$$

is displaced far in the direction of the sulfate at ordinary temperatures. Hence the apparent stability of galena is a result of the slowness of the oxidation, and in terms of equilibrium anglesite is the more stable mineral. This double usage of the word stability is a frequent source of confusion.

To avoid ambiguity in this book, "stability" will be used only in the sense of "stable with respect to equilibrium." Substances or mixtures of substances which seem stable because they react very slowly will be referred to as "metastable," or "apparently stable." Both stability and metastability will become clearer after we have discussed energy changes in equilibrium reactions (Chap. 8).

1-8 CONVENTIONS

Several arbitrary conventions have been introduced in this chapter for the handling of equilibrium constants. These conventions and a few important definitions are repeated here for easy reference.

1. For a reversible reaction $aA + bB + \cdots \rightleftharpoons yY + zZ + \cdots$, the equilibrium constant is defined as

$$K = \frac{[Y]^y[Z]^z \cdots}{[A]^a[B]^b \cdots}$$

2. K has a definite value for a particular chemical equation, not for a particular substance or a particular process.
3. Values of K are given for 25°C and 1 atm total pressure, unless other temperatures and pressures are specified.
4. Concentrations of gases in the equilibrium-constant formula are expressed as partial pressures in atmospheres. $\rightarrow m/l$
5. Concentrations of solutes in aqueous solution are given either as moles of solute per liter of solution (*molarity* or *molar concentration*, abbreviated M) or as moles of solute per kilogram of water (*molality* or *molal concentration*, abbreviated m). In dilute solutions the difference between the two is unimportant for geological calculations.
6. Concentrations of pure solids and pure liquids, and of water in liquid aqueous solutions, are included in the equilibrium constant and hence do not appear in the formula.
7. An equilibrium is called *homogeneous* if all the reacting substances are in a single *phase*, *heterogeneous* if they are in two or more phases. Most equilibria of interest in geology are heterogeneous.
8. A mixture is chemically *stable* either because it is at equilibrium or because reactions are slow. The first meaning is generally understood unless the second is specified.

1-9 A WORD ABOUT PROBLEM SOLVING

The solving of mathematical problems is not a main purpose of this book, but some facility in handling equilibrium constants is necessary for an understanding of geochemical arguments. The following general remarks about some common kinds of problems may be helpful:

1. Equilibrium constants and the concentrations of ions in geologically important solutions are generally small numbers, best represented by expressions with negative exponents. Such expressions may be written with integral exponents (for example, 2.5×10^{-5}) or with fractional exponents (for example, $10^{-4.6}$). Sometimes one form is more convenient, sometimes the other. To convert from one to the other, find logarithms and then antilogarithms. For example, 2.5×10^{-5} may be written with a fractional exponent as $10^{-4.6}$, because

$$\log 2.5 \times 10^{-5} = \log 2.5 + \log 10^{-5} = 0.40 - 5 = -4.6$$

and antilog $-4.6 = 10^{-4.6}$. For a second example, $10^{-8.2}$ is equivalent to 6.3×10^{-9}, because $\log 10^{-8.2} = -8.2 = 0.8 - 9$, of which the antilog is 6.3×10^{-9}. With a little practice these conversions become almost automatic. In examples in future chapters the two modes of expression will be used interchangeably.

2. Since equilibrium problems are based on chemical equations, the first requirement in setting up a problem is to be sure that the equation or equations are accurate. This means that each equation (1) must be balanced with respect to both numbers of atoms and numbers of charges, and (2) must be chemically reasonable for the environment specified in the problem. Unbalanced or unrealistic chemical equations are a common source of error in problem-solving.

3. For a problem involving a single equilibrium, generally the only equation needed is the expression for the equilibrium constant. This must be set up carefully, following the conventions summarized in the last section; if the conventions are neglected, endless confusion results. The statement of the problem then gives enough relations among the concentrations so that a single one can be picked out as the unknown and labeled x. For example, in the decomposition of HCl gas (Sec. 1-2), H_2 and Cl_2 must be produced in equal amounts and each may be labeled x; then the concentration of the remaining HCl is the initial pressure minus $2x$. As another example, when $CaSO_4$ is stirred in $0.1M$ $CaCl_2$ solution (Sec. 1-5), a convenient x is the solubility; from the stoichiometry this is equal to $[SO_4^{2-}]$, and the concentration of Ca^{2+} is x plus the 0.1 mole/liter already present from the $CaCl_2$.

4. For a problem involving several equilibria, it is often more convenient to write down several equations in several unknowns and solve them simultaneously. Such problems will be illustrated in Chap. 2.

5. It is characteristic of this kind of problem that the equation ultimately obtained for solution is difficult to handle mathematically, even when the original problem seems fairly simple. Practically always the equation is at least of second degree, and cubic and quartic equations are common. Algebraic methods of solving such equations exist, but they are tedious and cumbersome. Generally, however, approximate solutions can be found quickly by noting that some terms in an equation are necessarily smaller than others and so can be neglected. The approximate solutions can be refined by back-substitution, but for geological purposes, and for many chemical purposes, the refinements are

usually unnecessary. The real art in handling equilibrium-constant equations comes in sensing what approximations are reasonable, so that solutions can be arrived at quickly and easily.

6. The solution to a problem cannot be more accurate than the data that go into it. If, for example, the solubility product of anhydrite is 3.4×10^{-5}, mathematics unrestrained by common sense would give for the solubility in pure water a figure $5.832 \times 10^{-3} M$. Since the equilibrium constant is given only to two significant figures, all digits beyond the first two in the answer are meaningless. Not only must spurious mathematical accuracy be avoided; in geochemical work, it is fruitless also to seek accuracy beyond what natural conditions warrant. Suppose, for example, that the solubility product of anhydrite has been measured very accurately, so that the solubility in pure water at 25°C can be confidently fixed at 5.82×10^{-3} rather than $5.80 \times 10^{-3} M$. This would have no geologic significance, because any natural solution shows enough variation in temperature and contains enough other ions to change the solubility by more than this difference.

7. Since answers are generally arrived at by approximation and since the answers are expected to fit primarily geological rather than laboratory situations, it is more than usually important in problems of this kind that answers be checked for reasonableness. Once a problem has been solved, three questions should always be asked: (1) Is the answer grossly reasonable? (2) Does the solution fit the mathematical equations within a tolerable limit of accuracy? (3) Would the solution be significant in natural environments, where at least some uncontrolled variables are always present?

8. Problems in this book are approached on a basis of expediency rather than logic. Methods of attack are adapted to individual problems, and emphasis is kept as far as possible on physical relationships rather than mathematical elegance. Individuals differ in the kind of mathematical approach they can follow most easily, and some will find a more rigorous logical development preferable. For such readers two references from the list at the end of the chapter are particularly recommended: Garrels & Christ (1965) and Stumm and Morgan (1970).

PROBLEMS

Answers to some of the numerical problems are given in Appendix XII.

1 A mixture of hydrogen and oxygen at room temperature does not react appreciably but, when ignited by a flame or an electric spark, explodes violently as the two gases unite to form water. Water vapor is stable even at high temperatures, dissociating into its elements to an appreciable extent only above 1500°C. At 2000° and 1 atm, it is about 0.4% dissociated.

(a) Set up an expression for the equilibrium constant for the reaction

$$2H_2 + O_2 \rightleftharpoons 2H_2O$$

(b) Calculate the value of the constant at 2000°C.

(c) If additional hydrogen is added to an equilibrium mixture at 2000°, how is the equilibrium affected?

(d) If the total pressure on an equilibrium mixture at 2000° is increased, how is the equilibrium affected?

(e) Which reaction, forward or reverse, is exothermic? How, then, is the composition of an equilibrium mixture affected by a rise in temperature?

(f) In items (c), (d), and (e), how is the *equilibrium constant* affected by the suggested changes?

(g) Is a mixture of hydrogen and oxygen at room temperature stable?

2 Is the equilibrium established when excess CaF_2 is stirred in water heterogeneous or homogeneous? What phases are present?

3 The solubility product of calcium carbonate at 25°C is 4.5×10^{-9}.

(a) Calculate the solubility of $CaCO_3$ in pure water at 25°. Express the solubility as (1) moles per liter, (2) grams of $CaCO_3$ per 100 ml, (3) parts per million (ppm) of Ca.

(b) Calculate the solubility of $CaCO_3$ in a solution of $0.05M$ $CaCl_2$ at 25°.

(c) What is the ratio of SO_4^{2-} to CO_3^{2-} in a solution at equilibrium with both $CaSO_4$ and $CaCO_3$?

4 The solubility of Ag_2SO_4 at 25°C is 0.8 g/100g H_2O. Calculate the solubility product.

5 The equilibrium constant for Eq. (1-17) is 5×10^{43}. Calculate the pressure of oxygen in equilibrium with a mixture of hematite and magnetite at 25°. How many *molecules* per liter does this represent?

6 The equilibrium constant for the reaction

$$2Fe^{3+} + 2Cl^- \rightleftharpoons 2Fe^{2+} + Cl_2$$

is approximately 10^{-20}. Calculate the equilibrium ratio $[Fe^{2+}]/[Fe^{3+}]$ for (a) $[Cl_2] = 1$ atm and $[Cl^-] = 1M$, and (b) $[Cl_2] = 10^{-10}$ atm and $[Cl^-] = 1M$. Would you expect chlorine to oxidize Fe^{2+} appreciably at ordinary temperatures? Would you expect to be able to smell Cl_2 over a solution of $FeCl_3$?

7 The solubility product of PbS is $10^{-27.5}$ and that of ZnS is $10^{-24.7}$. What is the ratio of $[Pb^{2+}]$ to $[Zn^{2+}]$ in equilibrium with both galena and sphalerite? If a solution containing 100 times as much Zn^{2+} as Pb^{2+} percolates through a mixture of the sulfides, would galena be replaced by sphalerite or sphalerite by galena?

8 The solubility of H_2S in water is 2.3 liters per liter of solution at 25° and 1 atm. Using this figure, calculate the equilibrium constant for the reaction

$$H_2S(gas) \rightleftharpoons H_2S(aq)$$

9 From your general knowledge of the chemical behavior and geologic occurrence of the following substances, indicate which are stable and which are metastable at ordinary temperatures.

(a) Quartz exposed to air.

(b) Magnetite exposed to air.

(c) A mixture of olivine and silica.

(d) A mixture of kaolinite and calcite.

(e) Petroleum exposed to air.

(f) Cassiterite exposed to air.

(g) Calcium oxide exposed to air.

10 In Sec. 1-5, the ratio $[Ca^{2+}]/[Ba^{2+}]$ in equilibrium with both $CaSO_4$ and $BaSO_4$ was found to be 3.4×10^5. What are the actual concentrations of Ca^{2+} and Ba^{2+} in the solution? What is the concentration of SO_4^{2-}?

11 Suppose that a solution is just saturated with calcium sulfate, but no solid calcium sulfate is present. If a small amount of a more soluble calcium salt in fairly concentrated solution (say, $1M$ $CaCl_2$) is added, would you expect a precipitate to form? Explain.

REFERENCES AND SUGGESTIONS FOR FURTHER READING

Butler, J. N., "Ionic Equilibrium, a Mathematical Approach," Addison-Wesley Publishing Company, Inc., Reading, Mass., 1964. The mathematical background of equilibrium calculations, from a chemist's point of view, is presented for many kinds of reactions. The mathematics is not difficult, and less knowledge of physical chemistry is assumed than in Garrels and Christ. The same author and publisher have produced a briefer and far cheaper paperback volume, "Solubility and pH Calculations," which covers many of the topics in the hard-cover edition.

Garrels, R. M., and C. L. Christ, "Minerals, Solutions, and Equilibria," Harper & Row, Publishers, Incorporated, New York, 1965. The mathematical handling of equilibrium constants for geologically important reactions is treated in great detail, with more attention to formal rigor than was attempted in this chapter. A good background in physical chemistry is assumed.

Sienko, M. J., and R. A. Plane, "Chemistry," 4th ed., McGraw-Hill Book Company, New York, 1971. The material covered in this book or a similar textbook of modern chemistry should be familiar to the reader. Sienko and Plane give a particularly good elementary treatment of equilibrium and reaction rates, plus the necessary background of descriptive chemistry, methods of balancing equations, and the handling of simple chemical calculations.

Stumm, W., and J. J. Morgan, "Aquatic Chemistry," Wiley-Interscience, John Wiley & Sons, Inc., New York, 1970. A sophisticated treatment of equilibrium; requires a good background in physical chemistry.

ACIDS AND BASES

The three words acid, base, and alkali are old ones, commonly used in geologic and chemical writing even before the sciences took modern form at the end of the eighteenth century. As knowledge increased regarding rock origins on the one hand and chemical relationships on the other, the meanings of the three words underwent gradual changes. Unfortunately the changes have not been parallel in geology and chemistry. Present usages differ in several important respects, leading to much confusion in geochemistry, for it is often not clear whether an author is using the words with their chemical or their geological connotation.

One objective in this chapter will be to try to bring order out of this confusion of nomenclature. A second purpose is to extend the discussion of equilibrium begun in the first chapter, for the equilibrium concept is a powerful tool in systematizing our knowledge of acid-base reactions.

2-1 CHEMICAL DEFINITIONS

Acids, from the time of Robert Boyle (1663), have been described as substances with a sour taste and possessing the ability to dissolve many substances, to change the color of vegetable dyes like litmus, and to react with bases to form salts. Early in the nineteenth century it was proved that the essential element common to all acids is hydrogen. Toward the end of the century, as a part of his ionic theory, Arrhenius proposed that acids differ from other hydrogen-containing compounds in that they partially dissociate when dissolved in water to set free hydrogen ions, H^+. Following Arrhenius, present-day chemists commonly ascribe the acid properties of a solution to the presence of free hydrogen ion.

Brønsted and others in the present century have pointed out that this idea cannot be correct, because H^+ represents nothing more than an isolated proton, which could not possibly exist by itself in the presence of water. It would necessarily be hydrated, forming the ion H_3O^+ (hydronium ion), so that the dissociation of an acid would be represented by an equation like

$$HCl + H_2O \rightarrow H_3O^+ + Cl^- \tag{2-1}$$

rather than the simple formulation of Arrhenius

$$HCl \rightarrow H^+ + Cl^- \tag{2-2}$$

According to Brønsted, an acid is a molecule or ion capable of giving H^+ to another molecule or ion [as HCl gives it to H_2O in Eq. (2-1)]; in other words an acid is a "proton donor." Other chemists have objected that Brønsted's representation is still not quite correct, because water forms associated molecules like $(H_2O)_2$ and $(H_2O)_3$, so that "hydronium ion" includes $H_5O_2^+$ and $H_7O_3^+$ as well as H_3O^+. A further generalization has been suggested by chemists who would leave hydrogen out of the definition altogether and let "acid" refer to any substance whose molecules can react with unattached electron pairs in other compounds.

Strictly, then, there is no really correct chemical definition of acid that is simple enough for everyday use. Fortunately, despite their apparent differences, the various definitions give consistent and not very different interpretations of most common reactions. We can choose, therefore, the formulation that is most convenient for geologic purposes, realizing that it is not strictly accurate, but knowing that the inaccuracy is of too low an order to affect geologic predictions. So we pick, as the simplest representation, the type of reaction favored by Arrhenius [Eq. (2-2)]; we define an acid as *a substance containing hydrogen which gives free hydrogen ions when dissolved in water*, and we describe the characteristic properties of acids as the properties of hydrogen ion.

The term *base* has a similar history. For a long time it was used loosely to describe substances whose water solutions have a soapy feel, a bitter taste, the ability to neutralize acids, and the property of reversing the color changes that acids produce in vegetable dyes. Such properties are possessed by a variety of materials, seemingly not closely related—ammonia, metal oxides, carbonates, hydroxides, and several others. Since the time of Arrhenius the properties common to all bases have been ascribed to hydroxide ion, OH^-, and the term is generally restricted to compounds like NaOH and $Ca(OH)_2$ which dissociate to give this ion directly:

$$Ca(OH)_2 \rightarrow Ca^{2+} + 2OH^-$$

Brønsted refers to OH^- itself as a base and broadens the term to include all ions and molecules which, like OH^-, are capable of uniting with H^+ ("proton acceptors"). In the Brønsted terminology a neutralization reaction like

$$H_2CO_3 + OH^- \rightarrow HCO_3^- + H_2O$$

is described as transfer of a proton from the donor (acid) H_2CO_3 to the acceptor (base) OH^-; in the older terminology of Arrhenius, a soluble base like NaOH has supplied OH^- to react with the acid. Brønsted includes in his definition of base such substances as ammonia, NH_3, because it unites with H^+ to form NH_4^+; carbonate ion, CO_3^{2-}, because it readily forms HCO_3^-; and sulfide ion, S^{2-}, because it forms HS^-. As in the case of acids, the two modes of description are less different than they seem. For geologic purposes the Arrhenius formulation is more convenient; hence we shall define a base as *a substance containing the* OH *group that yields* OH^- *on dissolving in water*, and we shall describe the characteristic properties of bases as the properties of hydroxide ion.

Alkali is an Arabic word originally used for the bitter extract produced by leaching the ashes of a desert plant. It was extended to the bitter-tasting salts that sometimes collect in desert lakes (chiefly the substance we call sodium carbonate), and ultimately to other compounds prepared from these salts and from extracts of plant ashes. Today, as used in chemistry, it refers generally to the soluble strong bases like NaOH, KOH, and $Ba(OH)_2$, but usage is not entirely consistent. The adjective *alkaline* is practically a synonym for *basic*, referring to any solution containing appreciable OH^- or any substance capable of forming such a solution. The derived term "alkali metal" means any metal of the group sodium, potassium, lithium, rubidium, cesium, and the term "alkaline-earth metal" means one of the group calcium, strontium, barium (often including also magnesium, beryllium, and radium).

2-2 GEOLOGIC USAGE

The various shades of meaning given to acid, base, and alkali by chemists are confusing enough, but in geology the situation is worse.

Many oxides of nonmetals dissolve in water to form acids (CO_2 and SO_3 are familiar examples), and the term "acid oxide" or "acid anhydride" is often extended to any nonmetallic oxide whether or not it is appreciably soluble. Hence SiO_2 becomes an "acid oxide," and rocks containing a high percentage of SiO_2 are "acid" rocks. Similarly, "basic oxide" is used for any metal oxide, regardless of whether it dissolves enough to give appreciable OH^-, and "basic" rocks are those containing an abundance of metal oxides (especially MgO, FeO, and CaO). The oxides of sodium and potassium, in geologic parlance, are generally called "alkaline" rather than basic, and rocks having unusually large amounts of these oxides are described as alkaline or alkalic.

This divergence from chemical usage begins the confusion but is by no means the end of the story. A solution whose analysis shows an excess of the common anions Cl^-, SO_4^{2-}, HCO_3^- over the common cations Na^+, K^+, Mg^{2+}, Ca^{2+} is necessarily acid (unless unusual cations are present to make up the necessary positive charge), and from this fact it is often assumed that the mere presence of such ions as Cl^- and SO_4^{2-} (also F^-, Br^-, $H_2BO_3^-$) betokens an acid environment. Similarly a solution containing large amounts of Na^+ and K^+ is commonly referred to as alkaline, regardless of the amount of OH^- or H^+ present. The terms base and basic are subject to even broader interpretation. In rock analyses, "base" may refer to either oxides or metals; "base exchange" and "basic front" refer to movement of metal ions; and it is seldom quite clear whether "base" is restricted to iron and magnesium or includes also calcium, sodium, and potassium. Thus when a geologist speaks of an "alkaline solution," a "zone of acid alteration," or a "basic environment," his statements are not precise unless the terms are further delimited.

In geochemistry the words acid, base, and alkali are used very frequently, and it is desirable that they be given a more precise connotation than they often have in current geologic usage. We shall adhere strictly to the chemical definitions as given above, using "acid" only with reference to hydrogen ion and "base" only with reference to hydroxide ion. "Alkali" will mean a strong base, and "alkaline" will be a synonym for "basic," except that we shall retain the common names "alkali metal" and "alkaline-earth metal." Instead of calling a quartz-rich rock "acidic," we shall describe it as *felsic* or *silicic*. For "basic" rocks and minerals, we shall use the synonym *mafic*. Rocks rich in sodium and potassium we shall continue to call *alkalic* or *alkaline*, because no convenient alternative name has been proposed. These definitions are by no means entirely free of difficulties, but they do avoid the more glaring ambiguities in geologic literature.

2-3 THE pH

When HCl dissolves in water, evidence from conductivity, depression of the freezing point, and other properties indicates that the acid is completely dissociated into ions:

$$HCl \rightarrow H^+ + Cl^- \tag{2-2}$$

This equation can represent a chemical equilibrium only if the undissociated HCl on the left-hand side is regarded as HCl gas in the space above the solution. Within the solution itself no HCl molecules exist (or their concentration is negligibly small); the solution consists simply of independent hydrogen ions and chloride ions. At any concentrations of interest in geology, no equilibrium between ions and undissociated acid need be considered.

Similarly, when a strong base like NaOH dissolves in water, its dissociation is complete:

$$NaOH \rightarrow Na^+ + OH^- \tag{2-3}$$

Only in a very concentrated solution, in contact with solid NaOH, would this equation represent an equilibrium. In geologic situations a solution of NaOH may be regarded simply as independent Na^+ and OH^- ions.

The reaction between an acid and a base, called *neutralization*, is illustrated in its simplest form when dilute solutions of HCl and NaOH are mixed:

$$Na^+ + OH^- + H^+ + Cl^- \rightleftharpoons H_2O + Na^+ + Cl^-$$

Or more simply, since Na^+ and Cl^- are unaffected,

$$OH^- + H^+ \rightleftharpoons H_2O \qquad (2\text{-}4)$$

This same equation represents the neutralization of any strong acid by any strong base.

What happens if a solution containing exactly 1 equivalent of acid is carefully neutralized with exactly 1 equivalent of base? On first thought, one might guess that all of the H^+ and OH^- would be used up, so that the neutral solution would contain none of these ions at all. Experimentally we find that this does not happen, for water itself is slightly dissociated. The reaction shown by Eq. (2-4) proceeds until H^+ and OH^- have equal concentrations of $10^{-7}M$, at which point the solution, by definition, is neutral. In other words, the dissociation constant for water is

$$H_2O \rightleftharpoons H^+ + OH^- \qquad K_w = [H^+][OH^-] = 10^{-14} \qquad (2\text{-}5)$$

It is fortunate for simplifying calculations that the exponent, shown in Eq. (2-5) rounded off to -14, is actually so nearly a whole number at 25°C. A more precise value for the exponent obtained by careful measurements is -13.998, but for all ordinary purposes at usual temperatures the value of this important constant is taken as 10^{-14}. The variation of the constant with temperature is shown in Fig. 2-1.

The product $[H^+][OH^-]$ is constant for all water solutions, not for neutral ones only. This means that H^+ is present even in strongly basic solutions and OH^- in strong acids—in very minute concentrations, of course, but enough to keep the product of $[H^+]$ and $[OH^-]$ equal to 10^{-14}. It follows that the acidity or alkalinity of a solution can be specified by giving the concentration of either H^+ or OH^- alone. Thus a $1N$ solution of a strong acid has $1M$ H^+, a neutral solution has $10^{-7}M$ H^+, and a $1N$ solution of a strong base has $10^{-14}M$ H^+. (The symbol N, read "normal," means equivalents of acid or base per liter of solution.) The whole range from strong acids to strong bases can thus be expressed in terms of $[H^+]$, as a series of powers of 10. We can simplify the representation even further by discarding the 10's and using only the exponents, and by changing the sign of the exponents from negative to positive. The numbers so obtained are called pH values. In more formal language we define the pH of a solution as *the negative logarithm of the hydrogen-ion concentration.* A $1N$ solution of a strong acid has a pH of 0, a neutral solution a pH of 7, a $1N$ solution of a strong base a pH of 14. A solution containing $3 \times 10^{-4}M$ H^+ has a pH of 3.5; one containing $10^{-5.3}M$ OH^- has a pH of 8.7.

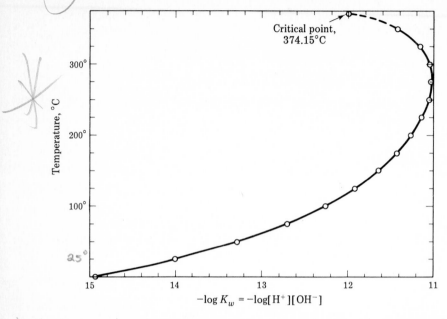

Figure 2-1 Change of the ionic dissociation constant of water with temperature, for water in equilibrium with its vapor. (*Source: Fisher, J. R., and H. L. Barnes, Jour. Physical Chemistry, vol. 76, pp. 90–99, 1972.*)

In nature, observed pH's lie mostly in the range 4 to 9. Streams in humid regions generally show values between 5 and 6.5, in arid regions between 7 and 8. Soil water, especially if decaying vegetation is abundant, may have pH's down to 4 or a little lower. Ocean water normally shows a pH range of 8.1 to 8.3. Soil water and playa-lake water in deserts may have pH's of 9 or even higher. The lowest recorded pH's in nature are found in solutions in contact with oxidizing pyrite; values even less than 0 have been recorded from such environments.

2-4 DISSOCIATION CONSTANTS OF WEAK ACIDS

If solutions of sulfuric and carbonic acids of similar total acid concentration are compared, the H_2SO_4 solution shows much greater activity as an acid. In comparison with $0.1M$ H_2CO_3, a solution of $0.1M$ H_2SO_4 has a sourer taste, a more pronounced effect on indicator dyes, and a greater solvent action on metals. Sulfuric acid is therefore described as a *strong* acid, carbonic acid as a *weak* acid. The difference in behavior is explained as a difference in extent of dissociation, H_2SO_4 in dilute solution being almost completely broken down into ions and H_2CO_3 only slightly so.

As a first guess, one might suppose that the dissociation of a weak acid like H_2CO_3 could be represented by an equilibrium reaction of the form

$$H_2CO_3 \rightleftharpoons 2H^+ + CO_3^{2-} \tag{2-6}$$

The manner in which the pH changes on gradual neutralization, however, shows that the reaction is more complicated. All weak acids containing more than one H atom per molecule dissociate in steps:

$$H_2CO_3 \rightleftharpoons H^+ + HCO_3^- \tag{2-7}$$

$$HCO_3^- \rightleftharpoons H^+ + CO_3^{2-} \tag{2-8}$$

Like other equilibria, these can be formulated in terms of constants representing quotients of concentrations:

$$\frac{[H^+][HCO_3^-]}{[H_2CO_3]} = K_1 = 10^{-6.4} = 4.2 \times 10^{-7} \tag{2-9}$$

$$\frac{[H^+][CO_3^{2-}]}{[HCO_3^-]} = K_2 = 10^{-10.3} = 5.0 \times 10^{-11} \tag{2-10}$$

This particular kind of equilibrium constant is called a *dissociation constant*. Note that here we cannot disregard the concentration of the undissociated substance, as we could disregard the undissolved solid in formulating a solubility product (Sec. 1-5). The undissociated molecules [H_2CO_3 in Eq. (2-9) and HCO_3^- in Eq. (2-10)] are present *in the solution*, and obviously their concentration affects the concentration of H^+. Dissociation constants (Table VII-2 in the appendix) are important in making comparisons of the strengths of different acids and in calculating the pH's of various solutions.

To illustrate the use of dissociation constants in finding hydrogen-ion concentrations, let us calculate the pH of a solution of $0.01M$ H_2CO_3. The figure for concentration, $0.01M$, in this sort of expression refers to *total* carbonate, i.e.,

$$\Sigma CO_3 = [H_2CO_3] + [HCO_3^-] + [CO_3^{2-}] = 0.01 \text{ mole/liter} \tag{2-11}$$

In other words, "$0.01M$ H_2CO_3" means "a solution obtained by dissolving 0.01 mole of CO_2 in a liter of water"; some of the acid will be present as ions, but the designation $0.01M$ H_2CO_3 refers to the total amount dissolved. The expression $[H_2CO_3]$ in Eq. (2-11), by contrast, is the concentration of undissociated acid molecules alone.

Another equation can be set up to express the fact that the solution is electrically neutral, i.e., that the concentration of positive charges must equal the concentration of negative charges:

$$[H^+] = [OH^-] + [HCO_3^-] + 2[CO_3^{2-}] \tag{2-12}$$

[The coefficient before $[CO_3^{2-}]$ is necessary because each mole of a divalent anion is equivalent to 2 moles of negative charge. Thus if x moles of CO_3^{2-} are present, they must be balanced by $2x$ moles of H^+.] Still another relation that must be satisfied in any aqueous solution is Eq. (2-5):

$$[H^+][OH^-] = 10^{-14} \qquad (2\text{-}5)$$

We now have a set of five equations [(2-9), (2-10), (2-11), (2-12), (2-5)] relating five unknown concentrations (concentrations of H^+, OH^-, H_2CO_3, HCO_3^-, CO_3^{2-}). The equations include three experimentally determined constants (K_1, K_2, and K_w) and the given total carbonate concentration, $0.01M$. To find the pH, we need only solve the five equations for $[H^+]$ by the ordinary rules of algebra.

A glance at the equations, however, shows that the necessary algebra is far from simple. As noted before (Sec. 1-9), this is a common situation in problems based on equilibrium constants: we set up simple equations relating the concentrations of substances in an equilibrium and then find that the mathematics required to solve the equations is rather formidable. Sometimes a complete formal solution is necessary, but as a rule we try to avoid the mathematical labor by making approximations. Especially in problems dealing with geologic environments, where great accuracy would be pointless anyway, approximate methods generally give satisfactory solutions.

Thus, in the present example, we note first that K_2 is almost 10,000 times smaller than K_1. This means that (H^+) contributed by the second dissociation is small compared with that from the first dissociation. Since the only source of CO_3^{2-} in the solution is the second dissociation, we could also guess that the concentration of this ion is small. As a first assumption, therefore, we guess that $[CO_3^{2-}]$ in Eqs. (2-11) and (2-12) is negligible in comparison with $[H_2CO_3]$ and $[HCO_3^-]$. A second reasonable assumption is that $[OH^-]$ is small relative to $[H^+]$, since we start by dissolving an acid in water. Using these assumptions, we set $[H^+]$ equal to $[HCO_3^-]$ [from Eq. (2-12)], and then Eqs. (2-9) and (2-11) become

$$\frac{[H^+]^2}{[H_2CO_3]} = 10^{-6.4} \qquad (2\text{-}13)$$

$$[H_2CO_3] + [H^+] = 0.01 \qquad (2\text{-}14)$$

The problem is thus reduced to two equations in two unknowns. These could be solved directly without much trouble, but we might first see if even this much mathematics can be avoided by trying a third assumption. The value of K_1 tells us that H_2CO_3 is a weak acid; therefore the $[H^+]$ from its dissociation is probably small in comparison with the concentration of undissociated molecules. We guess, therefore, that $[H^+]$ in Eq. (2-14) can be neglected, and $[H_2CO_3]$ can be set equal to 0.01. On this basis Eq. (2-13) can be solved immediately for $[H^+]$:

$$[H^+]^2 = 0.01 \times 10^{-6.4} \qquad [H^+] = 10^{-4.2}M \qquad pH = 4.2$$

Concentrations of other ions would then be, approximately,

$$[HCO_3^-] = [H^+] = 10^{-4.2}M$$

$$[OH^-] = \frac{10^{-14}}{[H^+]} = 10^{-9.8}M$$

$$[CO_3^{2-}] = \frac{10^{-10.3}[HCO_3^-]}{[H^+]} = 10^{-10.3}M$$

An answer obtained by making so many approximations must be checked by substituting in all the original equations:

$$10^{-4.2}\frac{10^{-4.2}}{0.01} = 10^{-6.4} \tag{2-9}$$

$$10^{-4.2}\frac{10^{-10.3}}{10^{-4.2}} = 10^{-10.3} \tag{2-10}$$

$$10^{-10.3} + 10^{-4.2} + 0.01 = 0.01 \tag{2-11}$$

$$10^{-4.2} = 10^{-9.8} + 10^{-4.2} + 2 \times 10^{-10.3} \tag{2-12}$$

The first two equations are exact equalities (as they must be from the nature of the assumptions), but the last two are obviously not quite true. Equation (2-11) shows the largest discrepancy; it amounts to saying that $0.01006 = 0.01$, or that $1,006 = 1,000$. The error is less than 1%, which for most geological purposes (and many chemical purposes as well) is inconsequential. The assumptions are therefore fully justified, and a more exact solution obtained by solving the five equations algebraically would give a wholly illusory accuracy.

Thus a solution of $0.01M$ H_2CO_3 has a pH of 4.2, or a hydrogen-ion concentration of $10^{-4.2}M$. This means that less than 1 percent of the acid molecules are broken up into ions.

Since it is a general rule that the first dissociation constant of a weak acid is greater than the second by a factor of 10^4 to 10^6, we may use the same reasoning for any weak acid:

$$H_nA \rightleftharpoons H^+ + H_{n-1}A^- \tag{2-15}$$

$$[H^+] = (\Sigma A \times K_1)^{1/2} \tag{2-16}$$

Thus for H_2S, with $K_1 = 10^{-7}$ and $K_2 = 10^{-13}$, the hydrogen-ion concentration of a $0.001M$ solution is

$$[H^+] = (0.001 \times 10^{-7})^{1/2} = 10^{-5}M$$

and the pH would be 5.0.

In using an approximate formula like this, one must be alert to its limitations. At best it is safe only for pH's accurate to two significant figures. If K_1 is fairly large, say 10^{-1} or 10^{-2}—in other words, if the acid is only moderately weak—$[H^+]$ can no longer be considered small compared with $[H_nA]$; the two equations corresponding to Eqs. (2-13) and (2-14) would then have to be solved by substitution. Again, if either K_1 or ΣA is very small, the answers will be unrealistic. For example, suppose that the previous calculation is attempted for $10^{-8}M$ H_2CO_3 instead of $10^{-2}M$. Equation (2-16) gives

$$[H^+] = (10^{-8} \times 10^{-6.4})^{1/2} = 10^{-7.2}M$$

This is clearly wrong, because the dissociation of water itself gives a larger value, $10^{-7.0}M$. The theoretical pH of such a solution could be evaluated by including Eqs. (2-12) and (2-5) in the calculation, but this would seldom be worthwhile. Any geologically important solution would contain enough additional dissolved material so that its pH would be determined by other reactions, not by this exceedingly minute quantity of H_2CO_3.

How small can ΣA become before the approximate answer is unusable? A good way to get a feeling for this limit is to calculate the pH of water in equilibrium with the CO_2 of ordinary air. The total dissolved CO_2, as calculated in the last chapter (Sec. 1-4), is about $10^{-5}M$. The problem is thus identical with the one above except that ΣCO_3 is now $10^{-5}M$ instead of $10^{-2}M$. From Eq. (2-16):

$$[H^+] = (10^{-5} \times 10^{-6.4})^{1/2} = 10^{-5.7}M$$

Checking back over the approximations to see if this number is reasonable, we note that $[CO_3^{2-}]$ and $[OH^-]$ are still small enough to be safely neglected, but the third assumption, that $[H^+]$ is small compared with $[H_2CO_3]$, is no longer obvious. If $[H^+]$ is $10^{-5.7}M$, then this much H_2CO_3 must have dissociated, and the remaining concentration of undissociated H_2CO_3 is $10^{-5} - 10^{-5.7}$, or $10^{-5.1}M$, rather than $10^{-5.0}M$. Then Eq. (2-9) would be

$$\frac{10^{-5.7} \times 10^{-5.7}}{10^{-5.1}} = 10^{-6.4} \qquad H_2CO_3 = [CO_3][H]$$

which is no longer quite true. If the third assumption is ignored and if Eq. (2-9) is solved as a quadratic,

$$\frac{[H]^2}{10^{-5} - [H^+]} = 10^{-6.4}$$

a more accurate value for $[H^+]$ is obtained, $10^{-5.74}M$. Mathematically this figure is better, but the apparent accuracy is suspect because the exponents in both K_1 and ΣCO_3 are given only to two significant figures. In any event, the approximate pH is within 1% of the more accurate one; even if the concentrations are expressed with integral exponents,

Approximate: $2.0 \times 10^{-6}M$

Accurate: $1.8 \times 10^{-6}M$

the discrepancy is only about 10% of the value. For most geologic purposes this is well within the limits of toleration. We conclude, therefore, that the approximate calculation using Eq. (2-16) in the case of carbonic acid is good down to concentrations of about $10^{-5}M$.

In addition to this mathematical conclusion, an important result of the computation is the demonstration that water exposed to atmospheric CO_2 at 25°C has a pH of 5.7. This assumes, of course, that the water is not in contact with other soluble material which could give a different pH. It is worth emphasizing that, although pure water has a theoretical neutral pH of 7.0, this pH is never encountered at the earth's surface unless the water contains dissolved material that neutralizes the dissolved CO_2. All water that we ordinarily consider "pure," either in nature or in the laboratory, acquires a pH near 5.7 if it stands for a short time exposed to air.

2-5 DISSOCIATION CONSTANTS OF HYDROXIDES

By analogy with acids, strong and weak bases can be distinguished on the basis of their degree of dissociation into ions. A strong base like KOH dissociates completely:

$$KOH \rightarrow K^+ + OH^-$$

At cation concentrations likely to be encountered in geology, the hydroxides of Li, Na, Rb, Cs, Sr, and Ba may also be considered soluble strong bases. An example of a soluble weak base is ammonia:

$$NH_3 + H_2O \text{ (or } NH_4OH) \rightleftharpoons NH_4^+ + OH^- \qquad K = 2 \times 10^{-5} \quad (2\text{-}17)$$

The analogy with acids cannot be carried very far, however, because all other hydroxides of geologic interest are only slightly soluble. To a good approximation, for many purposes the dissociation of an insoluble hydroxide may be considered complete insofar as the compound dissolves:

$$Mg(OH)_2(s) \rightleftharpoons Mg^{2+} + 2OH^- \qquad K = [Mg^{2+}][OH^-]^2 = 10^{-11.2} \quad (2\text{-}18)$$

[The symbol (s) following $Mg(OH)_2$ means "solid"; the equation shows equilibrium between ions in solution and the solid compound.] Magnesium hydroxide forms only a weakly basic solution, but the weakness is due to insolubility rather than failure of the dissolved compound to dissociate. Or, to say the same thing in another way, the dissociation constant of a slightly soluble base is the same as a solubility product.

Actually the situation is not this simple. Bases dissociate in steps, just as acids do, and the unraveling of the various dissociation reactions turns out to be a complicated branch of inorganic chemistry. Consider, for example, the behavior of copper hydroxide. If a base is added to a solution of a cupric salt, a blue precipitate of $Cu(OH)_2$ appears. Detailed study of copper concentrations at various pH's shows that this compound dissociates in steps:

$$Cu(OH)_2(s) \rightleftharpoons CuOH^+ + OH^- \qquad K = [CuOH^+][OH^-] = 10^{-13.0} \quad (2\text{-}19)$$

$$CuOH^+ \rightleftharpoons Cu^{2+} + OH^- \qquad K = \frac{[Cu^{2+}][OH^-]}{[CuOH^+]} = 10^{-6.3} \quad (2\text{-}20)$$

Furthermore, the solid hydroxide can react with OH^- to a slight extent (i.e., it is slightly amphoteric):

$$Cu(OH)_2(s) + OH^- \rightleftharpoons Cu(OH)_3^- \qquad K = \frac{[Cu(OH)_3^-]}{[OH^-]} = 10^{-2.9} \quad (2\text{-}21)$$

This does not end the story, for copper can form additional ions in solution containing more than one copper atom per ion, for example, $Cu_2(OH)_2^{2+}$. And if certain common anions are present, copper forms precipitates like $Cu_4(OH)_6SO_4$ and $Cu_2(OH)_3Cl$ rather than the simple $Cu(OH)_2$. As if this were not enough, the precipitated $Cu(OH)_2$ is not stable in contact with the solution, but changes over a period of a few days into the slightly less soluble oxide, CuO. Equilibrium constants are known, at least approximately, for all the pertinent reactions, so that concentrations of the various copper species can be calculated at given pH's, but the algebra becomes complex and tedious.

Not all metals have been studied in such detail, so that the necessary information to make detailed calculations is spotty and of widely varying quality. Some of the best current estimates are listed in the appendix, Table VII-3 (page 555), but many of the figures will doubtless be changed as research continues. In general, "total" dissociation constants [assuming complete dissociation without steps, as in Eq. (2-18)] are more accurately known than the constants for the step reactions.

Simple calculations from the total dissociation constants often give results, in situations of geologic interest, that are not very far different from results using all the individual constants. For example, suppose we are interested in the solubility of $Cu(OH)_2$ in a solution whose pH is 6.0. The "total" constant is

$$Cu(OH)_2(s) \rightleftharpoons Cu^{2+} + 2OH^- \qquad K = [Cu^{2+}][OH^-]^2 = 10^{-19.3} \quad (2\text{-}22)$$

At pH 6 the OH^- concentration is $10^{-8}M$; hence the amount of Cu^{2+} in solution would be calculated from this constant as $10^{-19.3}/10^{-16} = 10^{-3.3}M$. Now, for comparison, we try the same calculation using the separate constants for each ion. Equations (2-19) and (2-20) give

$$[CuOH^+] = \frac{10^{-13.0}}{10^{-8}} = 10^{-5.0}M$$

$$\frac{[Cu^{2+}]}{[CuOH^+]} = \frac{10^{-6.3}}{10^{-8.0}} = 10^{1.7} = 50$$

Hence $[CuOH^+]$ is much smaller than $[Cu^{2+}]$ and does not contribute appreciably to the solubility of the hydroxide. The contribution of $Cu(OH)_3^-$ is still smaller, as is evident from the value of K in Eq. (2-21). An ion like $Cu_2(OH)_2^{2+}$ is not likely to be important when $[OH^-]$ is so low, because its concentration would vary with the square of $[OH^-]$:

$$2Cu^{2+} + 2OH^- \rightleftharpoons Cu_2(OH)_2^{2+} \qquad (2-23)$$

Hence in this case the simple calculation from the total dissociation constant gives an accurate figure for the solubility of $Cu(OH)_2$.

In a solution with pH 8 the result is less satisfactory. A rough calculation from Eq. (2-22) now gives $[Cu^{2+}] = 10^{-7.5}M$. Equations (2-19) and (2-20), however, show that $[CuOH^+]$ is $10^{-7.0}M$, and the ratio $[CuOH^+]/[Cu^{2+}]$ is $10^{0.3}$ or about 2. The chief ion present is therefore $CuOH^+$, and the total solubility is given by the sum

$$[CuOH^+] + [Cu^{2+}] = 10^{-7.0} + 10^{-7.3} = 10^{-6.8}M$$

This is greater than the solubility estimated from the total ionization constant by a factor of about 5. For some geologic purposes a discrepancy of this magnitude would not be serious, but at pH's any higher than 8 the divergence becomes so great that the result of the simple calculation is misleading.

As a second example, consider the question, At what pH would $Cu(OH)_2$ precipitate if a base is added slowly to an acid solution of $10^{-4}M$ Cu^{2+}? Simple substitution in Eq. (2-22) gives at once a rough figure of $10^{-7.7}M$ for the OH^- concentration at which the solubility product would be reached, or a pH of 6.3. To consider the details of the precipitation process, we might guess that gradual addition of OH^- would lead first to formation of $CuOH^+$, then as a second step to precipitation of $Cu(OH)_2$. Equation (2-20) tells us that the two ions Cu^{2+} and $CuOH^+$ will have equal concentrations when $[OH^-]$ reaches $10^{-6.3}M$ (pH $= 7.7$). Evidently precipitation starts before this happens, or before much $CuOH^+$ has formed, so that consideration of the individual steps does not change the result appreciably. Here again the simple calculation from Eq. (2-22) is entirely adequate.

For other metal hydroxides there is unfortunately no easy way to tell, from casual inspection of the constants, just how satisfactory the calculations from the total constant will be. In general, the simple calculation gives better results for hydroxides of divalent metals than for those of trivalent metals, and better results at low pH's than at high pH's. The best rule is to use the constants for the step reactions when they are given, just as one does for acids, and to use the total constants for order-of-magnitude guesses when the separate constants are uncertain or unavailable.

2-6 DISSOCIATION CONSTANTS OF SALTS

Most salts are completely dissociated insofar as they dissolve. Solutions of soluble salts are treated in calculations as if they consist entirely of independent ions; for slightly soluble salts the dissociation constants are identical with the solubility products. Some salts are exceptions to this general rule, in that they ionize only partially in solution. A good example is lead chloride, for which the more important dissociation constants are

$$PbCl_2(s) \rightleftharpoons PbCl_2(aq) \qquad K = [PbCl_2(aq)] = 10^{-3.2}$$

$$PbCl_2(aq) \rightleftharpoons PbCl^+ + Cl^- \qquad K = \frac{[PbCl^+][Cl^-]}{[PbCl_2]} = 10^{-0.2}$$

$$PbCl^+ \rightleftharpoons Pb^{2+} + Cl^- \qquad K = \frac{[Pb^{2+}][Cl^-]}{[PbCl^+]} = 10^{-1.6}$$

Unless the concentration of chloride ion is unusually high, however (greater than about $0.01M$), these constants show that Pb^{2+} is far and away the most abundant lead species in solution, so that this complication can often be ignored. But in concentrated salt solutions, like seawater and the water responsible for deposition of many ore deposits, complexes like $PbCl^+$ and $PbCl_2(aq)$ may play an important role. Equilibrium constants for the partial dissociation of some salts are given in the appendix, Table VII-4.

2-7 HYDROLYSIS

Definition

Some salts dissolve in water to give neutral solutions, for example, NaCl, K_2SO_4, $Ba(NO_3)_2$. All of them have this in common, that they consist of the cation of a strong base and the anion of a strong acid. Other salts give distinctly acid or basic solutions, acid if the cation forms a weak base [for example, NH_4Cl, $Fe_2(SO_4)_3$] and basic if the anion forms a weak acid [for example, K_2CO_3, Na_2S]. The reaction leading to an excess of H^+ or OH^- in these solutions is called *hydrolysis*. We can formulate hydrolysis reactions and predict how far they will go by a further application of equilibrium reasoning.

Hydrolysis is an old word meaning literally "breakup by means of water." It is inherited from an earlier period in chemistry when water was supposed to split a salt into an acid and a base:

$$K_2CO_3 + 2H_2O \rightarrow 2KOH + H_2CO_3$$

We would now describe the reaction differently, first mentally dissecting the salt into ions, then noting that K^+ is indifferent to the water or to anything else in solution, and so focusing our attention on the CO_3^{2-}:

$$CO_3^{2-} + H_2O \rightleftharpoons HCO_3^- + OH^- \tag{2-24}$$

This is the essential reaction in the hydrolysis of carbonates. We no longer think of the water as breaking up the salt, but as uniting with one of the ions of the salt. The reaction does not go to completion but quickly establishes equilibrium; the position of equilibrium is determined by the extent of dissociation of the two substances H_2O and HCO_3^-. In effect, there is competition for H^+ between CO_3^{2-} and OH^-:

$$H^+ + OH^- \rightarrow H_2O$$

$$H^+ + CO_3^{2-} \rightarrow HCO_3^-$$

In the competition OH^- has a great advantage, since K for the first reaction is 10^{14} and that for the second is only $10^{10.3}$ [reciprocals of the K's in Eqs. (2-5) and (2-10)]. The CO_3^{2-} nevertheless gets enough H^+ away from OH^- to make the solution distinctly basic.

The hydrolysis of a salt like $Fe_2(SO_4)_3$ may be written

$$Fe^{3+} + H_2O \rightleftharpoons FeOH^{2+} + H^+$$

The SO_4^{2-} can be ignored because it is the anion of a strong acid. We can visualize the reaction as a partitioning of OH^- between Fe^{3+} and H^+, the H^+ having considerable advantage but the Fe^{3+} being able to attract enough OH^- to leave the solution distinctly acid. Thus, in modern language, *hydrolysis is the reaction between water and the ion of a weak acid or a weak base*. If only one ion of a pair hydrolyzes, as in the above examples, the solution becomes basic or acidic; if both ions hydrolyze, their effects may partly or completely cancel each other and leave the solution nearly neutral.

Hydrolysis of Na_2CO_3

The quantitative handling of hydrolysis is a simple extension of our work with dissociation constants. Suppose, for example, that we need to know the approximate pH of a $0.01M$ solution of Na_2CO_3. We set up the equilibrium constant for Eq. (2-24):

$$K = \frac{[HCO_3^-][OH^-]}{[CO_3^{2-}]} \tag{2-25}$$

To evaluate this constant, we resort to a trick. If both numerator and denominator are multiplied by $[H^+]$, we obtain

$$K = \frac{[HCO_3^-][OH^-][H^+]}{[CO_3^{2-}][H^+]} = \frac{[HCO_3^-]}{[CO_3^{2-}][H^+]}[OH^-][H^+]$$

This is the product of two constants we already know; the first is the reciprocal of the constant in Eq. (2-10), and the second is the constant in Eq. (2-5). Hence

$$K_{hydrolysis} = \frac{K_{water}}{K_{dissociation}} = \frac{10^{-14}}{5.0 \times 10^{-11}} = 2.0 \times 10^{-4}$$

Then if the solution contains no other carbonate besides Na_2CO_3 and no other HCO_3^- except that produced by hydrolysis, we can set $[OH^-]$ equal to $[HCO_3^-]$ and $[CO_3^{2-}]$ equal to $0.01 - [HCO_3^-]$ or $0.01 - [OH^-]$. Hence

$$K_{hydrolysis} = \frac{[OH^-]^2}{0.01 - [OH^-]} = 2.0 \times 10^{-4}$$

$$[OH^-]^2 + 2.0 \times 10^{-4}[OH^-] - 2.0 \times 10^{-6} = 0$$

This quadratic equation can be solved for $[OH^-]$, and from this the pH is readily obtained.

To solve the quadratic, let us first, as usual, try a shortcut. We move the third term to the right-hand side of the equation, and try eliminating in succession the second term and then the first term:

$[OH^-]^2 = 2.0 \times 10^{-6}$ from which $[OH^-] = 1.4 \times 10^{-3}M$

$2.0 \times 10^{-4}[OH^-] = 2.0 \times 10^{-6}$ from which $[OH^-] = 10^{-2}M$

The second answer is hardly probable, since it would mean that hydrolysis goes to completion. To prove that it is wrong, we may substitute:

$$(10^{-2})^2 + 2.0 \times 10^{-4} \times 10^{-2} - 2.0 \times 10^{-6} = 10^{-4} = 0$$

which is clearly not true. We now test the first answer:

$$(1.4 \times 10^{-3})^2 + 2.0 \times 10^{-4} \times 1.4 \times 10^{-3} - 2.0 \times 10^{-6} = 0$$

or

$$2.0 \times 10^{-6} + 2.8 \times 10^{-7} - 2.0 \times 10^{-6} = 0$$

This is better, but it is still not quite true. Inspection indicates that $[OH^-]$ should be made a trifle smaller than $1.4 \times 10^{-3}M$, so that the first two terms would add up to 2.0×10^{-6}. Trial and error shows that $1.3 \times 10^{-3}M$ is satisfactory. In this case the shortcut has not saved much time; the same answer could be obtained about as fast by using the familiar quadratic formula.

Note that, in solving this problem, we have tried a method of approximation different from the one used in calculating the pH of H_2CO_3 solutions (Sec. 2-3). We could have used the former method by guessing, to begin with, that the degree of hydrolysis is fairly small (since $K_{hydrolysis}$ is small), so that $[OH^-]$ should be small in comparison with the unhydrolyzed CO_3^{2-}. Then in Eq. (2-25), $[CO_3^{2-}]$ would be considered approximately equal to total carbonate, or $0.01M$, and the equation would be

$$\frac{[OH^-]^2}{0.01} = 2.0 \times 10^{-4}$$

This gives the same approximate answer as before, and the same trial-and-error correction would be needed. It makes no difference, of course, which method of approximation is used; as a general rule, in solving equilibrium-constant problems, there are several ways to make approximations that lead to the same result.

To go back to the problem we started with, the figure $1.3 \times 10^{-3}M$ for $[OH^-]$ means that carbonate ion at a concentration of $0.01M$ is about 13% hydrolyzed. Since $1.3 \times 10^{-3} = 10^{-2.9}$, the pH of the solution is $14 - 2.9$ or 11.1. Thus a dilute solution of a soluble carbonate is fairly strongly alkaline.

Hydrolysis of CuSO₄

As an example of hydrolysis of a cation, consider a $10^{-4}M$ solution of $CuSO_4$. The hydrolysis reaction is

$$Cu^{2+} + H_2O \rightleftharpoons CuOH^+ + H^+ \tag{2-26}$$

for which an equilibrium constant can be set up:

$$K = \frac{[CuOH^+][H^+]}{[Cu^{2+}]} = \frac{[CuOH^+]}{[Cu^{2+}][OH^-]}[H^+][OH^-]$$

In this expression, multiplication of both numerator and denominator of K by $[OH^-]$ makes it possible to evaluate the constant as a product of the reciprocal of the constant for Eq. (2-20) and the constant for the dissociation of water [Eq. (2-5)]:

$$K = \frac{1}{10^{-6.3}}10^{-14} = 10^{-7.7}$$

If no other source of Cu^{2+} or H^+ is present in the solution, we may set $[CuOH^+] = [H^+]$ and $[Cu^{2+}] = 10^{-4} - [H^+]$, whence

$$K = \frac{[H^+]^2}{10^{-4} - [H^+]} = 10^{-7.7}$$

and therefore

$$[H^+]^2 + 10^{-7.7}[H^+] - 10^{-11.7} = 0$$

An approximate solution is $[H^+] = 10^{-5.9}M$; hence the pH of $10^{-4}M$ $CuSO_4$ is about 5.9.

For this kind of problem it is preferable to use the stepwise dissociation of the hydroxide [Eq. (2-20) rather than Eq. (2-22)]. If the hydrolysis had been set up to show the formation of $Cu(OH)_2$ instead of $CuOH^+$,

$$Cu^{2+} + 2H_2O \rightleftharpoons Cu(OH)_2(s) + 2H^+$$

the calculated pH would be 6.4. The discrepancy between this and the more accurate value, 5.9, is uncomfortably large, and it is obviously more realistic to show the hydrolysis going only to $CuOH^+$, since no solid $Cu(OH)_2$ appears in copper sulfate solutions.

General Equation for Hydrolysis

From these examples comes the generalization that any hydrolysis constant may be obtained by dividing the dissociation constant of water (raised to a power if the coefficient of H^+ or OH^- is greater than 1) by the dissociation constant of the weak acid or weak base that is formed in the hydrolysis reaction:

$$K_h = \frac{K_w}{K_a} \quad \text{or} \quad K_h = \frac{K_w}{K_b} \tag{2-27}$$

One may generalize also that simple solutions of any soluble sulfide or soluble carbonate will necessarily be alkaline and that simple solutions of salts of the common heavy metals [for example, $FeCl_3$, $Pb(NO_3)_2$, $NiSO_4$] will necessarily be acidic, because of hydrolysis reactions analogous to Eqs. (2-24) and (2-26). We shall find these generalizations useful in discussing the many geologic processes in which hydrolysis plays an important role.

2-8 ESTIMATING IONIC CONCENTRATIONS

In a solution of given pH, if dissolved carbonate is known to be present, would it exist chiefly as H_2CO_3, HCO_3^-, or CO_3^{2-}? If the solution contains ferric iron, is it mainly Fe^{3+}, $FeOH^{2+}$, or $Fe(OH)_2^+$? Would dissolved zinc be present as positive ions (Zn^{2+} and $ZnOH^+$) or as negative ions [$Zn(OH)_3^-$ and $Zn(OH)_4^{2-}$]? This is a kind of question often encountered in geochemistry and easily answered if equilibrium constants are known.

The distribution of carbonate species will serve as a convenient example. We know to begin with, in a general qualitative way, that dissolved carbonate must exist chiefly as H_2CO_3 in acid solutions, as CO_3^{2-} in basic solutions, and as HCO_3^- in some intermediate range. To fix the limits, we write the equations for the two dissociation constants [Eqs. (2-9) and (2-10)] in the form

$$\frac{[HCO_3^-]}{[H_2CO_3]} = \frac{10^{-6.4}}{[H^+]} \quad \text{and} \quad \frac{[CO_3^{2-}]}{[HCO_3^-]} = \frac{10^{-10.3}}{[H^+]}$$

From these expressions, the concentrations of HCO_3^- and H_2CO_3 must be equal when $[H^+]$ has a numerical value equal to K_1, and the concentrations of CO_3^{2-} and HCO_3^- are equal when $[H^+] = K_2$. Hence we can say immediately that H_2CO_3 is the dominant carbonate species in all solutions with pH less than 6.4 [or $[H^+]$ greater than $10^{-6.4}M$], HCO_3^- is dominant in the pH range 6.4 to 10.3, and CO_3^{2-} is dominant at pH's above 10.3. These rules hold for any solution, regardless of how dilute or how concentrated it may be or what other solutes may be present.

Suppose now that we have given also a total analytical concentration of carbonate, say $0.001M$. In a solution whose pH is 6.4 the concentrations of both H_2CO_3 and HCO_3^- must be half this number, or $0.0005M$, and the concentration of CO_3^{2-} is very small; at pH 10.3 the concentrations of CO_3^{2-} and HCO_3^- are both

$0.0005M$ and $[H_2CO_3]$ is very small. At pH's well below 6.4, $[H_2CO_3]$ is effectively $0.001M$; at pH's well above 10.3, $[CO_3^{2-}]$ is $0.001M$; and over a good part of the intermediate range, $[HCO_3^-]$ must be $0.001M$.

It is often desirable to know the concentrations of all three carbonate species in a given solution, even though one or two may be very minor. For example, what are the concentrations of CO_3^{2-} and H_2CO_3 in a solution containing $0.001M$ total dissolved carbonate and having a pH of 8.0? This is in the intermediate range, where most of the dissolved carbonate exists as HCO_3^-, so that this ion may be assigned a concentration of approximately $0.001M$. Then the equation for K_2 [Eq. (2-10)] becomes

$$\frac{[CO_3^{2-}]}{0.001} = \frac{10^{-10.3}}{[H^+]}$$

and the equation for K_1 [Eq. (2-9)]:

$$\frac{0.001}{[H_2CO_3]} = \frac{10^{-6.4}}{[H^+]}$$

If $[H^+]$ is $10^{-8}M$, these equations give $[CO_3^{2-}] = 10^{-5.3}M$ and $[H_2CO_3] = 10^{-4.6}M$.

To generalize these results, it is convenient to rewrite the two equations in logarithmic form:

$$\log [CO_3^{2-}] - (-3) = -10.3 - \log [H^+]$$
$$(-3) - \log [H_2CO_3] = -6.4 - \log [H^+]$$

These may be simplified to

$$\log [CO_3^{2-}] = -13.3 + pH$$
$$\log [H_2CO_3] = 3.4 - pH$$

If now $\log [CO_3^{2-}]$ is plotted against pH, it should give a straight line with unit positive slope, and $\log [H_2CO_3]$ should give a straight line with unit negative slope. These relations hold over the pH range in which HCO_3^- has the approximate concentration $0.001M$, in other words from roughly 7.0 to 9.5. Similar equations can be set up for other pH ranges, giving a combined plot, shown in Fig. 2-2. The diagram is drawn for a total carbonate concentration of $0.001M$, but it can be used for any desired concentration by simply shifting the vertical scale up or down. From the diagram the concentrations of each of the three carbonate species at any pH can be read as intersections of the appropriate lines with a vertical line through the pH value. A similar diagram for sulfide solutions is given in Fig. 2-3.

The two acids H_2CO_3 and H_2S are the most important weak acids in geologic environments, and an understanding of the relations of the two acids and their ions, as summarized in Figs. 2-2 and 2-3, is essential to geochemical work with natural solutions. Similar diagrams may be constructed for other weak acids, and similar reasoning may be applied to other solutes that can exist in different species.

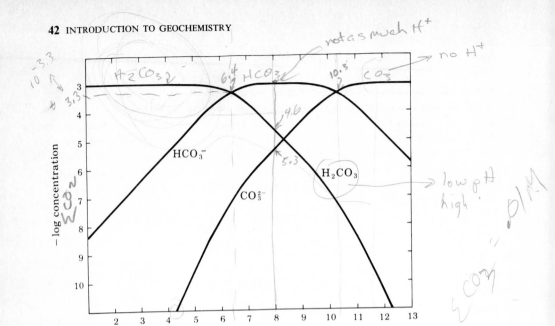

Figure 2-2 Concentrations of carbonate species at 25°C, in solutions with total dissolved carbonate = 0.001M.

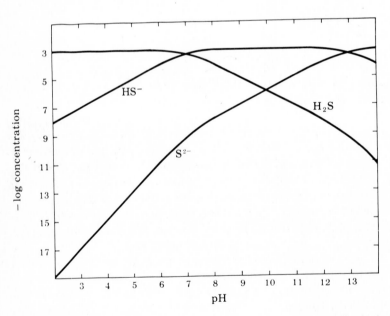

Figure 2-3 Concentrations of sulfide species at 25°C, in solutions with total dissolved sulfide = 0.001M.

2-9 CARBONATE EQUILIBRIA

As another illustration of the use of equilibrium reasoning, we consider next the relations between carbonic acid and carbonate minerals. These relations determine the conditions under which carbonate rocks are formed or dissolved and likewise the conditions of formation of carbonate gangue minerals in veins. The discussion in this section will be limited to qualitative reasoning, and to a single carbonate—the carbonate of calcium, which is by far the most abundant carbonate in nature. In the next chapter we shall make the treatment more quantitative and extend it to other common carbonate minerals.

A strong acid dissolves calcium carbonate by the familiar reaction

$$CaCO_3 + 2H^+ \rightarrow Ca^{2+} + H_2O + CO_2 \qquad \frac{[Ca][CO_2]}{[H]^2} \qquad (2\text{-}28)$$

If the concentration of acid is low, a more accurate equation would be

$$CaCO_3 + H^+ \rightarrow Ca^{2+} + HCO_3^- \qquad \frac{[Ca][HCO_2]}{[H]} \qquad (2\text{-}29)$$

showing that H^+ takes CO_3^{2-} away from Ca^{2+} to form the very weak (little dissociated) acid HCO_3^-. These reactions would take place in nature, for example, where strongly acid solutions from the weathering of pyrite encounter limestone. The reactions can be reversed by any process that uses up H^+; for example, if a base is added,

high pH $\rightarrow Ca^{2+} + HCO_3^- + OH^- \rightarrow CaCO_3 + H_2O \qquad \dfrac{1}{[Ca][HCO_3][OH]} \qquad (2\text{-}30)$

Quite evidently the solubility of $CaCO_3$ is determined in large part by the pH of its environment. By working out equilibrium constants for the above reactions, we could express this dependence quantitatively, but first it will be useful to get a feeling for the qualitative relationships.

Under natural conditions the dissolving of calcium carbonate is a little more complicated, because the acids involved are usually weak rather than strong. When limestone dissolves in carbonic acid, for example, the overall process may be summarized by the equation

$$CaCO_3 + H_2CO_3 \rightleftharpoons Ca^{2+} + 2HCO_3^- \qquad (2\text{-}31)$$

Note that the two HCO_3^- ions are from different sources: one is simply left over from the dissociation of H_2CO_3, and the other is formed by the reaction of H^+ from the acid with $CaCO_3$, as shown by Eq. (2-29). Equation (2-31) is the essential reaction for an understanding of carbonate behavior in nature. The forward reaction shows what happens when limestone weathers, when limestone is dissolved to form caves, or when marble is dissolved by ore-bearing solutions in the walls of a fissure. The reverse process represents the precipitation of calcium carbonate in the sea, as a cementing material in sedimentary rocks, or where droplets evaporate at the tip of a stalactite.

The effect of pH on solubility is shown as well by Eq. (2-31) as by the simpler equations preceding it. At low pH, where most dissolved carbonate exists as H_2CO_3 (Fig. 2-2), the forward reaction is favored; at high pH, the reverse reaction leading to precipitation is favored, because OH^- reacts preferentially with the stronger acid, H_2CO_3, rather than with the very weak HCO_3^-. The equation shows also that the position of equilibrium (and hence the solubility) depends on the pressure of CO_2 above the solution, since this pressure helps to determine the concentration of dissolved H_2CO_3:

$$H_2O + CO_2$$
$$CaCO_3 + H_2CO_3 \rightleftharpoons Ca^{2+} + 2HCO_3^-$$

Any process that increases the amount of CO_2 available to the solution makes more $CaCO_3$ dissolve; anything that decreases the amount of CO_2 causes $CaCO_3$ to precipitate. Some of the important natural processes that affect the solubility of $CaCO_3$ by changing the position of equilibrium in Eq. (2-31) are described in the following paragraphs.

Temperature Changes

The solubility of $CaCO_3$ in pure water decreases somewhat as the temperature rises. This is opposite to the behavior of most salts; the general result of increasing temperature is to give higher solubilities, but a number of carbonates and sulfates are exceptions. In addition to this effect, the solubility of $CaCO_3$ in natural waters decreases at higher temperatures because CO_2, like any other gas, is less soluble in hot water than in cold water. In general, the solubility of carbonates is much more influenced by this change in solubility of CO_2 than by the temperature coefficient of the solubility itself. As an illustration of the effect of temperature, $CaCO_3$ dissolves at great depths in the ocean, where the water is perennially cold, but precipitates near the surface, especially in the tropics, where the water is warm.

Changes in Pressure

The effect of pressure by itself, independent of its effect on CO_2, is to increase the solubility of $CaCO_3$ slightly. Where pressures are very large, the effect of pressure may be substantial; in the deep parts of the ocean, for example, pressure alone would increase the solubility to about twice its surface value. The major reason for an influence of pressure in near-surface environments, however, is the change in amount of dissolved CO_2 when the pressure of the gas changes in the surrounding atmosphere. Theoretically even day-to-day barometric changes should have a detectable effect on solubility, and the local production of CO_2 in abnormal amounts, say by a forest fire, an industrial plant, or a volcanic eruption, could cause a marked increase temporarily. But circulation of the atmosphere is so effective in keeping the partial pressure of CO_2 uniform that this factor is probably less important than the others.

Organic Activity

[handwritten: cause precip. of CaCO₃ by removing CO₂ from water by photosynthesis]

Many organisms use calcium carbonate in the construction of their shells. Just how they accomplish this is not certain, but they flourish in greatest numbers in water approximately saturated with $CaCO_3$, where only a minor change in pH is needed to cause precipitation. Green plants may cause precipitation of $CaCO_3$ indirectly, by removing CO_2 from water in the process of photosynthesis. Abundant green algae in the warm waters of the Bahama Banks, for example, aid in the precipitation of the limy mud and sand with which the banks are covered.

Decay

[handwritten: makes it more acidic, CaCO₃ dissolves gives CO₂]

Decay of organic matter in the presence of air or aerated water gives CO_2 in large amounts and hence makes $CaCO_3$ in the vicinity more soluble. If access of air is restricted or cut off entirely, the processes of decay are more complicated and the effect on solubility of $CaCO_3$ is not predictable. Any CO_2 or H_2S produced would make the water acid and lead to increased solubility; on the other hand, ammonia is a common product of decay which would have the opposite effect. Measurements of pH in stagnant waters suggest that anaerobic decay most often leads to increased acidity, but the effect may be different for different temperatures and different kinds of organic matter.

Carbonic acid, as these many examples show, is important in controlling the solubility of $CaCO_3$ in most natural environments, but the compound is somewhat soluble even in water containing no CO_2: *[handwritten: forms a base]*

$$CaCO_3 + H_2O \rightleftharpoons Ca^{2+} + HCO_3^- + OH^- \qquad (2\text{-}32)$$

[handwritten: ion of weak acid]

This is a hydrolysis reaction, possible because HCO_3^- is so weak an acid (or because the $H-CO_3$ bond is so strong). Even the small amount of CO_3^{2-} produced by the dissolving of $CaCO_3$ can take a little H^+ away from the OH^- of water. The reaction, of course, cannot go far in the forward direction, not nearly so far as the corresponding reaction for a soluble carbonate [Eq. (2-24)], because here the reverse reaction is aided by the insolubility of $CaCO_3$. But the reaction does go far enough to make water in contact with carbonates appreciably basic. In nature, most water solutions are exposed to CO_2 from the atmosphere, and this complicates the hydrolysis; experimentally it is found that water containing suspended $CaCO_3$ and left exposed to the air acquires a pH of approximately 8.

In summary, the solubility of carbonates in nature is controlled by fairly simple equilibria involving H_2CO_3, HCO_3^-, CO_3^{2-}, and water. The principal equilibrium [Eq. (2-31)] is very sensitive to changes in the amount of dissolved CO_2, and this is dependent on a variety of influences. Much surface water and ground water is approximately saturated with $CaCO_3$, and such waters can either dissolve or precipitate the carbonate, depending on slight alterations in external conditions. Hydrolysis of even fairly insoluble carbonates is sufficient to make solutions in contact with them slightly basic.

2-10 BUFFERS

The pH of ocean water sampled near the surface is almost always between the narrow limits 8.1 and 8.3. Locally and temporarily it may deviate from this range, but by and large the pH stays surprisingly constant. In the laboratory, a sample of seawater appears to resist a change in pH when acid or base is deliberately added. If, say, a liter of seawater and a liter of distilled water are set side by side and if a few drops of HCl are added to each, the distilled water will show a pH of 2 or 3 while that of the seawater remains close to 8. Whence comes this capacity for maintaining a nearly constant concentration of hydrogen and hydroxide ions?

For an answer we look once more to equilibrium reactions. We start, however, with a system much simpler than seawater, a solution containing roughly $0.01M$ H_2CO_3 and $0.01M$ $NaHCO_3$. This mixture, we would find experimentally, behaves like seawater in that it shows little tendency to change its pH when a few drops of acid or base are added. What are the principal substances present in the solution? One is Na^+, but it can be dropped from consideration because it undergoes no reaction with common acids and bases. The other two principal substances are HCO_3^- and undissociated H_2CO_3, both present at concentrations of about $0.01M$, because their dissociation and hydrolysis reactions are too slight to affect their concentrations appreciably. From Fig. 2-2 we see that a solution with equal amounts of H_2CO_3 and HCO_3^- should have a pH of 6.4. We could get the same number from Eq. (2-9):

$$[H^+] = \frac{[H_2CO_3]}{[HCO_3^-]} K_1 = \frac{0.01}{0.01} K_1 = K_1 = 10^{-6.4}M \tag{2-33}$$

Now when acid is added, the H^+ reacts with HCO_3^-:

$$H^+ + HCO_3^- \rightarrow H_2CO_3$$

This, of course, will change the ratio of $[H_2CO_3]$ to $[HCO_3^-]$ in Eq. (2-33), and thereby the $[H^+]$ also; but actually the pH is not very greatly affected unless the ratio changes a great deal. Suppose, for example, that enough acid is added to change the ratio from approximately 1 to approximately 2 (which could be accomplished by adding about 3 ml of $1M$ HCl to a liter of solution); this would give $[H^+] = 2 \times 10^{-6.4}M$ and pH = 6.1 instead of 6.4. By way of contrast, if distilled water were used instead of the H_2CO_3-$NaHCO_3$ solution, the same amount of acid would give a pH of 2.5. The HCO_3^-, in effect, takes up enough of the added H^+ so that the amount of free H^+ in the solution is only slightly increased.

Now note what happens if a strong base is added. The OH^- reacts with the H_2CO_3:

$$H_2CO_3 + OH^- \rightarrow HCO_3^- + H_2O \tag{2-34}$$

The ratio $[H_2CO_3]/[HCO_3^-]$ this time is decreased, and $[H^+]$ is correspondingly decreased, but again the effect on the pH is relatively slight. Addition of 3 ml of

$1M$ NaOH, for example, would make the pH about 6.7 instead of 6.4, whereas the same amount of base added to distilled water would give 11.5. The H_2CO_3 uses up most of the added OH^-, so that free hydroxide ion in the solution is increased only a trifle.

Solutions of this kind, capable of absorbing considerable H^+ or OH^- without showing much change in pH, are called *buffers*. In general, a buffer consists of approximately equal amounts of a weak acid (like H_2CO_3) and a salt of the acid (like $NaHCO_3$). The pH maintained by any particular buffer is determined by the dissociation constant of the acid. The maintenance of pH is, of course, possible only within limits. When enough strong acid or base has been added to a buffer to change the ratio of acid to ion by a factor of more than 10, the buffer loses its effectiveness, and the pH responds thereafter directly to increments of H^+ or OH^-.

Buffering action is not limited to acids. Mixtures of bases and the corresponding cations (say NH_3 and NH_4^+) would obviously be just as efficient. Or one could employ a slightly soluble salt as part of a buffer. A solution of calcium bicarbonate in contact with calcium carbonate, for example, tends to maintain its pH because H^+ is absorbed in the reaction

$$CaCO_3 + H^+ \rightarrow Ca^{2+} + HCO_3^- \tag{2-35}$$

and OH^- in the reaction

$$Ca^{2+} + HCO_3^- + OH^- \rightarrow CaCO_3 + H_2O \tag{2-36}$$

The intermediate ion of a dihydrogen acid has some ability to control changes in pH all by itself, for it undergoes reactions of either of the following types:

$$HCO_3^- + H^+ \rightarrow H_2CO_3$$
$$HCO_3^- + OH^- \rightarrow CO_3^{2-} + H_2O$$

Many natural solutions have the right combinations of solutes to serve as effective buffers. Seawater is an excellent example, but its buffering action is complex. Any attempt to make seawater more acid is countered by the reactions

$$H^+ + CO_3^{2-} \rightarrow HCO_3^- \quad \text{and} \quad H^+ + HCO_3^- \rightarrow H_2CO_3$$

If solid $CaCO_3$ is present, as it always is in many parts of the sea, excess H^+ is also cut down by the reaction shown in Eq. (2-35). The addition of OH^- leads to the counterreactions

$$OH^- + HCO_3^- \rightarrow CO_3^{2-} + H_2O$$

and

$$OH^- + H_2CO_3 \rightarrow HCO_3^- + H_2O$$

The H_2CO_3 for the second reaction is present only in minute quantities, but the CO_2 of the atmosphere forms a reservoir of additional H_2CO_3 if the amount of base to be handled is large. The precipitation of $CaCO_3$ according to Eq. (2-36) is another method by which seawater can reduce the concentration of OH^-. In addition to all these reactions involving carbonates and carbonate ions, seawater contains enough boron so that reactions like

$$H^+ + H_2BO_3^- \rightarrow H_3BO_3$$

and

$$OH^- + H_3BO_3 \rightarrow H_2BO_3^- + H_2O$$

play a significant secondary role.

These various processes serve to hold the pH of seawater in the neighborhood of 8 and probably have so held it for a long time in the geologic past. The ultimate controls, it should be noted, are CO_2 in the atmosphere and $CaCO_3$ in the bottom sediments. Any long-continued addition of acid, say as the result of large-scale production of HCl and CO_2 by volcanic activity, would lead to marked solution of the $CaCO_3$; any long-continued addition of base would mean a depletion of atmospheric CO_2. Since abundant calcium carbonate and carbon dioxide have been in contact with the oceans at least since the beginning of the Paleozoic era and probably in earlier periods also, it seems unlikely that the pH of seawater has varied appreciably during the latter part of geologic time.

Although the carbonate system is thus the probable immediate control on the pH of seawater, a complete explanation would have to go deeper. One could ask, for example, why the oceans at some time in the past acquired a pH suitable for the precipitation of $CaCO_3$. At least theoretically the water might be much more acid and still exist in equilibrium with atmospheric CO_2. The ultimate explanation would go back to the properties of silicate minerals, whose reactions are slower than those of carbonates but which, over long periods, must determine the inorganic character of seawater. We shall return to this question in a later chapter.

PROBLEMS

1 List the substances (ions and/or molecules) present in a solution made by stirring sodium sulfide (Na_2S) into water. Which are present in large concentrations and which in very minute concentrations? How would the concentration of each substance change as HCl is added to the solution? Is the solution alkaline to start with? Does it remain alkaline as the acid is added?

2 A nepheline basalt is a rock containing the minerals nepheline (approximately $NaAlSiO_4$), plagioclase, and pyroxene. In what sense is this rock alkaline? In what sense is it basic?

3 The analysis of a sample of ground water shows 44.9 g/liter of Na^+, 6.6 g/liter of Ca^{2+}, 81.9 g/liter of Cl^-, and 1.0 g/liter of SO_4^{2-}. No other ions are present in appreciable amounts except H^+ or OH^-. Is the solution acid or alkaline, in the chemical sense? Approximately what is its pH?

4 Calculate the hydrogen-ion concentration and the pH of a solution of $0.0001M$ H_2CO_3, and calculate also what fraction of the H_2CO_3 has dissociated. Compare this fraction with the fraction dissociated when the concentration is $0.01M$ (Sec. 2-4). What generalization can you make regarding the effect of dilution on degree of dissociation?

5 The solubility of amorphous silica in water is about 120 ppm of SiO_2 at 25°C. The solution contains silicic acid, H_4SiO_4, whose first dissociation constant is $10^{-9.9}$. What is the pH of a saturated silica solution?

6 If solid silver chloride is stirred in a solution of $1M$ NaCl until equilibrium is established, what are the concentrations of the different silver species present (Ag^+, AgCl(aq), $AgCl_2^-$)? What is the total concentration of dissolved silver? (Use data from Appendix VII.)

7 Find the pH of a solution of K_2CO_3 containing 20 g of the salt per liter.

8 Estimate the pH of a $0.01M$ solution of Na_2S.

9 From the data in the appendix, Table VII-3, find the concentrations of each of the important aluminum ions—Al^{3+}, $AlOH^{2+}$, $Al(OH)_2^+$, and $Al(OH)_4^-$—in a solution at equilibrium with amorphous $Al(OH)_3$, at pH's of 4, 5, 6, 7, 8, and 9. Which is the dominant aluminum species at each pH? What is the total concentration of aluminum in solution at each pH? How would the results be altered if the solid is gibbsite rather than amorphous $Al(OH)_3$? Explain how your figures bear out the statement that $Al(OH)_3$ is an amphoteric hydroxide. (Retain your answers to this question for use in Prob. 7-8).

10 Make calculations similar to those in Prob. 9 for $Fe(OH)_2$. How do the results corroborate the statement that $Fe(OH)_2$ is not amphoteric? On the basis of these two sets of calculations, what conclusions can you draw about the amphoteric character of other hydroxides in Table VII-3 in the appendix?

11 Again using the data in the appendix, Table VII-3, calculate the pH of a solution of $0.1M$ $ZnCl_2$. If a base is added slowly to a slightly acid solution of $0.001M$ $ZnCl_2$, at what pH will $Zn(OH)_2$ begin to precipitate? Use the figures for amorphous $Zn(OH)_2$, and make the calculations (a) by using the total dissociation constant and (b) by using the stepwise dissociation constants.

12 If a solution having a concentration of $0.001M$ each of Al^{3+}, Cu^{2+}, Fe^{2+}, Fe^{3+}, Mg^{2+}, and Zn^{2+} is gradually made alkaline, in what order would you expect the hydroxides to precipitate?

13 If a solution containing Ca^{2+}, Mg^{2+}, Fe^{2+}, Mn^{2+}, Ba^{2+}, and Zn^{2+}, each in $0.001M$ concentration, comes in contact with a solution containing CO_3^{2-}, in what order will precipitates form? Answer the same question for a solution containing SO_4^{2-}.

14 In the normal pH range of surface waters (4 to 9), what would be the dominant dissolved species (ions or molecules or both) containing boron? Silicon? Fluorine? Divalent selenium? Quinquevalent phosphorus? (Use dissociation constants from Appendix VII.)

15 In what part of the normal pH range would the dominant carbon-containing species be undissociated H_2CO_3? In what part would it be HCO_3^-? In what sort of geologic environment would CO_3^{2-} be the dominant species?

16 A solution contains $0.1M$ total carbonate ($H_2CO_3 + HCO_3^- + CO_3^{2-}$). If the pH is 6, find the concentration of each carbonate species present. Compare your answers with Fig. 2-2. If the solution contains also $10^{-4}M$ Ca^{2+}, is it saturated, unsaturated, or supersaturated with $CaCO_3$?

17 Is the solution of Prob. 16 a buffer? If 1 ml of $6M$ HCl is added to a liter of this solution, what is the principal reaction that takes place? What is the final pH?

18 In Fig. 2-3, what determines the location of the point where the lines for H_2S and HS^- cross? The lines for HS^- and S^{2-}? The lines for H_2S and S^{2-}? Can you frame a generalization for the form of such a diagram for any dihydrogen acid, H_2A? From Fig. 2-3, read the concentration of S^{2-} (a) in a solution with $\Sigma S = 10^{-3}M$ and a pH of 4.5; (b) in a solution with $\Sigma S = 10^{-5}M$ and a pH of 7.0.

19 Suppose that a solution with pH 11 stands in contact with air. Over a period of a day or so, before evaporation has reduced its volume appreciably, would you expect its pH to remain constant? Answer the same question for a solution with pH 4.

REFERENCES AND SUGGESTIONS FOR FURTHER READING

The books listed at the end of Chap. 1 are good references for further details, both qualitative and quantitative, on modern concepts of acids and bases, hydrolysis, and buffers.

Barnes, I., Field measurement of alkalinity and pH, *U.S. Geol. Survey Water-Supply Paper* 1535-H, 1964. A description of the precautions necessary to obtain accurate measurements of pH and carbonate-ion concentrations.

Smith, R. M., and Martell, A. E., "Critical Stability Constants, vol. 4: Inorganic Complexes," Plenum Press, New York and London, 1976. Tables of solubility products, dissociation constants of acids and bases, and stability constants of ionic and molecular complexes.

CARBONATE SEDIMENTS

As a first example of the application of quantitative chemical reasoning to geologic problems, we choose the behavior of carbonates in sedimentary environments. These compounds have the advantage that their reactions are for the most part simple, rapid, and easily reversible, so that the ideas of chemical equilibrium from the preceding two chapters may be used directly. Furthermore, the behavior of the carbonates has been exhaustively studied by both chemists and geologists, so that a wealth of data is available for study. Most of our attention will be focused on the carbonate of calcium, which is by far the most abundant chemical sediment.

3-1 CALCIUM CARBONATE: SOLUBILITY CALCULATIONS

The deposition of calcium carbonate is controlled primarily by equilibrium in the reaction

$$CaCO_3(s) + H_2CO_3 \rightleftharpoons Ca^{2+} + 2HCO_3^- \tag{2-31}$$

In the last chapter (Sec. 2-9) we discussed the qualitative effects on this equilibrium of changes in pressure, temperature, pH, and concentration. Our purpose now is to see how some of these changes can be treated quantitatively. We shall try, for example, to answer such questions as these: How much calcium carbonate can be dissolved in a solution of known composition and pH? What is the equilibrium pH of water standing in contact with limestone? If a base is added slowly to a solution containing calcium ion and carbonic acid, at what pH will calcium carbonate begin to precipitate?

For the solubility product of calcite, the most common form of calcium carbonate, we find in tables

$$CaCO_3 \rightleftharpoons Ca^{2+} + CO_3^{2-}$$

$$K = [Ca^{2+}][CO_3^{2-}] = 10^{-8.3} = 4.5 \times 10^{-9} \text{ at } 25°C \qquad (3\text{-}1)$$

The solubility in pure water, according to the methods of Chap. 1, should then be the square root of this number, or about $6.8 \times 10^{-5}M$. We recall, however, that CO_3^{2-} hydrolyzes in water, so that the process of solution is a bit more complicated:

$$CaCO_3 + H_2O \rightleftharpoons Ca^{2+} + OH^- + HCO_3^- \qquad (2\text{-}32)$$

For this reaction the equilibrium constant can be evaluated by combining other constants:

$$K = [Ca^{2+}][OH^-][HCO_3^-] = [Ca^{2+}][CO_3^{2-}] \times [OH^-][H^+] \times \frac{[HCO_3^-]}{[H^+][CO_3^{2-}]}$$

$$= \frac{10^{-8.3} \times 10^{-14}}{10^{-10.3}} = 10^{-12}$$

In pure water, according to Eq. (2-32), the three ions Ca^{2+}, OH^-, and HCO_3^- should be formed in equal amounts, so that

$$[Ca^{2+}] = [OH^-] = [HCO_3^-] = \sqrt[3]{10^{-12}} = 10^{-4} \text{ mole/liter}$$

Hence calcite should dissolve in CO_2-free water to a concentration of $0.0001M$, or about 0.01 g/liter, and the solution should acquire a pH of 10. This is not quite the end of the story, because at such a high pH the acid HCO_3^- is appreciably dissociated. From Fig. 2-2, in fact, we can read that the ratio $[CO_3^{2-}]/[HCO_3^-]$ should be about 0.5. This means that our assumption about the equality of the three ions is not valid and that more $CaCO_3$ must dissolve to maintain the ion product at 10^{-12}. Using the rough figure 0.5 for the carbonate-bicarbonate ratio, we could guess that the concentrations of HCO_3^- and OH^- would be about two-thirds of the concentration of Ca^{2+}, instead of equal to it, so that the last equation should be

$$[Ca^{2+}] = \tfrac{3}{2}[OH^-] = \tfrac{3}{2}[HCO_3^-]$$

$$= \sqrt[3]{\tfrac{9}{4} \times 10^{-12}} = 1.3 \times 10^{-4}M$$

Thus the solubility is raised to a figure about twice as large as the $6.8 \times 10^{-5}M$ that we obtained by simply taking the square root of the solubility product; at the same time the OH^- concentration is lowered slightly, so that the pH is approximately 9.9 instead of 10.0.

This problem of calculating concentrations when calcium carbonate is in equilibrium with pure water may be set up more formally as follows. We note that the reactions contributing to the equilibrium involve a total of six concentrations: $[H^+]$, $[OH^-]$, $[Ca^{2+}]$, $[CO_3^{2-}]$, $[HCO_3^-]$, and $[H_2CO_3]$. To solve a problem with six unknowns requires that we have six equations. Four of these equations are supplied by the dissociation constants of H_2O, $CaCO_3$, H_2CO_3, and HCO_3^- [Eqs. (2-5), (3-1), (2-9), and (2-10)]. Another equation expresses the fact that all carbonate in solution is supplied by dissociation of $CaCO_3$, so that the sum of all carbonate concentrations must equal the concentration of calcium:

$$[Ca^{2+}] = [CO_3^{2-}] + [HCO_3^-] + [H_2CO_3] \tag{3-2}$$

The sixth equation comes from the requirement that the solution remain electrically neutral; the total concentration of positively charged ions must equal the total for negatively charged ions, with doubly charged ions counted twice:

$$2[Ca^{2+}] + [H^+] = 2[CO_3^{2-}] + [HCO_3^-] + [OH^-] \tag{3-3}$$

The array of six equations can then be solved in any one of various ways (see Garrels and Christ, 1965, page 78). An advantage of the formal approach is that results of greater accuracy can be obtained, but a simpler calculation like that in the last paragraph often provides all the accuracy that is justified by the quality of the original data.

The corresponding calculation for water containing CO_2 must be based on Eq. (2-31), for which the equilibrium constant is

$$K = \frac{[Ca^{2+}][HCO_3^-]^2}{[H_2CO_3]}$$

$$= [Ca^{2+}][CO_3^{2-}]\frac{[HCO_3^-]}{[CO_3^{2-}][H^+]}\frac{[HCO_3^-][H^+]}{[H_2CO_3]}$$

$$= \frac{10^{-8.3} \times 10^{-6.4}}{10^{-10.3}} = 10^{-4.4}$$

In the reaction of Eq. (2-31), 2 moles of HCO_3^- are produced for every mole of Ca^{2+}. Hence, if there is no other source of these ions,

$$[Ca^{2+}] = \tfrac{1}{2}[HCO_3^-]$$

For equilibrium with CO_2 in the atmosphere, the concentration of dissolved CO_2 or H_2CO_3 is approximately $10^{-5}M$ (Sec. 1-4). We can then substitute in the expression for the equilibrium constant:

$$\frac{[Ca^{2+}][HCO_3^-]^2}{[H_2CO_3]} = \frac{\tfrac{1}{2}[HCO_3^-]^3}{10^{-5}} = 10^{-4.4}$$

whence

$$[HCO_3^-] = 10^{-3.03} = 9.3 \times 10^{-4}M$$

The concentration of Ca^{2+} in the solution is half this figure, or $4.7 \times 10^{-4}M$, more than three times the concentration in pure water. To find the pH of the

#6

solution, we substitute values for $[HCO_3^-]$ and $[H_2CO_3]$ in the expression for the dissociation constant of carbonic acid:

$$\frac{[H^+][HCO_3^-]}{[H_2CO_3]} = \frac{[H^+] \times 10^{-3.0}}{10^{-5}} = 10^{-6.4}$$

whence $[H^+] = 10^{-8.4}M$, and pH = 8.4.

This calculation, it should be noted, is only approximate because $[CO_3^{2-}]$ is assumed to be negligibly small. If the concentration of this ion were appreciable (in other words, if the simple dissolving of $CaCO_3$ to form Ca^{2+} and CO_3^{2-} is significant in comparison with the reaction with H_2CO_3), then it would no longer be true that $[Ca^{2+}] = \frac{1}{2}[HCO_3^-]$, and a more complicated procedure would be necessary. The magnitude of $[CO_3^{2-}]$ can be estimated from the relation [derived from Eq. (2-10)]

$$[CO_3^{2-}] = 10^{-10.3}\frac{[HCO_3^-]}{[H^+]} = 10^{-10.3}\frac{10^{-3.0}}{10^{-8.4}} = 10^{-4.9}M$$

This is only about $\frac{1}{100}$ of the concentration of HCO_3^-, so for all ordinary purposes the assumption is justified.

In surface water at temperatures near 25°, therefore, the solubility of calcite ranges from about 1.3 to $4.7 \times 10^{-4}M$, or 0.01 to 0.05 g/liter, depending on the degree of saturation with CO_2. The higher figure may be exceeded in colder water because CO_2 becomes more soluble; it will be exceeded also in places where CO_2 is unusually abundant or where some other source of acid keeps the pH low. In soils, for example, the decomposition of organic matter gives local concentrations of CO_2 often on the order of 0.1 atm, and sometimes as high as 1 atm. The pH's of solutions that have come to equilibrium with calcite should be in the range 8 to 10—close to the lower figure at the earth's surface, and close to the upper figure at a depth out of contact with the atmosphere.

These numbers give us a quantitative expression of the variation in solubility of $CaCO_3$ that we predicted from qualitative arguments in the last chapter. Our next move obviously should be to compare the theoretically derived numbers with actual measured concentrations of $CaCO_3$ in natural solutions. The comparison is easily made, but it turns out to be most disillusioning. Concentrations of $CaCO_3$ in natural waters are extremely variable and only rarely come close to the numbers predicted in the last few paragraphs. Low concentrations can be plausibly explained as the result of failure of solutions to reach equilibrium with solid carbonate. In many natural waters, however, the discrepancy is in the opposite direction: concentrations are embarrassingly high, much higher than can be accounted for even with generous assumptions about temperature, CO_2 pressure, and acidity.

Possible reasons for these higher concentrations are discussed in the following six sections. They are worth examining in some detail, because they illustrate beautifully the difficulties in trying to make theoretical predictions adaptable to complex natural environments.

3-2 CALCITE AND ARAGONITE

One complication affecting solubility is the fact that $CaCO_3$ is polymorphous. The two common naturally occurring crystal forms are calcite and aragonite; at least one other (vaterite) can be prepared artificially and is known as a very rare mineral in nature. Solubilities of the various forms are different, as is shown by the solubility products for calcite and aragonite:

$$[Ca^{2+}][CO_3^{2-}] = 10^{-8.35} = 4.5 \times 10^{-9} \text{ for calcite} \tag{3-1}$$

$$[Ca^{2+}][CO_3^{2-}] = 10^{-8.22} = 6.0 \times 10^{-9} \text{ for aragonite} \tag{3-4}$$

A similar difference in solubility products was noted earlier (Sec. 1-5) for the two common forms of calcium sulfate, gypsum and anhydrite.

The existence of two solubilities for the same chemical compound suggests an awkward question. Suppose we have established equilibrium between solid aragonite and its saturated solution. The product of the concentration of the two ions is larger than the equilibrium product for calcite. Why, then, doesn't calcite precipitate? Theoretically, it would seem that the excess Ca^{2+} and CO_3^{2-} should combine to form calcite; then more aragonite should dissociate to reestablish its own equilibrium, more calcite should precipitate, and so on—the net result being the slow conversion of solid aragonite into solid calcite.

Practically, this doesn't happen, or at least it doesn't happen rapidly enough to be observed. A plausible explanation is that nuclei are not present for calcite to precipitate around. The solution is supersaturated with respect to calcite, and like any supersaturated solution it is essentially unstable; but when supersaturation is very slight, as in this case, crystallization is extremely slow unless seed crystals are present. If finely ground calcite is added to the equilibrium solution, and if the mixture is warmed to speed up reactions, the change becomes appreciable.

Aragonite is therefore unstable with respect to calcite under ordinary conditions. It becomes the stable form of calcium carbonate at high pressures, as might be anticipated from its greater density (2.9 g/cm^3, in contrast to 2.7 for calcite). The necessary pressure is so high, however, that it cannot be a factor in environments of sedimentation near the earth's surface. This leads to another awkward question: Why does aragonite ever appear in sedimentary rocks? When conditions are right for calcium carbonate to precipitate, why doesn't the more stable calcite form in all cases?

There is no entirely satisfactory answer to this question. It is an empirical fact that many polymorphic substances show this same tendency to precipitate first in metastable forms, which change only slowly to the stable varieties. We can guess that the manner of crystallization depends at least in part on reaction rates, the metastable forms being able to crystallize more quickly. In the case of calcium carbonate, experiments show that calcite and aragonite often precipitate together, the proportion of aragonite being greater if the reaction is carried out at high temperatures. The presence of certain other ions in solution, particularly Mg^{2+} and Sr^{2+}, also aids the formation of aragonite. Living creatures that use calcium carbonate in their shells may precipitate either polymorph, some species favoring one and some the other; many pelecypods precipitate both in alternate layers. In whatever way aragonite may form, during the course of geologic time it gradually alters to calcite, so that in old rocks and shells aragonite is a rare mineral.

3-3 SUPERSATURATION

The fact that calcite does not precipitate immediately from solutions at equilibrium with aragonite is proof that calcium carbonate can exist for extended periods in supersaturated solution. Why do we not use this simple fact to explain all the anomalously high concentrations of calcium carbonate in natural solutions?

The question is a tricky one to handle, because the phenomenon of supersaturation is very imperfectly understood. Some compounds, it is well known, can remain in solution almost indefinitely at concentrations far above saturation equilibrium; others precipitate whenever the saturation value is exceeded by a very small amount; and there is little theoretical basis for predicting which way a given compound will behave. Furthermore, the onset of precipitation is influenced by a variety of factors that are difficult to control or predict: mechanical disturbances, dust particles in the solution, the material of the container, unevenness of surfaces in contact with the solution. Probably in any precipitation process at least momentary supersaturation exists before precipitation begins, but the persistence of supersaturation depends unpredictably on these factors as well as on the nature of the solute and on its concentration. In any given case predictions are possible only on the basis of empirical evidence.

For $CaCO_3$, laboratory experiments show that this compound ordinarily precipitates rapidly, without any lag ascribable to supersaturation. If the experiment is done carefully, so as to minimize external disturbance and to build up the concentrations of Ca^{2+} and CO_3^{2-} slowly, supersaturation up to about twice the normal solubility can be demonstrated and can be maintained for times that are long by laboratory standards. In natural waters, especially where CO_2 is being slowly removed by aquatic plants, apparent temporary supersaturation up to five times the solubility has been recorded. But such cases are rare, and in general the high concentrations of $CaCO_3$ sometimes found in nature can be more plausibly explained in other ways.

The suggestion that supersaturation might account for abnormal carbonate concentrations in nature is an example of a kind of explanation that often looks superficially attractive. It depends on a nonequilibrium phenomenon—on the hypothesis that carbonate solubilities in natural environments are determined by processes to which the laboratory-derived rules of chemical equilibrium do not apply. In effect, it appeals to a supposed slow rate of change from a metastable to a stable mixture. Now there is nothing basically wrong with this kind of assumption; certainly slow rates and nonequilibrium mixtures are commonplace in the laboratory as well as in nature. Here, however, the hypothesis of large and widespread deviation from equilibrium in carbonate solutions is not supported by empirical data, so that it becomes a pure assumption. The assumption cannot really be proved wrong, because quantitative information on reaction rates is still very meager, but it gives no insight into the geochemistry of carbonates, since no specific prediction can be made from it. This is the kind of hypothesis that at first glance seems satisfying, but on closer examination turns out to be so vague and general that it explains nothing. If equilibrium does not exist, or is not closely approached, then the field is wide open for almost any kind of speculation. In some cases this supposition is necessary, but ordinarily it is good policy to turn to nonequilibrium hypotheses only as a last resort, after explanations based on equilibrium have been tried and found wanting.

3-4 EFFECT OF GRAIN SIZE ON SOLUBILITY

Another factor that modifies the solubility relations of $CaCO_3$ is the size of crystals exposed to a saturated solution. Experimentally, very tiny grains show a greater solubility (and, of course, a greater solubility product) than do large crystals. The mechanism of dissolving furnishes a ready explanation: ions escape from a crystal most easily on exposed corners and edges, which are more numerous on small particles than on large ones. The effect is noticeable only when particle dimensions are extremely small, say of the order of a few millimicrons. Because of the greater solubility of small particles, ions in equilibrium with them constitute a supersaturated solution with respect to large particles. This means that the small ones should eventually dissolve, and the large one should grow at their expense. A precipitate on first forming necessarily consists of very tiny grains; on standing the grains grow in size, at first rapidly and then more slowly, so that with most substances the crystals will have reached stable dimensions within a few seconds or minutes. Thus the measured solubility is the equilibrium solubility of larger crystals, and ordinarily the higher solubility of tiny particles is only a temporary phenomenon.

The growth of large grains at the expense of small ones in contact with a saturated solution is a familiar process both in geologic environments and in chemical laboratories. The chemist, faced with the problem of filtering a very fine-grained precipitate like $BaSO_4$ or $Fe(OH)_3$, often "digests" the precipitate by holding it near the boiling point for a few minutes; at the high temperature the growth of particles is accelerated, and the crystals soon become large enough for a filter paper to hold them. A geologist sees in this same phenomenon an explanation for the recrystallization and increase in grain size of many rocks formed originally as fine chemical precipitates. The coarse grains of some limestones and the development of chalcedony or quartz from opal are common examples.

3-5 THE EFFECT OF OTHER ELECTROLYTES: ACTIVITY AND ACTIVITY COEFFICIENT

Neither the substitution of aragonite for calcite nor the substitution of little grains for big ones has a large effect on the solubility of calcium carbonate. A third factor, however—the presence of other ions in solution—may change the solubility by tens or hundreds of times.

The effect of a common ion we have talked about before (Sec. 1-5): addition of either Ca^{2+} or CO_3^{2-} to a saturated solution of $CaCO_3$ will make the solubility less, by an amount that is easily computed from the solubility product. Electrolytes which do not supply a common ion generally have the opposite effect, making the solubility greater. To account for this observation, let us consider some details of the precipitation process. In the immediate vicinity of a Ca^{2+} ion in water solution, we would expect to find the particle surrounded by water molecules; since the water molecules are polar, with one end relatively positive and the other negative, we would expect them to be oriented around the particle, with their negative ends close to the ion and their positive ends facing outward. The CO_3^{2-} ions would be similarly hydrated, with the water molecules in their vicinity showing opposite orientations. These loosely held retinues of molecules around the ions would serve to shield Ca^{2+} and CO_3^{2-} from coming close to each other and precipitating. The shielding is evidently not very efficient, because most of the ions precipitate out of the solution in spite of it. Now suppose an electrolyte, say NaCl, is added. The solution will contain abundant Na^+ and Cl^- ions, and we might expect that these ions as well as water molecules would be attracted to Ca^{2+} and CO_3^{2-}. Each Ca^{2+} we can picture as the center of a cluster of water molecules and Cl^- ions, each CO_3^{2-} as the center of a cluster of water molecules and Na^+ ions. The added ions will increase the shielding action of the water, and it will be harder for the Ca^{2+} and CO_3^{2-} to find each other and precipitate. The solubility will therefore be greater, and the amount of increase should depend, within limits, on the concentration of added salt.

We can carry this naïve qualitative reasoning a bit further. If increased solubility in electrolyte solutions results from protective shields of opposite charge around each ion, we could guess that the effect would be greater in solutions containing divalent ions than in those with only univalent ions, since the attraction for multiply charged ions should be stronger. Thus the solubility of $CaCO_3$ should be higher in $0.1M$ $MgSO_4$ than in $0.1M$ NaCl solution. By the same argument, the increase in solubility should be greater for salts made up of multivalent ions than for those with only univalent ions; for example, the solubility of $CaCO_3$ should be increased more than the solubility of AgCl if both are put in a $NaNO_3$ solution. We could predict also that the effect of electrolytes on the solubility of undissociated solutes like H_2CO_3 or H_4SiO_4 should be small, since ions would have little tendency to cluster around uncharged molecules. In general, predictions of this sort are borne out by experiment.

Somewhat greater precision can be given to these qualitative conclusions if the concentration of an electrolyte solution is expressed as ionic strength rather than simple molarity. The *ionic strength* of a solution is defined as

$$I = \frac{1}{2} \sum_i c_i z_i^2$$ (3-5)

where c_i is the concentration of an ion in moles per liter, z_i is its charge, and the sum is taken over all the ions in the solution. For a solution containing $0.1M$ $BaCl_2$ and $0.04M$ $NaNO_3$, for example, the ionic strength would be

$$I = \tfrac{1}{2}(0.1 \times 4 + 0.2 \times 1 + 0.04 \times 1 + 0.04 \times 1) = 0.34$$

The ionic strength differs from the sum of molar concentrations in a complex solution in that it emphasizes the effect of higher charges of multivalent ions. Empirically it is found that the ionic strength gives a good measure of the effect of electrolytes on solubility. For insoluble salts of a given type (univalent like $AgCl$; didivalent, like $CaSO_4$; unidivalent, like Ag_2SO_4; and so on), the effect on solubility of any solution of given ionic strength, regardless of its composition, is roughly the same. Furthermore, experiments show that the solubility of a given salt generally shows a rough proportionality to the square root of ionic strength, at least for fairly dilute solutions. These are useful rules, but they are only approximate and have many exceptions.

In order to attach numbers to the effect of electrolytes on solubility, we define a quantity called *activity*. Suppose that the solubility of a salt is measured in solutions of different ionic strengths. By plotting the solubility against ionic strength (or better, against its square root), we can extrapolate back to a hypothetical solubility at zero ionic strength (Fig. 3-1). Generally this will be nearly the

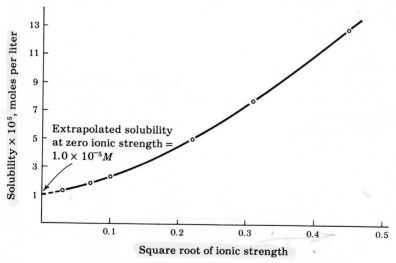

Figure 3-1 Change in solubility of $BaSO_4$ with ionic strength.

same as the solubility measured when the salt is stirred in pure water. The value of the solubility at this point is called the *activity* of the salt in a saturated solution, and the solubility product is called the *activity product*, or the product of the activities of the separate ions.

By definition, the activity and the activity product are constant for any saturated solution at a given temperature. In a solution with ionic strength higher than zero, then, the increased solubility may be expressed by saying that the measured concentration of $CaCO_3$ is greater than its activity. The relation between the two quantities may be shown by means of a factor called the *activity coefficient*:

$$a_{CaCO_3} = \gamma_{CaCO_3} \times [CaCO_3 \text{ in solution}] \tag{3-6}$$

where a is activity, γ is activity coefficient, and $[CaCO_3$ in solution] means the number of moles of $CaCO_3$ dissolved in a liter of solution. Similar expressions can be set up for activities and activity coefficients of the separate ions:

$$a_{Ca^{2+}} = \gamma_{Ca^{2+}}[Ca^{2+}] \quad \text{and} \quad a_{CO_3^{2-}} = \gamma_{CO_3^{2-}}[CO_3^{2-}] \tag{3-7}$$

In very dilute solutions (with ionic strength less than, say, 0.001) the activity coefficients are close to 1; this means that activities are practically equal to concentrations, and the activity product is practically equal to the solubility product which we have previously defined in terms of concentrations (Sec. 1-5):

$$a_{Ca^{2+}} a_{CO_3^{2-}} = K_a \cong [Ca^{2+}][CO_3^{2-}] \tag{3-8}$$

As the ionic strength of the solution increases, the activity coefficients decrease. For practically all ions the activity coefficient remains less than 1 for ionic strengths up to about 2. Some salts and some ions in very concentrated solutions, however, have activity coefficients much greater than 1, which means that, as a solution becomes concentrated, the solubility of a slightly soluble salt may pass through a maximum and then decrease abruptly.

Activities have a more general significance than the above definition would indicate. They need not be defined only for a saturated solution, but may be used in any solution. The activity of an ion, so to speak, is its *effective concentration*, the part of its analytical concentration that determines its behavior toward other ions with which it may react. At a given ionic strength the activity is proportional to the concentration, and the proportionality constant is the activity coefficient. In other words, Eqs. (3-6) and (3-7) remain true, whether the solution is saturated with $CaCO_3$ or not.

Activities, like concentrations, may be expressed as moles per liter or moles per kilogram. This usage will be followed in this book, but activities may also be treated as dimensionless quantities (see Sec. 8-14).

Now, how can we measure activities and activity coefficients? The subject is too complicated for detailed treatment here, but a few general remarks are in order. We have mentioned activity coefficients for both salts and separate ions; actually only those for the salts are directly measurable, and the coefficients for ions are estimated from those for salts by making reasonable assumptions. One useful method of determining activities experimentally was described above: measurement of solubilities of a salt in solutions of different ionic strengths, and extrapolation of the solubilities to infinite dilution ($I = 0$). Other experimental methods include measurements of electromotive force, of vapor pressures, and of distribution of solutes between different solvents. Such measurements are described in detail in some of the references at the end of this chapter.

For work with reasonably dilute solutions where great accuracy is not important, activity coefficients can be estimated on the basis of a theoretical treatment formulated by Debye and Hückel. In this theory, ions are regarded as point centers of charge in a medium with a dielectric constant equal to that of the pure solvent. These are obviously idealizations, but for solutions with ionic strengths up to about 0.1 they give results in reasonable agreement with experiment. The simplest form of the Debye-Hückel theory gives as an expression for the activity coefficient of an ion

$$-\log \gamma = Az^2I^{1/2} \tag{3-9}$$

where z is the charge on the ion, I is the ionic strength of the solution, and A is a constant depending on temperature and the dielectric constant of the solvent, equal approximately to 0.51 in water at 25°C. The ionic strength, whose importance was recognized empirically before the Debye-Hückel theory was proposed, appears in the theoretical derivation as an essential factor in determining the size of γ. A more satisfactory expression for $\log \gamma$, derived in part from an extension of the theory and in part empirically, has been given by Davies (1962):

$$-\log \gamma = Az^2\left(\frac{I^{1/2}}{1 + I^{1/2}} - 0.2I\right) \tag{3-10}$$

Values of γ calculated from this equation are given in Table 3-1. The values can be considerably refined by using more complicated expressions for particular kinds

Table 3-1 Activity coefficients calculated from Davies' equation

Ionic strength	Ionic charge		
	±1	±2	±3
0.001	0.97	0.87	0.73
0.005	0.93	0.74	0.51
0.01	0.90	0.66	0.40
0.05	0.82	0.45	0.16
0.1	0.78	0.36	0.10
0.2	0.73	0.28	0.06
0.5	0.69	0.23	0.04
0.7	0.69	0.23	0.04

of ions, but for most geological purposes the numbers in Table 3-1 are sufficiently accurate.

Some of the basic assumptions of the Debye-Hückel theory break down at ionic strengths greater than 0.1, and extension of Davies' equation beyond this point is purely empirical. At ionic strengths above about 0.5, the equation gives only order-of-magnitude estimates. Such estimates may be useful in geochemical speculations, but their possible gross inaccuracy should be kept in mind.

From Table 3-1 the approximate quantitative effect of electrolytes on the solubility of $CaCO_3$ can be easily read. Even at an ionic strength as low as 0.001 (for example, in a $0.001M$ NaCl solution), the activity of Ca^{2+} and CO_3^{2-} is reduced to 87% of their concentration. This means that the solubility is $\frac{100}{87}$, or 1.15, times as great as its value at infinite dilution. The solubility product, using concentrations, would be greater than the activity product by the square of this number, since from Eqs. (3-7) and (3-8):

$$K_c = [Ca^{2+}][CO_3^{2-}] = \frac{a_{Ca^{2+}} a_{CO_3^{2-}}}{\gamma_{Ca^{2+}} \gamma_{CO_3^{2-}}} = \frac{K_a}{0.87^2} = 1.3 K_a$$

If the ionic strength of the solution is 0.5, the solubility is increased to $\frac{100}{23}$, or more than 4 times the value in very dilute solution, and the concentration product is nearly 20 times the activity product. Clearly, the effect of other electrolytes in solution is one of the major reasons why the measured concentration of $CaCO_3$ in some natural waters can be much larger than would be predicted from the solubility product.

More generally, the numbers in Table 3-1 show the deviations to be expected in the solubilities of any slightly soluble compounds due to the presence of electrolytes in solution. The numbers we have used heretofore for solubility products, and the numbers given in Appendix VII, are actually activity products. We have considered them equivalent to concentration products, and we have used them to calculate concentrations in solutions of fairly high ionic strength. This procedure gives reasonably accurate results in solutions whose electrolyte concentration is less than $0.001M$ and gives order-of-magnitude accuracy at concentrations up to $0.1M$. For geologic purposes this kind of accuracy is often sufficient, and we shall continue to assume in many later discussions that activity coefficients can be set equal to 1 and disregarded. But for precise work, even in fairly dilute solutions, the activity corrections cannot be neglected.

3-6 ION ASSOCIATION: FORMATION OF COMPLEXES

Another factor that might conceivably enhance the solubility of $CaCO_3$ is the possible association of its ions to form complexes. Thus far we have assumed that Ca^{2+} and CO_3^{2-} have no tendency to associate in solution, except to form a solid precipitate when the product of their concentrations exceeds the solubility product. This is nearly correct, but not quite. The two ions join to a slight extent into uncharged dissolved $CaCO_3$ molecules:

$$Ca^{2+} + CO_3^{2-} \rightleftharpoons CaCO_3(aq) \qquad K = \frac{a_{CaCO_3(aq)}}{a_{Ca^{2+}} a_{CO_3^{2-}}} = 10^{3.2}$$

As long as $a_{CO_3^{2-}}$ remains less than $10^{-3.2} M$ $(0.0006M)$, the value of K tells us that most of the dissolved calcium remains as the free ion Ca^{2+}. Solutions with $a_{CO_3^{2-}}$ greater than this can exist in nature only in unusual, highly alkaline environments, so that ordinarily $CaCO_3(aq)$ plays a minor role. In most situations of geologic interest, the solubility of solid $CaCO_3$ would be increased no more than a percent or so by this factor.

Neutral complexes like $CaCO_3(aq)$ are common in the chemistry of didivalent salts. The K's for the association reactions, so far as they are known, lie in the range 10^2 to 10^4. This means that the complexes are important only in solutions where the anion concentration is unusually high, and in most geologic situations the possibility of association can therefore be safely neglected. A few divalent ions form larger and more stable complexes by uniting with two or more anions; thus the complex ions $UO_2(CO_3)_2^{2-}$ and $UO_2(CO_3)_3^{4-}$ play an important role in the geologic behavior of uranium.

The term *complex* will appear in many future discussions. Its most general definition is simply any combination of two or more atoms that can exist in solution. On this basis such familiar substances as OH^-, SO_4^{2-}, H_2CO_3 would be complexes. In common usage, however, the word refers to more unusual atom combinations, like $CaCO_3(aq)$ and $UO_2(CO_3)_2^{2-}$ in the preceding paragraphs. Particularly important in geologic discussions are chloride complexes like $PbCl^+$, $PbCl_2(aq)$, $AuCl_4^-$, and sulfide complexes like AsS_2^- and HgS_2^{2-}. (For dissociation constants of some complexes, see the appendix, Table VII-4.)

3-7 EFFECT OF ORGANISMS ON SOLUBILITY

Organic activity, as noted briefly in Sec. 2-9, affects solubility in various ways. Some inorganic compounds are used for the construction of shells and skeletons; the compound that is now the center of attention, calcium carbonate, is an excellent example. Some organisms in nearly every phylum of invertebrates have found calcium carbonate a useful structural material, and its use goes back in geologic time at least to the beginning of the Cambrian period. The mechanism of its precipitation in shells is not known, but apparently it can form only in water that is saturated or nearly so. In part at least, the precipitated shell substance is covered with a thin film of organic material that protects it from dissolving if the water becomes temporarily unsaturated, either by cooling or by dilution. Whether $CaCO_3$ can be made to precipitate by organisms in a solution not saturated with its ions, so that its apparent solubility is reduced below the equilibrium value, is an unanswered question. Certainly there is no evidence that any organism can cause its precipitation from solutions very far below the saturation concentration. Thus organisms can prevent supersaturation and perhaps can lower the equilibrium solubility slightly, but at least in their shell-building activities they do not influence the solubility very greatly. Their role is chiefly to make use of a process which, in their absence, would take place inorganically.

Another way in which organisms can affect precipitation is by using the precipitation reaction to provide energy for their life processes. This operates chiefly with slow reactions that are far from equilibrium. Ferrous compounds in solution, for example, react slowly with atmospheric oxygen to precipitate ferric oxide, and some kinds of bacteria take advantage of this reaction to obtain energy. The ultimate solubility of the ferric oxide is not affected by the bacteria; they simply promote a reaction which would take place slowly in their absence, in this case a reaction that gives out considerable energy. They serve, so to speak, as an organic catalyst. The precipitation of calcium carbonate would probably not be used in this manner for obtaining energy, because the reaction is rapid and in natural solutions is seldom far from equilibrium, so that the amount of energy available is small.

A third sort of influence exerted by organisms is the causing of precipitation simply as an incidental by-product of their life processes. Water plants, as we have noted earlier, use up dissolved CO_2 during photosynthesis, thereby reducing the amount of H_2CO_3 in solution and leading to precipitation of $CaCO_3$. The precipitation of $CaCO_3$ may also be favored by the decay of organisms, if the products of decay are alkaline, or impeded if the products are acid, but unfortunately the acidity or alkalinity to be expected from decay in particular circumstances is not predictable from present data.

Thus organisms affect the formation of insoluble compounds primarily by creating conditions favorable for their precipitation rather than by changing the solubility product.

3-8 PRECIPITATION OF $CaCO_3$ IN SEAWATER

By way of summarizing this long discussion of the factors that influence solubility, let us survey briefly the behavior of calcium carbonate in the sea.

Seawater is a concentrated and exceedingly complex solution, containing electrolytes in great variety plus an abundance of living and dead organic material. The ordinary laws of dilute solution cannot be applied, or at best need great modification. The chemistry of seawater can be described fairly satisfactorily in general terms, but details about the behavior of even so simple a substance as $CaCO_3$ remain obscure.

Straight analysis of water taken from near the sea surface gives for the concentration of Ca^{2+} about $0.01M$, and for the concentration of CO_3^{2-} about $0.0003M$. The ion product is therefore 3×10^{-6}, which is roughly 700 times the activity product of calcite, 4.5×10^{-9}. Does this mean that seawater is enormously supersaturated with $CaCO_3$, or can we explain the discrepancy by referring to some of the factors discussed above that influence solubility?

Using the activity product of aragonite rather than calcite (6.0×10^{-9} rather than 4.5×10^{-9}) would reduce the discrepancy to 500 instead of 700, but this doesn't help a great deal. Nor can we appeal to the enhanced solubility of very tiny grains, because a large part of the ocean is in contact with well-crystallized calcite and aragonite. The high concentration of electrolytes, however, offers hope of explaining at least a part of the apparent supersaturation. The ionic strength of seawater is about 0.7, and in such a solution the activity coefficient of a divalent ion should be about 0.23 (Table 3-1). More accurate figures given by Garrels and Thompson (1962) are 0.28 for Ca^{2+} and 0.20 for CO_3^{2-}. With these coefficients we can calculate for the activity product:

$$a_{Ca^{2+}} a_{CO_3^{2-}} = 0.01 \times 0.28 \times 0.0003 \times 0.20 = 1.7 \times 10^{-7}$$

This is still almost 30 times larger than the equilibrium activity product of aragonite. Garrels and Thompson have shown, however, that ion association cuts down the amount of free Ca^{2+} by about 10 percent and of CO_3^{2-} by about 90 percent, the former uniting especially with the very abundant anion SO_4^{2-} and the latter with the abundant cation Mg^{2+}. Thus the concentration of free Ca^{2+} is reduced to about 91 percent of its analytical value, and that of CO_3^{2-} to about 9 percent of its analytical value. This means that the activity product should be changed to

$$0.91 \times 0.09 \times 1.7 \times 10^{-7} = 14 \times 10^{-9}$$

The discrepancy is not wholly removed, but this is about as close as we can expect to come in working with so complex a solution. It is a fair guess, therefore, that seawater is approximately saturated with $CaCO_3$. This seems eminently reasonable, inasmuch as $CaCO_3$ is observed to be in process of dissolving in some parts of the ocean and precipitating in other parts.

The slight shifts in the marine environment that exercise major control over precipitation and dissolution of $CaCO_3$ are so obvious that they hardly need pointing out. Precipitation is favored where the water is warm and where CO_2 is being lost through evaporation or photosynthesis, as on the shallow banks off Florida and the Bahamas. Cold water and abundance of CO_2 promote dissolution; the scarcity of calcareous shells in parts of the deep ocean below about 4,000 meters, for example, can be plausibly explained by the presence at these depths of cold masses of CO_2-saturated water moving slowly equatorward from the polar regions. Tide pools in temperate regions show evidence of precipitation of $CaCO_3$ when the trapped seawater warms up during the day and of dissolution at night when the water becomes cold.

The $CaCO_3$ that precipitates from seawater may be either calcite or aragonite, the circumstances that favor one or the other being still not entirely clear. With time, either during or after diagenesis, precipitated aragonite changes to the more stable calcite. Precipitation may occur either inorganically or through the agency of organisms; organisms may accomplish the precipitation either in the building of their shells, or incidentally as they remove CO_2 from the water, or during decay if they supply alkaline materials to the water. The old argument as to the relative amounts of limestone formed by inorganic precipitation on the one hand and by organic processes on the other has never been settled, but it is probably not a matter of great importance. In modern seas the places where abundant inorganic precipitation might be expected are precisely those places where organisms flourish most luxuriantly. Investigators disagree as to whether the fine-grained carbonate precipitated indirectly by organisms can be distinguished from an inorganic precipitate; if the distinction is difficult in modern sediments, it is doubtless impossible in older rocks where later recrystallization would have obscured whatever minute textural differences might have been present originally.

Once deposited as bottom sediment, $CaCO_3$ gradually hardens into limestone, through a series of processes that collectively come under the heading of *diagenesis:* growth of large crystals at the expense of small ones, conversion of aragonite into calcite, replacement of other materials by calcite, and deposition of calcite between the grains by circulating solutions. The ease with which $CaCO_3$ recrystallizes even at low temperatures is attested by the coarse, intergrowing grains of many older limestones. An unexplained detail in carbonate geochemistry is the fact that some limestones from far back in the Paleozoic remain very fine grained, contrasting sharply with the recrystallized textures that are common even in much more recent limestones.

Thus the chemistry of calcium carbonate in geologic processes is fairly well understood. Much of it is embodied in simple displacements of the equilibrium with carbonic acid and bicarbonate, although additional concepts are needed to account for the relations of the two polymorphs, the behavior of very tiny grains, the effect of dissolved electrolytes, and the influence of organisms. Quantitative treatment is feasible only for relatively simple freshwater solutions, but numbers that show fair agreement with observation can be obtained even for solutions as concentrated as seawater. We shall find that the rules of solubility developed in this section apply to other chemical precipitates of geologic interest, and that other carbonates in particular are governed by equilibria similar to the one we have used for calcium carbonate.

3-9 OTHER SIMPLE CARBONATES

General

All the carbonates that precipitate as sediments are somewhat soluble in carbonic acid, and all of them crystallize out of solution when the concentration of carbonic acid diminishes. This means that other carbonates enter into equilibria like the $CaCO_3$ equilibrium [Eq. (2-16)], according to reactions that can be symbolized

$$MCO_3 + H_2CO_3 \rightleftharpoons M^{2+} + 2HCO_3^- \tag{3-11}$$

the M^{2+} standing for Fe^{2+}, Sr^{2+}, Mn^{2+}, and so on. Like $CaCO_3$, the other carbonates are more soluble when the concentration of dissolved CO_2 is high, whether because of low temperature or high pressure or decay of organic matter, and less soluble when CO_2 is removed from solution by heating, by decrease in pressure, by addition of alkali, or by the photosynthetic activity of plants. The chemistry of carbonates is essentially simple, and their solubilities can be referred to shifts in a few well-understood equilibria.

Solubility (activity) products of the carbonates that might be expected in sedimentary rocks are shown in Table 3-2. The table shows also the measured concentrations of the metals in seawater, and the amounts of the metals that could theoretically exist in the same solution with the CO_3^{2-} of seawater, assuming that the solubility products are increased by a factor of 70. This factor is a purely empirical number, obtained as the factor necessary to bring the calculated $[Ca^{2+}]$ in equilibrium with the $3 \times 10^{-5}M$ of free CO_3^{2-} in seawater into agreement with the measured concentration. It includes the various corrections mentioned in a preceding paragraph: activity coefficients, ion association, and the unresolved discrepancy between calculated and measured values for the solubility of $CaCO_3$. Other carbonates of doubly charged metal ions, it may be reasonably assumed, are affected by these factors to about the same extent as $CaCO_3$; hence their

Table 3-2 Solubility products of carbonates and concentrations of metal ions in seawater at 25°C

	Solubility product of carbonate in distilled water†	Solubility product × 70	Measured concentration of metal ion in seawater, moles/liter‡	Possible concentration of metal ion at equilibrium with CO_3^{2-} of seawater (free $CO_3^{2-} = 3 \times 10^{-5}M$)§
$MgCO_3 \cdot 3H_2O$	2.5×10^{-6}	1.8×10^{-4}	0.053	6
$CaCO_3$(calcite)	4.5×10^{-9}	3.2×10^{-7}	0.010	0.011
$SrCO_3$	1.0×10^{-9}	7.0×10^{-8}	9.0×10^{-5}	2.3×10^{-3}
$BaCO_3$	5.0×10^{-9}	3.5×10^{-7}	1.5×10^{-7}	0.012
$MnCO_3$	5.0×10^{-10}	3.5×10^{-8}	3.6×10^{-9}	1.2×10^{-3}
$FeCO_3$	2.0×10^{-11}	1.4×10^{-9}	3.5×10^{-8}	4.7×10^{-5}
$CoCO_3$	1.0×10^{-10}	7.0×10^{-8}	8.0×10^{-10}	2.3×10^{-3}
$NiCO_3$	1.3×10^{-7}	9.1×10^{-6}	2.8×10^{-8}	0.30
$CuCO_3$	2.5×10^{-10}	1.8×10^{-8}	8.0×10^{-9}	6.0×10^{-4}
$ZnCO_3$	1.0×10^{-10}	7.0×10^{-9}	7.6×10^{-8}	2.3×10^{-4}
$CdCO_3$	2.0×10^{-14}	1.4×10^{-12}	1.0×10^{-9}	4.7×10^{-8}
$PbCO_3$	8.0×10^{-14}	5.6×10^{-12}	2.0×10^{-10}	1.9×10^{-7}

† From Table VII-1, Appendix VII.

‡ From Brewer, P. G., Minor elements in sea water, chapter 7 in "Chemical Oceanography," 2d ed., vol. 1, pp. 415–497, Academic Press, London, 1975.

§ The measured total CO_3^{2-} in seawater is of the order of 0.0003M, but much of this is associated with other ions, especially Mg^{2+}, so that the concentration of free CO_3^{2-} is only about one-tenth of this figure.

activity products should be increased by roughly the same amount to represent solubility equilibrium in seawater. The modified solubility products have meaning only as orders of magnitude, of course, but even with this limitation they bring out one striking fact: the only simple carbonate besides calcium carbonate that even approaches saturation in seawater is strontium carbonate, $SrCO_3$. The concentration of Ba^{2+} is kept below the equilibrium figure because $BaSO_4$ is more insoluble than $BaCO_3$ and because the concentration of SO_4^{2-} in seawater is far greater than that of CO_3^{2-}. Magnesium is kept low by the slow formation of dolomite (see page 70) and probably by the formation of clay minerals. Iron and manganese cannot become concentrated in seawater because they are oxidized and precipitated as oxides of higher oxidation number. It is small wonder, then, that the only simple carbonate that forms a large amount of sedimentary rock is the carbonate of calcium.

Replacement of one carbonate by another is often observed, both in sedimentary rocks and in veins. The conditions under which replacement can occur may be partly reconstructed from the data of Table 3-2. For example, the replacement of calcite by siderite should follow the equation

$$CaCO_3 + Fe^{2+} \rightleftharpoons Ca^{2+} + FeCO_3 \qquad (3\text{-}12)$$

for which the equilibrium constant is

$$K = \frac{[Ca^{2+}]}{[Fe^{2+}]} = \frac{[Ca^{2+}][CO_3^{2-}]}{[Fe^{2+}][CO_3^{2-}]} = \frac{4.5 \times 10^{-9}}{2.0 \times 10^{-11}} = 225$$

This means that at ordinary temperatures calcite should be replaced by siderite if it is in contact with a solution containing Fe^{2+} in a concentration more than $\frac{1}{225}$ that of Ca^{2+}, and likewise that siderite can be replaced by calcite only if the solution has more than 225 times as much Ca^{2+} as Fe^{2+}.

Iron Carbonate

Siderite precipitates when a solution containing abundant Fe^{2+} mixes with a solution containing HCO_3^- or CO_3^{2-}, or when a solution containing Fe^{2+} and HCO_3^- evaporates or becomes alkaline—in other words, whenever the equilibrium of Eq. (3-11) is shifted to the left. It is not a common sediment, because much of the iron liberated by weathering is oxidized at once to ferric oxide, and the amount of Fe^{2+} in most surface waters is vanishingly small. Under unusual conditions, where either the supply of ferrous iron is large or a reducing environment is maintained by abundant organic matter, siderite can precipitate in great quantities—as evidently happened, for example, in some of the lakes and swamps associated with formation of coal beds (giving the familiar "clay ironstones"). Siderite is rare as a primary mineral in marine sediments, because reducing environments in the sea commonly have abundant hydrogen sulfide and the sulfide of iron (pyrite) precipitates rather than the carbonate. In the form of concretions siderite occurs in a variety of sedimentary rocks, probably deposited after the rock was at least partly consolidated, from ground-water solutions whose dissolved oxygen had been exhausted.

The relation of siderite to iron oxide can be expressed in the equation

$$2H_2O + 2FeCO_3 + \tfrac{1}{2}O_2 \rightleftharpoons Fe_2O_3 + 2H_2CO_3 \qquad (3\text{-}13)$$

This may be regarded as an equilibrium process, displaced toward the right in the presence of free oxygen and displaced toward the left when organic material is present to use up the oxygen. The position of equilibrium depends not only on the oxygen, but also in a complicated way on the pH, since both $FeCO_3$ and Fe_2O_3 react with excess acid. We could study these complications by setting up an expression for the equilibrium constant in the usual way, but the discussion is best deferred until other ways of handling equilibria have been described in later chapters.

Magnesium Carbonate

The solubility relations of magnesium carbonate are complicated by the existence of stable hydrates. When CO_3^{2-} is added to Mg^{2+} at ordinary temperatures, either the trihydrate $MgCO_3 \cdot 3H_2O$ (which occurs in nature as the rare mineral nesquehonite) or the hydrated basic carbonate $Mg_4(OH)_2(CO_3)_3 \cdot 3H_2O$ (the mineral hydromagnesite) is precipitated, depending on temperature and pH. The normal carbonate $MgCO_3$ (magnesite) probably forms only at higher temperatures, or by slow alteration of one of the hydrated carbonates. Under usual conditions, therefore, the precipitation of magnesium is controlled by the solubility product of a hydrate:

$$MgCO_3 \cdot 3H_2O \rightleftharpoons Mg^{2+} + CO_3^{2-} + 3H_2O \qquad K = [Mg^{2+}][CO_3^{2-}] = 10^{-5.6}$$

This product, about 500 times that of calcite, is probably never exceeded in normal seawater despite the high concentration of magnesium. Rarely, in brines obtained by evaporation of seawater or in the water of salt lakes, the ratio Mg/Ca becomes high enough for the primary precipitation of magnesium carbonate. This compound also appears locally as a product of weathering of rocks rich in magnesium, particularly the ultramafic rocks. Presumably the first material formed is a hydrated carbonate, which may later be altered to magnesite, but details of the process are not clear.

Some marine organisms, in building their calcareous shells, include a good deal of magnesium in the shell structure. Such shell material forms the so-called "high-magnesium limestone" on consolidation. How this is accomplished, and why some organisms are more partial to magnesium than others, remain unanswered questions.

Carbonates of Mn, Sr, and Ba

Of the remaining carbonates in Table 3-2, $MnCO_3$ (rhodochrosite) is the only one known definitely to occur as a primary sediment. Manganese, like iron, is readily oxidized on exposure to air, so that Mn^{2+} is seldom abundant in surface waters, and the carbonate is correspondingly rare. Strontianite $(SrCO_3)$ occurs in sedimentary rocks, but probably always as a replacement or as a product of enrichment by leaching away of other material. Calcium carbonate formed in the sea always contains some Sr, as might be expected from the fact that both $CaCO_3$ and $SrCO_3$ are near saturation in seawater, but the $SrCO_3$ remains a very minor constituent of limestone unless it is concentrated by solutions later. The sulfate of strontium, celestite, is more insoluble than the carbonate in slightly acid solutions, so that celestite is a commoner mineral in sediments than strontianite, although it also is probably never a primary precipitate. Witherite $(BaCO_3)$ is exclusively a vein mineral; it cannot occur in sediments because barite, $BaSO_4$, is more insoluble and the ratio $[SO_4^{2-}]/[CO_3^{2-}]$ is high enough in most ground water and surface water to prevent the carbonate from forming.

3-10 THE DOLOMITE PROBLEM

The most stubborn question in carbonate geochemistry is the origin of the double carbonate dolomite. The "dolomite problem" may be stated very simply: Dolomite rock is one of the commonest of sedimentary materials, appearing as thick and extensive beds in strata of all ages from the Precambrian to the Cenozoic; there is no geologic evidence to indicate that its formation took place under unusual conditions of temperature or pressure; yet efforts to prepare dolomite in the laboratory under simulated sedimentary conditions have failed, and no dolomite is observed to be forming in nature in ordinary sedimentary environments. It is true that the laboratory precipitation of dolomite at low temperatures has recently been reported, from solutions at pH's greater than 9.5 and containing high concentrations of SO_4^{2-} and NO_3^-; it is also true that a little dolomite has been observed, probably forming as a primary precipitate, in present-day deposits from hot springs, in sediments of salt lakes, and in muds from salt lagoons undergoing strong solar evaporation. But the extreme conditions represented by both the laboratory experiments and the field occurrences, compared with usual sedimentary environments, make the abundance of dolomite in the sedimentary record seem all the more mysterious. Why should this one chemical sediment be such a conspicuous anomaly?

For answer we look first to crystallography. The mineral dolomite is a double carbonate of magnesium and calcium, $CaMg(CO_3)_2$, with a structure that may be visualized as a distorted NaCl framework (Sec. 5-1) in which the anions are CO_3^{2-} groups and the cations are regularly alternating Ca^{2+} and Mg^{2+}. The regular alternation is important. This is a special, highly ordered crystal structure, which perhaps takes a long time to grow. When attempts are made to precipitate dolomite in the laboratory, the usual result is a mixture of calcite and hydromagnesite, simple compounds which can form rapidly from solution. The magnesium-rich calcium carbonates in the shells of some marine organisms, referred to above, are not dolomite but structures with a random distribution of Ca and Mg—again the sort of structure that might be expected to form fairly rapidly. Dolomite is readily prepared artificially at temperatures somewhat over 100°, the function of temperature probably being to speed up the movement of ions so that Ca and Mg can find their places in the ordered structure within a reasonable time. Attempts to prepare the compound at successively lower temperatures give precipitates whose x-ray diffraction patterns show the characteristic lines of dolomite becoming progressively less numerous and more fuzzy. A reasonable inference from these facts is that dolomite forms at ordinary temperatures so very slowly that we have no chance to observe the process in nature or to duplicate it in the laboratory. This slowness of formation, due to the necessity of attaining a highly ordered structure, seems a convincing answer to at least part of the dolomite riddle.

One might expect to gain information about the origin of dolomite from solubility data, as one can for other carbonates. The difficulty lies in the uncertainty about attainment of solubility equilibrium. The amount of dolomite that goes into solution can be measured readily enough, but we cannot be sure that the solution has reached saturation because the ions will not recombine to form solid dolomite. An approximation to the solubility can be obtained by stirring dolomite in water and following the changes in concentration of one of its ions, or of the pH, until the concentrations no longer change; if reproducible results are obtained in such experiments, it seems likely that a condition approaching equilibrium has been reached. From experiments of this sort, Garrels, Thompson, and Siever (1960) have found that the apparent solubility product

$$K = a_{Ca^{2+}} a_{Mg^{2+}} a^2_{CO_3^{2-}} \qquad (3\text{-}14)$$

is slightly less than the same product of ions obtained from a mixture of calcite and magnesite treated in the same manner. In other words, reactions like

$$CaCO_3 + Mg^{2+} + 2HCO_3^- \rightleftharpoons CaMg(CO_3)_2 + H_2CO_3 \qquad (3\text{-}15)$$

$$2CaCO_3 + Mg^{2+} \rightleftharpoons CaMg(CO_3)_2 + Ca^{2+} \qquad (3\text{-}16)$$

are displaced in the direction of dolomite at usual concentrations of Ca^{2+} and Mg^{2+}, but the difference in solubilities is so slight that there is little drive to make the two ions assume the ordered structure of dolomite.

Regarding the numerical value of K in Eq. (3-14) there is still wide disagreement. Published values range from 10^{-17} to 10^{-19}; Garrels et al. find a value near the lower figure, but most estimates by indirect means favor a number closer to 10^{-17}. Measurements of $a_{Ca^{2+}}$, $a_{Mg^{2+}}$, and $a_{CO_3^{2-}}$ in cave waters by Holland et al. (1964) give an activity product higher than 10^{-15}, with no precipitation of dolomite; whatever the true solubility product may be, this indicates that supersaturation with respect to dolomite can persist for long periods.

Another difficulty in measuring the solubility of dolomite, or of any similar double salt, stems from a question as to just how the dissolving takes place. Does the salt dissolve as a whole (*congruent* solution), to give equimolal quantities of

Mg^{2+} and Ca^{2+}, or does the more soluble part of the salt, the $MgCO_3$, dissolve in greater amount (*incongruent* solution)? The two possibilities may be symbolized

$$CaMg(CO_3)_2 \rightarrow Ca^{2+} + Mg^{2+} + 2CO_3^{2-} \text{ (congruent solution)}$$

$$CaMg(CO_3)_2 \rightarrow CaCO_3 + Mg^{2+} + CO_3^{2-} \text{ (incongruent solution)}$$

The experiments of Garrels et al. show that dolomite dissolves congruently at ordinary temperatures, so that use of the simple expression in Eq. (3-14) for the solubility product is justified. At higher temperatures, however, work by others indicates that the dissolution is at least partly incongruent.

Geologically there is not much evidence that dolomites in older strata formed as primary precipitates, except possibly the dolomites associated with evaporite deposits. Many dolomites contain structures, particularly fossils, which originally must have been calcium carbonate, so that these dolomites certainly were formed by a reaction between Mg^{2+} and a $CaCO_3$ sediment. The characteristic poor preservation of fossils in dolomite, the coarseness of grain, and the commonly observed cavities and pore spaces are all indications that dolomite forms by reactions like Eqs. (3-15) and (3-16). The Mg^{2+} may come from seawater in contact with the limy sediment or buried with it, from ions taken up in the original $CaCO_3$ structure (particularly in shells), or from later solutions moving through the sediment; the reaction may represent an addition of new material [Eq. (3-15)], a replacement of original Ca^{2+} [Eq. (3-16)], or even in part precipitation of small amounts of original dolomite. In general, for any particular dolomite, it is impossible to sort out the effects of these various reactions. The conversion of calcium carbonate to dolomite commonly takes place shortly after deposition of the original sediment (during diagenesis), as is shown by the replacement of entire beds by dolomite and by the lack of influence on dolomitization of later structures in the rock. On the other hand, partial dolomitization of some limestones along networks of veinlets must represent the work of later solutions acting on solid rock.

Thus details of the formation of dolomite remain obscure, but recent experimental studies plus geologic observations have furnished fairly convincing evidence that most dolomite is a product of slow reactions altering originally deposited calcium carbonate.

3-11 CALCIUM PHOSPHATE

The sedimentary geochemistry of the phosphates in many ways resembles that of the carbonates. Like the carbonates, most phosphates except those of the alkali metals are insoluble in neutral and alkaline solutions. In acid solutions they dissolve, because phosphoric acid, like carbonic acid, is a fairly weak acid. The reactions are a bit more complicated, because phosphoric acid has three hydrogens per molecule instead of two:

$$H_3PO_4(aq) \rightleftharpoons H^+ + H_2PO_4^- \qquad K_1 = 10^{-2.1} \qquad \text{(3-17)}$$

$$H_2PO_4^- \rightleftharpoons H^+ + HPO_4^{2-} \qquad K_2 = 10^{-7.2} \qquad \text{(3-18)}$$

$$HPO_4^{2-} \rightleftharpoons H^+ + PO_4^{3-} \qquad K_3 = 10^{-12.4} \qquad \text{(3-19)}$$

From these activity constants we can read at once that the acid itself is considerably stronger than carbonic acid, since K_1 is fairly large; we see also that the principal phosphate ions in geologic environments will be $H_2PO_4^-$ in acid solutions and HPO_4^{2-} in alkaline solutions. The simple phosphate ion PO_4^{3-} becomes dominant only at pH's over 12.4, hence in geologic situations is always minor; but the precise amount of this ion at different pH's is of critical importance in discussions of phosphate precipitation, because the common sedimentary phosphates form when PO_4^{3-} reacts with cations.

The dissolving of a simple phosphate in acid may be represented by equations of the form

$$Ca_3(PO_4)_2(s) + 2H^+ \rightarrow 3Ca^{2+} + 2HPO_4^{2-} \qquad \text{(3-20)}$$

or $\qquad Ca_3(PO_4)_2(s) + 4H^+ \rightarrow 3Ca^{2+} + 2H_2PO_4^- \qquad \text{(3-21)}$

These equations are exactly analogous to Eqs. (2-29) and (2-28), respectively, for the action of acids on calcium carbonate.

When a solution of Ca^{2+} is added to a phosphate solution, the immediate precipitate is $Ca_3(PO_4)_2$ or $CaHPO_4$, depending on the pH. Both of these compounds are known as minerals, but they are very rare. The chemistry of calcium phosphate is complicated by the ability of this substance to react with other materials in solution to form a still more insoluble compound called apatite, which is the most abundant of the phosphate minerals. Properly speaking, apatite is not a single compound or mineral but a group of closely similar substances whose relations are still not altogether clear. The most familiar of the apatites is fluorapatite, whose formula may be written either $[Ca_3(PO_4)_2]_3CaF_2$ or $Ca_5(PO_4)_3F$. This is the common apatite of igneous rocks and is also an important constituent of phosphatic sediments. Another simple apatite is hydroxylapatite, $Ca_5(PO_4)_3OH$; all gradations exist between this and fluorapatite, depending on how much OH substitutes for F in the crystal structure, and intermediate varieties are commoner in sediments than the end members. Many other substitutions are possible in the apatite structure: Cl for part of the F and OH, CO_3 and SO_4 for PO_4, and cations like Sr, Y, Mn for Ca. Particularly common in sedimentary phosphates is carbonate-apatite, with CO_3 substituting (probably) for a few percent of the PO_4 groups. The sedimentary apatites of various compositions are generally not distinguishable, either by eye or with high magnification; they often appear completely amorphous, and their crystalline structure is revealed only by x-rays.

The great insolubility of compounds with the apatite structure is indicated by the measured solubility products of fluorapatite and hydroxylapatite:

$$Ca_5(PO_4)_3F \rightleftharpoons 5Ca^{2+} + 3PO_4^{3-} + F^- \qquad K = a_{Ca^{2+}}^5 a_{PO_4^{3-}}^3 \cdot a_{F^-} = 10^{-60.4}$$
$$(3\text{-}22)$$

$$Ca_5(PO_4)_3OH \rightleftharpoons 5Ca^{2+} + 3PO_4^{3-} + OH^- \qquad K = a_{Ca^{2+}}^5 a_{PO_4^{3-}}^3 \cdot a_{OH^-} = 10^{-57.8}$$
$$(3\text{-}23)$$

Because calcium is the only one of the common metals in geologic environments that forms such structures, calcium phosphate in its many varieties is by far the most abundant inorganic compound of phosphorus. The inorganic geochemistry of phosphorus is largely a study of calcium phosphate, just as the inorganic geochemistry of carbon is largely a study of calcium carbonate.

Experimentally, apatite has been formed in a variety of ways: by slow direct precipitation, with careful control of pH and concentrations; by slow reaction of solution with freshly precipitated calcium phosphate; and by replacement of calcium carbonate by reaction with dissolved phosphate. Furthermore, one kind of apatite can be changed to another by slow reaction with appropriate solutions. All these processes probably go on in nature. The replacement of calcium carbonate is strikingly illustrated by phosphate beds consisting largely of shell fragments, the original calcareous shell material being now entirely converted to apatite. Direct precipitation is suggested by the occurrence of phosphate in thin, uniform, largely nonfossiliferous strata interbedded with such common sedimentary materials as shale and shaly limestone; the Permian Phosphoria formation of southern Idaho and adjacent states is a good example. The slow conversion of hydroxylapatite to fluorapatite is illustrated by the fluorine content of bones: fresh bone material has little fluorine, but buried bones exposed to ground water show increasing amounts with age, the increase being so regular that in favorable locations the analysis of bones for fluorine can give a rough indication of their antiquity.

The existence of thick beds of nearly pure sedimentary apatite (" phosphorite " or " phosphate rock ") has long been a geologic puzzle. Such beds are not common, but in a few places are very extensive, e.g., in the southeastern United States, the northwestern states, North Africa, Russia west of the Urals. The difficult thing to understand is not so much the mechanism of precipitation, as the accumulation of such large quantities of phosphorus over long periods of time in particular areas. The amount of dissolved phosphate in river water and ocean water is extremely small. In the ocean one might expect, on the basis of solubilities, that calcium phosphate would precipitate under the same general conditions as calcium carbonate, so that phosphate should be a minor constituent of most limestones, and this guess is borne out by analyses. One possible way to explain phosphate accumulations, then, would be to suppose that limestone beds are extensively dissolved by later solutions, their original small phosphate content being concentrated as residual material because of its greater insolubility. An explanation of this sort is reasonable for much of the phosphate in the southeastern states, where geologic evidence indicates extensive reworking, probably through many cycles, of calcareous sediments with only minor phosphate initially.

In the Phosphoria formation of the northwestern states, however, the well-preserved sedimentary structures suggest direct precipitation of phosphate out of the Permian sea, or possibly conversion of originally formed calcareous sediment to phosphate very shortly after deposition. Where could the phosphorus for this direct precipitation or replacement have come from? The only possible answer would seem to be organic activity, for it is only in the proteins and hard parts of organisms that phosphorus accumulates to any extent in natural processes. The long argument about the origin of phosphorite is largely a dispute as to whether the influence of organisms is direct or indirect.

Living things flourish most luxuriantly in the upper few tens of meters of the sea, where sunlight can penetrate. As dead organisms and organic debris sink below this zone and decay, the phosphorus from the organic matter changes fairly rapidly to dissolved phosphate. This means that parts of the sea below the illuminated zone are relatively rich in phosphate, the concentration remaining fairly constant to great depths. In areas of the sea, therefore, where upwelling takes place—where water from depth rises toward the surface—the phosphate-rich water should promote an unusual abundance of organic activity. This prediction is fulfilled in the present oceans, for areas of upwelling, chiefly on the west side of continents in subtropical latitudes, are noted for their organic "productivity." These are also parts of the ocean where phosphate crusts and nodules are found in recent sediments on the continental shelf.

One can suppose that the accumulation of phosphate on the shelf is a direct response to organic activity, in that organic matter from both pelagic and bottom-dwelling organisms piles up so rapidly that its phosphorus content is partly converted to apatite before it is entirely consumed by scavengers. This kind of mechanism seems likely for a phosphorite in which organic remains, in the form of bone, teeth, and shell fragments, are particularly abundant. It is less plausible for nonfossiliferous beds, although the assumption can always be added that the organic matter here accumulated mostly as fine-grained, soft material whose structure could not be preserved.

An alternative hypothesis with considerable appeal makes the role of organisms in phosphate accumulation largely indirect. Conditions in upwelling water moving onto the continental shelf are favorable not only for the flourishing of organisms, but also for the inorganic precipitation of apatite. Cold water moving from depth toward the surface is heated and tends to lose carbon dioxide both because pressure decreases and because plant activity in the zone of photosynthesis consumes it. Thus water unusually rich in phosphorus moves into a region of increasing pH, which should favor the deposition of calcium phosphate even without organisms acting as intermediaries. A cogent objection to this idea is the observation that the same set of postulated conditions—water becoming warmer and more alkaline—should also lead to the formation of calcium carbonate, so that the phosphate precipitate should appear as only a minor constituent of limy mud rather than as the principal material deposited. The hypothesis can be saved by noting that, according to calculations from the solubility products, phosphate can precipitate in a slightly lower pH range than calcium carbonate; as the pH of upwelling water gradually rises, apatite would be the sole precipitate as long as this narrow range is not overstepped.

Whether the influence of organisms on the precipitation of calcium phosphate is largely direct or indirect, and whether the formation of apatite is chiefly by direct precipitation or by rapid replacement of first-formed calcium carbonate or tricalcium phosphate, are questions that remain unsettled. Perhaps, as in so many geologic arguments, all the postulated processes of origin play a role, and local circumstances determine the one that predominates. In any event, all the current hypotheses suggest as the most likely place of phosphate accumulation the outer part of the continental shelf, where detrital material is scarce, in latitudes where upwelling water brings abundant phosphate from depth. These conditions are sufficiently unusual to account for the limitation of thick phosphate beds to a few geographic areas and to a few periods of the geologic column. How effective this general picture can be as a means of prediction was brilliantly shown by Sheldon (1964), who located a new deposit of phosphate rock in southeastern Turkey by noting that this area should have the same relation to the reconstructed Cretaceous-Paleocene shoreline as do the deposits of North Africa and the Middle East.

Summary The theoretical ideas about chemical equilibrium which we developed in the first two chapters have given us here a basis for understanding the geochemistry of carbonates and phosphates in sedimentary environments. The influence of acids and bases, of changes in temperature and pressure, and of other ions in solution can be related to simple equilibrium processes. Given a deposit of calcium carbonate, we can set limits on the geologic conditions under which it could have been deposited. Given a solution of known composition, we can predict the conditions under which calcium carbonate would precipitate from it. Given data on the geology of an area, we can suggest where calcium carbonate should be in process of solution and where in process of precipitation, and how much calcium the ground water would contain at various points. From the properties of other carbonates we can describe the conditions under which they might replace, or be replaced by, the carbonate of calcium. In these various senses, chemistry provides an "understanding" of the geology of carbonates.

Geochemical questions relating to the carbonates can be given particularly specific and clean-cut answers. These compounds have relatively simple structures, and processes of solution and precipitation involve for the most part rapid reactions in which ions in dilute solution play an important role. This is the kind of situation where the quantitative rules of chemical equilibrium are particularly effective as a basis for predictions. As we turn our attention to more complex, slower-reacting compounds, we shall find that our explanations necessarily become more general and less quantitative.

PROBLEMS

1 Explain why underground water in contact with limestone is alkaline. Why is the pH higher if the water is out of contact with air?

2 Would calcite be appreciably soluble in contact with a solution whose pH is maintained at 4? With a solution at pH 11? With a solution containing $0.1M$ H_2S?

3 When ferric hydroxide is precipitated with ammonia during the course of chemical analysis, it is common practice to heat the solution to boiling before filtering. What purpose does the heating serve?

4 Would the activity of Cl^- be greater or less in seawater than in water from the Mississippi River? Answer the same question for the activity coefficient.

5 What is the ionic strength of a solution containing $0.4M$ NaCl and $0.1M$ $MgSO_4$? If 0.01 mole of $CaCl_2$ is dissolved in a liter of this solution, what is the activity of Ca^{2+}?

6 Calculate the solubility of calcite in water standing in contact with a soil atmosphere containing CO_2 at a partial pressure of 0.1 atm, and find the pH of the saturated solution.

7 (a) Given that the solubility product of rhodochrosite, $MnCO_3$, is $10^{-9.3}$, calculate the equilibrium constant for the reaction

$$MnCO_3 + H_2CO_3 \rightleftharpoons Mn^{2+} + 2HCO_3^-$$

(b) Qualitatively, how is the equilibrium constant affected by a change in pH from 6 to 8?
(c) How is the solubility of $MnCO_3$ affected by this change?

8 Suggest possible explanations for the facts that (a) calcium carbonate is the most abundant carbonate of sedimentary rocks, and (b) calcium in sedimentary environments appears chiefly in the form of carbonate, rather than as calcium sulfide, chloride, fluoride, oxide, silicate, or phosphate.

9 The solubility product (activity product) of calcium fluoride, CaF_2, in pure water is $10^{-10.5}$ at 25°C. The measured concentration of Ca^{2+} in seawater is about 400 ppm, and of F^- about 1.3 ppm. Using Table 3-1 to calculate the ion activities, show that these figures are in agreement with the observation that fluorite practically never occurs as a primary precipitate in marine sediments. What common chemical sediment does contain notable amounts of fluorine?

10 From the solubility products for CaF_2 and $CaCO_3$ (calcite), find the equilibrium constant for the reaction

$$CaCO_3 + 2F^- \rightleftharpoons CaF_2 + CO_3^{2-}$$

This figure should enable you to make predictions about the kind of solutions which might cause replacement of calcite by fluorite, or of fluorite by calcite. For example, suppose that a solution $0.001M$ in CO_3^{2-} and $0.001M$ in F^- percolates through limestone. Would calcite be replaced by fluorite? If the same solution moves through a deposit of fluorite in a vein, would calcite replace fluorite?

11 Qualitatively, how would a slight change in pH affect the ability of a solution to replace calcite or fluorite? (See Table VII-2 in the appendix for dissociation constants of HF and HCO_3^-.)

12 The solubility products of $SrCO_3$ and $SrSO_4$ are 1.0×10^{-9} and 3.0×10^{-7}, respectively. The concentrations of Sr^{2+}, free CO_3^{2-}, and SO_4^{2-} in seawater are $9 \times 10^{-5}M$, $2 \times 10^{-5}M$, and $0.028M$, respectively. Assuming that the effect of seawater on the solubility products is about the same, predict whether celestite or strontianite would be the first to precipitate in places where solutions bring large amounts of Sr^{2+} into the sea.

13 Show that hydroxylapatite is more stable in contact with seawater than either of the two simple calcium phosphates, $Ca_3(PO_4)_2$ and $CaHPO_4$. This is most easily done by writing down the two solubility products:

$$CaHPO_4 \rightleftharpoons Ca^{2+} + HPO_4^{2-} \qquad K = \text{approximately } 10^{-7}$$

$$Ca_3(PO_4)_2 \rightleftharpoons 3Ca^{2+} + 2PO_4^{3-} \qquad K = \text{approximately } 10^{-26}$$

and combining these with the constants for Eqs. (3-19), (3-23), and (2-5) to obtain constants for the equations

$$Ca_5(PO_4)_3OH + 4H^+ \rightleftharpoons 3CaHPO_4 + 2Ca^{2+} + H_2O$$

$$2Ca_5(PO_4)_3OH + 2H^+ \rightleftharpoons 3Ca_3(PO_4)_2 + Ca^{2+} + 2H_2O$$

Calculate the two constants, and compare the equilibrium values of $a_{Ca^{2+}}/a_{H^+}^2$ that you obtain with the value of this quotient in seawater. In which direction must the two reactions go in order for the equilibrium values to be attained? Which of the three compounds, then, is most stable?

14 The mineral variscite, $AlPO_4 \cdot 2H_2O$, has a solubility product $10^{-22.1}$. Write equations showing the reaction of this mineral with acid, and calculate its solubility in a solution with pH 3.

15 Is seawater saturated, unsaturated, or supersaturated with respect to $BaSO_4$? (Use concentrations of barium and sulfur from Appendix III and activity coefficients from Table 3-1.)

REFERENCES AND SUGGESTIONS FOR FURTHER READING

Berner, R. A., "Principles of Chemical Sedimentology," McGraw-Hill Book Company, New York, 1971. The chapters on calcium carbonate and calcium-magnesium carbonates give an excellent review of recent experimental and theoretical work.

Davies, C. W., "Ion Association," Butterworth & Co. (Publishers), Ltd., London, 1962. This is the source of the approximate equation for activity coefficients in Sec. 3-5. The book gives details of theoretical and experimental approaches to the problems of ion complexes.

Garrels, R. M., and C. Christ, "Minerals, Solutions, and Equilibria," Harper & Row, Publishers, Inc., New York, 1965. Chapter 3 covers the calculation of carbonate equilibria in great detail, for a variety of situations of geologic importance. Activities and activity coefficients are well explained in chapter 2.

Garrels, R. M., and M. E. Thompson, A chemical model for seawater, *Am. Jour. Sci.*, vol. 260, pp. 57–66, 1962. Calculation of concentrations of important dissolved species, and amount of ion association, from dissociation constants and activity coefficients.

Garrels, R. M., M. E. Thompson, and R. Siever, Stability of some carbonates at 25° and one atmosphere total pressure, *Am. Jour. Sci.*, vol. 258, pp. 402–418, 1960. Solubility products and free energies of formation calculated from solubility measurements; good discussion of calcite-dolomite relations.

Holland, H. D., T. V. Kirsipu, J. S. Huebner, and U. M. Oxburgh, On some aspects of the chemical evolution of cave waters, *Jour. Geology*, vol. 72, pp. 36–67, 1964. An interesting study of changes in the composition of cave waters, and of the solubility in such waters of calcite, aragonite, and dolomite.

Kramer, J. R., Sea water: saturation with apatites and carbonates, *Science*, vol. 146, pp. 637–638, 1964. Equilibrium calculations by computer on 1,200 analyses of seawater show that the water is slightly supersaturated with apatite and approximately saturated with $CaCO_3$.

Land, L. S., M. R. I. Salem, and D. W. Morrow, Paleohydrology of ancient dolomites: geochemical evidence, *Amer. Assoc. Petroleum Geologists Bull.*, vol. 59, pp. 1602–1625, 1975. Evidence for formation of dolomite by early influx of meteoric water into marine carbonate deposits. Review and critique of current ideas about the formation of dolomite rock.

Langmuir, D., Geochemistry of some carbonate ground waters in central Pennsylvania, *Geochim. et Cosmochim. Acta*, vol. 35, pp. 1023–1046, 1971. Analyses of spring and well waters in a limestone-dolomite terrain show that many waters are approximately saturated. Good discussion of precautions needed for laboratory study of natural waters.

Sheldon, R. P., Exploration for phosphate in Turkey, *Econ. Geology*, vol. 59, pp. 1159–1175, 1964. Discovery of important new phosphate deposits by application of a theory about the origin of phosphorites.

Sillén, L. G., The physical chemistry of seawater, in M. Sears (ed.), Oceanography, *Am. Assoc. Adv. Sci., Pub. 67*, pp. 549–581, 1961. A classical paper describing the physicochemical factors that control the concentrations of dissolved substances in the sea.

FOUR

CHEMICAL WEATHERING

Another geologic process in which we might expect to find chemical ideas useful is the weathering of rocks at the earth's surface. The reactions of weathering are in a sense the most familiar geochemical reactions of all, because they go on all around us, under conditions of ordinary temperature and pressure. The raw materials, the products, the agents of weathering are all accessible to study in the field and can all be brought easily into the laboratory. Here certainly we are justified in thinking that chemical principles should give us an insight into the ways of nature.

4-1 GENERAL NATURE OF WEATHERING REACTIONS

As often happens, the situation turns out to be less simple than it first appears. For one thing, weathering involves mechanical processes as well as chemical reactions—expansion of water on freezing, growth of roots, swelling of minerals due to hydration. We shall say little about these processes, simply because the focus here is on chemical details, but we must keep in mind that chemical weathering is only one aspect of the whole phenomenon. A second and more serious complication in applying chemical ideas is the extreme slowness of rock decay. In contrast to the rapid reactions of the carbonates that we studied in the last chapter, the reactions of weathering are sluggish, incomplete, often irreversible. In the laboratory we can easily duplicate the materials and environments of weathering, but the times elude us. This is a common difficulty in geochemical experimentation: we need centuries, and we have only weeks or months at our disposal. The time limitation is especially troublesome when we are dealing with low-temperature processes, as we must in the study of weathering.

The chemical reactions of weathering are basically simple. Details may seem complex, but the overall processes involve nothing more abstruse than ionic dissociation, addition of water and carbon dioxide, hydrolysis, and oxidation. Difficulties arise largely because of the complex composition of minerals and mixtures of minerals that undergo weathering, not from the reactions themselves.

From a different point of view, weathering means the approach to equilibrium of a system involving rocks, air, and water. An equilibrium assemblage in this system is hard to define precisely, but it would include minerals like limonite, quartz, and kaolinite. Actually the position of final equilibrium is not of great importance because the reactions are so very slow that we must deal for the most part with processes only partly complete. For this reason our knowledge of weathering is largely qualitative rather than quantitative. We can decipher what happens chemically in the decay of a rock, but we have no means of predicting accurately what the state of the rock will be at a particular time in the future.

We shall not attempt to define or delimit the term "chemical weathering" with any exactness. Reactions of rocks and minerals with the constituents of air and water at or near the earth's surface—something of this sort is generally understood. But should weathering include the alteration of minerals in mines at depths of hundreds or even thousands of feet, evidently caused by ground water which is largely of surface origin? Should it cover the alteration of rocks by seawater at great depths in the ocean? Such semantic questions we shall sidestep, limiting the discussion here to ordinary rocks on land and within a few feet or tens of feet of the surface.

4-2 CHANGES IN ROCK COMPOSITION

One approach to the subject of weathering is to consider the overall changes in composition from the fresh rock through its various stages of decay. A series of analyses showing such changes is given in Table 4-1.

Before we look at the changes in detail, a few words are in order about the methods of stating analyses of rocks and rock-derived materials. The most common form of statement is that shown in Table 4-1, in which concentrations of elements are given as weight percent of their oxides. This kind of statement has a long tradition behind it, dating from a time when rocks and rock minerals were thought to be made up essentially of oxides combined in various ways, so that the analysis supposedly represented actual building blocks of the complex substances present. Minerals containing oxygen can still be formally regarded as combinations of oxides (for example, the composition of albite can be written either $NaAlSi_3O_8$ or $Na_2O \cdot Al_2O_3 \cdot 6SiO_2$), but we know now that this formalism has little relation to actual mineral structures. Oxides are still convenient in analytical statements, however, for other reasons than tradition: most rocks consist largely of oxygen-containing minerals, but oxygen is generally not determined separately; ideally an analysis should add up to 100% (unless abundant minerals without oxygen are present), and the accuracy of the analysis can be judged in part by the

Table 4-1 Analyses of quartz-feldspar-biotite gneiss and weathered material derived from it†

Column I gives the analysis of a sample of fresh rock, and columns II, III, and IV give analyses of weathered material. In general, the degree of weathering increases from II to IV, but there is no assurance that the original material was precisely the same or that IV represents a longer time of weathering than II or III

	Chemical composition, weight percent			
	(I)	(II)	(III)	(IV)
SiO_2	71.54	68.09	70.30	55.07
Al_2O_3	14.62	17.31	18.34	26.14
Fe_2O_3	0.69	3.86	1.55	3.72
FeO	1.64	0.36	0.22	2.53
MgO	0.77	0.46	0.21	0.33
CaO	2.08	0.06	0.10	0.16
Na_2O	3.84	0.12	0.09	0.05
K_2O	3.92	3.48	2.47	0.14
H_2O	0.32	5.61	5.88	10.39
Others	0.65	0.56	0.54	0.58
Total	100.07	99.91	99.70	100.11

	Approximate mineral composition, volume percent			
Quartz	30	40	43	25
K-feldspar	19	18	13	1
Plagioclase	40	1	1	?
Biotite (+ chlorite)	7	Trace	Trace	0.2
Hornblende	1	None	None	Trace
Magnetite, ilmenite, secondary oxides	1.5	5	2	6
Kaolinite	None	36	40	66

† *Source:* Goldich, 1938.

closeness of its total to 100; in most of the common chemical transformations of rock material, say by weathering, metamorphism, or metasomatism, the elements are generally not separated from oxygen. Since oxides so commonly appear in analyses, most of them are given shortened names; those in Table 4-1 are called, in order, silica, alumina, ferric oxide, ferrous oxide, magnesia, lime, soda, and potash. This particular order, incidentally, is also traditional for the major oxides and has the merit that it brings chemically related elements together. For special purposes, of course, analyses are often stated in other ways—as weight percent of elements, as mole percent of either elements or oxides, or as parts per thousand or parts per million.

The chemical analyses in Table 4-1 include as a last item "others," lumping together TiO_2, MnO, BaO, P_2O_5, CO_2, and S. These are given separately in the original reference, but they are all minor and are not pertinent to the present discussion. At the bottom of the table are approximate mineral (or "modal") compositions, in percent by volume.

From the table, the losses of some constituents are immediately evident: most rapid for sodium, magnesium, and calcium, somewhat slower for potassium and silicon. This order of loss is a common one for the weathering of all types of rock, but it is by no means universal. Just which constituents will disappear first in any particular case depends on a variety of factors—mineral composition, rock texture, climate, drainage, amount of exposure—so that no exception-proof generalization can be framed.

To determine *how much* of any one constituent has been lost from a series of analyses like those in Table 4-1 is a more difficult problem. Each analysis, of course, gives only the *relative* amounts of various elements present at a particular stage of weathering. If, for example, a weathered rock has lost most of its original sodium and calcium but only a little of its aluminum and iron, the analysis will show an apparent increase in the latter two constituents, as illustrated by analyses I and II in the table. To explain a pair of analyses showing a decrease in some constituents and an increase in others, we obviously can make a number of guesses. Possibly the increase and decrease are both real; in other words, weathering might involve loss of some elements and addition of others, the total mass of rock remaining essentially constant. Or, some one constituent may be unaffected by weathering while everything else decreases; here the constant constituent and others whose loss is only slight will show an apparent increase. Or, finally, weathering may bring about a decrease in all elements, those which decrease least showing an apparent increase. Nothing in the analyses themselves enables us to choose between these alternatives.

If we had independent evidence as to how the mass or volume of rock has changed during weathering, our problem would be solved. Ordinarily such evidence is not obtainable, and we can proceed only by making an arbitrary assumption. The assumption commonly used is that alumina does not change appreciably during weathering. This guess seems reasonable on the grounds that Al_2O_3 in analyses of weathered material generally shows the greatest *apparent* increase and that of all common rock constituents Al is least abundant in surface waters. Inasmuch as aluminum is not completely absent from stream and ground waters, however, the assumption cannot be strictly accurate.

The calculation involves the following steps, as illustrated in Table 4-2:

1. Recalculate analyses to 100.00 by distributing the analytical error (columns I and III).
2. Assume Al_2O_3 constant. During weathering, 100 g of fresh rock has decreased in weight so that Al_2O_3 has apparently increased from 14.61 to 18.40%. Hence the total weight has decreased in the ratio 14.61/18.40, or from 100 to 79.40 g. The amount of each constituent in the 79.40 g can be found by multiplying each number in column III by this same ratio. This gives the numbers in column A.

Table 4-2 Calculation of gains and losses during weathering

Columns I and III, giving composition in weight percent, are repeated from Table 4-1, except that the analytical error in each has been distributed so that the totals are 100.00. Column A shows the calculated weight in grams of each oxide remaining from the weathering of 100 g of fresh rock, on the assumption of constant Al_2O_3. Column B shows the gains and losses of the different oxides in grams, and column C shows the same gains and losses in percentages of the original amounts

	(I)	(III)	(A)	(B)	(C)
SiO_2	71.48	70.51	55.99	-15.49	-22
Al_2O_3	14.61	18.40	14.61	0	0
Fe_2O_3	0.69	1.55	1.23	$+0.54$	$+78$
FeO	1.64	0.22	0.17	-1.47	-90
MgO	0.77	0.21	0.17	-0.60	-78
CaO	2.08	0.10	0.08	-2.00	-96
Na_2O	3.84	0.09	0.07	-3.77	-98
K_2O	3.92	2.48	1.97	-1.95	-50
H_2O	0.32	5.90	4.68	$+4.36$	$+1,360$
Others	0.65	0.54	0.43	-0.22	-34
Total	100.00	100.00	79.40	-20.60	

3. The decrease (or increase) in each constituent is found by subtracting the numbers in column A from those in column I (treating the latter as grams per 100 g rather than percent). This gives the numbers in column B.
4. The percentage decrease or increase of each constituent is computed by dividing the numbers in column B by those in column I, giving the numbers in column C.

This same method of calculation is often used with analyses showing other kinds of rock alteration, e.g., hydrothermal alteration near veins and igneous intrusives. The assumption of constant aluminum is on shakier grounds here but is partly justified by the lack of evidence that aluminum migrates extensively in the formation of veins and contact metamorphic aureoles.

4-3 SEQUENCE OF MINERAL ALTERATION

The usually rapid decrease of sodium, calcium, and magnesium, the slower loss of potassium and silicon, and the still slower loss of aluminum and iron are, of course, reflections of the susceptibility of various minerals to weathering. Ordinarily mafic minerals decay more rapidly than felsic minerals (although by no means always), liberating magnesium, iron, and in lesser amounts calcium and the alkalis; iron is in large part oxidized immediately to insoluble ferric oxide, so remains with the weathered material instead of being carried off in solution.

Among the feldspars plagioclase weathers faster than K-feldspar, and calcic plagioclase faster than sodic plagioclase, thus liberating sodium and calcium more rapidly than potassium. When the cations are set free, the Al-Si-O frameworks of the original silicate minerals are in part decomposed, in part reconstituted into the frameworks of clay minerals, so that only a part of the silicon and very little of the aluminum find their way into solution. Typical changes in mineral composition during weathering are shown in the lower part of Table 4-1.

These conclusions about the weathering of various minerals are substantiated by examination of weathered material in hand specimen and in thin section, and by experimental work on artificial weathering of minerals and rocks. It is a common field observation, for example, that a weathered surface of granite shows its dark minerals largely converted to limonite while the feldspars remain comparatively fresh, and that plagioclase has a more chalky appearance than K-feldspar. The relatively fast decay of Na-Ca-feldspar is often conspicuous in thin section, where plagioclase crystals may be flecked with tiny grains of clay minerals and calcite while the orthoclase or microcline is almost as clear and fresh as quartz. Experiments performed by letting powdered rocks and minerals stand in contact with water and dilute acids likewise show the faster decay of mafic minerals and calcic plagioclase, and the tendency of sodium, calcium, and magnesium to dissolve in larger amounts than the other cations.

On the basis of such observations and experiments, the common minerals of igneous rocks can be arranged in a series, or better in two parallel series, according to their rates of weathering:

Mafic minerals	Felsic minerals
Olivine	Ca-Na plagioclase
Pyroxene	Na-Ca plagioclase
Amphibole	K-feldspar
Biotite	Muscovite
	Quartz

These series have a striking resemblance to the order of crystallization of minerals from igneous melts (Sec. 14-1), the minerals that form at highest temperatures (olivine and calcic plagioclase) being those most susceptible to weathering processes. In a general way this seems reasonable enough, although it is not a necessary conclusion theoretically.

A word of caution is essential here. The above remarks on rates of weathering are broad generalizations, applicable to a majority of rocks and a majority of weathering environments. But they must not be taken as universal rules, for exceptions are very common. Biotite may appear more decayed than hornblende, and the mafic minerals of a granite may look fresher than its feldspars. Rate of weathering is so dependent on such factors as grain size and amount of fracturing as well as on straight chemical susceptibility that generalizations to cover all cases are well-nigh impossible.

4-4 AGENTS OF CHEMICAL WEATHERING

Dry air causes rock to decay only very slowly, as is attested by the marvelous preservation of carved inscriptions dating from three and four thousand years ago in the arid climate of Egypt. Moisture speeds up the process enormously, both because water itself is an active agent of weathering and because water holds in solution, and therefore in intimate contact with the rock surface, several substances which react with rock minerals. The more important of these substances are free oxygen, carbon dioxide, organic acids, and nitrogen acids.

Free oxygen is important in the decay of all rocks containing oxidizable substances, particularly iron and sulfur. At ordinary temperatures reactions involving free oxygen are slow; any number of readily oxidizable materials, such as wood, cloth, most metals, coal, and petroleum, can remain in contact with the oxygen in air almost indefinitely if water is absent and if the temperature remains low. Water speeds up oxidation, probably by dissolving minute quantities of minerals or other materials, for the reactions of oxygen are faster with dissolved substances, particularly with ions, than with solids. Water may enter the reaction itself, as in the formation of hydrates, but its role is largely that of a catalyst— simply to provide a favorable environment for otherwise extremely sluggish processes.

Carbon dioxide aids decay primarily by forming carbonic acid when it dissolves in water. All natural waters exposed to air are dilute solutions of this acid—rainwater, stream water, most ground water. We have calculated previously that such a solution, in the absence of other sources of H^+ or OH^-, has a pH of about 5.7 (Sec. 2-4). This slight acidity makes natural water a better solvent than strictly neutral water would be, a conclusion amply demonstrated by much experimental work on the dissolving of minerals in various solvents.

Locally, natural waters contain other materials that increase their acidity. The decaying humus of soil, for example, adds substances to water that lower its pH very commonly to 4.5 or 5.0, sometimes even to values under 4. The precise nature of these substances remains unclear. Partly the acidity is due simply to an abnormal concentration of CO_2 released by the decay; partly it can be ascribed to minute amounts of well-defined simple organic acids (e.g., acetic acid, CH_3COOH). In geologic literature the role of *humic acids* is often emphasized, but the importance of these substances in this context is questionable. Strictly the term "humic acid" refers to one of an ill-defined group of high-molecular-weight compounds obtained by digesting wood in strong alkali and then neutralizing with acid; the humic acids appear as a gelatinous precipitate, which if carefully washed shows practically no acid properties at all. As commonly used in geology, the word has a wider and vaguer significance, referring to any indefinite, dark-colored, partly colloidal material derived from decaying organic matter and imparting an acidity to the solution. The acidity, however, is probably more correctly ascribed to carbonic acid and simple organic acids than to humic acids.

The nitrogen acids HNO_3 and HNO_2 must also play at least a minor role in most natural waters. These acids may be derived either from organic decay and bacterial action in soils, or by the dissolving in rainwater of nitrogen oxides formed during lightning discharges. How significantly these acids contribute to the acidity of natural waters is uncertain. Much more locally, particularly in volcanic regions and in the oxidized zones of sulfide ore deposits, the sulfur acids H_2SO_3 and H_2SO_4 become important, in some places lowering the pH even below 1.

Sorting out the agents of weathering in this fashion is a useful introduction to weathering reactions, but no inference should be drawn that the various agents function individually. In nature the processes of solution, hydration, acid attack, and oxidation all take place simultaneously, and separating them into specific reactions is only a convenient means of pigeonholing them for discussion.

4-5 SOLUTION AND HYDRATION

Simplest of the weathering reactions are the dissolving of soluble minerals and the addition of water to form hydrates. Dissolving commonly means the setting free of ions, as in the weathering of salt deposits and gypsum beds, and in the much slower solution of carbonate rocks. The dissolving of silica, on the other hand, forms the neutral molecule H_4SiO_4, without any appreciable ionic dissociation in the pH range of natural solutions.

Hydration and dehydration reactions are more mysterious. Perhaps best understood is the change from gypsum to anhydrite and vice versa:

$$CaSO_4 \cdot 2H_2O \rightleftharpoons CaSO_4 + 2H_2O \qquad (4\text{-}1)$$

The stability ranges of the two minerals are fairly well known from experiment, and geologic occurrences conform well enough to the experimental results to indicate that both hydration and dehydration take place readily in nature. The fact that both minerals can exist indefinitely as museum specimens must mean, of course, that by human standards the reactions are slow. Far slower must be the reactions involving ferric oxide and its hydrates, for both hematite and limonite can persist for geologic eras. Freshly precipitated ferric oxide, often written $Fe(OH)_3$, is known to be unstable with respect to the anhydrous oxide:

$$2Fe(OH)_3 \rightarrow Fe_2O_3 + 3H_2O \qquad (4\text{-}2)$$

The relative stability of limonite and hematite, as symbolized approximately in the equation:

$$Fe_2O_3 + H_2O \rightleftharpoons 2FeOOH \qquad (4\text{-}3)$$

is much less certain. The common occurrence of anhydrous and hydrated iron oxides in similar geologic environments suggests that the difference in stability is not large and that reactions to form one from the other are extremely slow (see Sec. 10-1). Similar uncertainty exists regarding the hydrated and anhydrous forms of aluminum oxide and the manganese oxides.

4-6 WEATHERING OF CARBONATES

Carbonates are subject to attack by acids because the carbonate group CO_3^{2-} so readily unites with hydrogen ion to form the stable bicarbonate ion, HCO_3^-. (Recall from Chap. 2 that HCO_3^- is at once the anion of carbonic acid, H_2CO_3, and also itself an acid, since it dissociates very slightly into H^+ and CO_3^{2-}.) As an acid, HCO_3^- has a dissociation constant of 5.0×10^{-11}; any acid more dissociated than this is capable of reacting with carbonates. Carbonic acid, with a K of 4.5×10^{-7}, fits this requirement and is the commonest solvent of carbonates in nature:

$$CaCO_3 + H_2CO_3 \rightleftharpoons Ca^{2+} + 2HCO_3^- \qquad (2\text{-}31)$$

This reaction we have discussed at great length before (Sec. 2-9).

The equilibrium constant for Eq. (2-31) (for calcite) is 4.4×10^{-5}, which means that a solution in equilibrium with atmospheric CO_2 at 25° and containing no HCO_3^- except that derived from the reaction itself can dissolve calcite up to a maximum Ca^{2+} concentration of $4.7 \times 10^{-4}M$ (Sec. 3-1). The amount would be greater at lower temperatures because more CO_2 can dissolve, giving higher concentrations of H_2CO_3; it would also be greater if the supply of CO_2 is unusually large, as is possible in spring water or in places where vegetation is rapidly decaying; it would be less at high temperatures and less also if the solution contains HCO_3^- from other sources. These are all familiar deductions from equilibrium reasoning. In nature the dissolving of limestone in carbonic acid is evidenced by the fluted and pockmarked surfaces of limestone outcrops, by the widening of cracks in limestone, and by the high concentration of Ca^{2+} (commonly 0.1 to 0.2M) in ground water in limestone regions. Despite this apparently rapid solution, limestone commonly forms prominent cliffs and ridges, especially in arid regions. A possible explanation is that much limestone is a dense, impermeable rock into which water cannot easily penetrate, so that weathering is confined to exposed surfaces and widely spaced cracks.

The dissolving of other simple carbonates in natural environments occurs by the same kind of reaction, the extent of solution being roughly predictable from the dissociation constants. For example, the double carbonate dolomite is slightly less soluble than calcite, according to laboratory measurements (Sec. 3-10), and this agrees with the common field observation that weathered limestone surfaces show calcitic areas to be more deeply etched than adjacent areas that have been dolomitized. The difference in amount of etching may also be related to the greater rate at which calcite dissolves.

4-7 OXIDATION

Among the products of weathering, iron oxides are the most conspicuous because of their bright colors. All the oxides formed in contact with air are ferric oxides; they include two forms of the anhydrous compound Fe_2O_3 (the common mineral hematite and the less common magnetic oxide maghemite) and at least two hydrates, goethite $(HFeO_2)$ and lepidocrocite $(FeOOH)$. The material called limonite is chiefly fine-grained goethite, commonly mixed with more or less clay. The ferric oxide precipitated in the laboratory by adding base to Fe^{3+} is conventionally written $Fe(OH)_3$, but a less definite formula like $Fe_2O_3 \cdot nH_2O$ would be more appropriate. The color of the simple oxide is characteristically red and of the hydrates yellow to brown, but the color is not a safe guide to composition because it depends at least as much on the state of subdivision and on minor impurities as on the degree of hydration. The precise conditions of formation of the different compounds and the conditions under which hydration or dehydration may take place are still very imperfectly known. At the moment we are concerned with the change of ferrous compounds to ferric oxide, and the particular form the ferric oxide takes is of secondary importance. We shall use the formula Fe_2O_3, with the understanding that in nature this compound is often hydrated.

Any ferrous compound on prolonged exposure to the air is oxidized, according to reactions of the form

$$Fe_2SiO_4 \text{ (fayalite)} + \tfrac{1}{2}O_2 + 2H_2O \rightarrow Fe_2O_3 + H_4SiO_4 \qquad (4\text{-}4)$$

$$2CaFeSi_2O_6 \text{ (hedenbergite)} + \tfrac{1}{2}O_2 + 10H_2O + 4CO_2 \rightarrow$$
$$Fe_2O_3 + 4H_4SiO_4 + 2Ca^{2+} + 4HCO_3^- \qquad (4\text{-}5)$$

$$2FeCO_3 \text{ (siderite)} + \tfrac{1}{2}O_2 + 2H_2O \rightarrow Fe_2O_3 + 2H_2CO_3 \qquad (4\text{-}6)$$

Equations of this sort express only the overall result of the oxidation process: the tying up of iron in ferric oxide, the setting free of silica as dissolved H_4SiO_4 or colloidal SiO_2, and the ionizing of nonoxidizable metals like calcium. Almost certainly the reactions take place in steps. Details are not known, but very likely the reactions involve progressive slight dissolving of the ferrous compounds by H_2CO_3:

$$Fe_2SiO_4 + 4H_2CO_3 \rightarrow 2Fe^{2+} + 4HCO_3^- + H_4SiO_4 \qquad (4\text{-}7)$$

followed by oxidation of the Fe^{2+}:

$$2Fe^{2+} + 4HCO_3^- + \tfrac{1}{2}O_2 + 2H_2O \rightarrow Fe_2O_3 + 4H_2CO_3 \qquad (4\text{-}8)$$

The two steps may be widely separated, in case the original dissolving of iron takes place under reducing conditions (for example, through the agency of solutions containing or in contact with organic matter); the second step would follow only when the solution has moved into an oxidizing environment. Where the oxidation takes place on exposed surfaces, the two steps would not be distinguishable, for the Fe^{2+} would oxidize so rapidly that the amount present at any one time could not be detected.

The common occurrence of such oxidation reactions is, of course, a reflection of the great stability and great insolubility of ferric oxide. So stable is this substance that other ferric compounds are relatively uncommon. Locally the very slightly soluble ferric phosphate (strengite) or ferric arsenate (scorodite) may appear, and in arid regions the more soluble ferric sulfate. Under mildly oxidizing conditions in shallow seas, ferric iron may go into glauconite [approximately $KMgFe(SiO_3)_3 \cdot 3H_2O$], and near hot springs (perhaps sometimes also as a result of ordinary weathering) into jarosite [$KFe_3(OH)_6(SO_4)_2$]. But certainly the greatest part of the iron exposed to the atmosphere eventually appears as ferric oxide in one of its various forms.

Closely parallel to the behavior of iron is the oxidation of manganese. The results of this oxidation are less conspicuous, because manganese is much less abundant than iron and because the products of oxidation are dark brown or black rather than bright red and yellow-brown. Nevertheless, films of manganese oxide on the walls of cracks and on rock surfaces in desert areas ("desert varnish") are very common.

The oxidation of manganese minerals is more complicated than that of iron minerals, because manganese has two oxidation states higher than that of manganous ion. The $+3$ state is represented by the mineral manganite $(MnOOH)$, and the $+4$ state by pyrolusite (MnO_2). In addition, there is a host of more complex oxide minerals in which more than one oxidation state is represented; examples are braunite $(3Mn_2O_3 \cdot MnSiO_3)$, hausmannite (Mn_3O_4), and psilomelane (approximately $BaMn_9O_{18} \cdot 2H_2O$). Of all these the one that seems most stable on prolonged exposure to the atmosphere is pyrolusite, so that overall oxidation of manganese minerals can be represented by reactions of the form

$$MnSiO_3 \text{ (rhodonite)} + \tfrac{1}{2}O_2 + 2H_2O \rightarrow MnO_2 + H_4SiO_4 \qquad (4\text{-}9)$$

$$MnCO_3 \text{ (rhodochrosite)} + \tfrac{1}{2}O_2 + H_2O \rightarrow MnO_2 + H_2CO_3 \qquad (4\text{-}10)$$

Again these reactions doubtless take place in steps, and again the first step is most likely the slight dissolving of the manganous minerals by carbonic acid. The second step, the oxidation of Mn^{2+}, may follow at once, if conditions are oxidizing, or may be delayed indefinitely, if the rock is in contact with a reducing solution. The second step may be more complicated than for iron, because the other oxides and mixed oxides may form as intermediates on the way to pyrolusite.

A third common element that is oxidized during weathering is sulfur. In igneous rocks and in veins, this element occurs chiefly in sulfides—compounds with metals in which the sulfur has an oxidation number of -2. Oxidation can change the number to any one of several higher values, but in contact with air, equilibrium is reached only when the sulfur has attained its highest possible oxidation state, $+6$, in the form of a sulfate. Equations for such reactions may be deceptively simple:

$$PbS + 2O_2 \rightarrow PbSO_4 \qquad (4\text{-}11)$$

$$ZnS + 2O_2 \rightarrow Zn^{2+} + SO_4^{2-} \qquad (4\text{-}12)$$

($PbSO_4$ is represented as undissociated because it is fairly insoluble and commonly appears as anglesite encrusting or embaying galena; $ZnSO_4$, on the other hand, is very soluble.) These reactions, like those for the oxidation of iron and manganese, take place with extreme slowness or not at all in the absence of water. The function of the water is probably to supply carbonic acid to dissolve minute amounts of the sulfides:

$$PbS + 2H_2CO_3 \rightleftharpoons Pb^{2+} + H_2S + 2HCO_3^-$$

after which the hydrogen sulfide is oxidized:

$$H_2S + 2O_2 + Pb^{2+} + 2HCO_3^- \rightarrow PbSO_4 + 2H_2CO_3$$

Solutions resulting from the oxidation of sulfides are acid because of hydrolysis of the dissolved metal ion. For example,

$$Zn^{2+} + H_2O \rightleftharpoons ZnOH^+ + H^+$$

The amount of acidity depends on the stability of the dissolved metal-hydroxy complex (Sec. 2-6). For metals that form very insoluble oxides and hydroxides, hydrolysis may lead to precipitation of the solid. This kind of reaction is particularly important, and the resulting acid particularly strong, in the oxidation of the common sulfide pyrite, FeS_2. The high acidity results from the formation of the very insoluble ferric oxide (or a hydrated oxide):

$$2FeS_2 + \tfrac{15}{2}O_2 + 4H_2O \rightarrow Fe_2O_3 + 4SO_4^{2-} + 8H^+ \tag{4-13}$$

The results of this reaction are evident at outcrops of pyritic veins, where the rocks are heavily stained with yellow and brown ferric oxide and where the ground water has a sharply acid taste. Note that the sulfur in pyrite has an apparent oxidation number of -1, instead of the usual -2 for sulfides; note also that Eq. (4-13) represents simultaneous oxidation of two elements, Fe and S. In arid regions oxidation of pyrite commonly gives iron sulfate minerals, both ferrous and ferric sulfates, in addition to limonite or hematite. This suggests that the reaction may take place in steps:

$$FeS_2 + \tfrac{7}{2}O_2 + H_2O \rightarrow Fe^{2+} + 2SO_4^{2-} + 2H^+ \tag{4-14}$$

$$2Fe^{2+} + \tfrac{1}{2}O_2 + 2H^+ \rightarrow 2Fe^{3+} + H_2O \tag{4-15}$$

or

$$2Fe^{2+} + \tfrac{1}{2}O_2 + 2H_2O \rightarrow Fe_2O_3 + 4H^+ \tag{4-16}$$

How much of the iron becomes Fe^{3+} or $FeOH^{2+}$ in solution [Eq. (4-15)] and how much is precipitated as the oxide [Eq. (4-16)] depends on the pH of the immediately adjacent solution. The dissolved iron, Fe^{2+} and Fe^{3+}, may unite with SO_4^{2-} to form one or more of the sulfate minerals, provided the region is arid enough for these soluble compounds to persist. In humid regions the iron is practically all oxidized and precipitated as the oxide.

Sulfur, iron, and manganese are the only abundant elements in common rocks for which oxidation is an important part of weathering. Many of the less common elements—copper, arsenic, uranium, for example—are oxidized when their minerals are exposed to the atmosphere, but these reactions are best considered later on in connection with the chemistry of ore deposits (Chap. 18).

4-8 HYDROLYSIS OF SILICATES

The weathering of silicates is primarily a process of hydrolysis (Sec. 2-7). As a simple example, the mineral forsterite (magnesium-rich olivine) hydrolyzes according to the equation

$$Mg_2SiO_4 + 4H_2O \rightarrow 2Mg^{2+} + 4OH^- + H_4SiO_4 \qquad (4\text{-}17)$$

the hydrogen ion from water uniting with the silicate group to form the very weak silicic acid. Surface waters generally contain a little more H^+ than would be present in pure water, because of their dissolved CO_2, and this additional H^+ aids the process of hydrolysis. The carbonic acid may be included in the equation:

$$Mg_2SiO_4 + 4H_2CO_3 \rightarrow 2Mg^{2+} + 4HCO_3^- + H_4SiO_4 \qquad (4\text{-}18)$$

Locally, where acids stronger than carbonic acid are present, e.g., near a vein containing pyrite, the reaction becomes simply

$$Mg_2SiO_4 + 4H^+ \rightarrow 2Mg^{2+} + H_4SiO_4 \qquad (4\text{-}19)$$

and the abundant hydrogen ions are so effective that weathering in such places is unusually deep and unusually complete. Thus the equation for hydrolysis may be written in various ways, depending on the local availability of hydrogen ion.

These reactions are almost exactly analogous to reactions that might be written for carbonates:

$$MgCO_3 + H_2O \rightarrow Mg^{2+} + OH^- + HCO_3^-$$

$$MgCO_3 + H_2CO_3 \rightarrow Mg^{2+} + HCO_3^- + HCO_3^-$$

$$MgCO_3 + 2H^+ \rightarrow Mg^{2+} + H_2CO_3$$

The only significant difference is that the weak acid formed in the first two equations consists of ions (HCO_3^-) rather than neutral molecules (H_4SiO_4), which reflects the fact that carbonic acid is stronger than silicic acid. Neither the carbonate nor the silicate reactions are strictly reversible, for at ordinary temperatures the addition of Mg^{2+} to a bicarbonate or silicic acid solution gives hydrated compounds rather than the anhydrous salts.

For silicates containing several cations the hydrolysis reaction is less simple than for forsterite, because in general the different cations go into solution at different rates. At any stage of partial weathering, silicate grains are coated with an outer shell from which some cations have been preferentially removed, hence which must have a different composition from the mineral as a whole. This outer shell serves to "armor" the interior of a grain and makes the dissolving of most silicates extremely slow.

Aluminum silicates involve the further complication that one product of their weathering is practically always a clay mineral, a compound in which some of the original aluminum and silicon apparently remain combined. We can symbolize the reaction, using K-feldspar as an example of an aluminum silicate and kaolinite as a clay mineral:

$$4KAlSi_3O_8 + 22H_2O \rightarrow 4K^+ + 4OH^- + Al_4Si_4O_{10}(OH)_8 + 8H_4SiO_4 \qquad (4\text{-}20)$$

This is an oversimplified equation, representing a process that has not yet been duplicated in the laboratory at low temperatures. One can show, indeed, by leaving water in contact with finely ground feldspar for hours or days, that the solution becomes faintly alkaline and that it contains a little K^+, alumina, and silica; but the solid residue is still feldspar and not clay. The formation of kaolin from orthoclase has been demonstrated at temperatures over 200°C, but at ordinary temperatures it must be a very slow process.

Speculations about the mechanism of the reaction in Eq. (4-20) are numerous. It is possible, for example, that the hydrolysis does not take place directly, but in a series of steps. Perhaps the feldspar structure breaks down first to gibbsite, $Al(OH)_3$, setting silica free as dissolved silicic acid, and then kaolinite forms by a later reaction between $Al(OH)_3$ and H_4SiO_4. Or perhaps the second reaction involves the minute amount of aluminum that goes into solution; or perhaps both alumina and silica are separated from the feldspar as colloidal particles, which join later to form kaolinite. Evidence bearing on such questions has come from recent experimental work showing that some of the more complex clay minerals (montmorillonite, chlorite, serpentine) can be prepared at ordinary temperatures in the laboratory by slow reactions involving very dilute solutions or suspensions of the constituent oxides and hydroxides. Kaolinite is the most difficult of the clays to prepare in this manner; experimental data regarding its formation are still inconclusive. The fact that at least some of the clay minerals can be synthesized from their oxides shows that a stepwise mechanism for Eq. (4-20) is possible. It does not prove that the reaction cannot take place directly. Some geologic evidence suggests that the formation of clay may be accomplished by either mechanism—by a direct conversion of aluminum silicates *or* by a preliminary breakdown to alumina and silica, depending on the environment in which the reaction occurs. We shall return to this question later, after a discussion of clay-mineral structures (Sec. 7-4).

Regardless of how the reaction proceeds, hydrolysis of silicates always leaves the solution more basic than it was to begin with. If pure water is the agent of weathering, the pH rises above 7 [Eqs. (4-17) and (4-20)]; if carbonic acid or another acid is the agent, the acidity is reduced [Eq. (4-18)]. This generalization holds for hot water at depth as well as cold water near the surface. *Any solution in contact with silicate minerals cannot long remain appreciably acid, and if contact is continued, the solution must eventually become alkaline.* How alkaline it becomes depends on the nature of the silicates. A restraint on the increase in alkalinity is set by reactions of various forms of silica with OH^-:

$$H_2O + SiO_2 \text{ (quartz)} + OH^- \rightarrow H_3SiO_4^- \qquad K = 10^{0.4}$$

$$H_4SiO_4(aq) + OH^- \rightarrow H_3SiO_4^- + H_2O \qquad K = 10^{4.2}$$

These processes ordinarily keep the pH from rising much above 9.

In all these reactions the silica set free by weathering is represented as going into solution as silicic acid, H_4SiO_4. Certainly a large part of the silica behaves in this manner, finding its way into ground water and surface water. Locally its history may be different. Where silica is set free in high concentrations or where weathering solutions locally become concentrated, some of it may separate as colloidal or amorphous SiO_2. Part of the silica, especially from minerals of the amphibole and pyroxene groups, may not dissolve at all but may remain as amorphous residues of the original crystal structure. Some plants are known to take up silica from weathering material and then to release it during decay in the form of solid amorphous particles. Thus considerable silica may be left behind during weathering, in part perhaps to unite eventually with alumina to form clay, and in part to recrystallize into minute grains of quartz.

In summary, the weathering of silicates is chiefly a hydrolysis reaction, making the weathering solution alkaline or at least reducing its acidity, and yielding as principal products cations and silicic acid in solution and clay minerals as residual solids.

4-9 ENVIRONMENTS OF WEATHERING

We have studied in some detail the chemical reactions involved in weathering. From a broader point of view, let us now look at the weathering process as it is affected by various environments.

The ultimate product of weathering is soil, and varieties of soil should give us a clue to the influence of environment on weathering. Soil unfortunately is a very complex material, so complex that we had best defer detailed discussion until later; a few general remarks must suffice here. Soils from different environments appear to have recognizable characteristics, especially with regard to the nature of their clay minerals, which can be ascribed to the environment rather than to the original rock. Kaolinite soils, for example, are common in temperate climates with moderate to heavy rainfall, whereas the montmorillonite and illite clays (clay minerals with a high content of cations such as magnesium, iron, and potassium) are most abundant in soils of semiarid regions. The most striking influence of environment on soils is the formation of conspicuous red soils in parts of the humid tropics and subtropics. These soils have more alumina and ferric oxide and less silica than soils of cooler climates; extreme soils of this type, called *laterites*, consist almost wholly of alumina and/or iron oxide with practically no silica at all. The chemical reason for the breakdown of clay and removal of silica under conditions of high temperature and high rainfall has occasioned much argument, but remains an unsolved riddle (Sec. 7-5).

Except for differences in the clay minerals of soils, real differences in the processes and products of chemical weathering ascribable solely to environment are hard to find. Decay goes faster and penetrates more deeply in warm, humid climates than in arid climates, but the chemical reactions are not detectably different. Weathered granite from South Carolina and weathered granite from Arizona look alike, provided that weathering has reached about the same stage in both specimens. Even where rocks are wet by salt spray on the seacoast, the essential weathering reactions remain the same.

How deep does the zone of weathering extend? We are accustomed to thinking of weathering as a surface process, but clearly it can take place at lower levels wherever air and water penetrate. Mines, tunnels, and drill holes often encounter evidences of recent weathering hundreds of feet beneath the surface, the depth in any one place being determined by the effectiveness of circulation of air and water. The boundary between weathered and unweathered rock may be extremely irregular, the more massive, impermeable parts of the rock being left as projections or isolated boulders of unweathered material in the weathered zone. The permanent water table of an area commonly coincides at least roughly with the top of the unoxidized zone, probably for the reason that access of oxygen and carbon dioxide is much less free where rock interstices are continuously filled with water. In places the position of old water tables high above the present one can be deciphered from abrupt changes with depth in the degree of weathering.

That weathering takes place in buried sediments is indicated by the corroded, "moth-eaten" appearance of the less durable ferromagnesian grains, and likewise by the fact that most Paleozoic sandstones contain practically no minerals except the most resistant ones like quartz, orthoclase, zircon, and tourmaline. The progressive decrease of the more susceptible minerals in older rocks indicates that weathering reactions may go on for a long time. This brings us back to the problem of defining weathering: part of the decay of susceptible minerals may take place under the influence of the same solutions that cause replacement and cementation in the sediment; where should the line be drawn between processes of weathering and processes of lithification? There can be no simple answer, and the line is not of great importance as long as the separate processes are recognized.

One final environment of weathering should be mentioned, the environment of the ocean bottom. Weathering here is given a special name, halmyrolysis, but despite the long word not much is known about details of the process. The available agents are not greatly different from those on land, inasmuch as seawater generally contains dissolved O_2 and CO_2; the water itself is slightly alkaline, but the pH may drop to 5 or 6 in the sediments a few centimeters below the bottom; the presence of considerable magnesium, potassium, and sodium in the water, and of sulfur from decaying organic matter in the bottom sediments, provides some differences from most land environments. Clay minerals (probably chiefly illite), ferric oxide, and manganese dioxide are common products of submarine weathering, just as they are on land. More distinctive of this different environment are such minerals as glauconite (silicate of K, Fe^{2+}, and Fe^{3+}) and phillipsite (a potassium zeolite). Details of submarine weathering processes are a promising field for geochemical research.

PROBLEMS

1 Attempts have been made to devise a "weathering index," by which the concept of "degree of weathering" can be made quantitative. For example, one suggested index is the ratio

$$\frac{\text{Moles of } CaO + MgO + Na_2O + K_2O - H_2O}{\text{Moles of } SiO_2 + Al_2O_3 + Fe_2O_3 + CaO + MgO + Na_2O + K_2O}$$

As the weathered material loses Ca, Mg, Na, and K and appears to gain Al and Fe, this fraction will obviously decrease. By calculating such ratios for a series of rocks, we can compare their degrees of decay more accurately than is possible by the usual qualitative terms "slightly weathered," "extensively weathered," and so on. Do you think that this is a good idea? What advantages and what difficulties in the calculation of weathering indices can you point out?

2 For the analyses of fresh and weathered diabase in Table 4-3, calculate the loss or gain in each constituent on the assumption that Al_2O_3 remains constant. Make the calculation in terms both of grams per 100 g and of percent of original amounts. Note that the behavior of Na and K in this example is different from that in the weathered gneiss of Sec. 4-2.

Table 4-3

	Fresh	Weathered
SiO_2	47.28	44.44
Al_2O_3	20.22	23.19
Fe_2O_3	3.66	12.70
FeO	8.89	
MgO	3.17	2.82
CaO	7.09	6.03
Na_2O	3.94	3.93
K_2O	2.16	1.75
H_2O	2.73	2.73
Others	1.45	1.22
Total	100.59	98.81

3 You are comparing the heavy-mineral content of two samples of sand, one taken from near the top of a terrace and one from near the bottom. The formation of the terrace required at least several hundred thousand years, and as nearly as you can tell, the source of the terrace material did not change during that time. Of the following minerals, which would you expect to be markedly depleted in the lower sample: zircon, hornblende, garnet, olivine, biotite, labradorite, augite, tourmaline, magnetite, apatite?

4 Write balanced equations for the following reactions. Be sure that your equations are geologically reasonable, in the sense that the products are stable and can coexist in natural environments in the pH range in which the reactions would occur. Assume that the reactions take place at ordinary temperatures and pressures. (a) Calcite dissolves in carbonic acid. (b) Grossularite ($Ca_3Al_2Si_3O_{12}$) reacts with carbonic acid. (c) Sphalerite dissolves in carbonic acid. (d) Nepheline hydrolyzes. (Assume that kaolinite is one product and that nepheline can be represented by the simple formula $NaAlSiO_4$.)

5 List the chief products of chemical weathering of (a) basalt, (b) dolomite.

6 The hydrolysis of silicate minerals during weathering leads to the formation of OH^- [Eq. (4-20)]. Explain why this process generally does not give solutions with a pH higher than 9 during the weathering of ordinary silicate rocks.

7 Why cannot the weathering of orthoclase be discussed in terms of an equilibrium constant for the dissolving of $KAlSi_3O_8$?

8 The dissolved material in 56 perennial springs in granitic rocks of the Sierra Nevada has been studied by Feth, Roberson, and Polzer (1964). The climate is temperate and fairly humid. The rock is chiefly granodiorite and quartz monzonite, with quartz, orthoclase, andesine, biotite, and hornblende as principal minerals. No average analysis of the rock is available, but the single analysis given in Table 4-4 is probably representative. The right-hand analysis gives the mean composition of the spring waters in parts per million. Suggest a source for each of the dissolved substances, and suggest reasons why the relative amounts are so different from the relative amounts of the corresponding oxides in the rock.

Table 4-4

Composition of rock, % by weight		Dissolved material in springs, ppm	
SiO_2	67.4	SiO_2	24.6
Al_2O_3	15.8	Al	0.02
Fe_2O_3	1.7 ⎫	Fe	0.03
FeO	2.2 ⎭		
MgO	1.3	Mg	1.7
CaO	3.1	Ca	10.4
Na_2O	3.5	Na	6.0
K_2O	4.2	K	1.6
		HCO_3	54.6
pH of springwater = 6.8		SO_4	2.4
		Cl	1.1

REFERENCES AND SUGGESTIONS FOR FURTHER READING

Engel, C. G., and R. P. Sharp, Chemical data on desert varnish, *Geol. Soc. America Bull.*, vol. 69, pp. 487–518, 1958. Analyses of manganese-iron oxide coatings on desert boulders, and discussion of origin during weathering.

Feth, J. H., C. H. Roberson, and W. L. Polzer, Sources of mineral constituents in water from granitic rocks, Sierra Nevada, *U.S. Geol. Survey Water-Supply Paper* 1535-I, 1964. Details of weathering of granite in a humid temperate climate.

Goldich, S. S., A study in rock weathering, *Jour. Geology*, vol. 46, pp. 17–58, 1938. A detailed discussion of weathering of many kinds of rocks, including analyses of fresh and weathered materials.

Harriss, R. C., and J. A. S. Adams, Geochemical and mineralogical studies on the weathering of granitic rocks, *Am. Jour. Sci.*, vol. 264, pp. 146–173, 1966. A detailed study of soil profiles developed on five different kinds of granitic rock, including mineralogical changes and changes in major and minor elements.

Huang, W. H., and W. C. Kiang, Laboratory dissolution of plagioclase feldspars in water and organic acids at room temperature, *Am. Mineralogist*, vol. 57, pp. 1849–1859, 1972. An example of many recent papers describing results of experiments designed to reproduce weathering reactions in the laboratory.

Livingstone, D. A., Chemical composition of rivers and lakes, chap. G in Data of Geochemistry, 6th ed., *U.S. Geol. Survey Prof. Paper* 440-G, 1963. A compilation of analyses of surface waters, with dissolved material chiefly from weathering, and a discussion of overall rates of weathering by solution in various parts of the world.

Loughnan, F. C., "Chemical Weathering of the Silicate Minerals," American Elsevier, New York, 1969.

White, D. E., J. D. Hem, and G. A. Waring, Chemical composition of subsurface waters, chap. F in Data of Geochemistry, 6th ed., *U.S. Geol. Survey Prof. Paper* 440-F, 1963. A compilation of analyses of ground water. Criteria are set up for distinguishing waters deriving their dissolved material from the weathering of ordinary rocks from waters with dissolved material from other sources.

FIVE

STRUCTURAL CHEMISTRY

The clay minerals, in the last chapter, were dismissed with the observation that they are a complicated group of minerals formed by hydrolysis of aluminum silicates. This is obviously too superficial a treatment, for the clay minerals are the most abundant of the solid products of weathering. For a real understanding of rock decay, of soil formation, and of the makeup of shaly sediments, we must look more carefully at the constituents of clay.

This requires in turn that we explore some of the general characteristics of mineral structures. Classical chemistry alone tells us little about the different clay minerals: their empirical formulas are complicated, variable in detail but all discouragingly alike, and their reactions are generally slow and incomplete. As with other silicate minerals, structure holds the key to an understanding of the clays, but elucidation of clay structures for a long time lagged behind that of other minerals because clay particles are so fine-grained and so difficult to separate from other materials. In recent years concerted attack by means of x-rays, the electron microscope, and differential thermal analysis has brought rapid progress in our understanding of these minerals.

Structural chemistry is a branch of science closely related to crystallography and to solid-state physics. It is important in geochemistry for an understanding of many natural materials besides clays; therefore we shall develop the subject from a general viewpoint. Fortunately we can limit the discussion to an elementary treatment of basic principles, since the more intricate details of structure are not pertinent to most geochemical problems. Following a general treatment of mineral structures in this chapter, and an introduction to colloids in the next chapter, we shall apply this background to an investigation of the clay minerals in Chap. 7.

5-1 STRUCTURE OF NaCl

The chemical structures of most interest in geology are those of crystalline solids, especially solids made up of ions. A few materials found in nature, for example, diamond and sulfur, are constructed of atoms rather than ions. A few others, like asphalt, are best regarded as built up of molecules; in these the structure of the molecule itself may be complex and may largely determine the properties of the substance. But the rocks, minerals, and soils with which a geologist commonly deals are composed almost wholly of ionic solids, and it is the structure of these that will be the object of our inquiry.

A familiar crystalline solid, ordinary table salt, has an especially simple structure that will serve as a good starting point. The arrangement of ions in the structure is shown in Fig. 5-1. Note that the pattern is cubic, that Na^+ ions alternate with Cl^-, that every Na^+ is surrounded by six Cl^- ions at the corners of an octahedron and every Cl^- by six Na^+ ions. The figure shows only the arrangement, not the sizes; actually the ions should be regarded as touching their nearest neighbors, so that the entire volume of the crystal is occupied except for spaces between the spheres (Fig. 5-2).

Now what are these objects called "ions," which we write as Na^+ and Cl^- and which we represent in diagrams as tiny spheres? In the familiar model, a simple ion like Na^+ or Cl^- consists of a compact positive nucleus made up of protons and neutrons, surrounded by negative electrons, both nucleus and electrons being very tiny compared with their distance apart. The positive charge of Na^+ means that its nucleus contains one positive charge (one proton) in excess of the number of surrounding electrons, and the negative charge of Cl^- means that it contains one electron in excess. All this is simple enough, but difficulties arise if we try to make the picture more definite by specifying the arrangement and

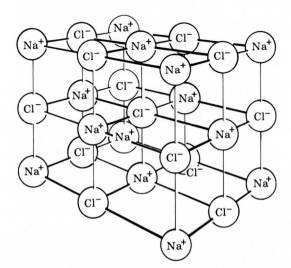

Figure 5-1 Arrangement of ions in the structure of NaCl. The relative sizes and distances are not to scale.

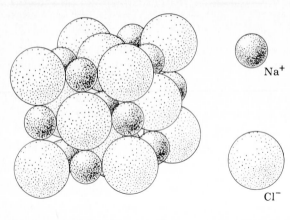

Na$^+$

Cl$^-$

Figure 5-2 Arrangement of ions in the structure of NaCl, with sizes drawn to scale. (*Source: Sienko and Plane, 1976, page 150.*)

movement of the electrons. For some purposes the electrons can be thought of as minute particles revolving around the nucleus in orbits, like planets around the sun, but this model is too simple to describe all aspects of electronic behavior. Actually the positions and motions of the electrons cannot be specified precisely; according to quantum theory, we can only indicate the probability of finding electrons at particular places within an ion or atom. This gives us a rather fuzzy and elusive sort of unit out of which to build our crystalline structures.

Fortunately the deeper questions about the nature of the electron are not of great moment in considering the elementary relations of ions in crystals. The important thing here is to remember that an ion consists essentially of widely spaced electric charges, so that its "boundaries" are actually boundaries of electric fields. We cannot expect, then, that ions will behave entirely like rigid spheres, nor can we expect to assign precise values to their radii or volumes.

The size of an ion, despite the fact that it cannot be defined or measured with great exactness, is a useful concept in discussing crystal structures. Some idea of the magnitude of ionic sizes can be gained by a very simple calculation. The molecular weight of sodium chloride tells us that 58.5 g must contain 6.02×10^{23} (Avogadro's number) "molecules" of NaCl, or 6.02×10^{23} Na$^+$ ions and the same number of Cl$^-$ ions. The density of NaCl is 2.16 g/cm^3, so that these ions occupy a volume of 58.5/2.16, or about 27 cm^3. If we assume that the two ions have roughly the same size, the volume of a single ion would be

$$\frac{27}{2 \times 6.02 \times 10^{23}} = 22 \times 10^{-24} \text{ cm}^3$$

The diameter of an ion would then be approximately the cube root of this figure or 2.8×10^{-8} cm, and the radius 1.4×10^{-8} cm (generally written 1.4 Å = 1.4 angstroms). Actually the radii are not equal, Cl$^-$ having a radius of about 1.7 Å and Na$^+$ of about 1.1 Å.

Better values of ionic sizes can be obtained by measurements of spacings of planes in crystal structures by means of x-rays, coupled with measurements of molar refractivities and with calculations from quantum mechanics. Methods of finding ionic sizes differ in detail, so that currently several sets of radii are available. Some of the differences are real, in that ionic radii for some purposes are defined differently, and therefore measured differently, than for other purposes. Some of the differences can be ascribed to experimental uncertainties and to differences in radii selected as standards. The discrepancies are mostly in the second decimal place and are not of great concern here.

The most widely used set of radii is given in Appendix IV, and values for some of the common ions are shown pictorially in Fig. 5-3. These values are the so-

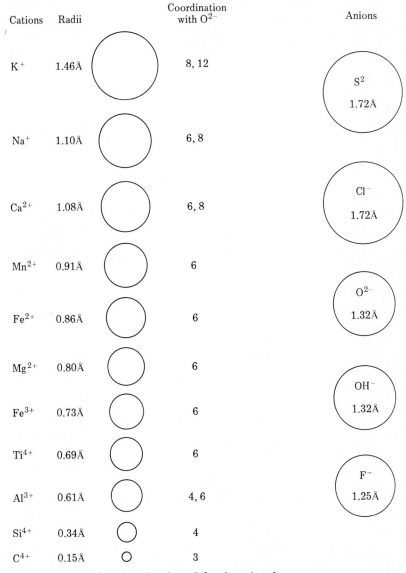

Figure 5-3 Radii of common ions in rock-forming minerals.

called "octahedral" radii, calculated on the assumption that the ion has octahedral coordination—in other words, that its closest neighbors in crystal structures are six other ions at the corners of an octahedron, like the six Cl^- ions around each Na^+ in the NaCl structure. For other types of coordination (for example, tetrahedral, as in diamond) the radii would be slightly different; thus, the "octahedral" radius of aluminum is 0.61 Å and the "tetrahedral" radius about 0.47 Å. Octahedral coordination is the commonest type for the cations of geologic interest and is a good average between the extremes found in crystals, hence the octahedral radii are the most generally useful.

An arrangement like the NaCl structure of Fig. 5-1 and 5-2 represents, for these particular ions, the most stable possible pattern, i.e., the arrangement in which the forces between adjacent Na^+ and Cl^- particles are a maximum, or for which the potential energy is a minimum. If we focus attention on the forces, we may speak of an *ionic bond* between Na^+ and Cl^-, and since NaCl is particularly stable, we would describe this bond as stronger than that, say, in a crystal of AgCl. Or if we think in terms of energy, we could say that the *crystal energy*—the amount of energy needed to tear apart a mole of NaCl into widely separated individual ions—of NaCl is larger than the crystal energy of AgCl.

The arrangement of ions in a crystal would be expected to have a close relation to some of the ordinary large-scale properties of the crystal, particularly its symmetry, cleavage, and effect on polarized light. The cubic pattern of ions in sodium chloride, for example, is reflected in the cubic habit of halite crystals, in the three mutually perpendicular planes of cleavage, and in the ability of light to traverse the crystal with the same speed in any direction. Historically these megascopic properties were studied in great detail long before anything was definitely known about the particles and the geometric patterns of crystal structures. It is one of the great triumphs of twentieth-century science that the inner structures elucidated by x-rays have been found to correlate with and to explain so elegantly the intricate relationships among geometric and optical properties discovered by nineteenth-century crystallographers.

5-2 OTHER IONIC SOLIDS

The pattern of ions in sodium chloride is very simple, and the correlation with elementary crystal properties is transparently obvious. Can we find similar patterns and similar correlations in other crystalline solids? To what extent can the properties of solids be related to geometric arrangements of ions treated simply as small, hard spheres?

We look first at compounds chemically similar to NaCl, consisting merely of an alkali metal joined to a halogen. These show a gratifying similarity, both in megascopic properties and in patterns of ions, except for a few salts of lithium, cesium, and fluorine. These three elements have ions of extreme sizes, small for lithium and fluorine and large for cesium; therefore we might guess that an explanation for the exceptions is somehow connected with different ionic volumes.

The simplest possible assumption is that any given combination of ions will form a structure in which the ions are packed as close together as possible, since this would make the forces between them a maximum. The packing arrangement to secure this closest approach would necessarily depend on the relative sizes of the anions and the cations. On this assumption our problem becomes the purely geometric one of how to arrange spheres of different sizes so as to give the densest possible packing. The problem is illustrated diagrammatically in two dimensions in Fig. 5-4.

From solid geometry the following rules can be worked out for three-dimensional structures:

1. If the *radius ratio* (radius of cation divided by radius of anion) is between 0.41 and 0.73, the pattern of closest packing has each ion surrounded by six ions of opposite sign. This is the sodium chloride structure (radius ratio for $Na^+/Cl^- = 0.64$).
2. If the radius ratio lies between 0.73 and 1, closest packing is secured when each ion has eight ions of opposite sign as closest neighbors. This is often described as the cesium chloride structure, although the radius ratio Cs^+/Cl^- (1.03) is slightly greater than the upper geometric limit.
3. If the radius ratio is in the range 0.22 to 0.41, each cation can surround itself with only four anions. For example, the radius ratio Si^{4+}/O^{2-} is 0.26, and the common Si-O structure in both silica minerals and silicates is the tetrahedron SiO_4^{4-}, in which Si^{4+} is linked to four O^{2-} ions (Sec. 5.6). This kind of structure is found also in sphalerite (ZnS), and is often called the sphalerite structure, although the ratio Zn^{2+}/S^{2-} (0.43) slightly exceeds the geometric limit.

The exceptions represented by cesium chloride and sphalerite are a warning that the radius-ratio rules do not hold with great exactness. In particular, the ratio is often greater than the upper limiting value, which means that the anions grouped around a cation do not touch each other—in other words, that the structure is not strictly close-packed. But for a great many simple compounds the

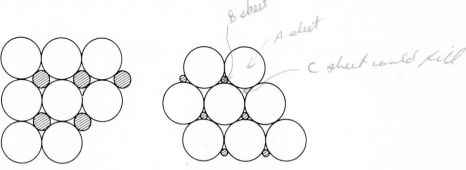

Radius ratio 0.41
4-coordination

Radius ratio 0.12
3-coordination

Figure 5-4 Relation between radius ratio and coordination number in two dimensions.

rules hold remarkably well, and crystal structures can be in large part predicted from geometry.

A further word is needed about nomenclature. In sodium chloride structures, with a pattern of six ions around each ion of opposite sign, the cation is said to have a *coordination number* of 6, or to be "6-coordinated." The anion could equally well be regarded as having a coordination number of 6, but conventionally the term is more commonly applied to the cation. The six ions surrounding a central ion are arranged at the corners of an octahedron, so that the sodium chloride structure is often spoken of as *octahedral.* Similarly the cesium chloride structure may be referred to as *cubic,* and the cesium ion may be assigned a coordination number of 8. The sphalerite structure is *tetrahedral,* and the zinc ion in sphalerite is 4-coordinated. In mineral structures the commonest anion by far is O^{2-}, with an ionic radius of 1.32 Å; most of the common cations have radii in the range 0.60 to 1.10 Å, giving radius ratios with oxygen of 0.45 to 0.83; hence the most frequent coordination number in minerals is 6. This is why ionic radii are generally tabulated as octahedral radii rather than tetrahedral or cubic.

The same sort of geometric argument can be used to make predictions about the crystal structures of more complex compounds, and for many of these the predictions fit observed facts just as admirably as they do for simple ones.

Is crystallography, then, no more than an extension of solid geometry? The answer, of course, is "No." A great deal more than geometry is involved, but we have so far looked only at specially selected examples for which geometric reason-

Table 5-1 Comparison of observed coordination numbers with numbers predicted from geometric radius ratios†

Ion	Radius ratio: $\dfrac{\text{Ionic radius}}{\text{Radius of } O^{2-}}$	Coordination predicted from ratio	Observed coordination numbers	Theoretical limiting radius ratios
Cs^+	1.35	12	12	
K^+	1.11	12	8–12	
				1.00
Sr^{2+}	0.92	8	8	
Na^+	0.83	8	6, 8	
Ca^{2+}	0.82	8	6, 8	
				0.73
Mn^{2+}	0.69	6	6	
Fe^{2+}	0.65	6	6	
Mg^{2+}	0.61	6	6	
Al^{3+}	0.46	6	4, 6	
				0.41
Si^{4+}	0.26	4	4	
				0.22
S^{6+}	0.15	3	4	
B^{3+}	0.15	3	3, 4	

† Data from Appendix IV. Radius of O^{2-} assumed = 1.32 Å.

ing works fairly well. In view of the complex electronic structure of ions, it is cause for wonder that the simple geometric model of ions as hard round spheres is actually so successful for a large number of substances.

The failure of geometric reasoning as a complete explanation is illustrated by Table 5-1, which shows a comparison between predicted and observed coordination numbers for several cations in compounds with oxygen. For cations whose radius ratios fall well within the theoretical limits, like Sr^{2+} and Fe^{2+}, the agreement of observed and predicted values is excellent. But cations like Ca^{2+}, Na^+, and Al^{3+}, whose ratios are close to the limiting figures, show variable coordination which the simple theory could not predict. A similar table for compounds containing sulfur would bring to light more serious discrepancies. Geometric reasoning breaks down also for complex anion groups like SO_4^{2-}, CO_3^{2-}, AsS_2^-; these groups show smaller interionic distances, and correspondingly greater stability, than would be expected from the sizes of the S^{6+}, C^{4+}, and As^{3+} ions alone. Our next step must be to inquire into the reasons for these deviations from strict geometric rules.

5-3 COVALENT BONDS

The nature of the deviation from radius-ratio predictions may be illustrated with some compounds of calcium and cadmium. These two metals have ions of almost identical radius: 1.08 Å for Ca^{2+} and 1.03 Å for Cd^{2+}. In their compounds with oxygen (ionic radius 1.32 Å) the interionic distances are 2.40 Å for CaO and 2.34 Å for CdO. There is nothing new here: the interionic distance for CdO is a trifle less than the sum of the radii, but the difference is hardly more than the uncertainty in the radii themselves. For the sulfides (ionic radius of $S^{2-} = 1.72$ Å), however, the interionic distances are 2.80 Å for CaS and 2.51 Å for CdS. Obviously something is amiss. It looks as if the ions in cadmium sulfide are squashed against each other, so that the interionic distance is considerably less than the sum of the radii. The crystal structures suggest the same thing: both compounds would be expected to have the NaCl structure (radius ratios approximately 0.61), but the prediction is fulfilled only for CaS; CdS has the sphalerite structure instead, showing that the ions are so deformed that only four rather than six can be grouped around an ion of opposite sign.

Deformation of ions is the simplest, most easily visualized model of deviations from the radius-ratio rule. We continue to regard the ions as spheres, but now as rubber balls rather than billiard balls. We think of one ion as deforming another, so that the distance between centers of the ions is shortened. One ion can be said to *polarize* the other, the implication being that the electrical symmetry of one or both is disturbed; thus, in the preceding example, the Cd^{2+} ion polarizes the large S^{2-} ion. This kind of description is common in older geochemical literature, but it is not well adapted to quantitative treatment. In recent work it has been largely superseded by other models based on the wave-mechanical picture of the atom.

The newer treatment focuses attention on the nature of the bond between atoms in a compound. In compounds of simple ionic character like NaCl, the formation of a bond between two atoms (an *ionic bond*) involves the complete removal of the outermost electron from one atom (Na) and its incorporation into the structure of the other (Cl), giving the latter a negative charge and leaving a positive charge on Na. Compounds like CdS have bonds of a different sort, in that electrons are partly *shared* between the atoms rather than completely transferred from one atom to another. The shared electrons are commonly grouped in pairs, which may be thought of as occupying positions about halfway between the atoms.

Chemical bonds of this latter kind, consisting of electron pairs held jointly between adjacent atoms, go by the name of *covalent* (or homopolar) bonds. Compounds with covalent bonds differ from ionic compounds, as a rule, in their slight solubilities in water and in their failure to conduct an electric current when melted. The formation of covalent bonds may lead to very stable molecules with little attraction for one another (for example Cl_2, CH_4, SO_2), in which case the compounds have low melting and boiling points and form solids with little strength or hardness. Alternatively such bonds may give continuous three-dimensional structures in which atoms are linked to others on all sides, as in diamond; in this case the substances have high melting points and exceptional hardness.

The ionic bonds in crystalline NaCl and the covalent bonds in Cl_2 or diamond are extreme types. In most substances the bonds are neither purely ionic nor purely covalent, but have an intermediate character; in other words, the electron pair between adjacent atoms may be regarded as somewhat displaced toward one of the atoms, but still attached to both. In crude symbols,

Na :Cl	H : Cl	Cl : Cl
Ionic bond, NaCl	Polar bond, HCl	Covalent bond, Cl_2

Bonds of intermediate type are called *polar-covalent* or simply *polar*, the term polar meaning that one end of such a bond is relatively more positive than the other. The partial separation of charge may give polar molecules (like HCl), or it may be compensated by oppositely directed bonds in the same molecule (for example, CCl_4). Water is an interesting case: although one might expect the polarity of the two bonds to compensate each other (H—O—H), the molecule is not linear but bent, Fig. 5-5. (The four electron pairs around the oxygen nucleus are approximately at the corners of a tetrahedron; the angle between the O—H

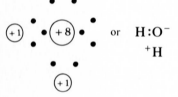

or   H:O$^-$
    $^+$H

Figure 5-5

bonds is 104.5°, close to the theoretical tetrahedral angle 109°28′.) Because of the bending, one side of the molecule has a net negative charge and the other side a positive charge.

Many polar compounds are soluble in water, but they generally do not dissociate appreciably (for example, sugar and alcohol). Some polar compounds, however, react with water to produce ions:

$$HCl + H_2O \rightleftharpoons H_3O^+ + Cl^-$$

This is a more accurate representation of the reaction we usually symbolize as $HCl \rightleftharpoons H^+ + Cl^-$ (Sec. 2-1).

Bonds in minerals, which cover the whole range from ionic to covalent, may be described as having so much "ionic character" or "covalent character." In CaS, for example, the bond is largely ionic, whereas the bond in CdS has considerable covalent character. The geometric radius-ratio rules of the last section apply strictly only to ionic compounds, and show increasing deviations as bonds become more covalent. Fortunately the bonds in many common minerals are sufficiently ionic so that the rules hold fairly well, and the effects of covalency can be treated as minor corrections. In other words, to answer a question posed several paragraphs back, mineralogical crystallography can indeed be regarded as largely an exercise in solid geometry, but with many complications arising from the partial covalent character of bonds.

The bonds in radicals like SO_4 and CO_3 are largely covalent and particularly strong, so that these groups are stable and remain intact through many chemical reactions. In the structures of crystals, such groups are so compact that to a first approximation they may be regarded as simple anionic units. For example, the structure of calcite may be pictured as a distorted NaCl structure, the Ca^{2+} and CO_3^{2-} ions playing the roles of Na^+ and Cl^-.

5-4 ELECTRONEGATIVITY

So far the discussion of bond type has been entirely qualitative. We have merely noted that bonds in some minerals have a partly covalent character, so that predictions about crystal structures from the radius-ratio rules are not wholly accurate. Obviously it would be desirable to make the concept quantitative—to express amount of covalent character by numbers, and to relate these numbers to other properties of elements so that the covalent character of a given bond could be predicted. This program has not yet been fully carried out, but several semiempirical schemes have been suggested by which the amount of covalent character can at least be estimated.

The most successful of these schemes is based on a concept called *electronegativity*. The general, qualitative meaning of this word is familiar from elementary chemistry: chlorine is an electronegative element because it readily forms negative ions in solution; sodium is an electropositive element because it forms positive ions; and copper is more electronegative (or less electropositive) than iron because Cu^{2+} will take electrons away from iron metal. The electronegativity of an element is clearly related to its ionization potential (for cations) and its electron affinity (for anions), but the relationship is not simple. Two examples of sets of numbers expressing electronegativity are given in Appendix IV and in abbreviated form in Table 5-2. One set (columns headed "Electronegativity") consists of values calculated by Pauling (1960) from bond strengths as measured by heats of formation, with an arbitrary range from 0.7 for Cs to 4.0 for F. The other set ("Percent ionic character") is based on electronegativities estimated by Povarennykh (1956) from ionization potentials and electron affinities; the numbers are not electronegativities as such, but percentages of ionic character of bonds with oxygen calculated from the electronegativities by Smith (1963).

Appendix IV gives an alphabetical listing for easy reference, but Table 5-2 is arranged so as to bring out more clearly the relations of the two sets of numbers with each other and with chemical properties of the elements. From Pauling's numbers one can generalize that a bond formed between any two atoms in the table is almost purely covalent if the electronegativities are similar, and largely ionic if the electronegativities are very different. Note, for example, that the electronegativity difference for NaCl is 2.1, for CaS 1.5, for CuS 0.5, and for CS_2 0.0, in agreement with the increasing covalent character of the bonds in this series of compounds. Smith's percentages express similar electronegativity differences for compounds of each cation with oxygen; thus the bonds Na—O, Ca—O, Cu—O, C—O have, respectively, 83, 79, 57, 23% ionic character. In general, as would be expected, Pauling's numbers are low and Smith's high for the active metals at the left side of Table 5-2, and Pauling's are high and Smith's low for the nonmetals at the right side. Since the two sets of numbers represent two different ways of

Table 5-2 Partial list of electronegativities and percentages of ionic character of bonds with oxygen†

Ion	Electronegativity	% ionic character	Ion	Electronegativity	% ionic character	Ion	Electronegativity	% ionic character
Cs^+	0.7	89	Mn^{2+}	1.5	72	Si^{4+}	1.8	48
K^+	0.8	87	Zn^{2+}	1.7	63	C^{4+}	2.5	23
Na^+	0.9	83	Sn^{2+}	1.8	73	P^{5+}	2.1	35
Li^+	1.0	82	Pb^{2+}	1.8	72	N^{5+}	3.0	9
Ba^{2+}	0.9	84	Fe^{2+}	1.8	69	Se	2.4	
Ca^{2+}	1.0	79	Fe^{3+}	1.9	54	S	2.5	
Mg^{2+}	1.2	71	Ag^+	1.9	71	O	3.5	
Be^{2+}	1.5	63	Cu^+	1.9	71	I	2.5	
Al^{3+}	1.5	60	Cu^{2+}	2.0	57	Cl	3.0	
B^{3+}	2.0	43	Au^+	2.4	62	F	4.0	

† Data from Appendix IV.

calculating electronegativity, however, the agreement is far from perfect. Discrepancies are particularly evident among the metals from the middle of the periodic system (*transition metals*) shown in the center of Table 5-2; as a single example, Pauling's numbers would make the Sn—O bond less ionic than the Zn—O bond, while Smith's numbers make it more ionic.

It is a general rule, illustrated by these two sets of numbers and also by several alternative sets that have been proposed as a measure of electronegativity, that numbers can be made to express very nicely the chemical properties of elements at the ends of the periods in the periodic table, but that unresolvable difficulties arise in trying to express the subtle and complicated relationships among the transition metals in the interior of the table. Electronegativity is a useful concept, but it cannot be depended on for wholly accurate predictions about character of bonding and coordination numbers in all kinds of compounds.

5-5 GENERAL RULES ABOUT BOND TYPE

The ionic radii and electronegativities in Appendix IV permit the formulation of a few useful rules about chemical bonds:

1. For a given cation and two different anions, the bond with the larger anion is more covalent (MgS is more covalent than MgO).
2. For a given anion and two different cations, the bond with the smaller cation is more covalent (MgO is more covalent than BaO).
3. For ions of similar size and different charge, the one with highest charge forms the most covalent bonds (Ca—O is more covalent than Na—O in Na_2O).
4. Ions of metals in the middle of the long periods of the periodic table form more covalent bonds with anions than do ions of similar size and charge in the first two or three groups of the table (CdS is more covalent than CaS, and FeO more covalent than MgO).

These rules have many exceptions, but they often help in making qualitative predictions about crystal structures and about the distribution of rare elements in geologic materials.

Such rules, it should be emphasized once again, are based on a crude model of electron-pair bonds linking spherical, somewhat deformable ions. The number of exceptions to the rules is a measure of the crudity of the model. Considerable refinement in predictions is possible by calculating bond energies and bond angles from quantum mechanics (see, for example, Fyfe, 1964), but for most purposes in geochemical arguments the simpler picture is adequate.

Covalent and ionic bonds are the only bond types of much interest in geology, but one other kind needs brief mention. The *metallic* bond, characteristic of all metals, is formed typically in substances whose atoms do not have sufficient valence electrons to combine into the usual stable shells of 8, either by sharing or by transfer. In this case the electrons may be pictured as largely free to wander from atom to atom, hence capable of moving along a metal under the influence of an electric potential difference. Some of the sulfide minerals, especially those with metallic luster, contain partly free electrons and so exhibit to some degree the characteristics of the metallic bond.

5-6 SILICATE STRUCTURES

The crystal structures of most interest in geology are also the most complicated—the structures of the multitudinous compounds called silicates.

We note first that the element silicon is a nonmetal of intermediate electronegativity, that it has an oxidation number of 4 and also a coordination number of 4, and that therefore the fundamental unit of silicate structures ought to be the anion SiO_4^{4-}, in which the silicon ion is surrounded by four oxygen ions at the corners of a tetrahedron. This expectation is borne out in nature, for silicon practically always occurs in SiO_4 tetrahedra. A few silicate minerals contain this group in the form of simple ions; for example, the common mineral olivine (Mg_2SiO_4) has a structure consisting of alternate Mg^{2+} and SiO_4^{4-} ions, much as magnesite consists of alternating Mg^{2+} and CO_3^{2-} ions. In such compounds silicon acts like a typical nonmetal, and its structural chemistry is no more complicated than that of sulfur or phosphorus or the carbon of carbonates. But compared with these other nonmetals, silicon displays astonishing versatility, for its simple ionic compounds are only the first step of an elaborate structural chemistry that includes rings, chains, sheets, and solid frameworks of interconnected silicon-oxygen groups. Why should this particular nonmetal form compounds of such enormous variety?

For an answer we recall that silicon, although a nonmetal, has properties in some measure intermediate between those of nonmetals and metals. A more metallic element like aluminum or magnesium would form a structure with oxygen in which metal ions and oxygen ions are linked together in a strong three-dimensional framework. Silicon exhibits this kind of behavior in quartz and the other silica minerals (tridymite, cristobalite, coesite, stishovite): here silicon remains at the center of SiO_4 tetrahedra, but each oxygen ion is linked with an adjacent silicon ion (Fig. 5-6). A more nonmetallic element like carbon or sulfur would form a volatile oxide consisting of self-contained molecules (like CO_2 or SO_3) having only very slight attraction for one another, and would also form ionic compounds with metals in whose structure the nonmetal is part of a compact anion (like CO_3^{2-} or SO_4^{2-}). Silicon forms no volatile oxide, but it does follow the behavior of nonmetals, as we saw in the last paragraph, by entering structures like that of olivine in the form of simple anions. Thus the behavior of silicon straddles the roles of metal and nonmetal. It forms not only structures typical of the extremes, but intermediate structures in which it is part of large silicon-oxygen units that are at once anions and more or less continuous frameworks. This dual

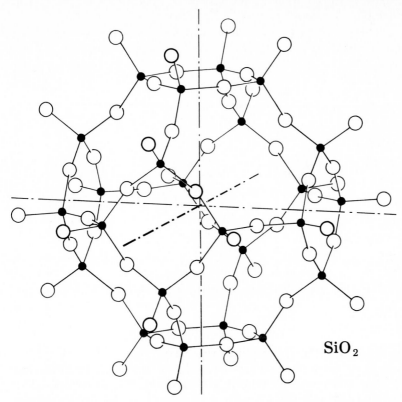

SiO_2

Figure 5-6 The arrangement of SiO_4 tetrahedra in the continuous-framework structure of the SiO_2 minerals. Open circles represent O^{2-}, small solid circles Si^{4+}. *(Source: Berry and Mason, 1959, page 466.)*

capacity of silicon, the ability to play the role of metal or nonmetal or anything in between, accounts for the diversity of silicate structures.

This explanation is largely in terms of chemical properties. We could say the same thing in more "structural" language by noting that the size of the Si^{4+} ion is intermediate between that of the smallest common metal ions (Ti^{4+} and Al^{3+}) and the largest nonmetal ions (P^{5+}, S^{6+}). If the silicon ion were smaller, it could polarize oxygen ions more effectively, deforming them to fit around it in compact, self-contained molecules or anions. If it were larger, its attraction for oxygen ions would be less, so that Si^{4+} and O^{2-} would simply be independent units in a framework like that of metal oxide crystals. The in-between size gives silicon the ability to perform either function in a crystal structure.

In silicates, then, the silicon ions occur always surrounded by oxygen ions at the corners of a tetrahedron. The tetrahedral groups may be independent anions, or they may be linked together in a variety of ways. The linkage simply means that one or more oxygens in a given tetrahedron are also part of adjacent tetrahedra

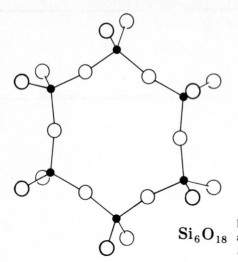

Si_6O_{18}

Figure 5-7 The arrangement of SiO_4 tetrahedra in a cyclosilicate: beryl. *(Source: Berry and Mason, 1959, page 463.)*

(Figs. 5-7 and 5-8). The possible varieties of structure permitted by these linkages are as follows:

Independent tetrahedral groups (*nesosilicates*). Silicon-oxygen tetrahedra as independent anions; the most familiar example is olivine.

Multiple tetrahedral groups (*sorosilicates*). Two to six tetrahedra linked together to form larger independent anions. A typical example is hemimorphite, $Zn_4Si_2O_7(OH)_2 \cdot H_2O$.

Ring structures of linked tetrahedra (*cyclosilicates*). The commonest example is beryl, $Be_3Al_2Si_6O_{18}$, with rings consisting of six tetrahedra (Fig. 5-7).

Chain structures (*inosilicates*). Tetrahedra linked to form linear chains of indefinite length. Two kinds of chains are found: single chains with a silicon-oxygen ratio of 1 : 3, characteristic of the pyroxenes (Fig. 5-8), and cross-linked double chains with a silicon-oxygen ratio of 4 : 11, characteristic of the amphiboles. The chains are bonded to one another by metal ions.

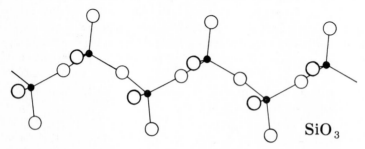

SiO_3

Figure 5-8 The arrangement of SiO_4 tetrahedra in a single-chain inosilicate: pyroxene. *(Source: Berry and Mason, 1959, page 464.)*

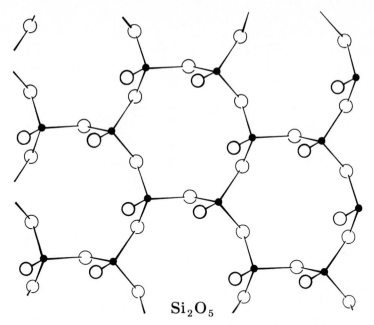

Si_2O_5

Figure 5-9 The arrangement of SiO_4 tetrahedra in the sheet structure of the phyllosilicates. *(Source: Berry and Mason, 1959, page 465.)*

Sheet structures (*phyllosilicates*). Three oxygens of each tetrahedron are linked with adjacent tetrahedra, forming flat sheets of indefinite extent. In effect this is the double-chain inosilicate structure extended in two dimensions instead of one. The silicon-oxygen ratio is 2 : 5. This structure is found in the micas and clay minerals; the sheet structures with a hexagonal pattern are reflected in the perfect basal cleavage and pseudohexagonal habit of these minerals (Fig. 5-9).

Framework structures (*tektosilicates*). Three-dimensional networks, each tetrahedron sharing all its oxygens with adjacent tetrahedra, thus giving a structure with silicon-oxygen ratio of 1 : 2. Quartz and the other silica minerals are good examples (Fig. 5-6). Other minerals may have this structure provided that some of the silicon ions are replaced by ions of lower charge; the commonest substitution is Al^{3+} for Si^{4+}, giving a negative charge to the framework which is balanced by positive metal ions. The feldspars and zeolites are familiar examples of this kind of structure.

5-7 ISOMORPHISM

The formula of olivine is customarily written $(Mg,Fe)_2SiO_4$. This signifies that olivine contains both magnesium and iron, that the ratio of the two metals is variable from one specimen to another but the ratio of total metal to silicon (in atoms or gram atoms) remains constant, and that magnesium ordinarily is present in greater amount than iron. For this mineral all ratios of magnesium to iron are possible, and the two *end members* Mg_2SiO_4 (forsterite) and Fe_2SiO_4 (fayalite) have similar crystal form and crystal structure. This relationship is described as *isomorphism,* and olivine is said to be an *isomorphous mixture* of its two end members. Alternatively olivine may be called a *solid solution* of fayalite and forsterite, the term referring to the fact that this solid resembles a liquid solution in that it remains homogeneous when its components are varied over a certain range of compositions.

The geometric basis of crystallography that we have discussed above gives at once a simple explanation of isomorphism. When two compounds differ in only a single constituent and when the two kinds of ions that play the role of this constituent have similar sizes, one might expect that either or both could fit into the same crystal pattern. Thus fayalite and forsterite differ only in the metal ions, and Mg^{2+} and Fe^{2+} have roughly the same radius (0.80 Å and 0.86 Å, respectively), so that the olivine structure can accommodate either or both. A similar isomorphous relationship between magnesium and iron is common in other silicate minerals; many of the pyroxenes [e.g., diopside, $Ca(Mg,Fe)Si_2O_6$] and amphiboles [e.g., actinolite, $Ca_2(Mg,Fe)_5(Si_4O_{11})_2(OH)_2$] are familiar examples.

It does not follow that all pairs of magnesium and iron compounds are isomorphous. Very little magnesium is found in pyrite (FeS_2), and very little iron in epsomite, $MgSO_4 \cdot 7H_2O$, for example. We might guess that the slight difference in ionic size prevents isomorphism in these compounds, or alternatively we recall that iron (a transition metal in the middle of the first long period of the periodic system) should form bonds of more covalent character than magnesium (near the beginning of a period), so that despite the similarity in size the ions can play somewhat different roles in crystal structures.

This example of iron-magnesium compounds leads to some general statements about isomorphism:

1. Two compounds are said to be isomorphous if they have the same, or nearly the same, crystal form.
2. The general requirement for isomorphism is that the two compounds contain ions of approximately the same size, or at least of the same relative size.
3. A pair of isomorphous compounds may show solid solution; i.e., homogeneous crystals may form containing the two end members in various ratios. All ratios between the end members may be possible, as they are between fayalite and forsterite (complete solid solution), or solid solution may be restricted to a few percent of one compound in the other (limited solid solution).
4. If isomorphism is due to similar *relative* sizes of ions and if absolute sizes are different, solid solution is generally not possible. For example, halite is isomorphous with galena, because the ratios of sizes Na^+/Cl^- and Pb^{2+}/S^{2-} are similar, but solid solution does not occur. Pairs of this kind are often called *isotypic.* (There is confusion in nomenclature here. We are following the practice of many authors who make isomorphism the general term, with isotypy a

special case; other authors use isotypy in a general sense and restrict isomorphism to pairs of compounds that show appreciable solid solution.)

5. Ionic size is the most important determiner of isomorphism, but by no means the only one. Isomorphism is most perfect, and solid solution is most complete, between compounds whose ions form bonds of similar covalent character. The structure of sphalerite is different from that of pyrrhotite, for example, even though Zn^{2+} and Fe^{2+} have nearly the same size, because the Zn—S bond is more covalent than the Fe—S bond; only limited solid solution is possible between sphalerite and a hypothetical compound FeS with the sphalerite structure.

1. same size
2. same covalent character
electronegativity

5-8 SUBSTITUTION

Another way of describing the relation between substances that form solid solutions is to say that one ion may *substitute* for another in the crystal structure. Thus iron substitutes for magnesium in the structure of olivine or pyroxene, but not in the structure of epsomite. To specify this particular kind of replacement, the more precise term *diadochic substitution* is sometimes used. Or one may say in the same sense that a given ion *proxies* another, for example, magnesium proxies iron in olivine. The possibilities of substitution in crystal structures become particularly important in the geochemistry of trace elements, because many of these elements substitute for major elements in common minerals instead of forming minerals of their own. Gallium, for example, is a major constituent in only a few exceedingly rare minerals but is always present in minor amounts as a substitute for aluminum in minerals like feldspar and mica, and for zinc in the mineral sphalerite.

From the previous discussion of isomorphism, it follows that the possibility of one ion substituting for another depends largely on size and secondarily on electronegativity (or polarizing ability). Beyond this qualitative statement it is difficult to go. Empirically we find that substitution is not extensive between elements whose ionic radii differ by more than 15%; for example, substitution of potassium for sodium in halite is very limited, despite the chemical similarity, because their ionic radii are far apart (1.46 Å and 1.10 Å). Empirically we find also that substitution is greater at high temperatures, presumably because added thermal energy serves to increase the vibratory motion of ions, so that the crystal structure expands and becomes more tolerant of foreign particles. The greater ease of substitution at high temperatures is often indicated by "exsolution textures" formed on cooling. Albite and orthoclase, for example, form a complete solid-solution series at temperatures of 500 to 600°C, but on cooling exsolve to produce the familiar perthite and antiperthite intergrowths. Another generalization about substitution is that some structures are more tolerant than others. The substitution of aluminum for silicon, as an example, is common in the feldspars and the clay minerals, but negligible in quartz. Rules of this sort are convenient to remember, but their qualitative form betrays a deficiency in present knowledge of just how the various factors control the substitution of one element for another.

The rules of substitution omit one ionic property that at first sight would seem important: they say nothing to forbid substitution between ions of different charge. Actually substitution of this sort is very common. A familiar illustration is the series of plagioclase feldspars, which are mixtures of the end members albite ($NaAlSi_3O_8$) and anorthite ($CaAl_2Si_2O_8$). The nature of the isomorphic substitution becomes clearer if we write, as a general formula for plagioclase, $(Na,Ca)(Si, Al)AlSi_2O_8$. In other words, Na and Ca are interchangeable, as might be expected from their ionic radii (1.10 Å and 1.08 Å, respectively), but in order to compensate for the charge difference some of the Si must also be interchangeable with Al. [Note that replacement of Si (0.34 Å) by Al (0.47 Å) is an exception—one of many, unfortunately—to the general rule that substitution is limited to pairs differing in ionic radius by less than 15%.] The substitution of Ga^{3+} for Zn^{2+} in sphalerite, of Li^+ for Mg^{2+} in mica, and of Ba^{2+} for K^+ in orthoclase are other examples of replacement by ions of different charge. Ordinarily this kind of replacement is limited to pairs whose charge difference is only one unit; there is little substitution, for example, of Se^{6+} for Si^{4+}, or of Y^{3+} for Na^+, despite resemblances in ionic radius.

Solid solution most commonly involves this sort of substitution of one ion by another, in other words the mixing of a pair of isomorphous compounds, but it may refer to other phenomena also. One kind of solid solution, called *interstitial solid solution*, depends on the entering of foreign ions into the open places of a crystal structure. It is particularly common among metals, whose crystals can take up large amounts of hydrogen, nitrogen, carbon, and other elements. Cristobalite, the high-temperature form of silica, has an open structure that can accommodate even the relatively large ions of sodium in its interstices. Another phenomenon that may be described as solid solution is the formation of a *defect structure*, or *defect lattice*, a crystal structure in which some of the lattice positions are unoccupied. Pyrrhotite, for example, has long been known to have an excess of sulfur over the amount required by the simple formula FeS; this was formerly ascribed to solid solution of sulfur in FeS, but investigation of the structure by x-rays shows that the crystals have a lattice in which some of the Fe^{2+} ions are lacking. Defect structures have been shown to be responsible for a good many stoichiometric anomalies in mineral formulas.

5-9 POLYMORPHISM

Two substances are *isomorphous* if they have similar crystal structures but different chemical formulas, *polymorphous* if they have similar formulas but different structures. Calcite and aragonite, for example, are polymorphs of $CaCO_3$; quartz, tridymite, and cristobalite are polymorphs of SiO_2. If a compound is known in only two crystalline modifications, it is often called *dimorphous* rather than polymorphous; thus pyrite is a dimorph of marcasite.

From a structural standpoint, polymorphism means that the ions making up a compound can be arranged in two or more different patterns. For example, rubidium ion, Rb^+, with an ionic radius intermediate between the radii of Cs^+ and Na^+, forms a chloride (RbCl) with the 8-coordinated cesium chloride structure at high pressures and the more open 6-coordinated structure of sodium chloride at ordinary pressures. Another example is the series of carbonates $MgCO_3$, $CaCO_3$, $SrCO_3$, $BaCO_3$, arranged in the order of increasing size of the cation; the compounds of the large ions Sr^{2+} and Ba^{2+} show the aragonite structure and that of the small ion Mg^{2+} the calcite structure, and only $CaCO_3$ with its cation of intermediate size is capable of crystallizing in either structure.

Polymorphism, however, is much more complicated than such a simple rule would indicate. Ideally we can refer it to geometry, but the factors influencing the geometry are numerous and inadequately understood. Temperature, of course, is one such factor, and pressure is another; in general, a particular structure is stable only over a certain range of temperature and pressure. Rhombic sulfur, for example, is stable at temperatures below 96°C, monoclinic above; and the temperature of the change, or *transition temperature*, is raised if the pressure is increased. Speaking still in generalities, we can say that high-temperature polymorphs have more open structures with higher symmetry than their low-temperature equivalents, and that high-pressure polymorphs have more closely packed structures with higher coordination numbers than low-pressure forms. The form in which a substance crystallizes, however, may be changed by impurities, even though these are sometimes present in very minor amounts. The form also depends on rate of crystallization: high-temperature forms sometimes appear in low-temperature environments, apparently because crystallization took place so rapidly that arrangement of ions into the more stable and more ordered arrangement of the low-temperature polymorph was impossible. Cristobalite, for example, although stable only at temperatures above 1470°C, is commonly found encrusting cavities in lava which could not have been at temperatures over a few hundred degrees, either because it contains enough impurities to modify its stability range profoundly or because it crystallized very rapidly.

A few polymorphs apparently have no true stability range. Marcasite, for example, is unstable with respect to pyrite at all temperatures and pressures. Such pairs are called *monotropic*, in contrast to *enantiotropic* pairs in which the change from one form to another is reversible at definite values of pressure and temperature. The existence of unstable monotropic modifications in nature must be ascribed to impurities or rate of crystallization or both.

Thus for polymorphism, as for isomorphism, we possess a considerable body of observations which can be summarized in a few useful but qualitative rules. We can make general predictions about what substances will be isomorphous and what compounds will show polymorphism, but the predictions often prove inaccurate because our knowledge of the effects of temperature, pressure, bond type, impurities, and reaction rates is still far from quantitative.

PROBLEMS

1 Why does zinc ion, Zn^{2+}, in many of its compounds show tetrahedral coordination, although its ionic radius is similar to that of ions like Mg^{2+} which show only octahedral coordination?

2 Why is the radius of Fe^{3+} smaller than the radius of Fe^{2+}? Would you expect, as a general rule, that the more highly charged ions of a multivalent element would have smaller radii than the ions of lower charge? Why?

3 The following substitutions are uncommon in minerals. Inspect each one, to see whether its infrequency of occurrence conforms to the general rules of isomorphous substitution.

$$Li^+ \text{ for } Na^+ \qquad C^{4+} \text{ for } Si^{4+} \qquad Cd^{2+} \text{ for } Na^+$$

$$Cu^+ \text{ for } Na^+ \qquad Sc^{3+} \text{ for } Li^+ \qquad Cl^- \text{ for } F^-$$

4 Why are intergrowth textures (perthite and antiperthite) common in alkali feldspars (albite and orthoclase) but not in the plagioclase feldspars?

5 A possible method of estimating the temperature of formation of sulfide veins is based on a determination of the iron content of sphalerite. Assuming that iron was present in excess in the solutions from which sphalerite crystallized, would you expect the iron content of sphalerite to be greater at high temperature or low? Why?

6 In each of the following pairs, choose the one in which the chemical bond would have more covalent character:

$$KCl \text{ and } KI \qquad Li_2S \text{ and } Cs_2S \qquad BaCl_2 \text{ and } HgCl_2$$

$$KCl \text{ and } BaCl_2 \qquad Cu_2O \text{ and } Cu_2S \qquad B_2O_3 \text{ and } Al_2O_3$$

7 Describe in general terms the crystal structure of diopside, forsterite, analcite, sericite, tremolite.

8 Why do we not find a variety of complex carbonate minerals with structures analogous to silicate structures?

9 If a cation can show more than one type of coordination, would you expect the higher coordination to be found in minerals formed at high temperatures or at low temperatures? Why? A good example is Al^{3+}, which shows 4-coordination in orthoclase, $KAlSi_3O_8$, and 6-coordination in kaolinite, $Al_4Si_4O_{10}(OH)_8$.

10 Consider a series of cations having the same electronic structure, for example the series Na^+, Mg^{2+}, ..., Cl^{7+}. Do ionic radii show a regular change through this sequence? Can you suggest a reason for the pattern of change?

REFERENCES AND SUGGESTIONS FOR FURTHER READING

Berry, L. G., and B. Mason, "Mineralogy." W. H. Freeman and Company, San Francisco, 1959.

Bloss, F. D., "Crystallography and Crystal Chemistry," Holt, Rinehart and Winston, Inc., New York, 1971. A standard textbook.

Fyfe, W. S., "Geochemistry of Solids," McGraw-Hill Book Company, New York, 1964. A sophisticated but clearly written treatment of the nature of chemical bonds and their role in crystal structures from the point of view of quantum mechanics.

Goldschmidt, V. M., "Geochemistry," Oxford University Press, Fair Lawn, N.J., 1954. Much of the foundation of modern crystal chemistry, and especially its applications in geochemistry, goes back to the work of Goldschmidt. Chapter 6 of this posthumous volume gives a good summary of Goldschmidt's ideas. Note that he speaks of the "polarization" of anions by cations rather than of the degree of covalent bonding between them.

Mason, B., "Principles of Geochemistry," 3d ed., John Wiley & Sons, Inc., New York, 1966. This is the best brief reference book on all aspects of geochemistry. Chapter 4 gives an excellent elementary survey of crystal chemistry.

Pauling, L., "The Nature of the Chemical Bond," 3d ed., Cornell University Press, Ithaca, N.Y., 1960. Methods of determining ionic radii are described in pages 511 to 519, and the development of an electronegativity scale in pages 88 to 102.

Povarennykh, A. S., O kolichestvennoi otsenke sostoyaniya khimicheskoi svyazi v mineralakh, *Dokl. Akad. Nauk SSSR*, vol. 109, pp. 993–996, 1956.

Smith, F. G., "Physical Geochemistry," Addison-Wesley Publishing Company, Inc., Reading, Mass., 1963. Chapter 2 is an excellent brief account of the development of ideas about coordination and chemical bonding. It includes a table of electronegativities as estimated by Povarennykh and percentages of ionic character of bonds in oxides and sulfides as calculated from the electronegativities by Smith.

Whittaker, E. J. W., and R. Muntus, Ionic radii for use in geochemistry, *Geochim. et Cosmochim. Acta*, vol. 34, pp. 945–956, 1970. Comprehensive table of ionic radii, and critical discussion of bases for estimating the radii.

diopside — inosilicate $CaMgSi_2O_6$
single chain

forsterite — Mg_2SiO_4 — nesosilicate   Mg form of olivine

analcite — tectosilicate

sericite — phyllosilicate $KAl_2(AlSi_3O_{10})(OH)_2$

tremolite — inosilicate — double chain
$Ca_2Mg_5Si_8O_{22}(OH)_2$

COLLOIDS

The structural concepts of the last chapter will help toward an understanding of clay minerals and the processes of weathering, but one part of the picture is still missing. An essential characteristic of most clays is their fine grain size, which gives them a set of properties that we could not predict from the crystal structure alone. In this chapter we set out to study the behavior of matter when it is finely subdivided, or in a *colloidal* state. As in the last chapter, we shall not confine the discussion to clay minerals, but rather look for generalizations that apply to colloids of all kinds. We shall meet colloids again later on in connection with the geochemistry of sedimentation and ore solutions.

6-1 DEFINITIONS

The term "colloid" goes back to a period in the history of chemistry when a distinction was attempted between "crystalloid" matter on the one hand and noncrystalline, or "colloid," matter on the other—the noncrystalline matter including substances like glue, soap, and gelatin which dissolved in water to give viscous solutions and which could not be obtained in the form of crystals. The distinction proved to be artificial, in that a great many substances could be prepared in both a crystalloidal and a colloidal form. The characteristics associated with colloids turned out to be simply the properties of matter in a finely divided state, and any substance that could be finely divided would exhibit these properties; on the other hand, many substances thought to be exclusively colloidal have been shown to have regularities of internal structure like those in crystals.

Hence there is no such thing as a "colloid" in the original sense of the word. In modern chemistry the term is used loosely to refer to a suspension of any fine-grained material, or to a material that can easily be made into such a suspension. Thus the finer-grained part of a sample of clay is a colloid, and the same word may be used to describe the liquid obtained by shaking the clay with water.

The adjective "colloidal" may be defined more precisely. A substance is colloidal, or in a colloidal state, if it consists of very tiny particles dispersed in another substance. Just how small the particles must be is not universally agreed upon, but a widely used definition puts the range of colloidal diameters at 10^{-3} to 10^{-6} mm (1 μm to 1 nm). The system including the colloidal particles and the medium in which they are suspended is called a *colloidal system*, or a *disperse system*.

6-2 KINDS OF COLLOIDAL SYSTEMS

Technically, colloidal particles may be either solid, liquid, or gas, and the dispersion medium likewise may be any one of the three states of matter. In geology we are concerned almost exclusively with water as the dispersion medium and with solids as the dispersed particles. The most notable exception is a colloidal system consisting of water suspended in petroleum, or petroleum in water; such a dispersion of one liquid in another is called an *emulsion*. In the geochemistry of the atmosphere, dispersions of fine solid particles and liquid droplets in the atmospheric gases, called *aerosols*, are sometimes important. Some naturally occurring crystals may contain inclusions of solid, liquid, or gas in the colloidal size range, thus forming systems with a solid dispersion medium, but the colloidal properties of such occurrences have only minor geologic significance.

Many solid-in-liquid systems may be prepared in two forms. If gelatin, for example, is dissolved in warm water, it forms a clear, transparent liquid which to all appearances is an ordinary solution. The colloidal nature of the particles in the "solution" may be established by some of the tests described in the next section. To specify such a liquid, actually a colloidal system but resembling a solution in its transparency and fluidity, we use the term *sol*. If the gelatin sol is cooled and allowed to stand, its character changes: we say it "sets," forming a translucent or transparent solid called a *gel*. Many colloids, like gelatin, may be obtained in either a sol or gel form, depending on concentration, methods of preparation, and time of standing. Just why a gel forms is still not completely understood, but it is natural to suppose that the colloidal particles form a network of fibers or chains in which the water is trapped.

Gelatin is an example of a large group of colloidal substances that dissolve directly to form disperse systems. The resulting systems are stable indefinitely, the suspended particles showing no tendency to settle out. Such colloids, provided the concentration is high enough, readily set to gels, and the process of gelation is reversible, in that the gel dissolves to form a sol if more water is added. The ease with which the gels form and dissolve, together with other less direct evidence, suggests that particles of these substances are accompanied in the sol by much adsorbed water. An appropriate name for such colloids is *hydrophilic*, which in Greek means "water-loving."

In contrast, a great many substances do not disperse themselves spontaneously but may be obtained as colloids by indirect means. If a reducing agent is added to a solution of a gold compound, for example, the resulting metallic gold may not precipitate, but may instead remain suspended in the form of colloidal particles, which color the solution bright red ("ruby gold") because of their ability to scatter light. Again, if one attempts to precipitate arsenic sulfide by bubbling hydrogen sulfide through an arsenic solution, the metal sulfide often remains suspended in colloidal form. Many other common precipitates form as colloids if the concentrations are in certain ranges, often to the embarrassment of the analyst who is trying to use the precipitation to separate one substance from another. Still another method of preparing colloids is by striking an electric arc under water: such an arc between platinum electrodes, for example, gives a platinum sol. Colloids formed by these indirect means are generally less stable than a colloid like gelatin, settling out partly or completely on standing or when disturbed. Some of them can be made to form gels, but temperatures and concentrations must be carefully controlled and the process of gelation is not reversible. Particles of these colloids are known to have much less adsorbed water, and the colloids are accordingly known as *hydrophobic,* or "water-fearing."

A few colloids are difficult to classify as either hydrophobic or hydrophilic. Silica, for example, behaves like a hydrophilic colloid in that a dilute silica sol is stable indefinitely and in that it readily forms a gel. On the other hand, silica does not spontaneously disperse itself in water; in this respect it exhibits the characteristic behavior of a hydrophobic colloid. Despite exceptions of this sort, the distinction between colloids with much adsorbed water and with little adsorbed water is a useful one.

6-3 PROPERTIES OF COLLOIDS

Colloidal particles lie in the size range between that of true solutions (particle diameters of the order of 10^{-7} mm) and that of suspensions whose particles quickly settle out (particle diameters greater than 10^{-3} mm). What are the characteristic properties that distinguish dispersions in this intermediate size range from true solutions on the one hand and coarse suspensions on the other?

Colloids differ from suspensions in their apparent homogeneity and in their stability. To the naked eye, and even in the field of an ordinary microscope, a colloid looks completely homogeneous, whereas the particles of a suspension are readily visible.

When a colloid is allowed to stand, it generally remains unchanged for an indefinite period, in contrast to the rapid settling of the particles in a suspension. Some colloids, it is true, are unstable in the sense that their particles slowly grow larger or coagulate and ultimately settle out, but the settling is much slower than that of a suspension and can be long delayed if the colloid is carefully prepared.

The distinction between colloids and true solutions is more difficult, because superficially they look alike. One indication that colloidal particles are present is the ability of a colloid to scatter light. This is readily demonstrated by passing a beam of light through a sol; the beam is outlined by a soft glow in the liquid, much as the path of light through a darkened room may be outlined by scattering from dust particles, whereas a light beam through a true solution is invisible from the side. This scattering of light is known as the Tyndall effect. In some colloids, especially colored ones or very concentrated ones, the Tyndall effect is so pronounced that the colloid shows opalescence—an apparent milkiness, or a difference in color when examined by reflected and transmitted light.

The ability of colloidal particles to scatter light may also be demonstrated by a microscope set up so that its field is illuminated from the side (an "ultramicroscope"). The particles themselves remain invisible, but they show up as bright points of light against a dark background. These points are in constant random motion (the "Brownian movement"), due to molecular bombardment; the particles are so small that collisions of water molecules against their sides do not balance out as they would for larger fragments, but push the particles first one way and then another. Study of particle motion with the ultramicroscope has given much information about the size and general behavior of colloidal particles.

The greater size of colloidal particles in comparison with the particles of a solution is also shown by an extreme slowness of diffusion. Diffusion can be prevented altogether by separating a colloid from pure water with a membrane of rubber or collodion: particles of water and of substances in true solution pass through the membrane without difficulty, but colloidal particles are too large for the tiny openings in the membrane. This provides a convenient means for purifying a colloid, since dissolved impurities in the colloidal system will diffuse out into the pure water on the other side of the membrane. The process is called *dialysis* (Fig. 6-1).

Still another characteristic of colloids is the lack of a predictable effect on the vapor pressure, boiling point, and melting point of water. A substance in true solution lowers the vapor pressure, depresses the freezing point, and raises the

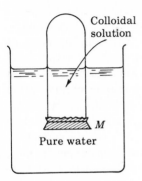

Colloidal solution

M

Pure water

Figure 6-1 Purification of a sol from electrolytes by dialysis. The sol is placed on one side of the membrane *M*, pure water on the other side. Ions from the sol diffuse through the membrane into the water, but colloidal particles remain behind.

boiling point; the effects on these properties follow simple, well-known rules that depend on the concentration and molecular weight of the dissolved material. Colloids, on the other hand, affect the properties of the liquid erratically. Hydrophobic colloids generally show very slight effects, whereas hydrophilic colloids may change the behavior of the liquid profoundly. These characteristics of colloids may again be referred to the sizes of particles, for the alteration in vapor pressure, freezing point, and boiling point is fundamentally an effect of the number of particles present. The tiny particles of a true solution are very numerous, whereas the larger particles of a colloid are relatively few even if the concentration is high. The erratic behavior of hydrophilic colloids is due to the adsorption of large amounts of water onto the particles.

Finally, a striking characteristic of colloids in the form of a gel is the development of colored diffusion bands, "Liesegang rings," when an electrolyte is allowed to diffuse into the gel. If a gel contains chromate ion, for example, and if a silver salt diffuses into it, regularly spaced bands of precipitated silver chromate develop in the gel. This phenomenon, first studied by the German chemist Liesegang, is still not entirely understood, but in general it must depend on a slow movement of silver ions leading to a depletion of chromate ions in some parts of the gel and supersaturation with silver chromate in other parts. The diffusion bands are strikingly similar to the rhythmic banding found in agate and in some ore deposits, suggesting that diffusion in gels may account for the origin of the natural color patterns.

6-4 ELECTRIC CHARGES ON COLLOIDAL PARTICLES

Some of the important characteristics of colloids are related to the fact that colloidal particles carry an electric charge. The existence of the charge is easily demonstrated, simply by passing a current through a sol: the colloidal particles migrate to one of the electrodes, and are coagulated as a precipitate when their charge is neutralized. The direction of motion of the particles is, of course, determined by the sign of their charge, which turns out to be positive for some colloids and negative for others.

The experimental facts are clear enough, but a completely satisfactory explanation for the facts is not yet possible. The charge is due largely to adsorption of ions on the particles, and a good reason for the adsorption is certainly the enormous total surface area which the particles expose to the solution. But why one kind of colloid has a greater charge than another, why one should be negative and another positive, why the amount and sign of the charge should depend on the nature and concentration of electrolytes in the solution—questions of this sort are largely in the realm of speculation. In talking about charges and their effects, we deal largely with a mass of empirical observations rather than with a simple and convincing theory.

The great surface area of colloids needs no more demonstration than an elementary calculation. If a cube 1 cm on a side is cut into eight equal parts, its surface area is increased from 6 to 12 cm^2; if each fragment is divided into eight smaller fragments 2.5 mm on a side, the area becomes 24 cm^2; and if this process of subdivision is continued until the diameter of each fragment is 10^{-5} mm, within the colloidal range, the surface area will have increased to 6 million cm^2, or 600 square meters. Now any surface in a solution will adsorb charged particles, if they are available—a phenomenon akin to contact potential, the development of opposite charges whenever two dissimilar substances are in contact. If the surface is the wall of a container, or the large grains of a suspension, the charge will be too small to observe. But the particles of a colloid, with their high ratio of surface area to volume, adsorb enough ions for the charge to be important.

The sign of the charge depends chiefly on the nature of the colloid, but there is no simple rule by which it may be predicted. Empirically it is found that the colloids of most interest in geology develop charges as shown in Table 6-1. A convenient rule is that sulfide sols and organic sols are negative, whereas oxide and hydroxide sols are positive. Two conspicuous exceptions are silica and manganese dioxide, both of which form negative sols. Some ions are more readily adsorbed than others and hence contribute more effectively to the charge on a colloid; two especially effective ions are H^+ and OH^-. Many colloids, of which ferric hydroxide is the most familiar example, may assume either a positive or negative charge, depending on whether the solution contains an excess of H^+ or OH^-. In nature, ferric oxide sols most commonly form in a faintly acid environment and so have a positive charge, but they are readily prepared with a negative charge in the laboratory by keeping the solution alkaline.

The electric charge on colloid particles is the principal reason that they remain dispersed indefinitely. Particles in the colloidal size range would stay suspended for a time simply because of the Brownian movement, but this buffeting by molecular bombardment would not be sufficient to prevent them from settling eventually. When the particles all have charges of the same sign, however, their mutual repulsion makes settling difficult. It follows that anything which will neutralize or diminish the charge should cause the particles to flocculate. This prediction is borne out by the fact that many colloids are readily flocculated simply by adding an electrolyte.

Table 6-1 Sign of charge on colloidal particles

Hydroxides and hydrated oxides	
Silica	−
Aluminum hydroxide	+
Ferric hydroxide	Usually +; may be −
Manganese dioxide	−
Titanium dioxide	+
Zirconium dioxide	+
Thorium dioxide	+
Sulfides	−
Carbonates	Usually +
Organic colloids	−

The precipitation of colloids by electrolytes is a complex phenomenon, for which theoretical explanations so far are very meager. It seems contradictory to learn that some electrolyte must be present in the first place for the colloid to form and be stabilized—because if electrolytes are completely absent, the colloidal particles would find no ions to adsorb—and then to learn that addition of more electrolyte has the opposite effect and causes the colloid to coagulate. The important consideration, apparently, is the amount and kind of ions present. Generally sols are more stable in very dilute electrolyte solutions than in concentrated ones, and are more easily precipitated by addition of some kinds of ions than by others. Two rules are useful in predicting the effect of electrolytes: (1) doubly and triply charged ions are more effective in coagulating colloids than singly charged ions, and (2) H^+ and OH^-, although univalent, are especially effective as coagulants.

These rules may be illustrated by experiments like the following: A stable arsenic sulfide sol is prepared by bubbling H_2S through a dilute solution of arsenious acid, H_3AsO_3. If solutions of NaCl and $CaCl_2$ of similar concentration are added to different portions of the sol, the calcium salt will prove much more effective in causing precipitation of the sulfide. This is because the sulfide particles are negatively charged, hence are flocculated by positive ions, and because the Ca^{2+} ions have a higher charge than Na^+ ions. If solutions of NaCl and Na_2SO_4 are used as electrolytes, no difference will be found in their effectiveness as coagulants because the positive ion in each is the same, Na^+. If HCl and NaOH are used, the former will be a better coagulant because H^+ is attracted to the negative sol particles. If the same electrolyte solutions are now tried with a positive sol, say $Al(OH)_3$, the NaCl and $CaCl_2$ will show equal effectiveness as precipitants, Na_2SO_4 will prove more active than NaCl, and NaOH will be better than HCl. Note that experiments of this sort provide a means for determining the sign of a colloid experimentally without having recourse to an electric current.

For a crude model of this kind of experiment, we can picture (Fig. 6-2) a colloidal sulfide particle as having a negative charge because of adsorbed S^{2-} ions. Positive ions are present in the solution but are not appreciably adsorbed. If a direct current is passed through such a sol, sulfide particles will migrate to the anode and positive ions to the cathode. In the immediate neighborhood of a colloidal particle positive ions form a vague layer adjacent to the layer of tightly held S^{2-} (an "electric double layer") simply because of the concentration of negative charge on the particle, but the effective negative charge is not greatly reduced by this diffuse positive halo. If now an electrolyte containing a divalent anion, say Ca^{2+}, is added, this ion is strongly attracted by the negative particle and neutralizes some of its charge. This means that adjacent particles no longer repel each other strongly, so that they can come together and settle out of the solution. A similar result would follow if H^+ is added in large amount, for this ion, like Ca^{2+}, is strongly attracted by the negative charges.

In simple cases of this sort the logic is reasonably straightforward, but many difficulties and exceptions are found when it is extended to other colloids and other electrolytes. For geologic applications the most important generalizations we can make are simply that different colloids may have charges of different sign,

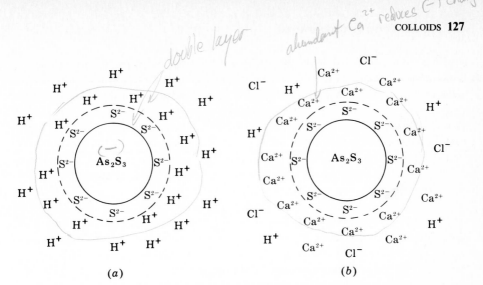

double layer

abundant Ca²⁺ reduces (−) charge

(a)   (b)

Figure 6-2 Diagram of a single colloidal particle in an As_2S_3 sol formed by bubbling H_2S through H_3AsO_3 solution. In (*a*) the particle is shown surrounded by a tightly held layer of adsorbed S^{2-} ions, which give the particle a negative charge. The concentration of negative charge attracts positive ions from the adjacent solution, in this case chiefly H^+, which are more loosely held in a diffuse outer layer. The combination of negative and positive charges makes up the "electric double layer" around the particle, which is largely responsible for the stability of the sol. In (*b*) a solution of $CaCl_2$ has been added; the doubly charged Ca^{2+} ions are more strongly attracted to the S^{2-} layer than the H^+, so that the charge is effectively neutralized and the sol flocculates.

that many colloids are unstable in fairly concentrated electrolyte solutions, and that, for any given colloid, some electrolytes are more effective than others in causing precipitation. Rules of this kind provide possible explanations for the partial separation of substances during sedimentation and during ore deposition.

6-5 ION EXCHANGE

Ions adsorbed on colloidal particles are held by forces that range from weak to very strong, depending on the kind of colloid and the kind of ion. In general the attachment is weak enough so that the ions are easily replaceable by others. If a colloid with one kind of ion adsorbed on its particles is added to an electrolyte solution containing different ions, some of the original ions will be set free in the solution and some of the new ions will be adsorbed in their place. This phenomenon goes by the name of *ion exchange*. It is often called, less precisely, *base exchange*, by the same generous extension of the meaning of "base" that we have discussed earlier (Sec. 2-2). Most often we are concerned with replacement of cations rather than anions, so that the term *cation exchange* is appropriate.

Ion exchange is not limited to colloids, but may be applied also to the substitution of one ion for another in crystals, on surfaces, or in particles of any dimension. The most familiar example of ion exchange is the process commonly used for softening water. If hard water containing Ca^{2+} ions is passed over crystals of a zeolite (usually an artificial zeolite) with Na^+ as part of its crystal structure, the Ca^{2+} takes the place of the Na^+ in the crystals and the relatively harmless Na^+ is added to the water. When most of the Na^+ has been replaced, the zeolite must be regenerated, which is accomplished by passing over it a concentrated NaCl solution. The ion exchange is now reversed because of the high concentration of Na^+, and the crystals become once more a sodium zeolite. In symbols the reaction may be thought of as an equilibrium,

$$Na\text{-zeolite} + Ca^{2+} \rightleftharpoons Ca\text{-zeolite} + Na^+ \qquad (6\text{-}1)$$

and the equilibrium may be displaced in one direction or the other by changing the relative concentrations of Na^+ and Ca^{2+}. This is a particularly clear and simple example of ion exchange, but more complicated cases can always be broken down into similar exchange equilibria.

The exchangeability of an adsorbed ion depends on how it is attached to the colloidal particle (or to a larger crystal). Several possible methods of attachment can be listed, although in practice it is seldom easy to distinguish one clearly from another. The weakest bonds between ions and surfaces are those due to "residual" forces (also called "physical" or "van der Waals" forces), forces that arise from electrical interaction between moving electrons in the ions and surfaces; these are the forces that lead to the adsorption of at least minor amounts of ions from gases or liquids on any exposed surface, however inert it may seem. A second kind of attachment is the sort that leads to formation of an electric "double layer" around a colloidal particle; either the more firmly held inner layer or the less well defined outer layer may be replaced. This kind of bonding grades into the attachment of ions by chemical valence forces—simple electrostatic forces arising from excess and deficiency of electrons—to the corners and edges of particles, where exposed parts of the crystal structure have unsatisfied charges. And attachment at the edge of particles is not very different from actual substitution in the crystal structure itself, the sort of substitution that is involved in water-softening by zeolites. Thus we encounter all gradations, from ions so loosely held to surfaces that they are readily exchanged for any others to ions that are integral parts of a crystal structure; these latter are exchangeable in some particularly open structures like the zeolites, but more commonly are exchangeable only when they are on corners or edges of crystal fragments.

In experiments on ion exchange it is seldom possible or even desirable to determine just what kinds of attachment of ions to surfaces are involved. Usually the important consideration is that some surfaces or some kinds of colloidal particles can be shown to hold adsorbed ions much more strongly, and to hold them in much greater amounts, than do others. For example, most cations are more effectively removed from solution by shaking the solution with montmorillonite than with kaolinite, even when the two clays are added in equal amount and have comparable grain size.

For any given type of particle or surface, different ions show a wide range of adsorbability. The behavior of zeolites in water softening is a good example: Ca^{2+}

is removed by Na-zeolite even from very dilute solution, but to regenerate the Na-zeolite requires that the Ca-zeolite be treated with a very concentrated Na^+ solution. In other words, equilibrium is established for Eq. (6-1) when the ratio $[Na^+]/[Ca^{2+}]$ is much greater than 1. Experiments of this kind, involving measurements of the amount of one ion released from an adsorbent by a given concentration of another, provide us with a measure of relative adsorbabilities. Unfortunately, such experiments do not lead to any simple rules about the behavior of ions toward adsorbents, apparently because adsorption depends on too many factors: the nature of the adsorbent (chemical composition, particle size, method of preparation, aging), the temperature, and the kind and amounts of other ions present in the solution.

As a first guess, we might suppose that the relative adsorbability of two ions would depend on such properties as their sizes, their charges, and their ability to form covalent bonds, the same properties that determine their roles in crystal structures. It would seem natural, for example, that (1) the smaller of two ions would be more firmly held to a surface than the larger, (2) a multivalent ion would be more firmly attached than a univalent ion, and (3) an ion whose bonds have strong covalent character would be more readily adsorbed than one whose bonds are dominantly ionic. In some experiments these expectations are fulfilled, but exceptions are so numerous and so serious that the generalizations are not very useful. The first one, in particular, is often violated. We would predict, for example, that Na^+ should be more readily adsorbed from solution than its larger relative K^+, but experimental results are usually just the reverse. The entire series of alkali metals, in fact, in most adsorption experiments shows a progressive *decrease* in adsorbability from Cs^+ to Li^+, in direct contradiction to the first rule above. One can explain away the discrepancy by noting that the smallest ion, Li^+, is the most highly hydrated and therefore effectively larger than Cs^+ in water solution, but this introduces yet another variable into adsorption experiments. The second and third rules above show fewer flagrant exceptions; one can expect, for example, that in most experimental situations Ba^{2+} will be adsorbed in preference to K^+, and Cd^{2+} in preference to Ca^{2+}.

In the geochemistry of colloids, ion exchange is important for two principal reasons. The most obvious one is the mechanism it provides for redistribution of metal ions between solution and sediments. Thus the high Na/K ratio in seawater can be partly explained by the greater ease of adsorption of K^+ on clays, organic matter, and chemical precipitates; and the high concentrations of metals like cobalt and lead often found in manganese ores can be accounted for by adsorption of the ions on manganese dioxide sols. The second important effect of ion exchange is the influence of different ions on the properties of the adsorbing substance. This is especially significant for clays; the outstanding example is the difference in properties between clays containing chiefly adsorbed Na^+ and clays containing chiefly Ca^{2+}, the former being sticky and impervious while the latter are granular, easily worked, and readily permeable. For agricultural purposes the calcium clays are more desirable, and gypsum is often added to convert soda clays into the calcium variety. For water reservoirs, on the other hand, it is sometimes necessary to convert calcium clays to soda clays by adding brine in order to make the bottom sediments less permeable.

Despite the importance of adsorption and ion exchange, our knowledge regarding them is still largely empirical. So many variables are involved, and some are so difficult to control, that experiments all too often lead only to broad generalizations rather than to specific rules. In discussions to follow we shall often mention adsorption and ion exchange as important geologic processes, but we shall seldom be able to predict just what their effects should be in particular situations.

6-6 STABILITY OF COLLOIDS

One of the important geological questions about colloids may be phrased like this: for a given metal, is there a compound that forms a sol readily, so that the sol can serve as a means of transporting the metal, either in surface waters or groundwaters? The effectiveness of the sol in carrying the metal must depend on its stability, for obviously a colloid that flocculates readily would not serve. The ultimate fate of the metal will depend on the conditions necessary to bring about flocculation. Iron, for example, in the ferric state is known to form a very insoluble hydroxide, which often remains dispersed as a stable $Fe(OH)_3$ sol; the amount of this sol in stream water is greater than that in the sea, an observation which suggests (although it does not prove) that the sol is flocculated on contact with electrolytes in the sea.

From the discussion of the last two sections, we could guess that the stability of a given colloid is difficult to predict. Any colloidal dispersion is essentially unstable; its particles are large enough to settle out under the influence of gravity, but are prevented from doing so by the accident of adsorbing enough ions to maintain an electric charge. Many colloids demonstrate their instability by gradually precipitating on standing, but some are stable enough to remain suspended indefinitely. Every chemical analyst knows how troublesome colloids can be, simply because they remain in suspension and therefore are not filterable, despite efforts to make them precipitate. But to know in advance which substances are likely to form sols and which form clean precipitates is something that an analyst must learn from experience and not by predictions from theory.

By and large, colloids form more readily in dilute solutions than in concentrated ones. When H_2S is added to a dissolved salt of mercury, for example, the precipitated HgS remains suspended as a sol if the concentration of mercury is only a few parts per million, but appears as a heavy black precipitate if the concentration is higher. This rule has many exceptions, however; a notable one is barium sulfate, which comes down as a crystalline precipitate even from very dilute solutions but can be prepared as a colloid in concentrated solutions. Another common observation is that metal hydroxides form colloids more readily than metal sulfides, but again this would have to be qualified by pointing out exceptions. The stability of a colloid also depends on details of its method of preparation; thus manganese dioxide can be prepared as a stable colloid in the laboratory by adding hydrogen peroxide slowly to dilute potassium permanganate, but it flocculates unless the concentration, the acidity, the temperature, and the rate of addition of peroxide are kept within narrow limits. Most such requirements are purely empirical knowledge, discovered by trial and error. So at present it is simply not possible to predict whether any particular compound of geologic interest will form a stable sol under specified conditions. One can only try to simulate geologic conditions in the laboratory, and see if the compound will flocculate or remain dispersed.

Suppose we have a sample of a stable sol: under what conditions will it flocculate? Here again useful generalizations are meager. We have noted that colloids are precipitated by electrolytes, and that some electrolytes are better precipitants than others. Yet this rule is hedged around with difficulties. For a colloid to form at all, at least a little electrolyte must be present to supply adsorbable ions; therefore only a fairly concentrated electrolyte will be effective in flocculation, and the meaning of "fairly concentrated" will differ from one colloid to the next and from one electrolyte to another. Furthermore, some colloids are stable both in dilute electrolytes and in concentrated ones, but not in an intermediate range. For example, the positively charged particles of a gold sol in dilute solution are flocculated by the addition of sodium chloride, but if the salt is added fast enough, the particles remain suspended and acquire a negative charge, apparently because Cl^- is adsorbed in sufficient quantity not only to neutralize the original positive charge but to make the particles negative. Nevertheless, the rule about colloids being precipitated by electrolytes has enough validity for us to state that particles remain suspended in freshwater longer than in seawater, and that the extraordinary clearness of the water in many salt lakes is due to the rapid coagulation of any suspended material brought into them.

The effect of heat on colloids is a problem on which chemical data are practically nonexistent, but which has great interest for geologists because of its bearing on the possible role of colloids in hydrothermal solutions. One might expect that high temperature in general would favor flocculation of colloids, because increased agitation would give the particles more opportunity to come into contact and because the growth of large particles at the expense of small particles (Sec. 3-4) would be accelerated. Certainly for many colloids this generalization holds; a favorite trick in chemical analysis, for example, is to speed up the flocculation of a partly colloidal precipitate by keeping it for a time near the boiling point of water. Yet iron hydroxide sols are best prepared in the laboratory in a boiling solution, and experiments have shown that mercury sulfide sols are stable for long periods at 100° and gold sols at temperatures as high as 150°C. Until more data become available, speculations about the behavior of colloids in hydrothermal solutions can have little guidance from physical chemistry.

One curious observation about the stability of colloids is the fact that a given colloid may be much more stable in the presence of a second colloid than it is by itself. When ammonium sulfide is added to a copper solution, for example, a precipitate of copper sulfide appears at once; if the solution contains a mere trace of gelatin, the copper sulfide does not precipitate but is stabilized as a sol. The sulfide is said to be "protected" by the gelatin, and the gelatin is called a *protective colloid*. In general, protective colloids are hydrophilic. One's first guess about the mechanism of protective action would be that the hydrophilic colloid simply coats the particles of the hydrophobic colloid and so keeps them apart, but experiments show that the amounts of hydrophilic colloid necessary are so very minute that all the particles of the protected colloid could not possibly be covered. Again we must deal with a phenomenon for which present theory is inadequate, but which empirically plays an important role in colloidal behavior. Geologically, protective colloids are of particular interest in surface waters containing an abundance of organic matter, for the hydrophilic organic colloids serve as efficient protectors of such inorganic materials as iron hydroxide and manganese dioxide, making these colloids more stable and hence more easily transportable than they otherwise would be. Silica is another important protective colloid in geologic environments. Its efficacy in stabilizing sols of gold and mercury sulfide has been demonstrated experimentally; its effect on sols of iron and manganese oxide is less clear, some experiments showing a stabilizing influence and others a mutual precipitation of the two colloids.

6-7 SILICA AS A CHEMICAL SEDIMENT

This background of colloidal chemistry gives us a basis for looking into the complex behavior of silica in sedimentary environments.

The common occurrence of silica in sediments as nodules with rounded or irregular shapes, the cryptocrystalline or amorphous nature of the silica, and the frequent presence of colored bands resembling Liesegang rings have long been interpreted as evidence for the importance of colloidal phenomena in silica deposition. In the laboratory, the ability of silica to form stable sols and gels is well known. Complications arise from the facts that different forms of silica have widely different solubilities, and that changes of silica from one of its solid forms to true solution or colloid, and from solution to colloid or precipitate, involve slow reactions.

The commonest form of silica, quartz, has a very low solubility at ordinary temperatures, about 6 ppm of SiO_2. Fresh silica gel, on the other hand, dissolves slowly to give an equilibrium concentration of roughly 120 ppm. Other forms of silica—opal, tridymite, cristobalite, chalcedony—have solubilities between these extremes. The different solubilities are another example of the same phenomenon we met in considering the solubilities of calcite and aragonite (Sec. 3-2), but here the differences are much larger. As with calcium carbonate, the solubilities reflect differences in stability, quartz being the most stable form of silica at ordinary conditions and silica gel the least.

Dissolved silica exists as silicic acid, H_4SiO_4:

$$SiO_2 + 2H_2O \rightleftharpoons H_4SiO_4$$

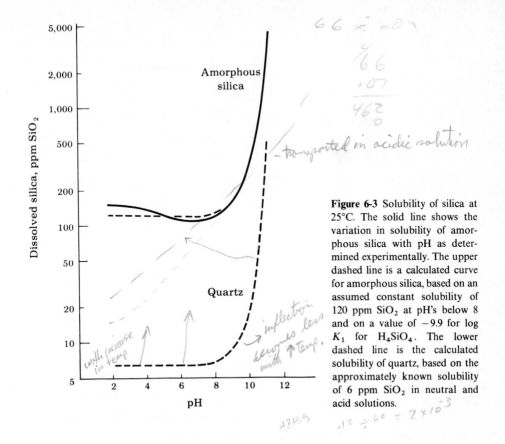

Figure **6-3** Solubility of silica at 25°C. The solid line shows the variation in solubility of amorphous silica with pH as determined experimentally. The upper dashed line is a calculated curve for amorphous silica, based on an assumed constant solubility of 120 ppm SiO_2 at pH's below 8 and on a value of -9.9 for log K_1 for H_4SiO_4. The lower dashed line is the calculated solubility of quartz, based on the approximately known solubility of 6 ppm SiO_2 in neutral and acid solutions.

Silicic acid is analogous to carbonic acid, H_2CO_3, except that the molecule contains an extra molecule of H_2O. The acid is much weaker than carbonic acid, its first dissociation constant being more than a thousand times smaller:

$$H_4SiO_4 \rightleftharpoons H^+ + H_3SiO_4^- \qquad K = 10^{-9.9}$$

Hence the solubility of silica should not be affected by pH at values below about 9. This is contrary to a common statement in older geologic literature; it was often assumed, simply because silica is very soluble in concentrated alkali, that the solubility decreases steadily with pH over the entire pH range. Laboratory experiment, however, amply bears out the prediction that the solubility remains constant at pH's less than 9 (Fig. 6-3).

Supersaturated solutions of H_4SiO_4 are readily prepared either by adding acid to a solution of sodium silicate or by cooling a solution of silica gel prepared at high temperature. From a supersaturated solution silica does not crystallize, but slowly forms a sol; over a period of days or weeks the concentration of dissolved H_4SiO_4 gradually decreases to the equilibrium value, and the remainder of the silica stays suspended as colloidal particles. A sol with a few hundred parts per million of excess silica is stable indefinitely, but a more concentrated sol eventually forms a gelatinous precipitate (in alkaline solution) or sets to a uniform gel (in acid solution). As with other colloids, the coagulation is aided by electrolytes. It is important to note that *only the silica in colloidal form coagulates*, the part in true solution remaining unaffected.

If a silica sol is diluted, the colloidal particles gradually dissociate into H_4SiO_4 molecules until the equilibrium solubility is reached. Because the breaking up is slow, sols with much less than 100 ppm of SiO_2 can be prepared and can exist for periods of days or weeks; from these dilute sols silica can be coagulated by adding electrolytes, but most of the precipitated silica goes back into solution on standing.

These laboratory observations give a basis for understanding the behavior of silica in nature. In stream water and ground water the concentration of silica ordinarily lies between 10 and 60 ppm SiO_2, well below the equilibrium solubility for amorphous silica. To explain the low figure, we can plausibly guess that the dissolving of quartz and silicates, which must supply the bulk of the silica, would not give as high a concentration as does dissolution of silica gel in the laboratory. The concentration may also be diminished by organisms that use silica. Most of the silica in surface waters is in true solution, but near hot springs some may be at least temporarily colloidal because the concentration in the original hot water may be high (up to about 350 ppm at 90°C). In seawater the silica concentration is very low, only 1 or 2 ppm. It is often supposed that the low concentration results from coagulation of colloidal silica by the electrolytes of seawater, but this is impossible, because most of the silica entering the sea is in true solution. The low concentration is more probably maintained by the activity of organisms like diatoms and radiolaria that use silica for shell material, and perhaps in part by the precipitation of silicates like glauconite, authigenic feldspar, chlorite, and illite.

Sedimentary silica is chiefly in the form of chert, and the origin of this material has been a much-debated geochemical problem. Where lava flows enter the sea, and where hot springs associated with volcanic activity supply water to the sea, silica may become so concentrated that direct precipitation of silica gel is possible; this is an easy explanation for the abundant chert often found with the products of submarine eruptions. But much chert has no obvious connection with volcanic activity and presumably must have formed as a sediment in contact with normal seawater. How is this possible, when seawater is apparently so undersaturated with silica? No entirely satisfactory answer has been proposed, but a possible explanation depends on the activity of silica-secreting organisms. These creatures are able to utilize silica even from very dilute solutions, perhaps protecting it from redissolving with a film of organic matter. If the hard parts of dead organisms accumulate, one might expect that reorganization of material would take place—solution of silica on exposed corners and edges, and precipitation on the larger shell fragments. This is the same process, of course, as the growth of large crystals at the expense of small ones in the case of calcium carbonate. Such a conversion of siliceous skeletons to opaline chert has been demonstrated for the Monterey diatomite of California (Bramlette, 1946), and a similar explanation is at least plausible for other cherts that contain abundant fragments of organic origin.

Whether it can be applied also to cherts with no sign of fossil fragments preserved is an open question.

Whatever may be an acceptable explanation for the origin of the silica in chert, geologic evidence suggests that during its formative process much chert passes through a stage when it behaves as a fairly coherent gel. The shape of chert nodules, the shrinkage cracks both in nodules and bedded chert, the common banded appearance so similar to the Liesegang rings of laboratory experiments, the intricate crumpling often observed in bedded cherts, all suggest gel behavior.

Another common geologic observation is that chert extensively replaces other rock materials, often preserving original textures and structures with marvelous fidelity. Petrified wood is an outstanding example. Some of the chert in sedimentary rocks is doubtless of replacement origin, formed after the rock solidified by silica-rich solutions circulating through it. Details of the replacement process are obscure. Why, for example, do silica-rich solutions in some rocks replace earlier minerals with opal or chalcedony, and in other rocks cause overgrowths of crystalline quartz around original sand grains? Perhaps the presence of quartz nuclei favors deposition of more quartz, or perhaps slightly higher temperatures lead to formation of the most stable form of silica. Another puzzle is the precise mechanism of replacement: what sort of solutions are responsible for the widespread replacement of calcite by silica, and the almost equally widespread replacement of silica by calcite? We can speculate that slight changes in pH and silica concentration might be responsible. A slightly acid solution supersaturated with silica could cause calcite to dissolve and silica to be deposited; a slightly alkaline solution undersaturated with silica could be responsible for the opposite change. But such statements are merely guesses in the absence of a laboratory demonstration.

6-8 CLAY MINERALS AS COLLOIDS

The clay minerals are often colloidal, and much in their behavior can be explained on the basis of colloid chemistry. Clay suspended in water, for example, can be flocculated by the addition of electrolytes, and the capacity of some clays for ion exchange is well known. Nevertheless, as colloids the clay minerals have a few peculiarities which should be noted.

Most clay samples are entirely, or almost entirely, crystalline. The crystal structures of their particles show clearly in x-ray-diffraction patterns and are often revealed by the geometric shapes of the grains in electron photomicrographs (see Fig. 7-1). For most clays these crystalline particles are flat plates like mica flakes, much smaller in one dimension than in the other two. In a given sample the grains may have an enormous range in size, from those easily visible with an optical microscope to those whose dimensions lie entirely within the colloidal range. A majority of the particles are often "semicolloidal," in the sense that their thickness is of the order of a few millimicrons whereas their longer dimensions exceed the upper limit of colloidal sizes.

The flattened shapes of clay particles, together with peculiarities of the crystal structure which we shall study in the next chapter, give opportunity for adsorbing ions in a variety of ways. Some ions are held to broken bonds on the edges of the flakes, some to the flat surfaces; some make their way into spaces between the layers of the crystal structure, and some enter into the structure of the crystal itself to take the place of one of its constituents. Ions in these different positions vary greatly in their exchangeability; some are replaced by others even from dilute solutions, but some are so firmly held that exchange is slight even from concentrated solutions. Exchangeability also depends on the kind of ions, some ions, by reason of charge or size, fitting into the clay structure better than others. Still another factor in determining ion exchange is the kind of clay mineral or minerals present, since the clays differ greatly among themselves in their capacity for picking up and holding foreign ions. The processes of ion exchange in clays are thus extremely complicated and only partly understood. This is a subject of much current research interest, for the properties of clays as raw materials for agriculture, ceramics, and engineering purposes depend in large part on the quantity and nature of their adsorbed ions.

The ability of a clay to adsorb water is another important determiner of its properties. Like colloids of all kinds, clay particles pick up and hold water to their surfaces, but different clays show enormous variation in the amount of water adsorbed. Kaolinite clays are relatively nonplastic, because their particles hold only a little water. A pure montmorillonite clay, on the other hand, behaves like a hydrophilic colloid, in that its particles adsorb water until the clay swells to several times its original volume and acquires a gel-like consistency. Most clay samples show behavior between these two extremes.

Thus we can correlate some properties of clays with the fact that their particles have dimensions entirely or partly in the colloidal range.

6-9 GEOLOGIC EVIDENCES OF FORMER COLLOIDS

The importance of colloids in present-day geologic processes is plain to see. The transport of insoluble metal compounds by streams, the precipitation of fine-grained material in seawater, the adsorption of ions by clays and organic matter in the soil are only a few examples of the role played by finely dispersed particles. What evidence do we have that colloids have affected geologic processes in the past as well?

The question is a difficult one because colloids are essentially unstable. In the course of geologic time they can be expected to flocculate and perhaps to crystallize, so that the materials we examine today may show little sign of their colloidal history.

Materials that are amorphous or very finely crystalline, that commonly occur with botryoidal or mammillary surfaces, that show indistinct color banding similar to Liesegang rings, and that consist of compounds known to form colloids readily in either natural or artificial environments, are often assigned a colloidal origin. Chert is a good example: its texture, its banding, its rounded surfaces resemble so closely gelatinous silica produced in the laboratory that the inference of origin as a mass of silica gel seems natural. There is one pitfall here, in that artificial gels contain 90% or more water, whereas chert seldom has more than 5%; furthermore, if artificial gels are allowed to dehydrate, they lose their gelatinous appearance and smooth surfaces and disintegrate into flaky opaque material. The objection may not be insuperable, because chert in its process of formation may pick up silica as it loses water, but the entire sequence of events has not yet been duplicated in the laboratory. Limonite and wad are also commonly supposed to represent hardened colloids, probably with good reason, although the same difficulty about water content presents itself in comparing the natural materials with artificial precipitates of the same general composition. A colloidal origin is often inferred even for materials that do not readily form gels in the laboratory, particularly calcite and siderite, when these substances occur in discrete bodies with microcrystalline texture and smooth, hummocky surfaces. Septarian nodules are a good example: the crystal-lined cracks in these odd structures can be made plausible on the colloid hypothesis by supposing that they represent shrinkage cracks produced by partial dehydration of the gel as it hardened.

The same general criteria for colloidal origin are also extended to textures of metallic ores in hydrothermal veins, despite the lack of experimental evidence on colloidal behavior at high temperatures. Finely crystalline ores with rounded surfaces in open cavities, particularly if they show bands of different composition, are commonly cited, and the assumption of colloidal origin is sometimes stretched even to finely banded ores without rounded surfaces. With our present knowledge of colloidal chemistry it is hardly worthwhile to debate the pros and cons of this hypothesis. The assumption that these ores were deposited from colloidal solutions is not unreasonable, but there is little positive evidence to support it. The converse statement, however, that coarse crystalline ores cannot be deposited from colloidal solutions, has little justification. It is true enough that coarse crystals do not form from colloids at low temperatures in the short times of laboratory experiments, but about the capacity of such crystals to grow from high-temperature colloids in the course of geologic time we simply have no present basis for guessing.

Compared with other branches of physical chemistry, the study of colloids is still in a primitive state, necessarily so because of the enormous complexity of colloidal phenomena. A mass of empirical data has been accumulated, but general principles do not go far beyond the data and are riddled with qualifications and exceptions. Regarding colloids we must be more than usually cautious, therefore, in trying to extrapolate from laboratory experiments to geologic processes.

PROBLEMS

1 By what observations or experiments could you prove that a red liquid containing gold is a sol of the metal rather than a true solution?

2 Describe two kinds of experiments by which you could establish that the particles of a gold sol carry a positive charge.

Fe⁰

precipitate because there is a mixture of (+) and (−) charge

3 What would you expect to happen when sols of ferric oxide and arsenic sulfide are mixed?

4 One method of determining the molecular weight of a substance in solution is to measure the depression of the freezing point caused by a known concentration of the dissolved substance. Would this be a feasible method for finding the average weight of the particles of a sol? Why or why not?

No because colloids aren't predictable in their effect

5 On an artificial island in San Francisco Bay an attempt was made to construct a fresh water pond, but the clay fill of the island was too permeable for the pond to hold water. The difficulty was solved by filling the pond with seawater and keeping it full for several days, after which the seawater was pumped out and replaced by fresh water. The water now shows little tendency to seep away. Suggest an explanation. *seawater contains high Na⁺, it gets absorbed onto colloids, they precipitate and the*

6 What is meant by each of the following terms? *"water loving" → solution passes through membrane*

Hydrophilic colloid
Emulsion~ *liquid in liquid*
Ion exchange

Dialysis — *separating of solution and colloid*
Liesegang bands — *colored diffusion bands*
Protective colloid

7 What is the concentration of $H_3SiO_4^-$ in a solution with pH 8.0 and a total silica content of 100 ppm?

8 Assume that quartz at any pH at 25°C dissolves to the extent of 6 ppm SiO_2, forming undissociated H_4SiO_4. At high pH's additional amounts dissolve to form $H_3SiO_4^-$. Using the dissociation constant of silicic acid, set up an equation expressing the total solubility in parts per million as a function of pH. Compare your equation with the lower dashed curve in Fig. 6-3.

9 Water taken from hot springs in Yellowstone Park at temperatures near 100°C has a pH of about 7 and contains dissolved silica at a concentration of approximately 350 ppm SiO_2. Hot-spring orifices are commonly surrounded by low mounds of opaline silica (geyserite). If a sample of the water is placed in a stoppered bottle and kept at room temperature, successive analyses over a period of weeks show diminishing amounts of dissolved silica, but the solution shows no visible change. After several years the sample is still clear and colorless, and analyses give a constant value of about 140 ppm SiO_2; if salt is added to the water, gelatinous flocs appear. Explain these observations.

10 Set up an equation showing a possible mechanism for the replacement of limestone by silica from a solution of H_4SiO_4 and, by using equilibrium constants, discuss conditions under which this replacement might take place. *high pH — silica dissolves, carbonate precipitates*
low pH — carbonate dissolves, silica precipitates

REFERENCES AND SUGGESTIONS FOR FURTHER READING

Bramlette, M. N., The Monterey formation of California and the origin of its siliceous rocks, *U.S. Geol. Survey Prof. Paper* 212, 1946. A classical paper giving detailed field and laboratory evidence for the conversion of diatomite into chert.

Carroll, D., Ion exchange in clays and other minerals, *Geol. Soc. America Bull.*, vol. 70, pp. 749–780, 1959. An excellent general review.

Kastner, M., J. B. Keene, and J. M. Gieskes, Diagenesis of siliceous oozes, *Geochim. et Cosmochim. Acta*, vol. 41, pp. 1041–1049, 1977. Experimental study of slow changes in silica sediments in various simulated sedimentary environments; suggestions about the origin of chert and replacement of carbonate rocks by silica.

Morey, G. W., R. O. Fournier, and J. J. Rowe, Solubility of amorphous silica at 25°C., *Jour. Geophys. Research*, vol. 69, pp. 1995–2002, 1964; Solubility of quartz in water in the temperature interval 25–300°C., *Geochim. et Cosmochim. Acta*, vol. 26, pp. 1029–1044, 1962. Detailed experimental work on the complicated and curious solubility relations of different forms of silica.

Murata, K. J., I. Friedman, and J. D. Gleason, Oxygen isotope relations between diagenetic silica minerals in Monterey shale, *Am. Jour. Sci.*, vol. 277, pp. 259–272, 1977. A reinvestigation of the siliceous rocks studied by Bramlette (1946, above), using ratios of oxygen isotopes to estimate temperatures at which the forms of silica change during diagenesis.

Murray, J. W., Interaction of metal ions at the manganese dioxide-solution interface, *Geochim. et Cosmochim. Acta*, vol. 39, pp. 505–519, 1975. An example of many recent papers on details of the adsorption behavior of various metal ions.

Robinson, B. P., Ion-exchange and disposal of radioactive wastes—a survey of literature, *U.S. Geol. Survey Water-Supply Paper* 1616, 1962. Review of the theory of adsorption and applications to disposal of radioactive wastes.

Roedder, E., The noncolloidal origin of "colloform" textures in sphalerite ores, *Econ. Geology*, vol. 63, pp. 451–471, 1968. A critical study of textures in metallic ores that are often thought to be evidence for colloidal origin.

Siever, R., K. C. Beck, and R. A. Berner, Composition of interstitial waters of modern sediments, *Jour. Geology*, vol. 73, pp. 39–73, 1965. A study of water squeezed from modern sediments, describing changes in pH, in ratios of cations, and in concentration of silicon.

SEVEN

CLAY MINERALS AND SOILS

After a long digression on structural chemistry and the chemistry of colloids, we come back in this chapter to the products of weathering. The ultimate solid product of weathering is the material called soil, and chief among the constituents of most soils are the clay minerals. We start, then, by investigating the structure and properties of the clays, and go on to consider the role these minerals play in the formation of soils.

7-1 STRUCTURE

Introduction

Geologists have used the word "clay" in two senses: as a size term, to refer to material of any composition whose average grain size is less than 0.004 mm (this figure varies somewhat with different authors), and as a mineralogic term, to refer to a group of minerals with a specific range of composition and a particular kind of crystallographic structure. The two meanings often overlap or coincide, because the fine-grained part of a soil or sediment most commonly consists largely of clay minerals. The second meaning is the commoner one in recent literature, and it is the only one that will concern us here.

Chemically the clay minerals are best described as hydrous aluminum silicates. This simple definition must be qualified by adding that many clays contain other metals in addition to aluminum, particularly magnesium and iron, and that some extreme types have almost no aluminum at all. Crude molecular formulas of the principal clays may be written $H_4Al_2Si_2O_9$ (kaolin) and $HAlSi_2O_6$ (montmorillonite), but these tell us practically nothing about the behavior of the minerals. Actual compositions are never this simple, but show variations in the Si/Al ratio, a variable quantity of water, and usually considerable amounts of magnesium, iron, calcium, and the alkali metals. Between composition and properties there is little correspondence. Two clays with similar proportions of the various elements may show great differences in plasticity and in capacity for ion exchange, while another pair with very different compositions may show strikingly similar properties. Chemical analyses by themselves obviously are of little help in understanding the clays; hence we turn instead to the structural considerations of Chap. 5.

Characteristically the clay minerals are extremely fine-grained, so fine that ordinary microscopes are inadequate for studies of crystal morphology. X-ray-diffraction cameras and electron microscopes are the instruments that have furnished most of our modern information about clays. By their aid the general patterns of clay structure have been deciphered, the apparent anomalies of chemical composition have been explained, and a classification of clay minerals has been established that fits fairly satisfactorily the observed macroscopic properties.

Octahedral and Tetrahedral Sheets

Clay minerals (with a few rare exceptions) are phyllosilicates, silicates with continuous sheet structures like the micas. This could be anticipated from the flakelike form of clay particles; the resemblance to micas is apparent also in the hexagonal shapes which the flakes of some clays assume (Fig. 7-1). Both micas and clays have a characteristic structure, made up of alternating layers or sheets of two kinds. One sheet consists of the ions Al^{3+}, O^{2-}, and OH^-; the negative ions form octahedra around Al^{3+}, the relative numbers of O^{2-} and OH^- being adjusted to satisfy the valences of the entire structure; the O^{2-} and OH^- are shared between adjacent octahedra, so that the structure is continuous in two dimensions (Fig. 7-2). This pattern by itself, not as part of a clay structure, is identical with the structure of gibbsite $[Al_2(OH)_6]$; it is often referred to as the "gibbsite sheet," or the "octahedral sheet," of the clay structure. The second kind of sheet is made up of Si^{4+}, O^{2-}, and OH^- ions; each Si^{4+} is in the center of a tetrahedron of oxygen ions; the tetrahedra all face the same direction, and the oxygens at their bases are linked so as to form hexagonal rings (Fig. 7-3). This sheet is the "silica sheet," or the "tetrahedral sheet," of the clay structure. The complete clay structure (or mica structure) consists of one of several possible combinations of the octahedral and tetrahedral sheets. The simplest combination, the layer structure of kaolinite, is shown in Fig. 7-4: a single octahedral sheet is linked to a single tetrahedral sheet by sharing some of the oxygen ions. The double layer extends indefinitely in two dimensions, and the clay crystal is built up of a succession of these layers one on top of another.

The diagrammatic representations in Figs. 7-2 to 7-4 show the geometry correctly but give a false impression of ionic sizes. The big units of the structure are, of course, O^{2-} and OH^-; these have almost identical sizes (radius 1.32 Å), since the

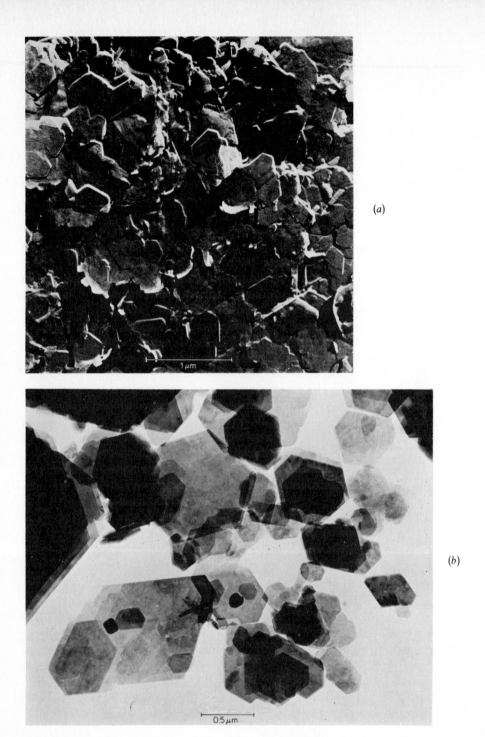

(a)

(b)

Figure 7-1 Electron photomicrographs of kaolinite: (*a*) replica; (*b*) transmission. Note the hexagonal shapes of many of the grains. The magnification is about 43,000 times. (*Source: Kenneth Towe, Smithsonian Institution, Washington, D.C.*)

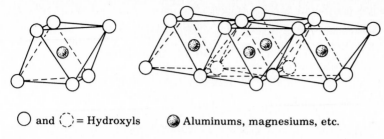

○ and ⟨⟩ = Hydroxyls   ◉ Aluminums, magnesiums, etc.

Figure 7-2 Diagram of the octahedral sheet, or gibbsite sheet, of phyllosilicate structures. (The left-hand diagram shows a single octahedral unit.) When Al^{3+} occupies the centers of the octahedra, only two-thirds of the possible sites are filled; when Mg^{2+} occupies these positions, all sites are filled. *(Source: Grim, 1968, page 52.)*

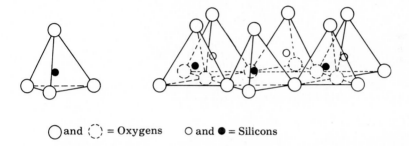

○ and ⟨⟩ = Oxygens   ○ and ● = Silicons

Figure 7-3 Diagram of the tetrahedral sheet, or silica sheet, of phyllosilicate structures. (The left-hand diagram is a single tetrahedral unit.) *(Source: Grim, 1968, page 52.)*

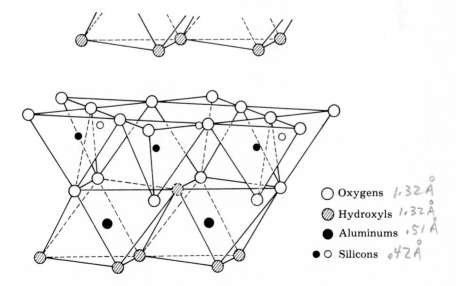

○ Oxygens $1.32 \overset{o}{A}$
◎ Hydroxyls $1.32 \overset{o}{A}$
● Aluminums $.51 \overset{o}{A}$
●○ Silicons $.42 \overset{o}{A}$

Figure 7-4 Diagram of the structure of kaolinite. *(Source: Grim, 1968, page 58.)*

size of the added proton in OH^- is negligible. These large anions should be thought of as in contact or nearly so, clustered around the smaller cations. Only four can fit around the tiny Si^{4+} (0.42 Å), but six can be accommodated around the larger Al^{3+} (0.51 Å).

Kaolinite and Montmorillonite

Another way in which the octahedral and tetrahedral sheets can be put together is shown in Fig. 7-5. Here, in montmorillonite, an octahedral sheet is sandwiched between two tetrahedral sheets, so that the theoretical Si/Al ratio is twice the ratio in kaolin. A convenient way to summarize the difference between the kaolinite and

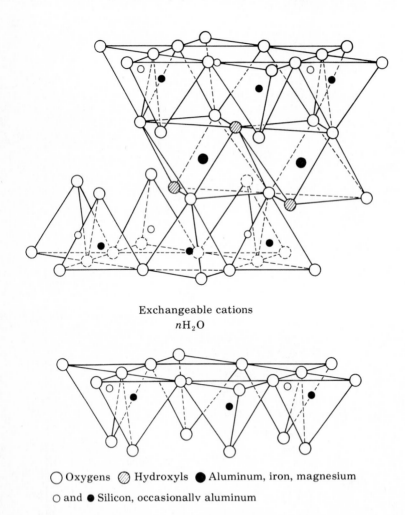

Exchangeable cations
$n\mathrm{H}_2\mathrm{O}$

○ Oxygens ⊘ Hydroxyls ● Aluminum, iron, magnesium

○ and ● Silicon, occasionally aluminum

Figure 7-5 Diagram of the structure of montmorillonite. *(Source: Grim, 1968, page 79.)*

montmorillonite structures is simply to tabulate the numbers of ions at different levels in a unit of structure:

Kaolinite	Montmorillonite
$6O^{2-}$	$6O^{2-}$
$4Si^{4+}$	$4Si^{4+}$
$4O^{2-} + 2OH^-$	$4O^{2-} + 2OH^-$
$4Al^{3+}$	$4Al^{3+}$
$6OH^-$	$4O^{2-} + 2OH^-$
	$4Si^{4+}$
	$6O^{2-}$

In this representation, the layers designated $4O^{2-} + 2OH^-$ are common to the octahedral and tetrahedral sheets. A kaolin-type clay is often said to have a two-layer structure (one gibbsite sheet and one silica sheet), whereas a clay with the montmorillonite pattern is said to have a three-layer structure. Other combinations of the unit sheets are possible, but these two are the most common.

The conventional formula for kaolinite, $Al_4Si_4O_{10}(OH)_8$, is clearly a condensation of the layer arrangement just described. The conventional formula for montmorillonite is $Al_4(Si_4O_{10})_2(OH)_4$; since adsorbed water is always present, a more accurate formula is $Al_4(Si_4O_{10})_2(OH)_4 \cdot nH_2O$.

Successive unit layers of the clay structure are stacked one on top of another. Differences in the geometry of stacking give rise to clays with different properties; thus the hydrothermal clay minerals dickite and nacrite have the same two-sheet structure as kaolinite, but differ in the position of one layer with respect to the next. Whatever the arrangement, the distance from the O^{2-} or OH^- ions of one layer to the O^{2-} or OH^- of the next is greater than that between the anions within each layer, so that the mineral splits between the layers much more easily than across them. This, of course, is reflected in the macroscopic (001) cleavage of the clay minerals and micas. Successive layers are more easily separated in the montmorillonite structure than in the kaolinite structure, the difference being correlated with the fact that adjacent layers in kaolinite have O^{2-} ions in one facing OH^- in the next, while in montmorillonite O^{2-} of one layer always faces O^{2-} of another. This difference is often expressed by saying that the kaolinite structure is asymmetric (gibbsite-silica, gibbsite-silica, ...) and the montmorillonite structure symmetric (silica-gibbsite-silica, silica-gibbsite-silica, ...). The attraction of O^{2-} for OH^- is much greater than that of O^{2-} for other O^{2-} ions; one can think of the H^+ from the OH^- as shared between the O^{2-} of adjacent layers of kaolinite (forming a "hydrogen bond"), and so linking the layers together. Much of the difference in macroscopic properties between the two types of clay can be ascribed to the greater ease with which the layers of the symmetric structure separate from one another.

Isomorphous Substitution

Theoretically one might expect a great deal of isomorphous substitution in the clay-mineral structures, since Si^{4+} and Al^{3+} seemingly should be replaceable by many other ions with coordination numbers 4 and 6. The expectation is abundantly fulfilled in montmorillonite clays, but not in kaolinite. This difference in the amount of isomorphous substitution is another of the important differences between the two kinds of structure: kaolinite *always* has a composition approximating the ideal formula $Al_4Si_4O_{10}(OH)_8$, but montmorillonite *never* has the ideal formula $Al_4(Si_4O_{10})_2(OH)_4$ because of extensive substitution. The common substitution in the tetrahedral sheets of montmorillonite is Al^{3+} for Si^{4+}, the amount of substitution being limited to about 15%. In the octahedral sheets a much greater variety of substitution is possible, and the amount may be anywhere from very small to 100%; the most common substitutions are Mg^{2+} and Fe^{3+}, and rarer ones are Zn^{2+}, Ni^{2+}, Li^+, and Cr^{3+}. A formula for montmorillonite more realistic than the ideal one would be $(Al, Mg, Fe^{3+})_4(Si, Al)_8O_{20}(OH)_4 \cdot nH_2O$.

The substitution of Al^{3+} for Si^{4+} and of Mg^{2+} for Al^{3+} leaves a deficiency of positive charge in the montmorillonite layers. The deficiency may be compensated in various ways: by replacement of O^{2-} by OH^-, by introduction of excess cations into the octahedral layer (Al^{3+} in the ideal structure fills only two-thirds of the available positions), and by adsorption of cations onto the surface of individual layers. Some of the compensation is always accomplished by the last method, so that a constant characteristic of montmorillonite clays is the presence of abundant adsorbed ions. These ions are held loosely to the clay structure, and are readily replaced by others, so that montmorillonite clays typically have a high capacity for ion exchange.

Illite and Muscovite

One of the ions most firmly held by montmorillonite is potassium, K^+, a big ion whose size enables it to fit snugly between layers of the clay structure. Clays which have a deficiency of positive charge due largely to substitution in the tetrahedral sheets (hence close to the surface of the layers) hold the K^+ especially tightly, so that only part of it is replaceable by other ions. These clays, with successive layers held together by K^+ ions, have properties different from montmorillonite and are given the name *illite*. (The term *hydromica*, often used in European literature, is practically a synonym.) The illite structure is transitional to that of muscovite, in which the amount of K^+ is greater and bears a constant relation to the amounts of Si and Al. The difference in structure between montmorillonite, illite, and muscovite may be symbolized by the ideal formulas

$$Al_4Si_8O_{20}(OH)_4 \quad \text{Montmorillonite}$$

$$K_{0\text{-}2}Al_4(Si_{8\text{-}6}Al_{0\text{-}2})O_{20}(OH)_4 \quad \text{Illite}$$

$$K_2Al_4(Si_6Al_2)O_{20}(OH)_4 \quad \text{Muscovite}$$

The term "illite," therefore, is less a name for a definite mineral than a name for a group of substances with compositions intermediate between montmorillonite and muscovite. Recent work, in fact, suggests that much so-called illite is a mechanical mixture of fine-grained montmorillonite and muscovite, or a clay containing alternate layers having a montmorillonite and a muscovite structure. The name is convenient, however, as a general term to designate clay materials with less potassium than that corresponding to the mica formula.

7-2 CLASSIFICATION OF CLAY MINERALS

The structural ideas of the last section provide a good basis for theoretical classification, and structures correlate well enough with macroscopic properties to make the classification a generally useful one.

We divide the clays first into two-layer and three-layer types: those (like kaolin) whose layers consist of one tetrahedral and one octahedral sheet, and those (like montmorillonite and illite) whose layers have an octahedral sheet between two tetrahedral sheets. In general, the kaolinite clays have layers bound more tightly together and permit less substitution of other ions for Al and Si. These structural differences are reflected in less ion-exchange capacity for the kaolinite clays and less plasticity because of a smaller capacity for adsorbing water.

The two-layer clays are further classified according to the degree of filling of the octahedral sites in the gibbsite sheet. In the kaolinite group of minerals Al^{3+} fills only two-thirds of the available positions (hence these are called "dioctahedral" clays); in the serpentine (or septechlorite) group Mg^{2+} and other cations fill all available positions (three per half unit cell, hence "trioctahedral" clays). The kaolinite group is further subdivided, according to the way successive layers are stacked, into kaolinite (by far the commonest mineral of this group), dickite, nacrite, and halloysite.

Among the three-layer clays, an important grouping is based on the ease with which layers are separated from one another. Thus clays whose layers are held together by the attraction of K^+ for strong negative charges in the tetrahedral sheets (illite and muscovite) can be distinguished from clays with more diffuse negative charges originating in the central octahedral sheets and therefore with less tendency for successive layers to be tightly bound by cations (montmorillonite). The latter clays (often called *smectite* as a group, with montmorillonite a principal variety) are commonly described as having expanding lattices, because their layers are readily pushed apart by adsorbed water. This property is strikingly shown by the ability of montmorillonite to swell when placed in water (and other polar liquids), and by its very large ion-exchange capacity. Varieties of smectite may be distinguished on the basis of composition, particularly by the nature of the principal cation in the octahedral sheets: montmorillonite (Al), nontronite (Fe^{3+}), saponite (Mg), sauconite (Zn), hectorite (Li). A related mineral is vermiculite, a three-layer clay with all octahedral positions occupied by Mg^{2+} and Fe^{2+} and with more substitution of Al for Si than is usual in smectite; it is much less expansible, but has a similar high ion-exchange capacity.

In a broad classification, several other hydrous-silicate minerals produced by weathering, or at least by near-surface processes, may be grouped with the clay minerals. Especially important are the chlorite minerals, minerals whose structure consists of mica-like sheets [with the composition $(Mg,Fe)_6(Si,Al)_8O_{20}(OH)_4$] alternating with sheets having the structure of brucite [composition $(Mg,Al)_6$-$(OH)_{12}$]. Another way of looking at this structure is to regard it as similar to illite, with successive layers linked by $(Mg,Al)(OH)^+$ ions instead of by K^+. Glauconite is a variety of illite, with considerable replacement of the Al of the octahedral sheet by Fe^{3+} and of the interlayer K^+ by Ca^{2+} and Na^+. Another group of hydrous silicates are referred to by the names palygorskite, sepiolite, and attapulgite; their structures are not completely known but are probably based on silicon-oxygen double chains, like those in the structure of amphiboles, rather than on mica-like sheets. The fibrous nature of these minerals is a reflection of the chain structure.

One further complexity of structure is shown by clay minerals whose crystals consist of alternating layers of different kinds, the *mixed-layer* clays. The interlayering may follow a regular pattern—so many layers of one kind, then so many of the other—or it may be completely random. The chlorite minerals may be considered mixed-layer structures with a regular alternation of mica and brucite layers, and some illite is a mixed-layer structure of mica and montmorillonite.

Finally, clays should be mentioned which appear amorphous even to x-rays, clays given the general name *allophane*. Even in these clays there is probably some ordering of Al and O into octahedral patterns and of Si and O into tetrahedral patterns, but the units are so small and so poorly oriented with respect to one another that x-ray reflections cannot be obtained.

7-3 RELATION OF STRUCTURE TO PROPERTIES

In discussing colloidal properties of clays in the last chapter, and details of structure in this chapter, we have already had occasion to point out some of the principal effects of crystal structure on megascopic properties. Here we need do little more than summarize the preceding discussions.

The properties of clays in general depend on their small grain size. In most samples of clay the particles have dimensions entirely or partially within the colloidal range, which means that they remain suspended in water for long periods, that they may be flocculated by electrolytes, that they adsorb ions out of solution, and that the adsorbed ions may be replaced by others when the concentration of the solution changes. Also, by virtue of their small size, clay particles have the ability to adsorb water and organic materials and so to become plastic, in other words, to remain coherent and capable of being molded when wet. These properties, although possessed in some degree by all clays, differ markedly from one clay mineral to another.

Minerals of the kaolinite group, because layers in the structure are held tightly together by the opposition of O^{2-} and OH^- and because little substitution of other cations for Al^{3+} and Si^{4+} is possible, show less capacity for adsorbing ions and water than do the other clays. Montmorillonite, with layers more easily separable and with extensive substitution, has abundant adsorbed ions and shows the capacity to adsorb water to an extreme degree. Montmorillonite therefore is much more plastic than kaolinite; its plasticity depends somewhat on the kind of adsorbed ions, being greatest for sodium and hydrogen montmorillonites and least for calcium montmorillonites. The extensive use of montmorillonite clays in drilling muds and in filters to remove undesirable ions and organic coloring materials from water and other liquids depends likewise on their adsorptive capacity. Illite clays and chlorites, with structural layers partly bound together by nonexchangeable cations, show intermediate plasticity and ion-exchange capacity. Vermiculite may have an ability to adsorb ions as great as that of montmorillonite, but its capacity for swelling by adsorption of water is more limited.

[margin notes: increasing plasticity; smectite; Ill.-to; Kaol]

Most clays in nature are mixtures of two or more clay minerals, and their properties are accordingly intermediate between the extremes. It is surprising, however, how greatly the properties of a clay sample may be influenced by a minor constituent; in particular, a small percentage of montmorillonite in a clay can radically change its plasticity. This ability of certain clays to modify the properties of mixtures makes difficult predictions about the behavior of a given sample from its composition alone. Much of our information about the behavior of clays, particularly about mixtures of different clay minerals, remains empirical. The more obvious properties we can link in general terms to crystal structures, but detailed correlation is still impossible.

7-4 FORMATION OF THE CLAY MINERALS *[margin note: ⊗ very important]*

The clay minerals are formed by alteration of aluminum silicates, both in weathering and in low-temperature hydrothermal processes. This much is obvious simply from geologic relationships, but details of the alteration process remain obscure. How does the chain structure of an amphibole, or the three-dimensional framework of a feldspar, convert itself into the hexagonal sheets of the clay minerals? Is it by a progressive modification of the preexisting structure, or do hydrated Al-O groups and Si-O groups detach themselves from the structure and later recombine into clays? If the second alternative is possible, can clays then precipitate from solution—say by reaction between dissolved silica and colloidal alumina in lakes or in the sea? These are some of the major conundrums about clay minerals which are still to be answered.

An obvious approach to such questions is through laboratory experiments. As so often happens in geochemistry, experiments carried out at low temperatures, under conditions duplicating as closely as possible those in nature, give unsatisfactory results because the reactions are so very slow. At higher temperatures reactions similar to those of weathering occur readily in the laboratory, but high-temperature results are always somewhat suspect because reactions may change their character over a range of a few hundred degrees. The best we can do is to combine the high-temperature findings with what hints we can get from low-temperature experiments, using the facts of geologic observation as a check on the extrapolation.

From qualitative experiments at high temperatures it has long been known that feldspar sealed in bombs with water and heated to temperatures of a few hundred degrees will form mica and various clay minerals. The nature of the product depends on the temperature, the amount of water, and the presence of other added materials such as HCl and metal oxides or chlorides. Kaolinite forms at temperatures under about 300°C, with or without added HCl, provided that metal compounds are low or absent; muscovite appears at higher temperatures and at higher ratios of KCl to HCl; montmorillonite is favored by oxides or chlorides of sodium and the alkaline-earth metals. Muscovite and the clays can also be produced under similar conditions if oxides are used as starting materials rather than feldspar. These results confirm, in general, what we might expect from the structures and compositions of the phyllosilicates.

These qualitative data have been refined in experiments by Hemley and his colleagues (1964). Instead of leaving the pressure of water vapor to be determined by the amount of H_2O placed in a sealed bomb at the beginning of an experiment, these workers arranged to have the bomb's interior connected with an outside source of water vapor, so that the pressure could be delicately controlled at all times. The amounts of HCl and alkali-metal salt were systematically varied, and the nature of the products was determined at various temperatures. As an example, Fig. 7-6 shows the products obtained in the system K_2O-Al_2O_3-SiO_2-H_2O under a pressure of 1,000 atm of water vapor. The two principal equilibria in this system are

$$3KAlSi_3O_8 + 2H^+ \rightleftharpoons KAl_3Si_3O_{10}(OH)_2 + 6SiO_2 + 2K^+ \qquad (7\text{-}1)$$
$$\text{K-feldspar} \qquad\qquad \text{Mica} \qquad\qquad \text{Quartz}$$

$$2KAl_3Si_3O_{10}(OH)_2 + 2H^+ + 3H_2O \rightleftharpoons 3Al_2Si_2O_5(OH)_4 + 2K^+ \qquad (7\text{-}2)$$
$$\text{Mica} \qquad\qquad\qquad\qquad \text{Kaolinite}$$

The equations show that equilibrium in both reactions depends on the ratio $[K^+]/[H^+]$. From the experimental results summarized in Fig. 7-6, the equilibrium ratio at a given temperature is higher for the first reaction than for the second, and for each reaction the ratio becomes larger as the temperature falls. Systems involving sodium and calcium compounds rather than potassium compounds give similar experimental results, except that mica has a smaller field of stability, and montmorillonite appears in addition to kaolinite in parts of the diagram where ratios of metal ion to hydrogen ion are large. The results agree with earlier work but are more useful, in that conditions of formation of clay minerals are defined more precisely and over a wider range of the important variables.

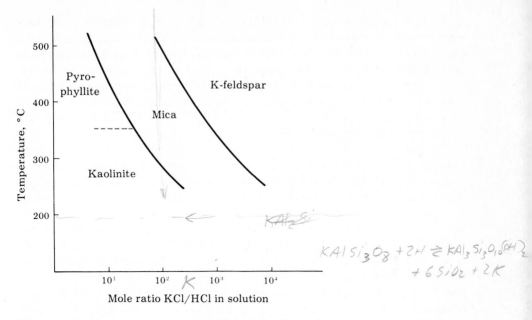

Figure 7-6 Reaction curves for the system $K_2O-Al_2O_3-SiO_2-H_2O$. Total pressure (chiefly water vapor) 15,000 psi (approximately 1,000 atm). Quartz present in the reaction mixture. Dashed line gives the experimentally determined temperature (but probably not the equilibrium temperature) of decomposition of kaolinite into pyrophyllite in 0.5m KCl solution. Upper solid line shows equilibrium between feldspar and mica [Eq. (7-1)]; lower solid line shows equilibrium between mica and kaolinite [Eq. (7-2)] in its lower part, and equilibrium between mica and pyrophyllite in its upper part. *(Source: Hemley and Jones, 1964, page 548.)*

Similar experiments at low temperatures are frustrating. When feldspar is stirred in water, a little of it dissolves, but no detectable clay minerals are formed. The dissolved material consists of minute concentrations of K^+, alumina, and silica, in proportions very different from those in the mineral; K^+, relative to Al and Si, is much higher in the solution. If the stirring is long-continued, the rate of dissolving becomes progressively slower. These facts mean that the feldspar grains must armor themselves with skins of a partly decomposed aluminosilicate framework which resists further attack. Low-temperature experiments of the opposite sort, involving attempts to produce clay minerals by mixing dissolved or colloidal silica and alumina, have been more encouraging. The slow formation of montmorillonite and chlorite by mixing very dilute solutions or colloidal suspensions of the constituent oxides at pH's in the range 8 to 9 is well established. To form kaolinite has proved more difficult, but small amounts are reported in recent experiments.

If we guess that the high-temperature results are similar to results that might be obtained at low temperatures by prolonging the experiments for geologic periods—in other words, if we assume that Eqs. (7-1) and (7-2) would eventually approach equilibrium at low temperatures—it is interesting to inquire whether the low-temperature equilibrium $[K^+]/[H^+]$ ratios might be estimated by extrapolating the curves in Fig. 7-6. The extrapolation is so gross that the ratios cannot be guessed with any certainty, but perhaps it is significant that the estimated values fall in the general range of measured $[K^+]/[H^+]$ ratios in ground and surface waters that have been long in contact with feldspar-bearing rocks. One may reasonably wonder, however, whether the existence or nonexistence of equilibrium has much importance in dealing with reactions so slow that equilibrium may not be reached even in geologic times.

To summarize currently available data, we conclude that clay minerals form from aluminosilicates by the action of water at temperatures up to a few hundred degrees, in reactions whose detailed mechanisms remain obscure and which at ordinary temperatures are extremely slow. Clay minerals can also form, again by slow reactions, from their constituent oxides or their solutions, under conditions that can be specified precisely only at high temperatures. These conclusions do not answer the question raised at the beginning of this section, whether clay minerals at ordinary temperatures actually form directly from aluminosilicate minerals, or whether the minerals must dissolve first into free silica and alumina, which thereafter combine to form clays. Geologic evidence can be cited in favor of either conjecture; the fact that aluminum and silicon dissolve separately in low-temperature experiments lends some support to the second. We shall return to this question in the discussion of laterites in the next section.

The suggestion from experimental work that acid solutions favor the formation of kaolinite and basic solutions the formation of montmorillonite at ordinary temperatures as well as at high temperatures is partly corroborated by data on the geologic occurrence of the two minerals. Kaolinite is commonly the chief clay mineral in soils of humid climates on well-drained slopes, where abundant vegetation makes soil solutions acid and cations are effectively leached away; montmorillonite is characteristic of soils in less humid climates, where soil solutions are slightly alkaline and cations are removed less rapidly. These generalizations seem plausible, but they are questioned by some geologists on the grounds that the clay minerals of a given area are often in large part inherited from earlier sediments rather than formed in place.

The effect of K^+ ions in favoring production of illite is indicated by the observation that this material is abundant in soils on which potash fertilizer has long been used, and by experiments in which montmorillonite is partly converted to illite by boiling in a solution of KCl and KOH. Illite is also the common clay mineral, as might be expected, in alkali soils of desert areas and in soils formed from igneous rocks especially rich in potassium.

7-5 STABILITY OF THE CLAY MINERALS

The clays, it is often said, are the stable products of weathering—the minerals that represent final attainment of equilibrium by the constituents of silicate rocks under normal conditions at the earth's surface. As a loose generalization this has some validity, but the multiplicity of clay minerals raises a further question: do the different clay minerals represent adjustment to slightly different weathering environments, or are some of the minerals merely metastable intermediates formed during the course of weathering but ultimately converted into others? No complete answer to this question is possible at present.

Environment of Formation versus Environment of Deposition

From the discussion in the last section it seems reasonable to suppose that kaolinite is the end product of weathering under conditions of acid soil solutions and good drainage in a temperate climate, that montmorillonite is the end product where solutions are alkaline, and that illite is the stable clay mineral where K^+ is abundant. We could also guess that clays formed in one environment might slowly change in character if the environment changes. In other words, clay minerals might be characteristic of particular environments of weathering, and a study of clays might make possible a reconstruction of the environment where they formed or were ultimately deposited. This is an attractive hypothesis, but there is no really convincing evidence to support it. It has not been demonstrated experimentally that one clay mineral can be changed into another at ordinary temperatures, except for the change from montmorillonite to illite mentioned above. Observational evidence is ambiguous, the clay minerals in some places seeming to confirm the hypothesis and in other places to contradict it. A major difficulty in trying to get observational evidence is that the clay minerals in a given soil may not have formed in place, but may be inherited from the weathering rock or may have been transported from elsewhere.

Some studies indicate that the clay minerals, once formed, do not change character readily but persist for times at least as long as geologic periods, even when their environment is altered. Frye et al. (1963), for example, have shown that layers of glacial till in Illinois can be identified by the clay minerals they contain, those from ice lobes to the west having much montmorillonite inherited from Cretaceous rocks, and those brought from the east containing largely illite from the Paleozoic rocks over which the glaciers moved. This suggests that clay minerals preserve a record of the environment from which they came rather than of the environment in which they were finally deposited. The argument is still not settled, and must await additional observational data.

Clay Minerals in Marine Sediments

The most common clay mineral of marine sedimentary rocks is illite, which suggests that montmorillonite is converted to illite by a slow reaction with the abundant K^+ of seawater. This seems a reasonable conclusion on the basis of evidence cited earlier, and more direct evidence is provided by data of Johns and Grim (1958) showing that illite (and chlorite) increase seaward at the expense of montmorillonite in sediments of the Mississippi delta. On the other hand, sediments with abundant montmorillonite are widespread in the modern ocean, and montmorillonite-rich marine sedimentary rocks (especially the bentonites) are found as far back as the Ordovician, suggesting that at least some montmorillonite is stable for geologically long times in a potassium-rich environment. Kaolinite is the dominant clay mineral of some marine sediments, both old and recent, an observation that speaks against the conversion of this mineral to either illite or montmorillonite even after long contact with alkaline solutions. Facts of this sort illustrate the difficulty, on the basis of present information, of trying to draw firm conclusions about the history of a sediment from the nature of its clay minerals.

Laterites

A further question about the stability of the clay minerals arises from data on the composition of tropical soils. Soils of the tropics show great variety, but in some places they contain only minor amounts of the clay minerals, their inorganic constituents being chiefly ferric oxide and aluminum oxide. Such soils are called *laterites*—ferruginous laterites if iron predominates, aluminous laterites, or *bauxites*, if aluminum predominates. Mineralogically these soils consist of various mixtures of hematite, goethite, gibbsite, boehmite, more rarely diaspore. Ferruginous laterites, as might be expected, are often thick and conspicuous where the underlying rock is an iron-rich variety like basalt or serpentine, while bauxite is well developed on aluminum-rich rocks like syenite and nepheline syenite. But lateritic soils are by no means limited to particular kinds of parent material; ferruginous laterites are reported from areas underlain by a great variety of rocks, even as iron-poor as granite, and bauxites are found as apparent weathering products of limestone and many kinds of volcanic rocks. Evidently weathering in parts of the tropics, regardless of the composition of the original rock, leads to removal from the soil not only of cations but also of silica. Does this mean that in warm climates the clay minerals are no longer stable, but break down into silica and alumina?

This question, the mechanism of origin of laterites, has long been argued by soil scientists and geologists. Thorough leaching is certainly involved, since laterites and bauxites are best developed where soils are well drained and where rainfall is heavy for at least part of the year. Beyond this, the geologic evidence is conflicting and chemical data are incomplete, so that the field is wide open for speculation. Some geologists have supposed that special processes of weathering must be operative in the tropics. Silica, for example, has been thought to be more soluble because soil solutions in the tropics are less acid than those of temperate climates; this guess has been proved wrong by measurements showing that silica is no more soluble in near-neutral solutions than in acids (Fig. 6-3). Or silica might be transported more easily in the tropics because organic matter is abundant to serve as a protective colloid; for this guess there is simply no evidence, either experimental or observational. Special kinds of vegetation and special varieties of bacteria have also been invoked as aids to tropical weathering, but again these are guesses unsupported by evidence. The most reasonable hypothesis is that tropical

de fowed from clays – from field data
clays from or ides – from lab data

weathering is not essentially different from weathering in temperate climates but simply goes farther toward completion. In other words, the clay minerals are not true end products of weathering but are metastable substances formed as intermediates in the slow breakdown of aluminosilicate minerals into their constituent oxides. All the reactions of weathering go faster in the tropics because of the higher temperature, and where conditions of rainfall and topography are particularly favorable for thorough leaching, the reactions may go beyond the clay-mineral stage. Field evidence for this hypothesis comes from the fact that normal clay soils are found in the tropics where leaching is less complete. Experimental support is furnished by the observation that feldspar stirred in water gives silica and a little alumina as separate substances in solution.

The problem of laterites is by no means solved in all its details by this hypothesis. One interesting detail, for example, concerns the mechanism by which the aluminum oxides form: do clay minerals form first, and thereafter break down into their oxides, or are alumina and silica formed directly from the original minerals, as they seem to be in laboratory experiments? If the second alternative is correct, does it follow that the clay minerals found in many tropical soils result from a recombination of the liberated alumina and silica? Field observations on laterites suggest that the reaction may go by either route, for in some places kaolinite is found close to fresh rock and bauxite farther away, while elsewhere bauxite rims the fresh material and kaolinite occurs at a distance. Furthermore, the frequently observed very sharp contacts between the three materials are at least suggestive of equilibrium in the reactions involved. It may be, then, that feldspar decomposes either to clay or directly to the oxides, and that clay may decompose into the oxides or be formed from them, depending on slight changes in such conditions as temperature, pH, and concentration of silica in the soil solution. Experimental data are not yet adequate to test these speculations.

Thermal Stability

The stability of clay minerals toward temperature change is not very great, as is shown by Hemley's experimental results (Sec. 7-4) and by the common field observation that, even at the lowest stage of metamorphism, muscovite occurs as an alteration of original kaolinite, montmorillonite, or illite. According to estimates of temperatures of metamorphism, this reaction must occur in nature not far above the boiling point of water, although higher temperatures are required to make it go at an appreciable rate in the laboratory. Recent work suggests that muscovite in very tiny flakes, or as layers in a mixed-layer clay structure, may form slowly in soils and shales even at ordinary temperatures.

Stability Diagrams

Reactions involving clay minerals, gibbsite, and primary aluminosilicates at ordinary temperatures are so slow that equilibrium cannot be assumed. It is nevertheless interesting to speculate as to what the mineral relations would be at a theoretical state of equilibrium. We might expect to find some approach to this

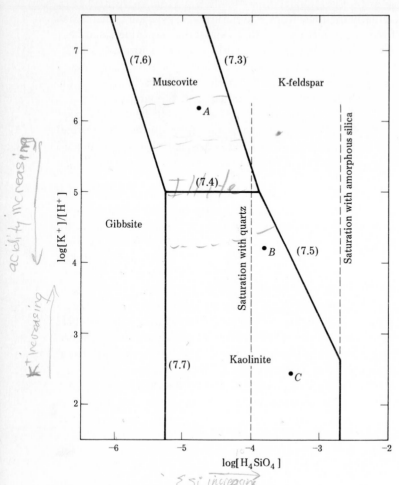

Figure 7-7 Stability relations of K-feldspar, mica, kaolinite, and gibbsite at 25°C and 1 atm, as functions of log $[K^+]/[H^+]$ and log $[H_4SiO_4]$. Numbers in parentheses refer to Eqs. 7-3 to 7-7.

state in geologic environments where weathering has gone on undisturbed for long periods.

A convenient way to display equilibrium relations is a diagram like Fig. 7-7, in which fields of stability of different minerals are plotted as functions of two variables. Four minerals are represented on this diagram: potash feldspar, $KAlSi_3O_8$ (as an example of a primary aluminosilicate); mica, $KAl_3Si_3O_{10}(OH)_2$; kaolinite, $Al_2Si_2O_5(OH)_4$; and gibbsite, $Al(OH)_3$. Theoretical equilibrium reactions among these substances may be written:

$$3KAlSi_3O_8 + 2H^+ + 12H_2O \rightleftharpoons KAl_3Si_3O_{10}(OH)_2 + 6H_4SiO_4 + 2K^+ \tag{7-3}$$

$$2KAl_3Si_3O_{10}(OH)_2 + 2H^+ + 3H_2O \rightleftharpoons 3Al_2Si_2O_5(OH)_4 + 2K^+ \tag{7-4}$$

$$2KAlSi_3O_8 + 2H^+ + 9H_2O \rightleftharpoons Al_2Si_2O_5(OH)_4 + 4H_4SiO_4 + 2K^+ \tag{7-5}$$

K-spar   *Kaolinite*   *silicic acid*

$$KAl_3Si_3O_{10}(OH)_2 + H^+ + 9H_2O \rightleftharpoons 3Al(OH)_3 + 3H_4SiO_4 + K^+ \tag{7-6}$$

muscovite   *gibbsite*   *silicic acid*

$$Al_2Si_2O_5(OH)_4 + 5H_2O \rightleftharpoons 2Al(OH)_3 + 2H_4SiO_4 \tag{7-7}$$

kaolinite   *gibbsite*   *silicic acid*

Note that Eq. (7-3) is the same as Eq. (4-20), except that it is written for a slightly acid solution; note also that Eqs. (7-3) and (7-4) are similar to Eqs. (7-1) and (7-2), except that for a low-temperature environment silica is written as H_4SiO_4 rather than SiO_2.

Equilibrium constants for the five reactions are:

$$\frac{[K^+]^2[H_4SiO_4]^6}{[H^+]^2} = 10^{-13.4} \qquad \text{or} \qquad 2\log\frac{[K^+]}{[H^+]} + 6\log[H_4SiO_4] = -13.4$$

$$\frac{[K^+]^2}{[H^+]^2} = 10^{10.0} \qquad \text{or} \qquad 2\log\frac{[K^+]}{[H^+]} = 10.0$$

$$\frac{[K^+]^2[H_4SiO_4]^4}{[H^+]^2} = 10^{-5.6} \qquad \text{or} \qquad 2\log\frac{[K^+]}{[H^+]} + 4\log[H_4SiO_4] = -5.6$$

$$\frac{[K^+][H_4SiO_4]^3}{[H^+]} = 10^{-10.8} \qquad \text{or} \qquad \log\frac{[K^+]}{[H^+]} + 3\log[H_4SiO_4] = -10.8$$

$$[H_4SiO_4]^2 = 10^{-10.5} \qquad \text{or} \qquad 2\log[H_4SiO_4] = -10.5$$

The numerical constants are not obtained by measuring actual equilibrium concentrations, for the obvious reason that equilibrium cannot be established in laboratory experiments. The numbers are calculated indirectly, by methods we will encounter in the next chapter.

In logarithmic form, each constant is expressed in terms of two variables only, $\log[K^+]/[H^+]$ and $\log[H_4SiO_4]$. Thus the five equations can be represented on a plot with these quantities as axes. On Fig. 7-7, the lines representing the five constants are identified by the numbers of the corresponding equations. Each of the four fields on the diagram shows the range of variables where one of the four minerals is stable, and each line shows the combination where a given pair would be at equilibrium. As might be expected, gibbsite is stable in solutions with low concentrations of silica, muscovite and feldspar in solutions with high concentrations of potassium, and kaolinite in solutions of relatively high acidity. Or, one can read from the diagram that feldspar becomes unstable as solutions become more acid (as $[K^+]/[H^+]$ decreases), changing first to mica and then to kaolinite (like the change at higher temperatures shown by Fig. 7-6); or that feldspar is altered in contact with solutions of decreasing silica content, changing first to mica or clay and then to gibbsite. Qualitatively, these predictions from the diagram agree with observations of weathering environments in nature.

The diagram is obviously incomplete, in that only a single clay mineral is represented. Illite, with a composition roughly between that of muscovite and kaolinite, would have a stability field overlapping the fields of these two minerals, above and below the horizontal line (7.4) on Fig. 7-7; but precise boundaries of the field cannot be given because illite has a widely variable composition. Montmorillonite could not be realistically shown on the diagram, because its stability depends on other variables, particularly the concentrations of Na^+ and Mg^{2+}.

In addition to its qualitative agreement with observation, Fig. 7-7 fits nicely some actual analytical data. The approximate composition of seawater, for example, is shown by point A; this is in the part of the diagram where illite would be stable, in agreement with the observed prevalence of this clay mineral in marine shales. Points B and C are representative compositions of soil water in contact with granite debris in a humid temperate climate, where kaolinite is the common clay mineral. This part of the diagram, in the kaolinite field with silica concentrations between saturation with quartz and saturation with amorphous silica, includes also the compositions of many stream waters in humid temperate regions. Such analytical data suggest that equilibrium may indeed be approached in some weathering environments. Numerous other analyses, however, would have anomalous positions on the diagram, indicating that attainment of equilibrium cannot be counted on when reactions are so very slow.

A diagram like Fig. 7-7 is in a sense dangerous, because it gives a false impression of quantitative accuracy in a situation where the plotted data are only approximate and approach to equilibrium is uncertain. But if its limitations are kept in mind, the diagram is a useful device for showing basic chemical relationships.

Summary The weathering of an aluminosilicate mineral is a slow and complicated hydrolysis reaction, in which the overall effect is removal of the cations and part of the silica in solution, while the aluminum and remaining silica stay behind as a clay mineral. The nature of the clay mineral originally formed probably depends on the environment of weathering, kaolinite being favored by acid solutions and good drainage, montmorillonite by alkaline solutions containing Ca^{2+}, Mg^{2+}, and Fe^{2+}, and illite by abundant K^+. In some parts of the tropics weathering leads to more complete removal of silica, leaving hydrated oxides of iron and aluminum as the principal solid inorganic residues. Clays of any kind carried to the sea are flocculated on contact with the electrolytes of seawater and become part of the sediment. During the early stages of metamorphism most clay minerals are converted into muscovite. The outstanding unsolved problems in this series of events are: (1) What is the mechanism of clay-mineral formation during hydrolysis? (2) Is there a real difference in stability with respect to decomposition into alumina and silica between tropical and temperate climates, or do clay minerals simply decompose faster in the tropics because of the higher temperatures? (3) How stable are the clay minerals with respect to changes from one to another in soils and sediments? (4) Is it possible for clay minerals to form at ordinary temperatures by a slow reaction between dissolved Al and Si, or between colloidal alumina and silica? (5) What are the conditions of formation of the less common clay minerals, especially halloysite, anauxite, and the palygorskite group?

7-6 FORMATION OF SOILS

Soils have figured prominently in the preceding discussions, for the obvious reason that most soils contain clay minerals as major constituents. In the remainder of this chapter soils will be the focus of attention.

The partly decomposed debris of rock weathering is subject to continued attack by the atmosphere and by rainwater percolating through it. Movement of water causes dissolved material and colloidal particles to be carried from one layer to another. Vegetation plays a role, particularly as it decays and so furnishes acids to the water; some of the organic matter may also be transported by the water and adsorbed by clays, thereby modifying their properties. Bacteria aid in decomposing the organic material and very probably serve as catalysts for inorganic reactions also, particularly for the oxidation of iron and manganese. The result of this complex of reactions is the material called soil. Soil formation is not sharply distinguished from weathering, but is rather just the last stage in the weathering process.

The essential thing about soil formation is the transposition of material from one level in the weathered debris to another. This gives rise to layers of different composition called *horizons*, and the vertical sequence of horizons from the surface down to fresh rock is called the *soil profile*. Horizons are designated by letters: the uppermost or A horizon is the soil layer from which downward-moving water has removed much of the soluble material, the B horizon is the intermediate layer in which some of the soluble and colloidal material is deposited, and the C horizon is the zone of fragmented but still largely unaltered debris that grades down into bedrock. Each horizon may be further subdivided by adding numbers; thus A_1 specifies a layer at the top of the A horizon containing abundant decaying vegetation. Some soils have well-developed profiles (meaning much transposition of material from the A to the B horizon, so that the horizons are clearly distinct from one another), others only a suggestion of profile formation.

The extent of profile development depends on a number of obvious variables. One is the kind of bedrock: clearly horizons will be more distinct in soils formed on granite or basalt than on a bed of kaolinitic clay or a pure quartz sandstone, because in the latter cases soil solutions will find little that can be moved from one horizon to another. A second variable is climate, particularly amount of rainfall: the quantity of material moved, other things being equal, must be greater where rainfall is higher and the amount of soil water is larger. Topography is a third factor, for on steep slopes soil is washed away before the A and B horizons can be clearly differentiated, and in swampy areas the movement of water is so slight that not much vertical transposition can take place. Other possible factors would be seasonal distribution of rainfall, length of the period of freezing, average temperature, and local sources of unusually acidic or basic solutions.

Observation indicates that soils developed under favorable conditions—on gentle slopes with adequate drainage, in a region where the climate has remained constant for a long time—acquire characteristics that depend chiefly on the climate. The nature of the bedrock makes little difference in the kind of soil ultimately formed (except for rocks of extreme compositions, such as ultramafic rocks or salt beds). In the early stages of weathering, of course, soils from a granite, an andesite, and a gabbro will be very different, but as time goes on, their characteristics converge to a soil type determined by the climate. Once the profile is fully developed and the soil is adjusted to the climate, it should be stable indefinitely. Slow erosion of the surface soil will be compensated by progressive weathering of the bedrock beneath, and the soil profile will gradually move downward while maintaining the characteristics of its various horizons. Such a soil is described as a *zonal* soil, meaning that it is typical of a certain climatic zone; alternatively it may be called a *mature* soil, to indicate that it has passed through a sequence of developmental stages and is now no longer changing. In contrast, soils on steep slopes or soils in areas of recent orogenic movement, where the nature of the bedrock still markedly influences the character of the soil, are spoken of as *azonal* or *immature* soils. How long it takes in years to convert an immature soil to a mature soil is unknown. Doubtless it varies greatly from one kind of bedrock to another and from one climatic regime to another; it is certainly measured at least in hundreds of years, probably often in thousands of years.

The concept of maturity in soil development is an extremely hazy one. "Zonal" is a somewhat more objective term and has become more common in recent literature. Serious objections have been raised even to this term, particularly on the grounds that a soil hardly ever completely loses the characteristics imposed on it by its parent material. Arguments over terminology, however, are unimportant for our purposes here. We need only recognize that climate is a dominant influence in soil development and that soils of stable, well-drained areas show major characteristics that we can predict more successfully from climatic data than from knowledge of bedrock petrography.

7-7 CLASSIFICATION OF SOILS

Soils pose a particularly troublesome problem in classification. They differ in so many ways: grain size, composition, plasticity, "workability," mechanical strength, color, fertility, permeability, parent material, degree of maturity, nature of the profile. Which of these are the most important characteristics, the ones on which a usable classification should be based? Obviously there is room here for several classifications, one suited for engineering uses of soils, one for agricultural uses, one for geologists interested in soil genesis, and so on. But the general problem of classification is not our concern, and we shall limit ourselves to the breakdown that has most geochemical significance. This turns out to be a classification based primarily on the climatic factors of soil development.

The major break is between soils of humid regions and soils of arid regions. Typical examples of the former are characterized by a concentration of iron oxide and aluminum silicates in the B horizon, and these soils are accordingly often referred to as *pedalfer* soils, pedalfer being a hybrid word from the Greek root for "soil" and the Latin words for the two elements. Soils of arid regions typically show concentrations of calcium salts and are designated *pedocal* soils. The zonal humid or pedalfer soils are further subdivided according to temperature, and the zonal aridic soils according to degree of aridity, as shown in Table 7-1. The dividing line between humid and aridic soils is commonly set at a rainfall of 25 in. per year, which in the United States coincides very roughly with a north-south line along the Mississippi River.

Most widespread of the zonal humid soils are the podzols and their relatives, soils typical of the humid parts of the temperate zones. In its most narrow sense, the word *podzol* refers to forest soils in the northern part of the temperate zones. Abundant vegetation makes the soil water acid, pH values reaching 4.0 to 4.5 in the clayey part of the soil and as low as 3.5 in the humus. This acid water effectively leaches alkali and alkaline-earth ions from the A horizon and is strong enough to remove much of the iron and aluminum also. The A horizon is left rich in silica and gray-white in color; the name "podzol," Russian for "under-ash," comes from the conspicuous light color of this upper layer. Ions of the alkalies and alkaline earths are almost completely removed in solution, but the iron and aluminum carried down from the A horizon are deposited in the B horizon as hydrates of iron oxide and colloidal clay particles, presumably in response to a slight rise in pH. The resulting iron-and-clay-rich B horizon is the "clay hardpan" typical of these soils. The precise role of organic matter in these processes is uncertain, but very likely organic material serves as a protective colloid in moving some of the oxides and clays downward. The clay in podzols is typically kaolinite, as would be expected from the acid environment and the depletion of lime and magnesia.

Soils in which leaching is less extreme and in which iron and aluminum are less concentrated in the B horizon are referred to as "podzolic" soils. They are characteristic of the central part of the temperate zones, for example most of eastern United States north of Tennessee and the Carolinas.

North of the forested areas of podzol are the treeless plains of the Arctic, the tundra, which gives its name to the typical cold-humid soils. Such soils have a great abundance of organic matter because the cold climate slows decay. Most commonly the subsoil is perpetually frozen, and the surface soil is frozen much of

Table 7-1 Climatic classification of soils

Pedalfer soils (average rainfall over 25 in./year) *[lacks clays, rich in oxides]*
    Laterite (tropics) *[hot]*
    Podzol (temperate zone) *[moderate — kaolinite]*
    Tundra (arctic zone) *[cold]*
Pedocal soils (average rainfall less than 25 in./year)
    Chernozem (rainfall 12–25 in./year, cool)
    Chestnut-brown (rainfall 10–15 in./year, warm)
    Desert and saline (rainfall less than 10 in./year, warm)

[handwritten: aridity increasing]

the year; this means that the soil is boggy and poorly drained, with little chance for profile development.

At the other extreme of temperature, in the humid tropics, a very different kind of soil is formed. This is the red soil called laterite, which we have discussed at length in a previous section. Chemically the chief characteristics are an enrichment of iron oxide and aluminum oxide throughout the profile, a depletion in silica, and removal of most of the alkali and alkaline-earth ions. The process is probably similar to podzolization but is more complete because weathering is more rapid and more intense. It should be recalled from the earlier discussion that laterites are by no means the only soils of the tropics and that soils containing kaolinite, in particular, are widespread.

Soils of humid areas near the edges of the tropics commonly have clay minerals mixed with the iron and aluminum oxides and are called "lateritic soils" to distinguish them from typical laterites. Such soils retain enough iron oxide distributed through the profile to color them red, red-brown, and yellow-brown; the famous red soils of Georgia and the Carolinas are a good example. Northward lateritic soils and podzolic soils intergrade.

Zonal aridic soils, in contrast to the humid soils, are characterized by concentrations of calcium salts somewhere in the profile. The reason is simply that water from rains is only rarely abundant enough to move completely through the soil and into drainage channels; much of the time it simply soaks into the ground and later rises back toward the surface because of capillary action. This means that only the most soluble soil constituents can be leached, particularly compounds of sodium, potassium, and magnesium, and that calcium is left behind to form salts as the capillary water rises and evaporates. Calcium carbonate is the commonest deposit, but gypsum may accompany or replace it locally. In areas where the rainfall is just under the 25-in. limit the deposition of these salts takes place far down in the B horizon, at depths of several feet or tens of feet. As aridity increases, the deposit moves closer to the surface and becomes a better-defined layer, forming the well-known "caliche" crusts so common at or near the surface in soils of the western states. Vegetation is neither so abundant nor so rapidly decayed as in more humid regions, so that soil water retains the slight alkalinity acquired by hydrolysis of silicate and carbonate minerals. The common clay mineral of arid soils, as might be expected, is montmorillonite; where potassium is abundant, illite may take its place.

The best known of the aridic or pedocal soils is the type characteristic of areas near the 25-in. rainfall boundary, a deep soil colored black by an abundance of organic matter and having an ill-defined zone of lime enrichment at depth—the soil called *chernozem*, or "black earth," perhaps the best of all agricultural soils because of its content of organic matter and unleached cations. With increasing dryness the soil becomes thinner and lighter-colored; varieties are often designated simply by color terms, such as "chestnut" and "red-brown." When dryness is extreme, the soil may retain very soluble salts like sodium chloride, and is accordingly described as a "saline" soil, or solonchak.

This brief review of a complex subject suffers from the usual handicap of superficial treatments, in that it gives a false illusion of simplicity. The classification of soils is so thorny a problem that experts still disagree, and schemes of classification have changed radically over the years. The simple break-down just described is not the one in current use by soil scientists, but the accepted classification is so enormously complex that a separate chapter would be needed to introduce it. For our purposes this seems too long a detour, and a rough division of soils according to their major chemical differences will be an adequate background for future discussions. An excellent description of the recent more detailed classification is given in the book by Birkeland listed in the bibliography for this chapter.

PROBLEMS

1 The chemical composition of a typical montmorillonite may be expressed by the formula $(Al,Mg, Fe^{3+},Zn^{2+})_4(Si,Al)_8O_{20}(OH)_4$. Explain in detail what this formula means.

2 Montmorillonite differs from kaolinite in its greater plasticity, its greater ability to adsorb water, and its greater ion-exchange capacity. How are these differences related to differences in the structure of the two minerals?

3 What would you expect to find as the dominant clay mineral in:
   (a) A mature soil developed on limestone in western Iowa?
   (b) An Ordovician shale containing graptolites?
   (c) A forest soil in the Laurentian highlands of Quebec?
   (d) An immature soil developed on gabbro in Virginia?
   (e) A soil developed on till in Minnesota?
   (f) A soil developed on trachyte in Colorado?

4 Why is a classification of soils based primarily on the nature of parent materials not a useful one for most purposes?

5 What are the important compositional and structural differences between chlorite and montmoril-lonite? Between illite and sericite? Between kaolinite and vermiculite?

6 In what ways is the origin of laterite a difficult geochemical problem? Suppose that you were setting out to study this problem: what sort of observational and experimental data would you look for?

7 What is meant by a clay with an "expanding lattice"? What peculiarity of structure gives rise to an expanding lattice?

8 In Prob. 2-9 you have calculated the total amount of dissolved aluminum in equilibrium with amorphous $Al(OH)_3$ as a function of pH. Make the same calculations for amorphous $Fe(OH)_3$, extending the pH range to 2 and to 11. Plot the figures for total Al and total Fe on a graph; this is most easily done by plotting logarithms of total metal concentration as ordinates against pH values as abscissas. On the curves indicate points that represent 10 ppm, 1 ppm, and 0.1 ppm of each metal. Add to the graph curves showing the solubility of quartz and amorphous silica, taken from Fig. 6-3 but with units changed to molar concentrations. How would the curves for Fe and Al be changed if equilibrium were established with goethite and boehmite rather than with the amorphous hydroxides? Roughly where would a curve for $Fe(OH)_2$ be located on the graph? (Recall Prob. 2-10.)

9 Use the graph of Prob. 8 as a basis for discussing the mobility of iron, aluminum, and silicon in various kinds of soil.

10 Describe what happens to the K-feldspar of a granite if the rock is permeated by hot water at a pressure of 1,000 atm, under the two following conditions:
   (a) The temperature remains constant at about 200°C and the acidity gradually increases.
   (b) The ratio $[K^+]/[H^+]$ remains constant at about 100, and the temperature falls.
What is the evidence that changes of this sort occur in nature?

REFERENCES AND SUGGESTIONS FOR FURTHER READING

Berner, R. A., "Principles of Chemical Sedimentology," McGraw-Hill Book Company, New York, 1971. Chapter 9 describes recent theoretical and experimental work on the formation and alteration of the clay minerals.

Birkeland, P. W., "Pedology, Weathering, and Geomorphological Research," Oxford University Press, 1974. Excellent description of soil-forming processes. Review of soil classification problems, and good exposition of modern classification schemes.

Elderfield, H., Hydrogenous material in marine sediments, chapter 27 in "Chemical Oceanography," edited by J. P. Riley and R. Chester, vol. 5, 2d ed., Academic Press, New York, 1976. Review of recent work on changes in clay minerals in contact with seawater.

Frye, J. C., H. D. Glass, and H. B. Willman, Mineralogy of glacial tills and their weathering profiles in Illinois, *Illinois Geol. Survey Circ.* 347, 1963. Clay minerals from various sources are so stable that they serve to identify till from glaciers that moved into Illinois from different directions.

Grim, R. E., "Clay Mineralogy," 2d ed., McGraw-Hill Book Company, New York, 1968. A standard reference book on the composition, structure, and genetic relations of the clay minerals. Chapter 13 is a good brief discussion of soil types and soil formation.

Grubb, P. L. C., Phase changes in aged sesquioxide gels and some analogies with katamorphic processes, *Mineralium Deposita*, vol. 4, pp. 30–51, 1969, and vol. 5, pp. 248–272, 1970. Experimental changes in mixed-oxide gels, with special reference to clay-mineral formation and to conditions of bauxite genesis. Reports formation of kaolinite at low temperatures.

Hem, J. D., and C. J. Lind, Kaolinite synthesis at 25°C, *Science*, vol. 184, pp. 1171–1173, 1974.

Hemley, J. J., and W. R. Jones, Chemical aspects of hydrothermal alteration with emphasis on hydrogen metasomatism, *Econ. Geology*, vol. 59, pp. 538–569, 1964. Review of experimental work on the K_2O-Na_2O-Al_2O_3-SiO_2-H_2O system and applications to alteration processes.

Johns, W. D., and R. E. Grim, Clay mineral composition of recent sediments from the Mississippi delta, *Journ. of Sedimentary Petrology*, vol. 28, pp. 186–199, 1958. Seaward change of montmorillonite to illite and chlorite in delta sediments.

Keller, W. D., Diagenesis in clay minerals—a review, *Clays and Clay Minerals, 11th Conf.*, pp. 136–157, 1963. A good review of the argument regarding stability versus diagenetic alteration of the clay minerals.

Siever, R., and N. Woodford, Sorption of silica by clay minerals, *Geochim. et Cosmochim. Acta*, vol. 37, pp. 1851–1880, 1973. Experimental study of silica-clay equilibria in water of various salinities.

EIGHT

FREE ENERGY

In this chapter we return to a formal study of chemical equilibrium, but our approach will be from a new direction. We look now at the energy evolved and consumed in chemical reactions and at the relation of this energy to equilibrium. This is a branch of physical chemistry called thermodynamics, a basic subject which permeates much of modern chemistry. As a predictive tool, thermodynamics has its greatest usefulness in the closed, carefully controlled systems of the chemical laboratory, but we shall find it an aid also in setting limits on what is possible and what is impossible in the more complex systems with which a geologist must deal.

Thermodynamics is an impressive logical structure built on a very few fundamental postulates. The derivation of its formulas from the basic postulates is not difficult mathematically, but it is too time-consuming for us to follow in this chapter. We shall limit the discussion here to a demonstration that the formulas are in reasonable agreement with experience, and leave their derivation to Appendix XI.

8-1 CHANGE IN ENTHALPY

The simple and obvious way to measure the energy of a chemical reaction is to put the ingredients in a calorimeter and record the amount of heat energy absorbed or liberated as the reaction takes place. To standardize the results so that one reaction can be compared with another, we specify that the process shall take place at constant temperature and pressure, or at least that the products of reaction shall be brought back to the initial temperature and pressure before the heat change is measured. Under these conditions we speak of the heat given out or taken up as the *heat of reaction*. It is measured in calories, either gram calories or kilogram calories, and is often written as part of the equation:

$$H_2 + \tfrac{1}{2}O_2 \rightarrow H_2O + 68.3 \text{ kcal}$$

This means that 68.3 kcal of heat are evolved for each mole of H_2O formed, when the reaction is carried out at 25°C and 1 atm pressure.

From a slightly different point of view, we may think of the reactants as possessing a certain amount of chemical energy and the products a different amount. In the example just cited, the product H_2O has less energy than does the combination H_2 and O_2, the excess being given up in the form of heat when the reaction takes place. In symbols, letting H stand for energy, we could write:

$$H_{\text{products}} - H_{\text{reactants}} = \Delta H = -68.3 \text{ kcal}$$

The symbol ΔH, read "delta H," means the change in H of the system when the reactants are transformed into the products. Here the H for the system becomes less during the reaction, so ΔH must have a negative sign. The particular kind of energy represented by H and ΔH goes by the name of *enthalpy*, or *heat content*; H itself refers to the enthalpy of the products or reactants, and ΔH is the *change in enthalpy* of the reaction.

It turns out that the absolute enthalpy of a substance or a mixture of substances cannot be measured. The quantity that concerns us in discussing a reaction is not the individual H's but the ΔH, the enthalpy change. This is easily measured as the heat change when the reaction takes place at constant temperature and pressure and is obviously the same number as the heat of reaction with the sign reversed. This odd procedure—introducing hypothetical quantities that cannot be measured, and using a negative sign for heat evolved—at first seems to make a simple subject needlessly complicated, but we shall find that it fits well into a larger logical framework.

For the hydrogen-oxygen reaction, then, we write

$$H_2 + \tfrac{1}{2}O_2 \rightarrow H_2O \qquad \Delta H = -68.3 \text{ kcal} \tag{8-1}$$

The enthalpy change is different at different temperatures and pressures, so we specify that -68.3 kcal refers to the reaction when it is carried out at 25° and 1 atm pressure (or when the resulting water is condensed and brought to these conditions). All enthalpy changes, unless otherwise specified, refer to these same conditions. Furthermore, ΔH obviously must be measured for a particular amount of substance; in our example the -68.3 kcal is the enthalpy change when 1 mole of hydrogen gas reacts with $\tfrac{1}{2}$ mole of oxygen. We could write equally well

$$2H_2 + O_2 \rightarrow 2H_2O \qquad \Delta H = -136.6 \text{ kcal}$$

In other words, a specific value of ΔH applies only to a particular formulation of the chemical equation. If the equation is multiplied by a number, the ΔH must be multiplied also.

For all exothermic reactions, like the hydrogen-oxygen reaction, ΔH is negative. In endothermic reactions ΔH is positive; for example:

$$N_2 + O_2 \rightleftharpoons 2NO \qquad \Delta H = +43.2 \text{ kcal}$$

$$NaCl \rightleftharpoons Na^+ + Cl^- \qquad \Delta H = +0.9 \text{ kcal}$$

Obviously, if an equation is reversed, the sign of ΔH changes:

$$2H_2O \rightleftharpoons 2H_2 + O_2 \qquad \Delta H = +136.6 \text{ kcal}$$

meaning simply that energy must be supplied to break water down into its elements.

For any compound, the enthalpy change involved in forming it from its elements is called its *heat of formation.* Thus the heat of formation of water is -68.3 kcal/mole and that of nitric oxide is $+21.6$ kcal/mole. Heats of formation always refer to 25°C and 1 atm pressure and to a single mole of the compound, unless conditions are specified otherwise. Such heats of formation have been measured for many simple compounds; those most useful in geochemistry are given in Appendix VIII.

By adding and subtracting heats of formation, the enthalpy change for any reaction can be computed. For example, the burning of hydrogen sulfide may be written

$$2H_2S + 3O_2 \rightarrow 2H_2O + 2SO_2$$

The heats of formation are

$$H_2 + S \rightarrow H_2S \qquad \Delta H = -4.9 \text{ kcal}$$

$$H_2 + \tfrac{1}{2}O_2 \rightarrow H_2O \qquad \Delta H = -68.3 \text{ kcal}$$

$$S + O_2 \rightarrow SO_2 \qquad \Delta H = -70.9 \text{ kcal}$$

Now if these three equations are doubled, if the second and third are added together, and if the first is subtracted from the sum, we obtain the original equation. Hence we find ΔH for the overall reaction by combining ΔH's for the part reactions:

$$\Delta H = (-136.6) + (-141.8) - (-9.8) = -269.6 \text{ kcal}$$

A simpler formulation is merely to write down the heat of formation and the coefficient beneath each substance in the equation:

$$2H_2S + 3O_2 \rightarrow 2H_2O + 2SO_2$$
$$2(-4.9) \quad 3 \times 0 \quad 2(-68.3) \quad 2(-70.9)$$

Then we get ΔH by adding the figures for the products and subtracting the sum of the figures for the reactants. This is a general rule for obtaining enthalpy changes.

8-2 CHANGE IN EQUILIBRIUM CONSTANTS WITH TEMPERATURE

Enthalpy differences can be used to express quantitatively the way an equilibrium constant changes as the temperature rises or falls. The relationship is given by the van't Hoff equation (see Appendix XI for derivation):

$$\frac{d \ln K}{dT} = \frac{\Delta H}{RT^2} \tag{8-2}$$

The expression $\ln K$ is the natural logarithm of the equilibrium constant, and T is the absolute temperature. R is a constant, the same as the "gas constant" in the expression $PV = nRT$ relating pressure, volume, and temperature for a perfect gas, a constant whose value is approximately 1.99 cal/mole (or 0.00199 kcal/mole). It is by no means evident why the equation should have this particular form, but at least we can see immediately that it agrees with qualitative predictions from Le Chatelier's rule: when ΔH is positive, meaning that the reaction as written is endothermic, the differential coefficient is also positive; in other words, K increases as T increases, so that the equilibrium is displaced in the direction which shows absorption of heat.

For numerical calculations it is convenient to restate Eq. (8-2) in integrated form. If ΔH is assumed constant, the integration is very simple:

$$\int d \ln K = \ln K = \frac{\Delta H}{R} \int \frac{dT}{T^2} = -\frac{\Delta H}{RT} + C$$

where C is the constant of integration. Ordinarily decimal logarithms are more convenient than natural logarithms, the conversion from one to the other following the equation

$$2.303 \log_{10} K = \ln_e K$$

Hence the integrated equation becomes

$$\log K = -\frac{\Delta H}{2.303RT} + C \tag{8-3}$$

The expression in the denominator on the right side, $2.303RT$, is very often encountered in thermodynamics. It is useful to commit to memory that $2.303R = 2.303 \times 1.99 = 4.58$ and that, when T is 25°C, the expression $2.303RT$ is

$$4.58 \times 298 = 1,364$$

When ΔH is expressed in kilocalories, these numbers become 0.00458 and 1.364, respectively.

As an example of how this equation is used, consider the reaction

$$CO + H_2O \rightleftharpoons CO_2 + H_2$$

From tables, the equilibrium constant at 417°C is 10, and the enthalpy change is − 10.2 kcal. From these figures the integration constant can be evaluated:

$$\log 10 = \frac{+10.2}{0.00458 \times 690} + C$$

whence

$$C = -2.24$$

Now to find K at another temperature, say 25°, the equation is used with this value for C:

$$\log K = \frac{10.2}{0.00458 \times 298} - 2.24 = 5.26$$

and

$$K = 10^{5.26} = 1.82 \times 10^5$$

In other words, the equilibrium constant decreases from 182,000 to 10 as the temperature rises from 25 to 417°. This is the direction of change we would predict qualitatively from Le Chatelier's principle for a reaction with negative ΔH.

Alternatively Eq. (8-2) may be integrated between limits:

$$\log \frac{K_2}{K_1} = \frac{\Delta H}{2.303R} \left(\frac{T_2 - T_1}{T_2 T_1} \right) \qquad (8\text{-}4)$$

where K_2 is the constant at temperature T_2 and K_1 the constant at T_1. In this formulation the example of the last paragraph would be set up as

$$\log \frac{10}{x} = \frac{-10,200}{2.303 \times 1.99} \left(\frac{690 - 298}{690 \times 298} \right)$$

As a second example of the use of van't Hoff's equation, let us try the reverse problem: given values of the equilibrium constant over a range of temperature, to calculate the enthalpy change for the reaction. We shall use a simple reaction, the dissolving of quartz in water (Sec. 6-7):

$$SiO_2(qz) + 2H_2O \rightleftharpoons H_4SiO_4(aq) \qquad K = a_{H_4SiO_4}$$

The equilibrium constant is merely the activity of silicic acid, since the activities of solid quartz and liquid water are 1 by convention. In dilute solutions the activity of H_4SiO_4 may be taken as equal to its concentration, and this may be expressed as the measured solubility of quartz. Values for the solubility of quartz over a range 25 to 200°C have been assembled by Siever (1962), from his own measurements and from the literature; Siever's graph is reproduced as Fig. 8-1. Solubilities are plotted as logarithms, and temperatures as reciprocals of the absolute temperature. The reason for this procedure lies in the form of Eq. (8-3):

$$\log K = -\frac{\Delta H}{2.303R} \left(\frac{1}{T} \right) + C = -A\frac{1}{T} + C$$

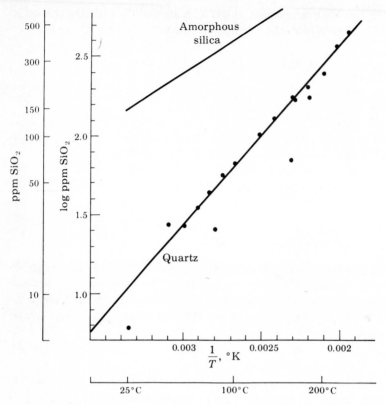

Figure 8-1 Solubility of silica as a function of temperature, plotted as log ppm SiO_2 against reciprocal of absolute temperature. (*Data from various sources, as compiled by Siever, Jour. Geology, vol. 70, pp. 127–150, 1962.*)

where A is a constant equal to $\Delta H/2.303R$. This is the equation of a straight line with negative slope, if log K and $1/T$ are plotted as variables. Siever's data, with the exception of a few points, clearly fit this requirement.

When the data are plotted and the best straight line is fitted to the points, the slope of the line is equal to $-\Delta H/2.303R$. Hence ΔH in kilocalories per mole can be found by multiplying the slope by 0.00458. In Fig. 8-1 the slope is 1,132, and ΔH for the dissolving of quartz is therefore

$$1{,}132 \times 0.00458 = 5.18 \text{ kcal/mole}$$

This number can be used to obtain the enthalpy of formation of $H_4SiO_4(aq)$: from Appendix VIII, the value of $\Delta H°$ for $SiO_2(qz)$ is -217.6 kcal, and for $2H_2O$ is -136.6; the figure for $H_4SiO_4(aq)$ must then be 5.18 kcal more positive than the sum of these, or -349.1 kcal.

This is a widely used method for obtaining enthalpy changes for reactions and enthalpies of formation that would be difficult to measure directly.

8-3 CHANGE OF K WITH TEMPERATURE WHEN ΔH IS NOT CONSTANT

Up to this point we have assumed that ΔH remains constant as the temperature changes. For most reactions the assumption is satisfactory if the temperature change is no more than a few tens of degrees, and for many reactions it holds within a percent or two even for changes of a few hundred degrees. Sometimes, however, the variation of ΔH with temperature becomes important: when greater accuracy is needed, when a temperature range of hundreds of degrees is considered, or when a reaction shows unusual variation in ΔH over small temperature ranges. For this purpose we must return to Eq. (8-2) and integrate it with ΔH under the integral sign instead of outside:

$$\int d \ln K = \frac{1}{R} \int \frac{\Delta H}{T^2}\, dT$$

To evaluate the integral, we need a way to express ΔH as a function of T.

To accomplish this, we recall the definition of heat capacity: the heat required to change the temperature of 1 mole of a substance by 1°C. The substance absorbs the heat as its temperature rises, and its enthalpy is thereby increased; hence we can recast the definition to read: heat capacity (at constant pressure) is the rate of change of enthalpy with temperature. In symbols:

$$c_P = \frac{dH}{dT}$$

The subscript P means that this is heat capacity at constant pressure. In a reaction mixture, we can think of the heat capacity of the reactants as added together on one side of the equation, and that of the products as added together on the other; then the latter sum minus the former is the change in heat capacity as the reaction takes place, which we will represent by Δc_P. It follows that

$$\Delta c_P = \frac{d\, \Delta H}{dT} \tag{8-5}$$

From this equation ΔH can be found by integration,

$$\Delta H = \int \Delta c_P\, dT \tag{8-6}$$

provided that Δc_P can be expressed as a function of temperature. Fortunately heat capacities are known for many substances and are commonly reported as power series in T with empirically determined coefficients. The general form is

$$c_P = a + bT + cT^2 + \cdots \tag{8-7}$$

where a, b, and c are simply constant numbers determined by experiment. Combining equations of this sort for the substances in a reaction mixture gives

$$\Delta c_P = \Delta a + \Delta bT + \Delta cT^2 + \cdots \tag{8-8}$$

where Δa, Δb, and Δc are the differences obtained by adding up the constants for the products and subtracting those for the reactants. Equation (8-8) can now be substituted in Eq. (8-6) and the integration performed:

$$\Delta H = A + \Delta aT + \tfrac{1}{2} \Delta bT^2 + \cdots \qquad (8\text{-}9)$$

where A is the constant of integration. Finally, this expression for ΔH can be used in Eq. (8-2), giving for the integral

$$\log K = \frac{1}{2.303R} \int \frac{\Delta H}{T^2} \, dT$$

$$= \frac{1}{4.58} \left(-\frac{A}{T} + \Delta a \ln T + \tfrac{1}{2} \Delta bT + \cdots - C \right) \qquad (8\text{-}10)$$

where C is the constant of the second integration.

Equation (8-10) is the general expression for the change of K with temperature. In order to use it, three things are necessary: (1) heat capacities must be known for all substances involved; (2) ΔH must be determined experimentally at one temperature to evaluate the constant A; (3) K must be determined experimentally at one temperature in order to evaluate the constant C. For many reactions of geochemical interest these quantities are not all known, and one must have recourse to the simpler but less accurate Eq. (8-3).

8-4 ENTHALPY CHANGE AS A MEASURE OF REACTIVITY

In Chap. 1 we set up as a major goal of geochemistry the ability to predict what will happen when one substance is mixed with another. Will a reaction take place, or will the substances exist side by side in chemical equilibrium? At first glance it looks as if enthalpy changes might be a key to answer such questions, for exothermic reactions often take place spontaneously while endothermic processes generally do not occur unless energy is supplied. Further, the more exothermic a reaction is (the higher its negative ΔH), the more energetically we expect it to proceed. The hydrogen-oxygen reaction, for example, is strongly exothermic and occurs with explosive violence if it is touched off with a flame or spark. The reaction between water and carbon dioxide to form carbonic acid, on the other hand, is only mildly exothermic and reaches equilibrium with no obvious indication that a reaction is taking place at all.

To say that the energy available in a reaction mixture determines how readily and how violently the reaction will take place seems entirely natural. It conforms to our intuitive prejudices, derived from experience with mechanical systems, where, in general, energy available does determine whether a given process will go of its own accord and how much work or heat we can expect to get out. A little reflection, however, will suggest that the generalization cannot be strictly true for chemical processes. Some reactions take place spontaneously despite the fact that they are endothermic; of their own accord they extract the necessary heat from their surroundings, cooling adjacent objects below the general temperature level. A good example is the dissociation of potassium nitrate: when this salt is stirred in water, the container quickly becomes cold because the reaction

$$KNO_3(s) \rightarrow K^+ + NO_3^- \qquad \Delta H = +8.4 \text{ kcal}$$

absorbs heat from its surroundings. Many other common solution processes are similarly endothermic. The enthalpy change, therefore, is by no means an infallible measure of the tendency of a reaction to take place.

It remains true that for very many reactions the size of the enthalpy change provides at least a gross indication of reactivity. In the absence of other data, enthalpy changes can be used fairly satisfactorily to make predictions about the behavior of unfamiliar substances. We might expect, therefore, that a quantity related to enthalpy would give us the necessary refinement to make the predictions exact. This quantity is called the free energy.

8-5 FREE ENERGY: DEFINITION

To see why enthalpy is not the only factor of importance in making a reaction take place, let us imagine a process in which there is no enthalpy change at all. We have two gases in containers separated by a partition; we remove the partition and let the gases mix. If we symbolize one gas by A, the other by B, and the mixture by A,B, the "equation" for the process is

$$A + B \rightarrow A,B$$

The experiment is imaginary because we assume that A and B behave as perfect gases; this is not a serious assumption as long as the pressure is low. The mixing of two perfect gases involves no energy change, so we may write for the reaction $\Delta H = 0$. Despite the lack of any evolved energy, the reaction takes place readily of its own accord.

Why does it take place? Because the mixture represents a state of greater disorder or randomness, or a state of higher probability, than the two separate pure gases. From ordinary experience we know that natural processes tend to produce disordered arrangements from ordered arrangements; think, for example, how quickly a neat pile of papers is scattered by a gust of wind, how readily the ordered arrangement of particles in a crystal disappears when salt dissolves in water, how effectively the processes of decay destroy the complexly organized structures of a dead animal. The mixing of two gases is a simple example of a general pattern in nature. Quite apart from energy changes, natural processes go spontaneously from states of order to states of disorder, and this tendency toward disorder is the other factor besides enthalpy that makes chemical reactions take place.

To give the tendency toward disorder symbolic expression, let us suppose tentatively that any element or compound possesses a certain "degree of disorder," which we symbolize as D. Then in a chemical reaction the degree of disorder of the products may be greater or less than that of the reactants; in other words, we think of D as increasing or decreasing when the reaction takes place. Since D increases for the mixing of two gases (or for any other spontaneous process that does not involve enthalpy change), we could write

$$A + B \rightarrow A,B \qquad \Delta D = +X$$

where X is some number to which we have not yet assigned units. Note that a spontaneous process is characterized by a *positive* value of ΔD, whereas a spontaneous process that involves change in enthalpy but not much change in disorder is characterized by a *negative* value of ΔH.

Now consider a reaction in which both enthalpy and degree of disorder undergo substantial change. If the change in enthalpy is negative ($\Delta H = -Y$) and the change in degree of disorder is positive ($\Delta D = +X$), both factors tend to make the reaction go spontaneously (the burning of gasoline is a good example). To put this in symbols, we need a combination of ΔH and ΔD that will express a summing of the effects of the two factors. Since we are accustomed to using a negative value of ΔH to indicate an exothermic process, we might think of ΔD as increasing this negative value when the two factors work together:

$$\text{Tendency to react} = \Delta H - \Delta D = -Y - (+X) = -Y - X$$

In a reaction with a negative enthalpy change and also a negative change in degree of disorder, the two factors work against each other. If the enthalpy change is large enough, the reaction can take place despite the decrease in disorder (the reaction of H_2 and O_2 to form H_2O vapor is an example):

$$\text{Tendency to react} = \Delta H - \Delta D = -Y - (-X)$$
$$= -Y + X \qquad \text{where } Y > X$$

For a process with positive enthalpy change (an endothermic reaction, for example the dissolving of salt in water), the reaction may still be spontaneous provided that ΔD is larger than ΔH:

$$\text{Tendency to react} = \Delta H - \Delta D = +Y - X \qquad \text{where } X > Y$$

Thus a combination of ΔH and a negative value of ΔD gives a quantity "tendency to react" which is negative for *any* spontaneous process, whatever the combination of enthalpy change and change in degree of disorder may be.

At first sight this all seems very roundabout. If we need an expression for tendency to react as a combination of two quantities, why not simply add the two quantities and adopt a positive sign for both when they work together to make a reaction go? It would be entirely reasonable to do so, but by historical accident the subject didn't develop this way. The negative signs in the last paragraph were introduced to make the language conform with the way thermodynamic equations are traditionally written. To repeat: we seek a quantity that will express the tendency for a reaction to take place; we choose a negative sign for the quantity when the reaction is spontaneous; we represent the quantity as a combination of enthalpy change, which is negative for a spontaneous process, and change in degree of disorder, which is positive for a spontaneous process; so we write the sum as $\Delta H - \Delta D$ rather than $\Delta H + \Delta D$.

Now to translate the above expressions into the usual language of thermodynamics. The quantity called "tendency to react" is the change in free energy, symbolized ΔG. It is a sort of modified ΔH—a quantity that includes degree of disorder as well as energy change and hence provides an accurate rather than approximate means of predicting whether and how far a reaction will go. Like ΔH, ΔG is a quantity of energy, measured in calories or kilocalories.

The quantity ΔD is harder to translate. How can we attach numbers to a concept as elusive as "degree of disorder"? More specifically, how can this concept be represented as a quantity of energy—as it must be, since both ΔG and ΔH are energy terms? The answer is too long for full discussion here; a partial explanation is given in Appendix XI. All we need do for the present is to note that the measure of degree of disorder in thermodynamics is a quantity called *entropy*, which is not itself an energy term but a quotient of energy divided by absolute temperature. The entropy change in a reaction is symbolized ΔS; like ΔG and ΔH, ΔS is a difference in sums of individual entropies for products and reactants. To convert entropy change into a quantity of energy, we need only multiply by absolute temperature:

$$\Delta D = T \, \Delta S$$

(The quantity ΔD we can henceforth forget; the concept it represents is universally symbolized by the product $T \, \Delta S$, and ΔD was introduced here only as a temporary crutch.)

The equations for "tendency to react" may now be rewritten in their usual form:

$$\Delta G = \Delta H - T \, \Delta S \tag{8-11}$$

The hypothetical quantity G for a single substance may be expressed similarly:

$$G = H - TS \tag{8-12}$$

In formal presentations of the subject this is the *defining equation* for free energy. Starting with this equation, we might have shown that a quantity so defined would have the desirable properties pointed out above; but it seems a little clearer, even if less logical, to begin by showing the need for the concept and then to work backward to the equation.

The quantity ΔS, the change in entropy, has great importance in thermodynamics. Numbers are assigned to it for a given reaction by supposing that the reaction can be set up to run reversibly—very slowly, and in such a way that the reaction can be reversed at any time by a slight change in external conditions. Under these idealized circumstances the entropy change is defined as the heat absorbed in the reaction divided by the absolute temperature:

$$\Delta S = \frac{q}{T}$$

When a reaction runs to some extent irreversibly, as all actual reactions do, the heat absorbed is less, but the entropy change remains the same. Thus the entropy change is a property of the reaction mixture and does not depend on how the reaction is carried out. The magnitude of the entropy change (like the magnitude of ΔD in preceding paragraphs) is a measure of the change in degree of disorder produced by the reaction. We shall not have occasion to use entropy extensively in this book, so we will leave its definition in this incomplete form. Methods of measuring entropies and of using entropies in calculations are discussed in Appendix XI.

The quantity ΔG, on the other hand, will find extensive use in this and subsequent chapters. Since it expresses the ability of substances to react and the extent to which reactions will go, it plays an important role in summarizing and in making predictions about a great variety of geochemical processes.

8-6 FREE ENERGY: CONVENTIONS AND QUALIFICATIONS

As we shall use the term, free energy will refer only to processes taking place at constant temperature and constant pressure. It may be defined equally well for a constant-volume process, and free energies so defined are often useful (see Helmholtz free energy, Appendix XI). One could also refer the definition to adiabatic rather than isothermal processes, although this is not generally done. Most geochemical processes for which the concept of free energy is useful take place under conditions of approximately constant pressure, and free energies given in tables always refer to constant temperature, so that these are appropriate conditions for our investigation.

Like enthalpies, numerical free energies are known only as relative quantities, never as absolute ones. All we ever get from experiment is a *difference* of enthalpy and a *difference* of free energy between the products and the reactants of a chemical process. We are concerned only with ΔG and ΔH, never with G and H themselves.

Free energies, like enthalpies, are expressed in calories or kilocalories. The convention of sign is similar: negative free energy means energy evolved; positive free energy means energy absorbed. Thus the free-energy change for the reaction between hydrogen and oxygen at 25°C and 1 atm is

$$H_2 + \tfrac{1}{2}O_2 \rightleftharpoons H_2O \qquad \Delta G = -56.7 \text{ kcal}$$

and the free-energy change for the separation of water into its elements is

$$H_2O \rightleftharpoons H_2 + \tfrac{1}{2}O_2 \qquad \Delta G = +56.7 \text{ kcal}$$

Again like enthalpies, free-energy changes depend on the coefficients in a reaction; thus the preceding equation could be written

$$2H_2O \rightleftharpoons 2H_2 + O_2 \qquad \Delta G = +113.4 \text{ kcal}$$

8-7 FREE ENERGIES OF FORMATION

The free energy of formation of a compound is the free-energy change accompanying the formation of 1 mole from its elements. Such free energies are important because they can be added and subtracted to determine the free-energy changes of reactions, just as heats of formation can be used to calculate enthalpy changes (Sec. 8-1). For example, the free energies of formation of the four substances involved in the following reaction are written below the formulas:

$$SO_2 + CO_2 \rightleftharpoons SO_3 + CO$$

$$\begin{array}{cccc} -71.7 & -94.3 & -88.7 & -32.8 \end{array}$$

and the free-energy change for the reaction is calculated by adding the numbers for the products and subtracting those for the reactants:

$$\Delta G = -88.7 - 32.8 + 71.7 + 94.3 = +44.5 \text{ kcal}$$

By convention, unless otherwise specified, the free energy of formation of a compound refers to a reaction of its elements carried out at 25°C and 1 atm pressure, and the elements to start with are assumed to be in their most stable forms under these conditions. Thus the number -71.7 kcal is the free energy of formation of 1 mole of SO_2 gas in a reaction carried out at 25° and 1 atm between O_2 gas and orthorhombic sulfur; the reaction does not actually take place at an appreciable rate under these conditions, but the energy can be calculated from determinations made at higher temperatures. The oxygen from which the compound forms is assumed to be O_2 gas, and the sulfur to be orthorhombic (rather than monoclinic), since these are the most stable states of the two elements at the specified conditions. For reactions involving mercury, the element would be taken as the liquid; iron would be the solid (in the crystal form called alpha-iron), bromine would be the liquid, nitrogen the diatomic gas N_2, and so on. Elements in these stable forms are said to be in their *standard states*, and free energies of compounds thus rigidly defined are called *standard free energies of formation*. These energies, symbolized ΔG_f°, are the ones commonly listed in tables (Appendix VIII).

For substances in solution, free energies of formation are referred conventionally to activities of 1 mole/liter. Thus, for H_2S as a pure gas and in aqueous solution, we write

$$S_{rh} + H_2(g) \rightarrow H_2S(g) \qquad \Delta G_f^\circ = -8.0 \text{ kcal}$$

$$S_{rh} + H_2(g) \rightarrow H_2S(aq) \qquad \Delta G_f^\circ = -6.7 \text{ kcal}$$

The symbol S_{rh} means rhombic sulfur, $H_2(g)$ means hydrogen gas at 1 atm, $H_2S(aq)$ means dissolved H_2S at an activity of $1M$, ΔG_f° means standard free energy of formation, and a temperature of 25°C is understood. For ions in solution a further convention is needed, because ions are never produced singly. We can determine, for example, the free-energy change in the reaction $Na + \frac{1}{2}Cl_2 \rightarrow$

$Na^+ + Cl^-$, but not for the separate processes $Na \rightarrow Na^+$ or $\frac{1}{2}Cl_2 \rightarrow Cl^-$. This means that some one dissociation reaction must be arbitrarily assigned a free-energy value, and other dissociation processes can then be referred to it. The convention universally adopted is

$$\frac{1}{2}H_2(g) \rightarrow H^+ + e^- \qquad \Delta G^\circ = 0$$

The symbol H^+ means hydrogen ion at an activity of $1M$ and the symbol e^- refers to the electron removed from the hydrogen atom.

In effect, then, we regard all elements in their standard states, and hydrogen ion at unit activity, as having zero free energies of formation. In an absolute sense this is wrong, but since we deal only with relative free energies, it is a workable convention. The standard free energies of formation in Appendix VIII are based on this rule.

8-8 FREE ENERGY AS A CRITERION OF EQUILIBRIUM

Unlike enthalpies, free-energy changes can be used to determine accurately how far a given reaction mixture is from equilibrium—in other words, whether the substances will react, and how far the reaction will go. The rules are very simple:

If $\Delta G < 0$, the reaction will take place spontaneously (although the rate may be so slow that no reaction is apparent)

If $\Delta G > 0$, the reaction cannot take place unless energy is supplied from an external source

If $\Delta G = 0$, the reaction mixture is at equilibrium

Similar rules cannot hold for enthalpy changes, because some endothermic reactions (for which $\Delta H > 0$) take place spontaneously. Such endothermic reactions would have a negative ΔG, but a positive $T \Delta S$ term [Eq. (8-11)] larger numerically than ΔG, so that ΔH becomes positive. These reactions are not common, and by and large a negative ΔG means a negative ΔH, especially if the ΔG is a large number, say more than 10 kcal. But ΔG has the advantage over ΔH that it gives an exact measure of the position of a reaction with respect to equilibrium, whereas ΔH gives only a rough indication.

Two examples will illustrate the use of free energy as a criterion of equilibrium. The reaction between sulfur and oxygen has a large negative ΔG,

$$S + O_2 \rightleftharpoons SO_2 \qquad \Delta G = -71{,}750 \text{ cal at } 25^\circ C$$

This means that a mixture of sulfur and oxygen is far from equilibrium; once ignited, the mixture will react until one or the other substance is almost entirely converted to SO_2. The reaction between hydrogen and iodine, on the other hand, has a small positive free-energy change:

$$\frac{1}{2}H_2 + \frac{1}{2}I_2 \rightleftharpoons HI \qquad \Delta G = +300 \text{ cal at } 25^\circ C$$

Hence a mixture of hydrogen gas at 1 atm, hydrogen iodide gas at 1 atm, and solid iodine, would be approximately at equilibrium; a little of the hydrogen iodide would tend to break down into its elements, but much would still be present when true equilibrium is established.

8-9 RELATION OF FREE-ENERGY CHANGE TO EQUILIBRIUM CONSTANT

If the magnitude of ΔG serves to measure how far a given mixture is from equilibrium, it obviously must have some simple relation to the equilibrium constant, for the constant is designed to yield the same sort of information. The relation is expressed in another formula whose derivation is given in Appendix XI:

$$\Delta G = -RT \ln K_a + RT \ln \frac{a_Y^y a_Z^z \cdots}{a_B^b a_D^d \cdots} \tag{8-13}$$

where the quotient in the last term refers to a general chemical equation:

$$bB + dD + \cdots \rightleftharpoons yY + zZ + \cdots \tag{8-14}$$

B, D, and so on, are chemical formulas and b, d, and so on, are coefficients; a is activity, and K_a is the equilibrium constant expressed in terms of activities (rather than concentrations). For the special case in which all substances are present at unit activity—$1M$ for substances in solution, 1 atm pressure for gases, and solids and liquids regarded as pure compounds—the equation assumes a simpler form. The quotient is now 1 and its logarithm is zero, so that

$$\Delta G° = -RT \ln K_a \tag{8-15}$$

At 25°C,

$$\Delta G° = -2.303 \times 1.99 \times 298 \times \log K_a$$
$$= -1,364 \log K_a \text{ for } \Delta G° \text{ in cal}$$
$$= -1.364 \log K_a \text{ for } \Delta G° \text{ in kcal}$$

The superscript in $\Delta G°$ indicates that this is the free-energy change for the reaction that would occur when all substances are present at unit activity; it is called the *standard free-energy change* for the reaction.

As an example, let us calculate the standard free-energy change at 25° for the reaction

$$CaSO_4 \rightleftharpoons Ca^{2+} + SO_4^{2-}$$

The equilibrium constant we have worked with before (Sec. 1-5):

$$K = a_{Ca^{2+}} a_{SO_4^{2-}} = 3.4 \times 10^{-5}$$

In the earlier discussion this constant was treated as a product of concentrations; actually the number refers to activities. If we are dealing with a solution of $CaSO_4$ in pure water, the concentrations of Ca^{2+} and SO_4^{2-} are so small that the difference between concentrations and activities is unimportant. The standard free-energy change is

$$\Delta G° = -RT \ln K_a = -1.364 \log 3.4 \times 10^{-5}$$

$$= -1.364(-4.47) = +6.1 \text{ kcal}$$

The positive number means that the reaction tends to go to the left when all substances are at unit activities, i.e., with solid $CaSO_4$ and the two ions Ca^{2+} and SO_4^{2-} each at an activity of $1M$. This, of course, agrees with laboratory experience: in a solution that contains such a high concentration of Ca^{2+} and SO_4^{2-}, solid $CaSO_4$ (in the form of gypsum) will precipitate rapidly until equilibrium is established.

For another example, consider the oxidation of galena to anglesite:

$$PbS + 2O_2 \rightleftharpoons PbSO_4 \qquad \Delta G° = -170.8 \text{ kcal} \qquad (8\text{-}16)$$

$$_{-23.6} \quad _{0} \qquad _{-194.4}$$

Here the $\Delta G°$ is calculated by subtracting the free energy of formation of PbS from that of $PbSO_4$. To find the equilibrium constant,

$$\Delta G° = -RT \ln K = -1.364 \log K = -170.8$$

so

$$\log K = \frac{+170.8}{1.364} = +125$$

and

$$K = 10^{125}$$

Both the high negative $\Delta G°$ and the enormous exponent in K indicate that this reaction goes practically to completion toward the right. For the reaction

$$PbS + SO_4^{2-} \rightleftharpoons PbSO_4 + S^{2-} \qquad \Delta G° = +27.7 \text{ kcal} \qquad (8\text{-}17)$$

$$_{-23.6} \quad _{-178.0} \qquad _{-194.4} \quad _{+20.5}$$

on the other hand, $\log K$ is $-27,700/1,364$, or -20.3, and K is $10^{-20.3}$; here the high positive $\Delta G°$ and the large negative exponent in K both mean that this reaction is displaced far to the left. For the reaction

$$PbS + 2H^+ \rightleftharpoons H_2S + Pb^{2+} \qquad \Delta G° = +11.1 \text{ kcal} \qquad (8\text{-}18)$$

$$_{-23.6} \quad _{0} \qquad _{-6.7} \quad _{-5.8}$$

$\log K$ is $-11.1/1.364$, or -8.1, and K is $10^{-8.1}$. Here the small positive ΔG^0 and the small exponent in K mean that the reaction is displaced toward the left, but that all four substances can exist together in appreciable amounts at equilibrium—as is shown by the experimental observations that lead sulfide is precipitated when H_2S is passed through an acid solution of a lead salt, and that a perceptible amount of H_2S is liberated when a fairly strong acid is dropped on galena. Generalizing from these figures, we can say that a reaction is fairly close to equilibrium as long as $\Delta G°$ is less than about 12 kcal.

8-10 FREE-ENERGY CHANGES FOR ACTIVITIES OTHER THAN UNITY

To calculate ΔG for reaction mixtures with any arbitrary activities, we go back to Eq. (8-13). A convenient example is the last reaction considered in the preceding paragraph, for which Eq. (8-13) would be:

$$\Delta G = -RT \ln K + RT \ln \frac{a_{H_2S} a_{Pb^{2+}}}{a_{H^+}^2}$$

$$= \Delta G° + 1{,}364 \log \frac{a_{H_2S} a_{Pb^{2+}}}{a_{H^+}^2}$$

We try first a calculation for $a_{H^+} = 10^{-3}M$, leaving a_{H_2S} and $a_{Pb^{2+}}$ at $1M$:

$$\Delta G = 11{,}100 + 1{,}364 \log \frac{1 \times 1}{10^{-6}} = 11{,}100 + 6 \times 1{,}364 = +19{,}300 \text{ cal}$$

In other words, decreasing the acidity makes the reaction go farther to the left, as we would predict, of course, from Le Chatelier's rule. We try next the effect of making both a_{H_2S} and $a_{Pb^{2+}}$ $10^{-4.5}M$, while a_{H^+} is left at $1M$:

$$\Delta G = 11{,}100 + 1{,}364 \log 10^{-9} = 11{,}100 - 9 \times 1{,}364 = -1200 \text{ cal}$$

Hence by making the concentration of the products fairly small, we give the reaction a negative free energy, so that it tends to go to the right, as we discover by the odor of H_2S when we put a drop of HCl on galena.

8-11 FREE-ENERGY CHANGES AT OTHER TEMPERATURES

Given the standard free-energy change for a reaction at 25°, how can we find the change at 100°? At first glance it seems that Eq. (8-15) should give this directly:

$$\Delta G = -4.58T \log K$$

The difficulty is that here ΔG is expressed as a function of log K as well as T, and log K changes just as rapidly as ΔG does when the temperature rises. This equation, therefore, *cannot be used directly* in calculating the change of ΔG with temperature.

If, however, we can find an expression for log K at various temperatures, the equation would enable us to convert this into a numerical value for ΔG. Such an expression we have worked with previously: Eq. (8-3) if ΔH can be considered constant, or Eq. (8-10) if ΔH varies with temperature. From Eq. (8-10), the most general expression for ΔG as a function of temperature would be

$$\frac{\Delta G}{T} = -4.58 \log K = -\int \frac{\Delta H}{T^2} \, dT$$

or

$$\Delta G = A - \Delta aT \ln T - \tfrac{1}{2} \Delta b T^2 - \cdots + CT \qquad (8\text{-}19)$$

For many reactions the T^2 term is negligibly small, so that the equation reduces to a simpler form:

$$\Delta G = A + BT \log T + CT \qquad (8\text{-}20)$$

in which B stands for $-2.303 \Delta a$. In tables (see references at the end of this chapter), ΔG's are often recorded as values of A, B, and C to be used with this equation.

An alternative method (actually equivalent to the preceding for the case of constant ΔH) is to use Eq. (8-11). From entropies recorded in tables (for example, Appendix VIII), the value of ΔS for a reaction can be calculated, and the difference between ΔH and $T \Delta S$ then gives the free energy for any value of T. This is a simple and rapid method for making rough estimates, but it is not very accurate, especially for large temperature ranges, because ΔH and ΔS may themselves show considerable variation with temperature.

As an example of such calculations, consider the reaction

$$Cu_2O(s) \rightleftharpoons 2Cu(s) + \tfrac{1}{2}O_2(g)$$

For the standard free energy of reaction, Kubaschewski, Evans, and Alcock (1967) give

$$A = 40{,}500 \qquad B = 3.92 \qquad C = -29.5$$

for the temperature range 298 to 1356K. Substitution in Eq. (8-20) gives

$$\Delta G = 40{,}500 + 3.92T \log T - 29.5T$$

Hence, at 25°C:

$$\Delta G = 40{,}500 + 3.92 \times 298 \times 2.474 - 29.5 \times 298$$

$$= 34{,}600 \text{ cal}$$

and at 400°C:

$$\Delta G = 40{,}500 + 3.92 \times 673 \times 2.828 - 29.5 \times 673$$

$$= 28{,}100 \text{ cal}$$

Alternatively, to make the calculation with Eq. (8-11), we find values of ΔH and S for the three substances from Appendix VIII:

	Cu_2O	$\rightleftharpoons$	$2Cu$	$+$	$\tfrac{1}{2}O_2$	
ΔH:	-40.3		0		0	$\Delta H_{reac} = +40.3$ kcal
S:	22.3		15.8		24.5	$\Delta S_{reac} = +18.0$ cal

Hence

$$\Delta G = \Delta H - T \Delta S = 40.3 - 0.018T$$

At 25°C:

$$\Delta G = 40.3 - 0.018 \times 298 = 34.9 \, \text{kcal}$$

and at 400°C:

$$\Delta G = 40.3 - 0.018 \times 673 = 28.2 \, \text{kcal}$$

In this case the simpler calculation neglecting the change of ΔH and ΔS with temperature gives a value for ΔG only 100 cal (28,200 − 28,100) different from the more accurate figure calculated from Eq. (8-20).

One caution is needed about the use of the equation $\Delta G = \Delta H - T \, \Delta S$: the equation can be used only for *reactions*, never for individual substances. This is because the values for entropy given in Appendix VIII (or similar tables), unlike the values for free energy and enthalpy, are *absolute* values (Appendix XI, p. 585). For example, one cannot find ΔG for the decomposition of Cu_2O at 400°C by simply looking up values for $\Delta H°$ (-40.0 kcal) and $S°$ (22.4 cal/degree) and putting them in the equation; one must write the complete reaction and find $\Delta S°$ by subtracting $S°$ for the compound from the $S°$ values for the elements, as was done in the calculation in the last paragraph. In other words, $\Delta G°$ and $\Delta H°$ of formation for a compound like Cu_2O are *relative* values, referred to assumed values of zero for the elements in their standard states, but the entropies for both compound and elements are *absolute* values.

8-12 FREE-ENERGY CHANGES AT OTHER PRESSURES

The change in the free energy of a reaction with pressure is obtained from a simple relation,

$$\frac{d \, \Delta G}{dP} = \Delta V \tag{8-21}$$

where ΔV is the overall change in volume during the reaction. Qualitatively, if the volume decreases (ΔV negative), this equation means that the free-energy change must become smaller (less positive, or more negative) as the pressure increases. In other words the reaction is displaced to the right, in the direction of decreasing volume, as Le Chatelier's rule requires. To use the equation quantitatively, we should have to find some way to express ΔV as a function of P and integrate.

8-13 USES AND LIMITATIONS OF FREE-ENERGY CALCULATIONS

In effect, by introducing the concept of free-energy change we have developed a second method of describing the tendency of a mixture to react. The first device was the equilibrium constant, from which we could deduce the concentration ratios that exist at equilibrium, and thus determine in what direction a given mixture must react to attain equilibrium. The standard free-energy change gives us the same sort of information, for it shows, by its sign and its magnitude, whether a mixture with unit concentrations will react one way or the other and to what extent. What have we gained by introducing this second method of description?

Free energies have two great advantages: they are additive, and they require less space in tables. To combine free energies for part reactions, or for reactions that follow one another, we need only add or subtract, while equilibrium constants must be multiplied or divided. In tables a separate equilibrium constant must be listed for each reaction, whereas free energies of formation in a relatively brief list can be used to compute free-energy changes (and hence equilibrium constants) for a large number of reactions.

Free energies give us a quantitative measure of *stability*, a concept we have discussed at some length earlier (Sec. 1-7). A mixture of substances all of whose possible reactions have zero free energy could undergo no change, hence would be stable; a mixture permitting reactions with negative free-energy changes would be unstable, and the magnitude of the possible free-energy changes would indicate the degree of instability. As in the earlier discussion, it should be emphasized that this kind of stability refers only to *possible* energy changes, not to the rate at which these energy changes actually occur. A piece of coal exposed to the air is apparently stable, but the stability is due only to the slowness with which coal reacts with oxygen at ordinary temperatures. A piece of quartz, on the other hand, is stable in contact with air for a different reason, because no conceivable reaction between quartz and the constituents of air would give a negative free-energy change. In this book "stability" or "true stability" will always refer to stability with respect to energy changes, and the apparent stability of some mixtures due to slow reaction rates will be called metastability.

The inability to give any information on reaction rates is one of the severe limitations on the application of free-energy reasoning in geochemistry. A second limitation is imposed simply by lack of data: free energies are just not available for many of the silicates with which a geologist must work, and a calculation of free energies at high temperatures is severely handicapped by scarcity of heat-capacity measurements on substances of geologic interest. A third very serious limitation is the fact that many geologic processes take place in "open" systems, systems in which matter is continually being added from the outside or is continually flowing away, so that real chemical equilibrium is not attained; consider the movement of fluid through a vein, for example, or the rush of gases through a volcanic orifice. Even when flow is not an important factor, the slowness of reactions and incompleteness of mixing may prevent attainment of equilibrium, as is obvious for immature soils and for half-digested inclusions in granites. Despite these restrictions, we shall find that reasoning on the basis of free energy and equilibrium will help in setting limits to geologic processes, in telling us at least which reactions are possible and which are completely out of the question. In restricted areas of geology, particularly in chemical sedimentation and in the crystallization of magmas, equilibrium reasoning can be carried farther and leads to exact predictions which can be checked against natural occurrences.

8-14 CHEMICAL POTENTIAL, FUGACITY, AND ACTIVITY

The idea embodied in free energy as a measure of equilibrium may be expressed in a variety of ways. Some of the nomenclature commonly used for particular situations is discussed in this section.

Escaping Tendency

Consider a very simple kind of equilibrium, the equilibrium between liquid water and water vapor in a closed container. The condition that equilibrium exists we can state by saying that, if a very small amount of liquid is vaporized or if a small amount of vapor is condensed, the change in free energy is zero. This is a useful formulation of the requirement for equilibrium, but for various purposes other kinds of statement are often more convenient.

In qualitative language, one may think of the liquid water as capable of vaporizing, or as tending to vaporize, or as seeking to "escape" from the liquid into the vapor phase. Thus we can assign to the water a certain *escaping tendency* with respect to the vapor. Similarly we can describe the vapor as having a certain escaping tendency with respect to the liquid. The condition of equilibrium would then be simply that the two escaping tendencies are equal. If the temperature is raised, the escaping tendency of the liquid increases more than that of the vapor, so that additional liquid must vaporize before equilibrium can be reestablished.

As a measure of the escaping tendency, we may use the free energy per mole, or *molal free energy*, of the water and its vapor. We designate molal free energy with a bar under the letter symbol:

$$\underline{G} = \frac{G}{n}$$

where G represents the free energy of n moles. The condition of equilibrium is

$$\underline{G}_{\text{liquid}} = \underline{G}_{\text{vapor}} \tag{8-22}$$

Alternatively, we may express this condition by thinking of the vaporization of an infinitesimal amount of water—the escape of dn moles of water from the liquid phase into the gas phase. The free energy lost by the liquid is $\underline{G}_{\text{liq}}\, dn$, and that gained by the vapor is $\underline{G}_{\text{vap}}\, dn$. If equilibrium exists, the net free-energy change must be zero:

$$dG = \underline{G}_{\text{vap}}\, dn - \underline{G}_{\text{liq}}\, dn = 0 \tag{8-23}$$

This is clearly equivalent to Eq. (8-22). Molal free energies so defined are not a very practical measure of escaping tendency, since numerical values of absolute free energies are unknown. For actual use we would substitute free energies of formation assigned on the basis of conventions described in Sec. 8-6.

Chemical Potential

The molal free energy may be given another name, *chemical potential*, suggesting that this quantity represents a sort of energy level of a substance in one phase of a system. An actual transfer of energy attendant on vaporization or condensation is expressed by the chemical potential multiplied by the number of moles transferred—in other words, by the total free-energy change. Analogously in mechanics an object suspended above the earth can be said to have a mechanical potential, measured by its height times the acceleration of gravity, gh; energy released when the object falls is then the product of this potential multiplied by its mass, mgh. Or in electricity: an electric charge has a certain electric potential (measured in volts) with respect to another charged object, and electrical energy released when the charge moves is the product of electric potential and the amount of charge. Chemical potential is less easily visualized, but it plays the same role in chemical reactions that these more familiar potentials play in mechanical and electrical processes. We may rephrase our condition of chemical equilibrium by saying that the chemical potentials of water in the liquid and vapor states must be equal.

Partial Molal Free Energy

So far this discussion is little more than a play on words, because we have been in essence saying the same thing over and over again in slightly different ways. The idea of chemical potential becomes more useful when we consider more complicated systems involving solutions. Suppose we change our simple example by stirring some salt into the liquid water: the equilibrium will readjust itself, and we inquire how the necessary condition for equilibrium between liquid and vapor can best be expressed. If we try to use equality of the molal free energies of water in the two phases, we meet the difficulty that molal free energy of water in a solution is not yet defined. It cannot be simply the free energy of the water present divided by the number of moles, because dissolving the salt involves an energy change; in other words, the free energy of the water is now a complex function of concentration as well as of temperature and pressure. We can still set up a number to represent the molal free energy, however, if we consider the change in free energy of the solution produced by an infinitesimal change in the amount of water. Instead of a quotient of macroscopic quantities, G/n, which we can use for a pure substance, we set up a differential coefficient:

$$\lim_{\Delta n_1 \to 0} \left(\frac{\Delta G_1}{\Delta n_1}\right) = \left(\frac{\partial G_1}{\partial n_1}\right)_{P,\,T,\,n_2,\,n_3\,\dots} = \bar{G}_1 \tag{8-24}$$

The subscripts mean that the differentiation is to be carried out with pressure, temperature, and concentrations of all other components of the solution held constant. In mathematical language this is a partial derivative, and the quantity it represents is therefore called the *partial molal free energy*. The term chemical potential is a synonym for this expression, as well as for the molal free energy of pure substances; it is often designated by the Greek letter mu, μ. A general statement of the condition of equilibrium can now be formulated for any system, however complex: *the chemical potential of each component must be the same in all phases of the system.*

Fugacity

Mathematically the partial molal free energy has the drawback that it approaches an infinite negative value as the concentration becomes indefinitely small. To circumvent this difficulty, it is often convenient to use still another measure of escaping tendency, called *fugacity*, defined by the equation

$$\left(\frac{\partial \ln f_1}{\partial \bar{G}_1}\right) = \frac{1}{RT} \tag{8-25}$$

where f_1 is fugacity and $\bar{G}_1$ is partial molal free energy. Alternatively this may be expressed in the integrated form

$$\bar{G}_1 - \bar{G}_1' = RT \ln \frac{f_1}{f_1'} \tag{8-26}$$

where $\bar{G}_1$ and $\bar{G}_1'$ are the partial molal free energies, and f_1 and f_1' the fugacities, of one substance at two different concentrations (or, for gases, at two different pressures). The fugacity, in effect, is a sort of idealized vapor pressure, and it is equal to the vapor pressure when the vapor behaves as a perfect gas.

The relation between fugacity and vapor pressure can be made clear by an argument based on the equation for the change of free energy with pressure [Eq. (8-21)]. A more general expression of this change is

$$\left(\frac{\partial G}{\partial P}\right)_T = V \tag{8-27}$$

which describes the change for a single substance rather than a reaction. For the case of 1 mole of a perfect gas, RT/P may be substituted for V:

$$\left(\frac{\partial \bar{G}}{\partial P}\right)_T = \frac{RT}{P} \tag{8-28}$$

At constant temperature this may be treated as an ordinary derivative; if dP is moved to the right side of the equation, integration gives

$$\int d\bar{G} = \int \frac{RT}{P}\, dP$$

$$\bar{G} - \bar{G}' = RT \ln \frac{P}{P'} \tag{8-29}$$

This has the same form as Eq. (8-26), indicating that vapor pressure may be used as an approximation for fugacity to the same extent that a perfect gas approximates the behavior of a real gas.

The criterion for equilibrium can be stated in terms of fugacities just as well as with the other measures of escaping tendency: *a state of equilibrium exists when for each component the fugacity (or the partial molal free energy, or the chemical potential) is the same in all phases of a system.*

Activity and Activity Coefficient

In very dilute solutions the fugacity of a volatile solute becomes equal to its vapor pressure, and this in turn is proportional to its concentration in the solution (Henry's law). To express the deviation of more concentrated solutions from Henry's law, it is convenient to use *relative fugacities*, fugacities referred to the fugacity in some assumed standard state:

$$\bar{G} - \bar{G}^\circ = RT \ln \frac{f}{f^\circ}$$

This equation is identical with Eq. (8-26), if $\bar{G}'_1$ and f'_1 are defined as the values of these quantities in a standard state. The standard state may be selected in different ways for different kinds of problems. For solutes a convenient standard state is a hypothetical solution of unit concentration that obeys Henry's law. (Practically all real solutions deviate considerably from Henry's law at concentrations near $1M$.) Then the ratio f/f° is called the *activity* of the solute. Evidently the activity will be closely related to the concentration; it is, in fact, a sort of idealized concentration, the concentration that would be expected if the solution obeyed Henry's law at all concentrations. The ratio of activity to concentration,

$$\frac{a}{c} = \gamma$$

expresses the extent to which the solution deviates from Henry's law; it is given the name *activity coefficient*. Although the concepts of activity and activity coefficient have been described here with relation to a volatile solute, they may be generalized to apply to nonvolatile solutes and to any kind of solution. We have met the terms before in a discussion of solubility equilibria (Sec. 3-5).

The best units to use for activities are not agreed upon by all writers. If the activity of a solute in a saturated solution is defined as in Sec. 3-5, as the equilibrium concentration extrapolated to zero ionic strength, the appropriate unit would be the same as the unit used to express concentration. With the more general definition given above, however, activity becomes the ratio of two fugacities, hence should be a dimensionless quantity. There is no discrepancy in the definitions; they can easily be reconciled by adjusting the definitions of standard fugacity or of activity coefficient. The way these related quantities are defined is largely a matter of choice, and practically it makes little difference, since numerical values of activity are the same with either usage. In this book activities will be expressed in the familiar units of concentration, most often as moles per liter. From a theoretical standpoint this is the poorer choice, but in geological problems it can seldom lead to difficulties, and moles per liter seem easier to visualize than abstract numbers.

Summary The various concepts we have introduced in this section are treated at length in textbooks of physical chemistry and thermodynamics, and numerical values are recorded in tables (see references at the end of this chapter). For some problems in geochemistry the numerical values are useful, especially values for activity coefficients, but for most discussions in this book the qualitative meanings will be sufficient. The important considerations are, in summary: the idea of chemical equilibrium may be visualized in terms of escaping tendencies, the tendency of each constituent to escape from one phase into another; escaping tendencies may be expressed quantitatively by chemical potentials (molal free energies for pure substances, partial molal free energies for constituents of solutions) or by fugacities; and the deviation of solutions from ideal behavior may be expressed by activity coefficients.

PROBLEMS

1 Using the data of Appendix VIII, find the standard enthalpy changes and free-energy changes for the following reactions at 25°C:

$$Sn + O_2 \rightleftharpoons SnO_2$$

$$KCl(s) \rightleftharpoons K^+ + Cl^-$$

$$2FeCO_3 + \tfrac{1}{2}O_2 + 2H_2O \rightleftharpoons Fe_2O_3 + 2H_2CO_3(aq)$$

In order to reach equilibrium, which of these reactions would go as written, i.e., from left to right? Which is more stable at ordinary temperatures in contact with air, tin or tin oxide? Hematite or siderite?

2 For the "water-gas reaction,"

$$H_2O(g) + C \rightleftharpoons H_2 + CO$$

$\Delta H°$ at 25°C is $+31.4$ kcal and the equilibrium constant K is $10^{-16.0}$. Calculate K for 100°C and for 200°C on the assumption that ΔH is constant. What is the ratio $[H_2]/[H_2O]$ in equilibrium with 0.1 atm of CO at these two temperatures?

3 From the data of Appendix VIII, calculate the equilibrium constant at 25° and at 250°C for the reaction

$$N_2 + 3H_2 \rightleftharpoons 2NH_3$$

In volcanic gas escaping from a fumarole at 600°C, would you expect the nitrogen to exist largely in the form of ammonia or largely as the free element?

4 From the data of Appendix VIII, calculate the solubility product at 25° for the reaction

$$PbS \rightleftharpoons Pb^{2+} + S^{2-}$$

What would be the solubility of PbS in pure water if this were the only reaction taking place? The measured solubility is 3×10^{-4} g/liter. Explain why the two values differ so markedly.

5 For the reaction $\frac{1}{2}H_2 + \frac{1}{2}Cl_2 \rightleftharpoons HCl$, the free-energy change as a function of temperature is given by the equation

$$\Delta G(cal) = -21,770 + 0.99T \log T - 5.22T$$

Calculate the free-energy change and the equilibrium constant at 100°C and at 1000°C, and find how much H_2 and Cl_2 would be in equilibrium with 1 atm of HCl at these temperatures.

6 In theories of formation of sulfide veins, a critical factor is the amount of free S^{2-} present in the vein solutions. Would the concentration of this ion resulting from dissociation of H_2S be greater or less at 100° than at 25°? Calculate the change in the two dissociation constants of H_2S between these temperatures.

7 It is often suggested that solutions which transport and deposit metallic sulfides are alkaline. At ordinary temperatures would galena dissolve appreciably in an alkaline solution to form $Pb(OH)_3^-$? (In this and similar questions, assume that "appreciable" solution means a concentration of at least $10^{-5}M$, which for most common metals is roughly equal to 1 ppm.)

8 In Fig. 7-7, page 156, lines are plotted representing equilibrium conditions for five reactions [Eqs. (7-3) to (7-7)]. To locate the lines, equilibrium constants were necessary for the reactions, but the constants could not be determined by direct experiment. For any two of the five equations, show how the equilibrium constants can be determined from the free-energy data in Appendix VIII.

REFERENCES AND SUGGESTIONS FOR FURTHER READING

Lewis, G. N., and M. Randall, "Thermodynamics," 2d ed., revised by K. S. Pitzer and L. Brewer, McGraw-Hill Book Company, New York, 1961. A standard text, giving the philosophical background of thermodynamics, derivation of formulas, and many applications.

MacWood, G. E., and F. H. Verhoek, How can you tell whether a reaction will occur? *Jour. Chem. Educ.*, vol. 38, pp. 334–337, 1961. An excellent elementary discussion of the meaning of the common thermodynamic functions; the explanation of the difficult concept of entropy is especially good.

Tables of free energies, enthalpies, and entropies:

Garrels, R. M., and C. L. Christ, "Solutions, Minerals, and Equilibria," Harper & Row, Publishers, Incorporated, New York, 1965.

Kubaschewski, O., E. L. Evans, and C. B. Alcock, "Metallurgical Thermochemistry," 4th ed., Pergamon Press, New York, 1967.

Robie, R. A., and D. R. Waldbaum, Thermodynamic properties of minerals and related substances at 298.15°K and one atmosphere pressure and at higher temperatures, *U.S. Geological Survey Bull.* 1259, 1968. (See note at end of Appendix VIII, page 565.)

Wagmann, D. D., W. H. Evans, V. B. Parker, I. Halow, S. M. Bailey, and R. H. Schumm, Selected values of chemical thermodynamic properties. U.S. National Bureau of Standards Technical Notes 270-3 and 270-4, 1968 and 1969.

OXIDATION POTENTIALS
AND Eh-pH DIAGRAMS

In addition to equilibrium constants and free energy, a third device is available for describing the tendency of one substance to react with another. This device, called the oxidation potential or oxidation-reduction potential, is usable only for reactions involving oxidation and reduction processes. For these reactions it is often the most convenient method of getting the desired quantitative information.

9-1 OXIDATION POTENTIALS

Oxidation and Reduction

Oxidation means an increase in oxidation number and reduction means a decrease. For example, when zinc displaces copper from a solution of copper sulfate,

$$Zn + Cu^{2+} \rightarrow Zn^{2+} + Cu \tag{9-1}$$

the zinc is oxidized (change in oxidation number from 0 to $+2$) and the copper is reduced ($+2$ to 0). When chlorine displaces bromine from a solution of sodium bromide,

$$Cl_2 + 2Br^- \rightarrow 2Cl^- + Br_2 \tag{9-2}$$

chlorine is reduced (0 to -1) and bromine is oxidized (-1 to 0). And when gold is dissolved by the action of MnO_2 in hydrochloric acid solution,

$$3MnO_2 + 2Au + 12H^+ + 8Cl^- \rightarrow 3Mn^{2+} + 2AuCl_4^- + 6H_2O \tag{9-3}$$

manganese is reduced ($+4$ to $+2$) and gold is oxidized (0 to $+3$). In the language of electrons, oxidation may alternatively be described as a loss of electrons and reduction as a gain; in the zinc-copper reaction, for example, each Zn atom loses two electrons to a copper ion. Note that any reaction of this sort must involve both an oxidation and a reduction, and that the total changes in oxidation number must balance.

Electrode Reactions

Any oxidation-reduction reaction, theoretically at least, can be set up so that the transfer of electrons from one element to another will take place along a wire. For the zinc-copper reaction the arrangement is very simple: pieces of the two metals are connected by a wire and submerged in copper sulfate solution. The piece of zinc slowly dissolves, fresh copper from the solution plates out on the copper metal, and a current flows through the wire. The process occurring at the zinc electrode may be symbolized

$$Zn \rightarrow Zn^{2+} + 2e^{-} \quad \text{oxidized} \quad \text{(loses electrons)} \tag{9-4}$$

where e^{-} indicates an electron. The liberated electrons move along the wire to the copper electrode, where they are used in the reaction

$$Cu^{2+} + 2e^{-} \rightarrow Cu \quad \text{reduced} \quad \text{(loses/gains electrons)} \tag{9-5}$$

Reactions of this kind, showing the processes that occur as electrons are produced or consumed at an electrode, are called *half-reactions*, or *electrode reactions*. Addition of two half-reactions gives the complete oxidation-reduction reaction; in this example Eq. (9-5) added to Eq. (9-4) gives Eq. (9-1). For a more complicated illustration, the electrode reactions corresponding to Eq. (9-3) are:

$$3MnO_2 + 12H^+ + 6e^- \rightarrow 3Mn^{2+} + 6H_2O$$

$$2Au + 8Cl^- \rightarrow 2AuCl_4^- + 6e^-$$

The potential difference between the electrodes of our zinc-copper cell can be measured if we add a galvanometer to the circuit. The amount of the potential difference depends on a great many variables, but we arrange to keep most of these constant. Thus we fix the concentration of both Cu^{2+} and Zn^{2+} at $1M$; we make sure that the metal of the electrodes is pure and has a clean surface; we hold the temperature at 25°C and the pressure at 1 atm; and we arrange to have as small a flow of current as possible. Under such conditions the measurement of potential difference is reproducible and may be compared with potential differences measured similarly for other oxidation-reduction reactions. (We pass over the technical difficulties in such measurements, which for some reactions are very troublesome.)

If we set up a number of cells similar to the zinc-copper cell, with various metals in contact with solutions of metal ions, we find that the metals can be arranged in a series according to their ability to displace one another from solution and according to the size of the potential difference produced by different pairs. Thus zinc displaces copper, copper displaces silver, and silver displaces gold; and the potential difference of a zinc-silver or zinc-gold cell is greater than the potential difference of a zinc-copper cell. On the other hand, the reverse reactions do not take place appreciably: silver placed in copper sulfate solution or in zinc sulfate solution causes no detectable reaction. Experiments of this sort give us the familiar electromotive series of metals, according to which we express the chemical activities of various metals with respect to one another.

Standard Potentials

In order to make the electromotive series quantitative, it is convenient to assign a potential difference to each half-reaction. This can be done by choosing some one half-reaction as a standard, giving it an arbitrary potential of zero, and then measuring other half-reactions against it. A convenient choice is the hydrogen couple,

$$\tfrac{1}{2}H_2 \rightarrow H^+ + e^- \qquad E° = 0.00 \text{ volt}$$

If we arrange a cell with zinc as one electrode and hydrogen as the other (by letting hydrogen at 1 atm pressure bubble over a platinum rod), and use a solution containing $1M$ H^+ and $1M$ Zn^{2+}, we obtain the potential difference for the overall reaction

$$Zn + 2H^+ \rightarrow Zn^{2+} + H_2 \qquad E° = 0.76 \text{ volt}$$

and we use this number as the potential for the zinc electrode reaction

$$Zn \rightarrow Zn^{2+} + 2e^- \qquad E° = 0.76 \text{ volt}$$

There is no way of measuring potentials for half-reactions independently. We get them only as differences between pairs of half-reactions, so that the actual numbers are no more than relative voltages compared with the hydrogen electrode.

Electrode potentials and potential differences for complete reactions are both designated by the symbol E. The symbol $E°$ refers to *standard* potentials or potential differences, for reactions that take place at 25°C with all substances at unit activities, i.e., with gases at 1 atm pressure and dissolved substances at 1 mole/liter.

Standard electrode potentials for reactions of geologic interest are tabulated in Appendix IX. Several details about the arrangement in this table should be noted. Each half-reaction is given with the reduced form of the element on the left and the oxidized form on the right. Strong reducing agents appear toward the top of the table, strong oxidizing agents near the bottom. Some reactions follow different courses in acid and basic solutions, because certain precipitates and complex ions are stable in one kind of solution but not in the other; hence it is necessary to include an auxiliary table for those reactions that occur only in an alkaline environment. The $+$ and $-$ signs given to the voltages are purely arbitrary, and unfortunately are not uniform from one reference to another. In this book we follow the usual practice in geochemical literature, making voltages more reducing than the hydrogen electrode negative and those more oxidizing than the hydrogen electrode positive.

Use of the Table of Oxidation Potentials

Qualitatively, the table tells us at a glance what reactions are possible and what are not. The reduced form of any couple will react with the oxidized form of any couple *below* it, but not with the oxidized form of a couple above it. Thus Pb will reduce Ag^+ but not Al^{3+}. Two reactions close together in the table will reach equilibrium with all substances present in appreciable amounts; thus metallic lead and tin are stable in contact with a solution containing fairly large amounts of both Pb^{2+} and Sn^{2+}. Technically, of course, all possible reactions indicated by the table are equilibrium processes. When we say " Pb will not reduce Al^{3+}," we mean more precisely that Pb will react until a very small amount of Pb^{2+} is formed, but that the ratio $(Al^{3+})/(Pb^{2+})$ is very large at equilibrium.

To use the table quantitatively to find the potential difference for a particular reaction, we need only subtract one half-reaction from another and subtract the corresponding voltages. Each reaction must be multiplied by a coefficient that will make the electron changes the same, for no free electrons can appear in the overall reaction. *The voltages, however, are not multiplied by the coefficients;* unlike free energies and enthalpies, the voltages are measurements of potential only and do not change with amount of substance present. For example, to find the potential for the oxidation of Fe^{2+} by MnO_2 in acid solution, we look up the two electrode reactions

$$Fe^{2+} \to Fe^{3+} + e^- \qquad E° = +0.77 \text{ volt}$$

$$Mn^{2+} + 2H_2O \to MnO_2 + 4H^+ + 2e^- \qquad E° = +1.23 \text{ volts}$$

In order to make the electron changes balance, we multiply the iron half-reaction by 2 and then subtract the manganese half-reaction. The standard potential for the overall reaction is found by subtracting the half-reaction potentials without multiplication:

$$MnO_2 + 4H^+ + 2Fe^{2+} \rightleftharpoons Mn^{2+} + 2H_2O + 2Fe^{3+} \qquad E° = -0.46 \text{ volt}$$

$$(9\text{-}6)$$

The convention regarding $+$ and $-$ signs means that a reaction which takes place spontaneously must have a negative voltage, and one that requires outside energy has a positive voltage.

9-2 RELATION OF OXIDATION POTENTIAL TO FREE ENERGY

Since the potential difference for a reaction computed from the data of Appendix IX is a measure of how far the reaction mixture is from equilibrium, it must clearly be related to the free-energy change. The relationship is a very simple one,

$$\Delta G = nfE \tag{9-7}$$

where n is the number of electrons that the equation shows shifting from one kind of atom to another and f is the Faraday constant. This constant is a number which, when multiplied by voltage, gives energy; the most common expression for it is 96,500 coulombs, the corresponding energies being expressed in volt-coulombs, or joules. Here we need energies expressed in calories, and the appropriate value of f is 23,061 cal/volt (or 23.1 kcal/volt). For example, the free-energy change in the manganese-iron reaction just discussed is

$$\Delta G° = nf E° = 2 \times 23.1(-0.46) = -21.2 \text{ kcal}$$

This checks well with the figure -21.1 kcal, which is obtained by adding up free energies of formation in the usual way. Note that a potential difference of half a volt corresponds to a fairly large free-energy change.

It should be emphasized again that free energies depend on how the equation is written—i.e., on whether the coefficients are doubled or tripled or halved—while the potential does not. The difference is taken care of in Eq. (9-7) by the factor n, which, of course, changes if the coefficients change.

So far we have talked exclusively about standard potentials, which are potentials for unit activities and a temperature of 25°C. To find the potential difference for a reaction under different conditions, we make use of the relations developed earlier for free energies. By combining Eqs. (9-7) and (8-13), we obtain

$$E = \frac{\Delta G}{nf} = \frac{\Delta G°}{nf} + \frac{RT}{nf} \ln \frac{a_Y^y a_Z^z \cdots}{a_B^b a_D^d \cdots}$$

$$= E° + \frac{2.303RT}{nf} \log \frac{a_Y^y a_Z^z \cdots}{a_B^b a_D^d \cdots} \tag{9-8}$$

an expression called the Nernst equation. For reactions at 25° the combination of constants before the logarithm is $0.059/n$, so that

$$E = E° + \frac{0.059}{n} \log \frac{a_Y^y a_Z^z \cdots}{a_B^b a_D^d \cdots} \tag{9-9}$$

For example, to find the potential for the manganese-iron reaction [Eq. (9-6)] in a solution with pH 3 and unit concentrations of other ions:

$$E = -0.46 + \frac{0.059}{2} \log \frac{a_{Mn^{2+}} a_{Fe^{3+}}^2}{a_{H^+}^4 a_{Fe^{2+}}^2}$$

$$= -0.46 + 0.03 \log \frac{1}{(10^{-3})^4}$$

$$= -0.46 + 12 \times 0.03 = -0.10 \text{ volt}$$

The smaller negative value for E means that the reaction would have less tendency to take place, as one would expect when the concentration of one of the reactants is reduced.

The relation of the oxidation potential to the equilibrium constant for a reaction may be formulated by combining Eqs. (9-7) and (8-15):

$$E^\circ = \frac{\Delta G^\circ}{nf} = -\frac{2.303RT \log K}{nf} = -\frac{0.059}{n} \log K \qquad (9\text{-}10)$$

An equation for calculating oxidation potentials at temperatures other than 25°C may be set up by combining Eqs. (9-7) and (8-11):

$$E = \frac{\Delta H - T \, \Delta S}{nf} \qquad (9\text{-}11)$$

The equation is usable only for temperature ranges in which ΔH and ΔS may be considered constant. To take account of variable enthalpy change, Eq. (9-7) may be combined with Eq. (8-19), but the resulting equation has limited use because few data on heat capacities of ions are available.

What advantages can be claimed for oxidation potentials as opposed to free energies in handling problems of equilibrium? Their chief merit is convenience: from a table of potentials like Appendix IX one can see at a glance, with no calculation except a mental note of how far apart two half-reactions are in the table, whether a given oxidation-reduction process can be expected to take place and approximately how far the reaction will go. If a more quantitative estimate is needed, the calculation involves only two figures rather than several. These advantages are offset by the facts that the half-reaction table is more cumbersome than a table of free energies, and that use of electrode potentials is limited to oxidation-reduction reactions.

9-3 REDOX POTENTIALS

One further advantage of oxidation potentials is their usefulness in treating problems which concern not specific reactions but the general oxidizing or reducing characteristics of a geologic environment. It is common knowledge, for example, that dissolved sulfur is largely in the form of SO_4^{2-} in water of the open sea, where conditions are oxidizing, but chiefly in the form of H_2S in the stagnant bottom waters of enclosed basins. Oxidation potentials make it possible to refine such qualitative statements and to estimate semiquantitatively just what concentrations of the various ions and compounds of sulfur can exist in these environments or in environments of intermediate characteristics.

The ability of a natural environment to oxidize sulfur, or bring about any other oxidation or reduction process, is measured by a quantity called its *redox potential*. Experimentally this is determined by immersing an inert electrode, usually platinum, in the environment—say in a sample of seawater, swamp muck, or soil—and determining the potential difference between the platinum and a hydrogen electrode or some other electrode of known potential. Measured redox potentials of seawater, for example, range between $+0.3$ volt for aerated water to -0.6 volt for water from bottom sediments containing organic matter. The term redox potential is used by some geochemists also as a synonym for oxidation potential, applicable to potentials of individual half-reactions as well as to potentials of environments. In either usage it is commonly given the symbol Eh.

As an example, suppose we find the redox potential of a sample of water to be 0.5 volt, and inquire as to the dominant form of dissolved iron in this environment. If the solution is acid, the choice lies between Fe^{2+} and Fe^{3+} (neglecting possible complexes). For the standard potential of the Fe^{2+}-Fe^{3+} couple we read from Appendix IX a value 0.77 volt. The measured potential is more reducing than this; hence we would expect qualitatively to find Fe^{2+} the chief ion. To get a quantitative value for the Fe^{2+}/Fe^{3+} ratio, we substitute in Eq. (9-9), using 0.77 as $E°$ and 0.5 as E:

$$0.5 = 0.77 + \frac{0.059}{1} \log \frac{[Fe^{3+}]}{[Fe^{2+}]}$$

$$\log \frac{[Fe^{3+}]}{[Fe^{2+}]} = -\frac{0.27}{0.059} = -4.58$$

$$\frac{[Fe^{3+}]}{[Fe^{2+}]} = 10^{-4.58} = 2.6 \times 10^{-5}$$

In this water, therefore, the concentration of Fe^{2+} is nearly 40,000 times that of Fe^{3+}.

Redox potential in many ways is analogous to pH. It measures the ability of an environment to supply electrons to an oxidizing agent, or to take up electrons from a reducing agent, just as the pH of an environment measures its ability to supply protons (hydrogen ions) to a base or to take up protons from an acid. In a complex solution like seawater or water in a soil the redox potential is determined by a number of reactions, just as pH is determined by the combined effects of the carbon dioxide system, the boric acid system, and various organic acids. The particular reactions are difficult to identify, and are less important than the overall ability of the environment to maintain its Eh and pH constant when small amounts of foreign material are added.

Unfortunately, redox potentials in nature cannot be determined as simply and unambiguously as this discussion has implied. The difficulty is that some of the reactions that determine redox potentials are slow, so that instantaneous readings with a platinum electrode do not give true equilibrium potential differences. This is particularly true for reactions involving oxygen, which, of course, include a great many of the most important oxidation reactions in nature. Most reactions in which oxygen plays a role take place by a series of steps, and one of the steps is very slow. Hence redox potentials measured in oxygen-containing environments are generally lower than equilibrium values, and there is no simple way to apply a correction factor. This means that most redox-potential measurements in nature give us only qualitative or semiquantitative information. It is useful nevertheless to make calculations based on the measured values and on theoretical potentials, since such calculations can at least set limits to the processes we may expect in natural environments.

9-4 LIMITS OF pH AND Eh IN NATURE

In order to make predictions about geologic processes, we need at least a rough idea as to the ranges of natural Eh and pH values. The ranges will obviously be more restricted than those with which the chemist is accustomed to deal in the laboratory.

The limits of pH we have mentioned in earlier discussions. The solutions of highest acidity found in nature are those formed by the dissolving of volcanic gases and by the weathering of ores containing pyrite. Locally such solutions may attain pH's less than zero (acidities greater than $1N$). Acidities of this magnitude are quickly lowered by reaction with adjacent rocks, and the rocks are thereby drastically altered, as is evident in the bleached and porous zones commonly found near fumaroles, hot springs, and pyritic ore deposits. Given enough time, contact with ordinary silicate or carbonate rocks would neutralize the solutions. Complete neutralization is generally prevented by solution of carbon dioxide from the atmosphere and of organic acids formed by decaying organic matter; these two are the source of acidity in most near-surface waters, giving pH's commonly in the range 5 to 6. Lower pH's are found in the A horizons of pedalfer soils, especially in podzols, where values as low as 3.5 are sometimes recorded. Disregarding the possible extremes, we can reasonably select a figure of 4 as the usual lower limit of pH's in natural environments.

At the other end of the scale, CO_2-free water in contact with carbonate rocks can acquire by hydrolysis a pH of about 10, and in contact with the silicates of ultramafic rocks a pH of nearly 12. Similar high values may be found in desert basins, where fractional crystallization and fractional solution have segregated alkaline salts like sodium carbonate and sodium borate. But most surface waters have sufficient contact with the atmosphere that such high alkalinities are not attained, and a reasonable upper limit of pH in most near-surface environments is about 9.

The strongest oxidizing agent commonly found in nature is the oxygen of the atmosphere. Stronger agents than this cannot persist, for the reason that they would react with water to liberate oxygen. Thus the upper limit of redox potentials is defined by the reaction

$$H_2O \rightleftharpoons \tfrac{1}{2}O_2 + 2H^+ + 2e^- \qquad E° = +1.23 \text{ volts}$$

The potential of this half-reaction clearly depends on the pH, as shown by the equation

$$E = +1.23 + 0.03 \log [O_2]^{1/2}[H^+]^2$$

For the usual concentration of O_2 we may use 0.2 atm, since oxygen makes up about one-fifth of the atmosphere by volume. Hence

$$Eh = +1.23 + 0.03 \log (0.2)^{1/2} + 0.059 \log [H^+]$$

$$= +1.22 - 0.059 \text{ pH} \tag{9-12}$$

Slowness of reaction ("overvoltage effects") should make it possible for stronger oxidizing agents to exist locally and temporarily. Actually, however, measured oxidizing potentials in nature are always well below this limit, so that the empirical equation

$$Eh = 1.04 - 0.059 \text{ pH}$$

is a more realistic upper boundary (Baas Becking et al., 1960). The discrepancy probably means that oxidation reactions involving O_2 have complicated mechanisms, possibly with a slow step in which traces of hydrogen peroxide act as an intermediate (Sato, 1960).

Reducing agents likewise are limited to substances that do not react with water, the reaction this time resulting in liberation of hydrogen. The limiting redox potential is that of the hydrogen electrode reaction,

$$H_2 \rightleftharpoons 2H^+ + 2e^- \qquad E° = 0.00 \text{ volt}$$

for which

$$E = 0.00 + 0.03 \log [H^+]^2 - 0.03 \log [H_2]$$

$$= -0.059 \text{ pH} - 0.03 \log [H_2]$$

Since the pressure of hydrogen gas in near-surface environments cannot exceed 1 atm, the maximum possible reducing potential in the presence of water would be

$$Eh = -0.059 \text{ pH} - 0.03 \log (1) = -0.059 \text{ pH} \tag{9-13}$$

Conceivably local conditions might permit stronger reducing reactions, particularly within bodies of organic material (coal or petroleum) out of contact with water.

9-5 Eh-pH DIAGRAMS

The usual limits of Eh and pH that we have just discussed may be conveniently plotted on a graph with Eh as ordinate and pH as abscissa (Fig. 9-1). A more accurate representation of Eh-pH limits has been constructed by Baas Becking et al. (1960), on the basis of measurements in many kinds of surface and near-surface environments (Fig. 9-2). This diagram includes rare environments with extreme

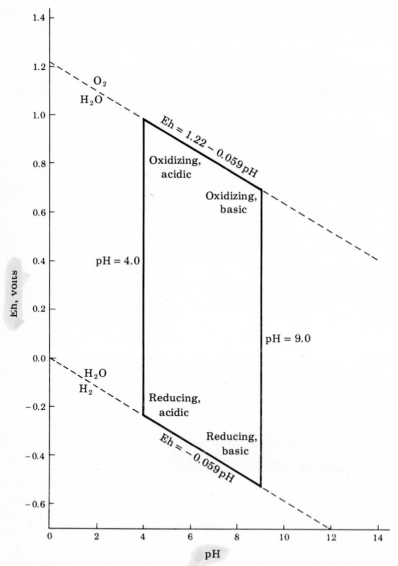

Figure 9-1 Framework of Eh-pH diagrams. The parallelogram outlines the usual limits of Eh and pH found in near-surface environments.

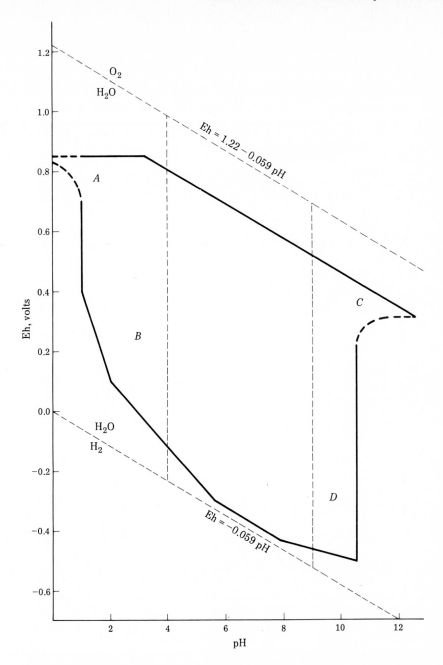

Figure 9-2 Measured limits of Eh and pH in natural environments, including extreme conditions, shown by heavy lines. *(Source: Baas Becking et al., 1960.)* Usual approximate limits (parallelogram of Fig. 9-1) shown by light lines. Letters *A*, *B*, *C*, and *D* refer to Problem 9 at end of chapter.

Eh's and pH's that are not shown by the simple parallelogram of Fig. 9-1. Since the parallelogram is easier to keep in mind and covers most situations in nature, it will be used in the following discussion.

The graph in Fig. 9-1 may be used to plot the potentials for various oxidation-reduction processes, in order to show the conditions under which these processes may be expected to occur in nature. For example, Fig. 9-3 shows one method of diagraming some of the oxidation reactions of iron. Consider first the couple

$$Fe^{2+} \rightleftharpoons Fe^{3+} + e^- \qquad E° = +0.77 \text{ volt}$$

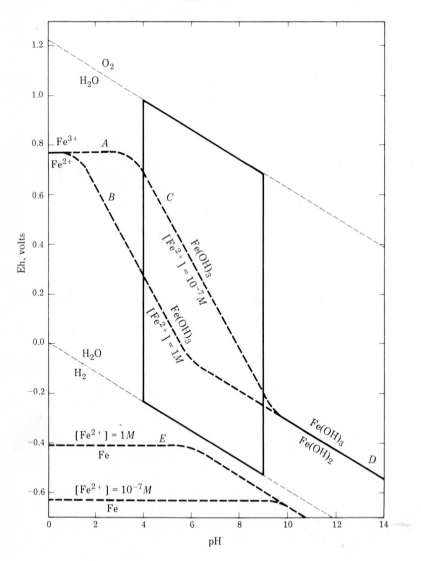

Figure 9-3 Eh-pH diagram for the simple ions and hydroxides of iron at 25°C.

This reaction is independent of pH, since neither H^+ nor OH^- appears in the equation, so that its potential may be plotted as a horizontal line (A). At any pair of Eh-pH values above the line, the ratio $[Fe^{3+}]/[Fe^{2+}]$ is greater than 1, at any pair below the line less than 1. The line cannot be continued far into the diagram because $Fe(OH)_3$ precipitates at pH's near 3 (the exact value depending on the total iron concentration). For pH's greater than 3 the simple electrode reaction must be replaced by

$$3H_2O + Fe^{2+} \rightleftharpoons Fe(OH)_3 + 3H^+ + e^- \qquad E° = +0.98 \text{ volt}$$

Here the relation of Eh to pH is given by

$$E = E° + 0.059 \log \frac{a_{H^+}^3}{a_{Fe^{2+}}}$$

$$Eh = 0.98 - 0.177 \text{ pH for } Fe^{2+} \text{ activity of } 1M$$

$$= 1.39 - 0.177 \text{ pH for } Fe^{2+} \text{ activity of } 10^{-7}M$$

These equations are plotted as the dashed lines (B and C) on Fig. 9-3; more than one line is needed because the potential depends on $[Fe^{2+}]$ as well as on $[H^+]$. Finally, in basic solutions the principal half-reaction is

$$Fe(OH)_2 + OH^- \rightleftharpoons Fe(OH)_3 + e^- \qquad E° = -0.56 \text{ volt}$$

$$Eh = -0.56 + 0.059 \log \frac{1}{[OH^-]}$$

$$Eh = +0.27 - 0.059 \text{ pH}$$

The corresponding line (D) is plotted on the right-hand side of Fig. 9-3.

The diagram shows that the change from ferrous to ferric iron falls approximately in the middle of the field representing conditions in nature, so that we would expect to find changes from one to the other very frequent, depending on slight shifts in the pH or Eh of the environment. This obviously corresponds with everyday experience. We find iron compounds reduced in the surface layer of a soil and oxidized beneath, reduced in bottom sediments of the sea and oxidized in seawater itself, and so on. The diagram also indicates that oxidation of iron takes place much more completely in alkaline solution than in acid, which fits the observation that larger amounts of dissolved iron are commonly present in slightly acid stream waters than in the faintly alkaline water of the oceans. The diagram includes also (line E) the potential of the reaction

$$Fe \rightleftharpoons Fe^{2+} + 2e^-$$

which falls below the field delimiting natural conditions. In other words, metallic iron is not to be expected in sedimentary environments because its presence would require too low a redox potential.

Potentials of a few other reactions are plotted in a similar manner on Fig. 9-4. Note that the line for $Au-AuCl_4^-$ is near the top of the field, indicating that gold

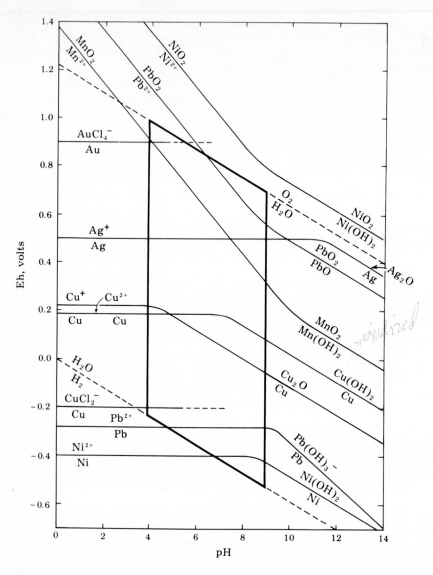

Figure 9-4 Eh-pH diagram for several metals. Activity of Cl^- assumed $1M$, and of metal ions $10^{-5}M$.

should remain as the metal except in strongly acid and strongly oxidizing environments; that copper may occur either in solution or as the native metal in sedimentary environments; and that manganese is more difficult to oxidize than iron. These deductions are all corroborated by observations in nature.

Eh-pH diagrams may be made geologically more realistic by plotting fields of stability for specific minerals. Suppose we inquire, for example, as to the Eh-pH conditions under which siderite would be deposited in preference to hematite. The equation relating the two minerals is

$$2FeCO_3 + 3H_2O \rightleftharpoons Fe_2O_3 + 2H_2CO_3 + 2H^+ + 2e^-$$

From free energies we calculate the $E°$ value, $+0.30$ volt, and then set up the equation for E at various concentrations:

$$E = E° + 0.03 \log a^2_{H_2CO_3} a^2_{H^+}$$

$$Eh = 0.30 + 0.059 \log a_{H_2CO_3} - 0.059 \text{ pH}$$

For any given total concentration of dissolved CO_2, $a_{H_2CO_3}$ is a function of pH alone (Fig. 2-2). Hence Eh can be expressed as a function (albeit a rather complicated one) of pH, and the corresponding line can be drawn on an Eh-pH graph to express equilibrium between siderite and hematite (line A, Fig. 9-5). Other iron minerals can be included by setting up similar equations and making reasonable assumptions about total dissolved sulfur, total dissolved silica, and so on. Figure 9-5 includes hematite, siderite, magnetite, and pyrite, for assumed concentrations of total sulfur (H_2S, HS^-, S^{2-}, SO_4^{2-}) equal to $10^{-6}M$ and of total carbonate (H_2CO_3, HCO_3^-, CO_3^{2-}) equal to $1M$.

One other kind of information can be given on the diagram: the concentration of Fe^{2+} or of total $Fe^{2+} + Fe^{3+}$ in equilibrium with the various minerals at different Eh-pH conditions. This is accomplished by setting up equations for the electrode potentials of reactions like

$$2Fe^{2+} + 3H_2O \rightleftharpoons Fe_2O_3 + 6H^+ + 2e^-$$

$$E = E° + 0.03 \log \frac{a^6_{H^+}}{a^2_{Fe^{2+}}}$$

The value of $E°$ may be calculated from free energies or looked up in tables, and turns out to be 0.65 volt. Hence

$$Eh = 0.65 - 0.17 \text{ pH} - 0.059 \log a_{Fe^{2+}}$$

Rearrangement gives $\log a_{Fe^{2+}}$ as a function of pH and Eh:

$$\log a_{Fe^{2+}} = \frac{0.65 - 0.17 \text{ pH} - Eh}{0.059}$$

In the field where hematite is the stable iron mineral, hematite cannot precipitate unless the activity of Fe^{2+} exceeds values given by this equation. Two solutions of the equation, for total activity of dissolved iron equal to $10^{-6}M$ and $10^{-4}M$, are shown by lines B and C in Fig. 9-5. For the siderite field the appropriate equation would be

$$FeCO_3 \rightleftharpoons Fe^{2+} + CO_3^{2-} \qquad K = 10^{-10.7}$$

whence

$$a_{Fe^{2+}} = \frac{10^{-10.7}}{a_{CO_3^{2-}}}$$

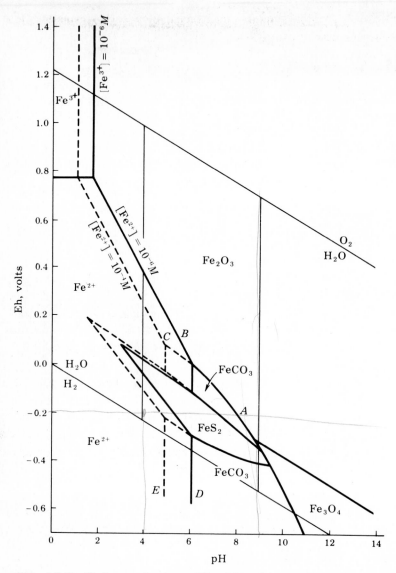

Figure 9-5 Eh-pH diagram showing stability fields of common iron minerals. Total activity of dissolved carbonate, $1M$, of dissolved sulfur, $10^{-6}M$. Solid field boundaries on left side of diagram are for total dissolved iron $= 10^{-6}M$, dashed lines for $10^{-4}M$. (*After Garrels and Christ, 1965, page 224.*)

The concentration of CO_3^{2-} would then be expressed as a function of total carbonate and pH, and substitution in this equation gives values of $a_{Fe^{2+}}$. Lines D and E in Fig. 9-5, showing equilibrium Fe^{2+} activities of $10^{-6}M$ and $10^{-4}M$ in the siderite field, are parallel to the Eh axis, since the equation involves no oxidation or reduction.

The Eh-pH relations in Fig. 9-5 may be interpreted as follows: Hematite is the stable mineral of iron in all moderately and strongly oxidizing environments. In reducing environments the stable mineral may be pyrite, siderite, or magnetite, depending on concentrations of sulfur and carbonate in the solution. For the conditions of high total carbonate ($1M$) and low total sulfur ($10^{-6}M$) shown in Fig. 9-5, siderite has two fields of stability separated by the field of pyrite, and magnetite is stable only in contact with strongly basic solutions. If dissolved carbonate is smaller and dissolved sulfur is higher, the field of pyrite expands until it fills nearly all the lower part of the diagram; a small field in which pyrrhotite is stable may appear at the extreme lower edge of the natural Eh range. If both carbonate and sulfur are very low, the field of magnetite extends into near-neutral environments. All the mineral transitions involving oxidation are favored by basic solutions, so that, for example, hematite may form from siderite in response to an increase in either pH or Eh. The occurrence of siderite is practically restricted to neutral and basic solutions; it can precipitate from weakly acid solutions only if the concentration of dissolved iron is abnormally high. Within narrow ranges of Eh and pH most pairs of iron minerals are stable together—magnetite-hematite, hematite-siderite, magnetite-siderite, siderite-pyrite. Even the pair hematite-pyrite, although forbidden by the low total sulfur of Fig. 9-5, has a stable existence in environments with higher sulfur. Most of these conclusions are familiar enough as geologic deductions based on field associations of iron minerals, but the diagram displays the underlying chemistry and makes the conclusions more quantitative.

Possible modifications of Fig. 9-5, constructed by varying the concentrations of sulfur, carbonate, and silicate, are explored in great detail in Garrels and Christ (1965), from which Fig. 9-5 is taken. One feature of the diagrams is particularly noteworthy, since it conflicts with older geologic literature: the prediction that magnetite and pyrrhotite under some conditions can occur as sedimentary minerals. Several recent papers have presented good evidence for the rare occurrence of these minerals as primary constituents of sediments—a striking confirmation of predictions from the thermodynamic data.

Thus an Eh-pH diagram is a convenient device for summarizing in quantitative fashion a body of chemical information, and for making predictions about reactions and associations among sedimentary minerals. Use of the diagrams is subject to the usual limitations encountered in applying quantitative chemistry to geologic problems, limitations imposed by (1) the greater number of variables in geologic situations and (2) the assumption of equilibrium. The nature of the limitations is nicely illustrated by Fig. 9-5. This figure displays relations among iron minerals in terms of two variables, acidity and oxidation potential. Two other geologically important variables, the total concentrations of carbonate and sulfur species in solution, are given arbitrary fixed values. To show the variation with either one of these, a three-dimensional diagram could be constructed with total sulfur or total carbon as the third axis; but three-dimensional diagrams are not easy to use, and relations are more clearly exhibited by constructing several two-dimensional plots with different fixed values for carbon and sulfur. For a complete chemical picture of iron in sediments we would need still other diagrams, to show the effects of such variables as dissolved silicate and phosphate. A notable omission from the diagram is a field for goethite (a major constituent of limonite), which is certainly an important iron mineral in sediments; its relation to hematite is shown by the reaction

$$2FeOOH \rightleftharpoons Fe_2O_3 + H_2O$$

The relative stability of the two minerals is evidently dependent on the vapor pressure (or activity) of water, which should therefore be yet another variable on our overloaded diagrams. Even if this variable is represented, however, the results will not be realistic because the hematite-goethite reaction is very slow, and attainment of equilibrium is not assured; it seems best to ignore this reaction and simply to note that the field of limonite on an Eh-pH diagram is approximately the same as that of hematite. In addition to all these variables must be added temperature and pressure. The usual Eh-pH diagram is drawn for conditions of 25°C and 1 atm total pressure, and of course its field boundaries will shift as either of these variables changes.

The list of provisos regarding the use of Eh-pH diagrams is still not ended. We should note, for example, that the stability fields in Fig. 9-5 are for pure iron compounds. Isomorphous substitution of other metals for iron is common, particularly magnesium and manganese in siderite and magnetite, and these impurities would change the field boundaries. The concentrations of Fe^{2+} and Fe^{3+} are correct, but for strict accuracy they should be supplemented by other iron ions: $FeOH^+$, $FeOH^{2+}$, $Fe(OH)_2^+$; it just happens that in this case these other substances are negligible in comparison with the simple ions. And finally, changes from one iron mineral to another are slow enough that we must expect occasionally to find a compound persisting metastably under conditions where the equilibrium relations shown on the diagram would not permit it.

With all these qualifications and restrictions, are Eh-pH diagrams for geologic purposes worth the time and effort they take to construct? The same question can be asked about any quantitative representation of geologic relations based on laboratory data for simple compounds and solutions. The answer must come from experience: Do the diagrams in fact summarize geologic observations realistically, and do they lead to predictions that can be tested against field occurrences? For Eh-pH diagrams the answer is a qualified "yes." This kind of plot is used increasingly in recent literature to display chemical relations in weathering and sedimentation, but correspondence with geologic observation is far from perfect.

PROBLEMS

1 Arrange the following, insofar as possible, in order of (a) decreasing pH and (b) decreasing Eh. In places where the order is ambiguous, explain why.

  A. Seawater from near the surface in the tropics.
  B. Soil water from the A horizon of a podzol.
  C. Water from a temporary playa lake in Death Valley.

D. Soil water from the A horizon of a chernozem.

E. Water from a lake in New York State.

F. Water in a small stream draining an area of schist containing abundant pyrite.

G. Water from a swamp in northern Canada.

2 Which of the following can be formed and continue to exist in near-surface environments? For those which cannot, give reasons to justify your answer.

$$Ca(OH)_2 \qquad MnO_4^- \qquad CoCl_3 \qquad Ag$$

$$UO_2^{2+} \qquad K_2S \qquad H_2CO_3 \qquad SiO$$

$$SO_4^{2-} \qquad Zn \qquad Al^{3+} \qquad Mn$$

3 On an Eh-pH diagram like Fig. 9-4, plot curves for the following half-reactions:

(*a*) $I^- - IO_3^-$

(*b*) $Cr^{3+} - Cr_2O_7^{2-}$ and $Cr(OH)_3 - CrO_4^{2-}$

(*c*) $Sn-Sn^{2+}$ and $Sn-Sn(OH)_3^-$

(*d*) $V^{3+} - VO^{2+}$ and $V(OH)_3 - VO(OH)_2$

(*e*) H_2S-S, $HS^- - S$, and $S^{2-} - S$

(*f*) $H_2SO_3 - SO_4^{2-}$ and $SO_3^{2-} - SO_4^{2-}$

Where it is necessary to assume a concentration for some ion other than H^+ and OH^-, use $10^{-5}M$.

4 Complete and balance the following oxidation-reduction equations. (One method for balancing such equations is outlined in Appendix X.)

$$Fe^{2+} + UO_2^{2+} \rightleftharpoons Fe^{3+} + U^{4+} \qquad \text{(strongly acid solution)}$$

$$Cl^- + MnO_4^- \rightleftharpoons Cl_2 + MnO_2 \qquad \text{(weakly acid solution)}$$

$$V(OH)_3 + O_2 \rightleftharpoons VO_4^{-3} \qquad \text{(alkaline solution)}$$

$$CH_4 + SO_4^{2-} \rightleftharpoons CO_2 + S \qquad \text{(acid solution)}$$

5 From the data of Appendix IX, find $E°$ and calculate $\Delta G°$ for the following reactions:

(*a*) $Mn + Cu^{2+} \rightleftharpoons Mn^{2+} + Cu$

(*b*) $MnO_2 + PbO + H_2O \rightleftharpoons Mn(OH)_2 + PbO_2$

(*c*) $3Cu + 2NO_3^- + 8H^+ \rightleftharpoons 3Cu^{2+} + 4H_2O + 2NO$

(*d*) $3HgS + 2NO_3^- + 8H^+ \rightleftharpoons 3Hg^{2+} + 4H_2O + 2NO + 3S$

(*e*) $2Ag + \frac{1}{2}O_2 + 2Cl^- + 2H^+ \rightleftharpoons 2AgCl + H_2O$

6 In lake water at a pH of 5 and a redox potential of $+0.30$ volt, what concentration of cupric ion could exist in contact with metallic copper?

7 What redox potential must an environment possess in order for the activities of Fe^{2+} and Fe^{3+} to be equal?

8 The following questions refer to Fig. 9-5:

(*a*) Describe the sequence of iron minerals that would form from a solution at a constant Eh of -0.2 volt, if its pH increases slowly from an initial value of 4.

(*b*) Describe the sequence of oxidation products of siderite in contact with a solution whose pH is maintained at 9 while oxidation occurs.

(*c*) Does the simple ferric ion, Fe^{3+}, play a role in ordinary near-surface solutions? In what sort of geologic environment might its concentration become appreciable?

9 The letters *A*, *B*, *C*, and *D* on Fig. 9-2 represent extreme conditions of Eh and pH sometimes found in natural environments. For each, suggest geologic situations in which these extreme conditions might be encountered.

REFERENCES AND SUGGESTIONS FOR FURTHER READING

Baas Becking, L. G. M., I. R. Kaplan, and D. Moore, Limits of the natural environment in terms of pH and oxidation-reduction potentials, *Jour. Geology*, vol. 68, pp. 243–284, 1960. A tabulation of recorded measurements of Eh and pH in natural environments, and a discussion of the factors that limit these variables in different geologic situations.

Barnes, I., and W. Back, Geochemistry of iron-rich ground water of southern Maryland, *Jour. Geology*, vol. 72, pp. 435–447, 1964. A good example of careful measurements of Eh and pH in natural environments.

Garrels, R. M., and C. L. Christ, "Solutions, Minerals, and Equilibria," Harper & Row, Publishers, Incorporated, New York, 1965. This is the standard reference on the measurement and use of oxidation potentials for geologic applications. The book has a great many Eh-pH diagrams, and gives detailed examples to show how the diagrams are constructed and interpreted.

Latimer, W. M., "Oxidation Potentials," 2d ed., Prentice-Hall, Inc. Englewood Cliffs, N.J., 1952. Although now badly out of date, this book remains a convenient source of data on oxidation-reduction potentials.

Sato, M., Geochemical environments in terms of Eh and pH, *Econ. Geology*, vol. 55, pp. 928–961, 1960. Experimental evidence that the upper limit of oxidation potentials in nature is determined by a rate-controlling step involving hydrogen peroxide rather than by the oxidation of water.

Sillén, L. G., Stability constants of metal-ion complexes, Sec. I: Inorganic ligands, *Chem. Soc. London Spec. Pub.* 17, 1964, and Supplement 1, *Chem. Soc. London Spec. Pub.* 25, 1971. Contains a compilation of oxidation-reduction potentials.

OXIDATION AND REDUCTION
IN SEDIMENTATION

With a background understanding of energy relationships in chemical reactions, we return now to some questions left unanswered in earlier discussions of sedimentary processes. Our concern here will be with the elements that undergo oxidation and reduction during sedimentation, in addition to the hydrolysis and precipitation reactions we have considered previously. The three most abundant of these elements, and the ones with which this chapter is primarily concerned, are iron, manganese, and sulfur. We might add carbon to this list, for it also is oxidized and reduced in a variety of processes during the formation of sediments, but the reactions of carbon are so complex and so unique that they will require a special chapter all to themselves (Chap. 11).

10-1 IRON SEDIMENTS

The general outlines of the sedimentary geochemistry of iron are clear from the Eh-pH diagrams in the last chapter. We shall review briefly the interpretation of these diagrams, and then consider other aspects of the behavior of iron which are not so easily expressed in quantitative terms.

Iron in minerals of igneous rocks is partly oxidized during weathering, giving the familiar brown, yellow, and red colors of rock surfaces, and partly dissolved as ferrous ion, Fe^{2+}. This ion can be transported long distances if the solution stays reducing and slightly acidic, and if it is not admixed with other ions that form insoluble compounds. A great variety of ions may cause precipitation, the most common ones being carbonate, sulfide, and silicate; the precipitation reaction for any of these is aided by an increase in alkalinity of the solution. Iron sulfide precipitates only under very reducing conditions, or from solutions in which the concentration of sulfide is unusually high. The precipitation of siderite, an iron silicate (for example, chamosite), or magnetite depends on relative concentrations of anions. Oxidation to hematite or goethite may take place at any time during the transportation, in solutions of any acidity within the normal range, but the reaction occurs most readily under alkaline conditions. These are the geologically useful conclusions that can be read from Fig. 9-5 and similar diagrams.

Thus the essential chemistry of iron deposition is fairly simple. The metal stays in solution as long as a delicate balance of conditions is maintained, and precipitates as one of a half-dozen possible compounds when conditions change. The geologic situations in which precipitation occurs are well known: by oxidation at the site of the original iron mineral, before any transportation has taken place; by oxidation in soil derived from the original rock, after only minor movement within the soil layer; by oxidation in streams, lakes, or swamps, when the water is aerated and loses its contained organic matter; in seawater, as oxide if the water is aerated, as hydrous silicate if the water is mildly reducing, as sulfide if the redox potential is low and sulfur is abundant. Ferrous carbonate precipitates in reducing freshwater environments, but probably not in present-day marine environments because seawater contains too much dissolved sulfur. The characteristics of modern sites of iron deposition make possible a plausible reconstruction of the environments where iron-bearing sediments of the past have accumulated. For example, facies of the Precambrian "iron formation" of the Lake Superior district have been identified with sediments formed at different depths in an ancient sea: hematite-bearing rocks deposited in an oxidizing environment near shore, magnetite and greenalite (a hydrous ferrous silicate) in deeper water under mildly reducing conditions, and pyrite in still deeper water where organic matter was abundant. Siderite-bearing rocks of the iron formation have been interpreted in two ways, either as swamp deposits from freshwater basins near the old shoreline, or as marine deposits formed at moderate depths in an ocean with less dissolved sulfur and more dissolved carbon dioxide than the ocean of modern times.

Several questions remain to be clarified. One concerns the relation of the anhydrous oxide hematite to the hydrated oxide goethite, which is the chief constituent of limonite. The free-energy change for the reaction

$$Fe_2O_3 + H_2O(liq) \rightleftharpoons 2FeOOH \qquad (10\text{-}1)$$

is small; reported values are a few hundred calories plus or minus, the exact number and the sign depending on whether one mineral or the other is very finely divided (and hence less stable). Hematite is ordinarily more coarsely crystalline than limonite, and this fact together with the geologic observation that limonite is scarce in older rocks suggests that the equilibrium in most natural environments is displaced to the left. Both forward and reverse reactions are slow, however, so that limonite can persist for geologically long times. The equation indicates that the activity of water should play an important role in determining relative stability, but geologic evidence on this point is ambiguous. It is true that limonite often appears in humid environments, for example in spring and swamp deposits, as might be expected, and that the red color of hematite is common in ancient sediments formed under arid conditions; yet limonite is the usual weathering product in modern deserts, and hematite forms at least locally in the humid tropics. Impurities in the two minerals would also necessarily affect their stabilities. The reaction provides a good illustration of the possible pitfalls in using free energies to make geologic predictions, especially for equilibria in which the free-energy change is near zero and both reactions are sluggish.

A second item that needs attention is the role of ferric oxide as a colloid. When conditions are right for the oxide to precipitate, it often forms instead a stable sol, and as a sol may be carried long distances; this is, in fact, the chief method of transportation of iron in surface waters. The sol should seemingly be stabilized by organic matter acting as a protective colloid, but experiments designed to test the protective action of organic materials derived from decaying vegetation have given ambiguous results. The iron colloid is flocculated by electrolytes, particularly where streams enter the sea. Probably most of the iron brought to the sea is precipitated first as hydrous ferric oxide by flocculation of the colloid in oxygen-rich near-shore waters, and is only later reduced to a ferrous compound by contact with organic matter in or near the bottom sediments.

A third gap in the previous discussion is a consideration of bacteria as possible precipitating agents. Certain species of bacteria have the ability to use the slow oxidation of iron compounds as an energy-producing reaction for their life processes. They probably do not cause any reaction to occur which would not happen anyway if given time enough, but they serve as efficient catalysts in speeding up a process like

$$2Fe^{2+} + \tfrac{1}{2}O_2 + 2H_2O \rightarrow Fe_2O_3 + 4H^+$$

Bacteria are active in forming "bog iron ores," deposits of limonite which grow in swamps and lakes, and also in causing precipitation of ferric oxide around springs. Bacterial deposition of ferric oxide can be very troublesome when it occurs on the walls of pipes or on turbine blades. Other species of bacteria probably aid in the formation of pyrite in reducing environments.

Precipitation of iron may be hindered by the formation of complexes (of both Fe^{2+} and Fe^{3+}) with certain organic compounds, particularly in fairly acid waters containing abundant organic material. Such complexes help to explain the apparently anomalous concentrations of dissolved iron sometimes reported in lakes and swamps of north temperate and subarctic areas.

The most serious of the remaining questions about iron is the mechanism by which iron sediments become segregated from other sedimentary materials. The various processes we have considered explain clearly how iron is transported and precipitated in nature, and we may reasonably suppose that such processes operate ceaselessly in surface waters of the present day; but most surface waters carry very little iron, and hence the precipitated iron is obscured by the bulk of other more abundant sediments. At some times in the past, however, iron-rich sediments have accumulated in large amounts and over long periods, without admixture of much other material. The Clinton ores of Silurian age in the southeastern states and the Precambrian Lake Superior iron formation are good examples. How was it possible for so much iron to be concentrated in one place?

A convincing answer is difficult to give, because we have no example in the present world of such large-scale segregation and precipitation of iron compounds. Explanations have been attempted along two lines. According to the first, a source of iron in greater abundance than usual is necessary. The source commonly suggested is volcanoes—submarine lava flows, submarine springs, or perhaps just the rapid weathering of mafic lavas and tuffs. As corroboratory evidence the existence of iron-rich springs near modern volcanoes can be cited. The second kind of explanation supposes that ordinary weathering, given a low-lying land mass and long-continued absence of orogenic movement, might serve to concentrate iron. The dissolving of iron in larger than normal amounts would be aided if the climate was warm and humid and if the land supported lush vegetation, so that the Eh and pH of stream water and ground water would be low. Precipitation of iron from such solutions, without much accompanying clastic material, can best be imagined in a restricted basin or arm of the sea, where the effect of waves and currents would be minor. Such a set of conditions is not unreasonable geologically. The assumed persistence of the favorable environment for long periods without tectonic disturbance may seem to stretch credibility, but other evidence for extended intervals of tectonic quiet is not lacking in the geologic record.

In a critical discussion of conditions necessary for large-scale accumulation of iron, James (1966) notes that the volcanic hypothesis and the weathering hypothesis are not really contradictory, and that each may provide an explanation for some deposits. Since an intimate association with volcanic rocks can be demonstrated for only a few iron ores, James tends to favor weathering and deposition in a restricted basin as a more general mechanism of concentration. To account for the very abundant iron ores of the Precambrian, when thick vegetation could not have aided the dissolving of iron from the lands, James notes that the atmosphere in Precambrian times probably contained more carbon dioxide than at present, making surface waters more acid and hence more effective in breaking down silicate minerals. Since silica as well as iron would be set free and dissolved by such waters, this suggestion has the merit of explaining also the large amounts of chert and quartz that are characteristically interlayered with iron minerals in the Precambrian ores.

10-2 MANGANESE SEDIMENTS

The chemistry of manganese resembles that of iron very closely. In solutions with low redox potential and low pH both manganese and iron are stable as the divalent ions. Both form carbonates, sulfides, and silicates that are fairly insoluble in neutral or basic solution. Both are readily oxidized under surface conditions to give very insoluble oxides. Details of the chemistry are different: manganese in nature shows two higher oxidation states, $+3$ and $+4$, whereas iron has only $+3$; oxidation of manganous compounds requires higher potentials than does oxidation of ferrous compounds; the simple sulfide of manganese, MnS, is more soluble than FeS, and the double sulfide MnS_2 is far less stable than FeS_2 (pyrite or marcasite). In the laboratory, where higher redox potentials are easily obtained, manganese also differs from iron in its ability to form stable compounds in which it has oxidation numbers of $+6$ (manganates, for example, Na_2MnO_4) and $+7$ (permanganates, for example, $KMnO_4$).

An Eh-pH diagram for manganese in simplified form is shown in Fig. 10-1. To facilitate comparison of manganese and iron, the diagram is drawn with the same fixed concentrations of total dissolved carbonate and sulfur as those used in Fig. 9-5: $1M$ for carbonate and $10^{-6}M$ for sulfur. The diagram shows that MnO_2 is the stable compound of manganese at high redox potentials, regardless of pH; that other oxides can form at lower redox potentials in basic solutions; that $MnCO_3$ is stable over a wide range of Eh and pH if dissolved carbonate is high; and that MnS does not form unless sulfide is much more concentrated and carbonate less concentrated (in fact, not unless total dissolved sulfur exceeds total carbonate by a factor of at least 100). Manganese silicates (for example, rhodonite, tephroite, neotocite) would occupy a prominent place on the right-hand side of the diagram if dissolved silica were high and carbonate low.

Even more than for iron compounds, the stability fields of manganese minerals may be modified by variations in composition. The simple formula $MnCO_3$ is reasonably accurate for rhodochrosite, and Mn_3O_4 for hausmannite, but Mn_2O_3 is a poor representation of manganite ($Mn_2O_3 \cdot H_2O$) and braunite ($Mn_2O_3 \cdot MnSiO_3$). MnO_2 is adequate for pyrolusite, but hardly for psilomelane or wad. Because the minerals are obviously the most stable forms under natural conditions, stability fields for the minerals must be somewhat larger than those shown for the simple compounds. Thus the diagram indicates that psilomelane (as represented by MnO_2) should not form as a primary mineral together with hausmannite, but this combination has been reported from low-temperature deposits; evidently the psilomelane field would be larger than that shown for MnO_2. Except for the psilomelane-hausmannite relation, the associations of primary manganese minerals found in nature fit the diagram well.

The relationship between manganese and iron minerals may be studied by superimposing Fig. 10-1 over Fig. 9-5. The most immediately obvious difference is the enormously greater field of the carbonate in the case of manganese, and of the highest oxide in the case of iron. This is a pictorial representation of the fact that iron compounds oxidize more readily in natural environments. The size of the $MnCO_3$ field, of course, is a consequence of the high assumed total carbonate; to indicate how the field would shrink at lower carbonate concentrations, the light dashed lines in the $MnCO_3$ field show its outline for total carbonate equal to $0.001M$. At this low value for carbonate the oxides would have larger fields, and over much of the normal Eh-pH range no manganese precipitate could form at all

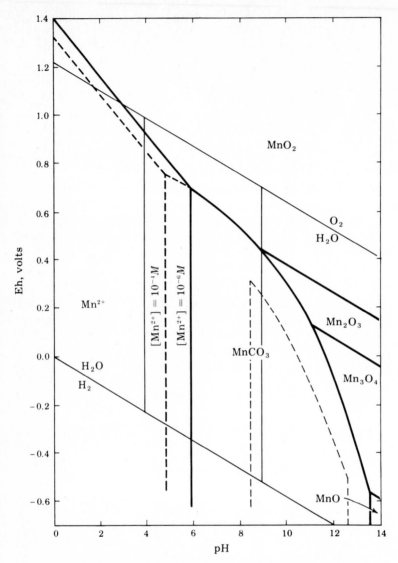

Figure 10-1 Eh-pH diagram showing stability fields of common manganese minerals. Assumed concentrations are the same as in Fig. 9-5: total carbonate, $1M$; total sulfur, $10^{-6}M$.

unless the concentration of Mn^{2+} is abnormally high. Note that hematite is stable in the presence of manganous compounds over a considerable Eh-pH range and that, in a large area of acid and moderately oxidizing conditions, hematite could precipitate while manganese remains in solution as Mn^{2+}. The great stability of pyrite is shown by the appearance of an FeS_2 field in Fig. 9-5, even when total sulfur has a concentration as low as $10^{-6}M$, while no corresponding manganese mineral can form even at much higher sulfur concentrations.

The sedimentary behavior of manganese, as reconstructed from Fig. 10-1, is similar to that of iron. The metal is dissolved from its compounds in igneous rocks as Mn^{2+} and remains in this form as long as the solution is slightly acid and not too oxidizing. Precipitation may occur whenever the pH increases, provided carbonate or silicate is present in sufficient concentration; conceivably the sulfide or hydroxide may form instead if the Eh is very low. If the solution becomes more oxidizing, say by prolonged exposure to air so that any organic material is oxidized, the manganese precipitates as one of the oxide minerals. With sufficient exposure to atmospheric oxygen, pyrolusite should be ultimately the most stable mineral. This general sequence of events can be observed wherever manganese deposits are forming at the present time, and can be reasonably assumed for sedimentary manganese deposits of the past. The readiness of the dioxide to precipitate on exposure of manganese solutions to air is evident in the black films, often showing intricate dendritic forms, deposited on the surfaces and in cracks of all kinds of rocks exposed to the atmosphere.

Again like iron, the higher manganese oxides commonly precipitate first as colloids and may be transported long distances as sols. These are unusual among oxide sols in that their particles carry a negative charge over a large pH range and therefore preferentially adsorb cations out of the solution. The particular assortment of cations commonly found with manganese oxide sediments is an odd one —K^+, Ni^{2+}, Co^{2+}, Pb^{2+}, Ba^{2+}, Cu^{2+} are especially abundant—and probably indicates that adsorption is supplemented by the formation of definite compounds.

Manganese resembles iron in the ability of its divalent ion to form complexes with organic compounds, and also in the speeding up of its oxidation by bacteria. Some of the same bacteria that help to precipitate iron can also derive energy from the oxidation of manganese, and at least two species apparently prefer manganese. Again the function of the bacteria is merely to speed up a reaction that is thermodynamically feasible, hence that would take place anyway if given a long enough time. The precipitation of manganese oxide by bacteria has proved a great nuisance in pipes for industrial and municipal water supplies, and in nature is probably responsible for the formation of manganese ores in bogs and lakes.

10-3 THE SEPARATION OF MANGANESE FROM IRON

For manganese as for iron, the great puzzle in its sedimentary geochemistry is not the chemical processes as such—these are understood in all but minor details— but the mechanism by which manganese compounds become separated from other sediments. Manganese compounds occur in small amount in nearly all sediments, and this we would expect from the general behavior outlined above; but locally manganese compounds are segregated by sedimentary processes into deposits of sufficient size and purity to be excellent ore. The mystery here is even more troublesome than for iron deposits, because we have to explain not only how manganese compounds come to be isolated from other clastic and chemical sediments, but also how they are separated from the very similar and much more abundant compounds of iron.

We may start, as we did for iron, either by assuming solutions unusually rich in manganese or by assuming the action of normal erosional processes subject to rigid control of Eh and pH over a long period of time. Manganese-rich solutions are a plausible enough postulate, because such solutions can be observed in springs in some volcanic areas. It is not very satisfying, however, simply to assume a solution containing manganese and not much else, produced by some mysterious volcanic process underground, because the assumption can be neither verified nor employed to make useful predictions. Such an assumption may be necessary, but before adopting it, we should explore all possible alternatives. Long-continued erosion can also be made plausible, but the behavior of manganese and iron is so similar that conditions under which they may be quantitatively separated are restricted.

Solutions obtained directly by the weathering of igneous rocks, or by the action of hot acids on igneous rocks, contain much more iron than manganese, roughly fifty times as much, which is the average ratio of iron to manganese in the earth's crust. This is established both by analysis of solutions in nature and by laboratory experiment. If such solutions are exposed to the air and made basic suddenly, both metals will precipitate, resulting in the formation of iron minerals with a small admixture of manganese; this corresponds to one type of iron ore. If, on the other hand, the pH increases very slowly, iron compounds reach the limit of solubility before manganese compounds and so can precipitate while manganese is left in solution. Most commonly conditions are oxidizing, so that the first compound to precipitate is ferric oxide. This process of isolating manganese in solution can be demonstrated in the laboratory and is also a commonplace in nature. For example, spring deposits are often reported where an iron-rich precipitate accumulates near the spring and a manganese-rich precipitate farther away. The precipitation of manganese after the iron has separated out can be effected in any of the usual ways: as an oxide if conditions are oxidizing and the solution becomes still more alkaline, or as the carbonate or silicate if conditions become reducing.

So effective is this separation process that one is tempted to postulate it as a general method of isolating manganese in near-surface solutions. Even where manganese-rich solutions appear in volcanic regions, one can suppose that iron initially present has precipitated out before the waters reached the surface. This would require, of course, that the formation of manganese deposits should be accompanied by precipitation of much greater quantities of iron minerals somewhere in the vicinity, either on the surface or underground. Field evidence, however, has not confirmed that precipitated iron minerals commonly appear in great abundance near sedimentary manganese deposits, so that this suggested mechanism for the isolation of manganese in large quantities is still no more than a plausible hypothesis.

Differences in the behavior of the colloidal oxides of the two metals have also been called upon as an explanation for their separation in nature. Data on the behavior of the two colloids are conflicting, the manganese colloid appearing more stable under some conditions and the iron colloid under others. Very probably differences between the colloids can lead to local slight enrichment of one metal or the other, but no evidence supports the hypothesis that deposits of nearly pure manganese oxides might be formed in this manner. Bacteria also may account for partial separation of iron and manganese locally, inasmuch as different species affect the two metals differently, but again neither experiment nor geologic observation suggests that anything like complete isolation of manganese can be so produced.

Still another possible method of separation depends on relative stabilities of organic complexes. In general the iron complexes are more stable, so that from solutions containing the two metals manganese would precipitate before iron. This reversal in the usual order of precipitation has been observed in deposits from organic-rich waters in Finland (Carlson et al., 1977).

Although other mechanisms of separation may be effective locally, the best general hypothesis for explaining how relatively pure iron deposits and pure manganese deposits can form in sedimentary processes is separation by differential oxidation and differential solubility.

The differences in solubilities of compounds and in oxidation potential likewise provide a possible explanation for the differences in behavior of iron and manganese in the sea. Some of the iron carried to the sea is precipitated in shallow water, as ferric oxide where the water is oxidizing and as pyrite in the more reducing water near or within the bottom sediments. Manganese can be precipitated only locally near shore, for the reason that its oxides are too easily reduced in the presence of organic matter and its sulfide is too soluble to remain long in contact with seawater except where Mn^{2+} is unusually concentrated. Hence most of the manganese should be widely dispersed in the sea as Mn^{2+} (or a complex of this ion), and perhaps is ultimately precipitated as oxide in deep water where dissolved oxygen is high and organic matter is scarce. Some of it may appear in manganese nodules, which are rounded masses of manganese dioxide up to a meter or so in diameter that form conspicuous parts of the present-day sediment over large areas of the ocean floor. So abundant are the nodules in places that commercial exploitation seems a real possibility. The nodules always contain iron oxide as well as manganese oxide, in amounts up to 50%, and commonly also have a percent or so of rare metals like copper, cobalt, and nickel. The origin of the manganese in the nodules is still a matter of dispute. The mechanism just described for transporting manganese from shallow water to deep water is a plausible hypothesis, but the common association of abundant nodules with areas of mafic volcanic rocks on the sea floor suggests that another source of the metal may be local submarine weathering of the volcanic material.

10-4 SULFUR SEDIMENTS

Sulfur is a third common element whose sedimentary geochemistry involves oxidation-reduction reactions. Its most common oxidation states in sedimentary environments are -2 (sulfides), 0 (the native element), and $+6$ (sulfates). Several other "formal" states are possible, because of sulfur's unique abilities to substitute for oxygen in anions and to form divalent anions consisting of groups of two or more sulfur atoms. Both of these capacities have their origin in the electron structure of the atom: the neutral atom has 6 valence electrons, permitting it to play the same role as oxygen,

$$
\begin{array}{cc}
:\!O\!: \quad {}^{2-} & :\!O\!: \quad {}^{2-} \\
:\!O\!:\!S\!:\!O\!: & :\!O\!:\!S\!:\!O\!: \\
:\!O\!: & :\!S\!: \\
\text{Sulfate ion} & \text{Thiosulfate ion}
\end{array}
$$

and also permitting one atom to complete a stable shell of 8 electrons by joining itself to another,

$$
\begin{array}{ccc}
:\!S\!: \quad {}^{2-} & :\!S\!:\!S\!: \quad {}^{2-} & :\!S\!:\!S\!:\!S\!: \quad {}^{2-} \\
\text{Sulfide ion} & \text{Disulfide ion} & \text{Trisulfide ion, etc.}
\end{array}
$$

The thiosulfate ion $(S_2O_3^{2-})$, in which the sulfur has a formal oxidation number of $+2$, is not known in minerals but has been detected in solutions in nature. Of the polysulfides the only one known to form stable minerals is the disulfide, which appears in pyrite (FeS_2), marcasite (FeS_2), and several rare species. The sulfur in pyrite and marcasite may be assigned a formal oxidation number of -1, or one sulfur atom may be considered to have a number of -2 and the other a number of 0.

The stable sulfur-containing species in solutions at ordinary temperatures and pressures are shown on an Eh-pH diagram in Fig. 10-2. Thiosulfate and polysulfide ions do not appear in the diagram, since they have only a metastable existence under usual conditions. The narrow wedge marked S_{rh} on the left-hand side of the diagram shows that native sulfur has a small field of stability in contact with near-surface solutions. The other fields are labeled with the formulas of the principal ions or molecules present. Over most of the area of any one field, the designated species makes up more than 99% of the total sulfur concentration in solution, hence has a concentration of approximately $0.001M$ for the particular conditions shown by Fig. 10-2. Near any boundary line the concentration diminishes; the line itself is drawn through points where the two species on either side have equal concentrations, approximately $0.0005M$. Note that dissolved sulfur in solutions at the earth's surface has only two principal oxidation states, $+6$ and -2; that the only important sulfur-containing ion in oxidizing solutions is SO_4^{2-}; that reducing solutions contain chiefly H_2S at pH's less than 7, and HS^- at pH's greater than 7; and that S^{2-} is never a major constituent of any geologically important solution. This does not mean, of course, that the geological role of S^{2-}

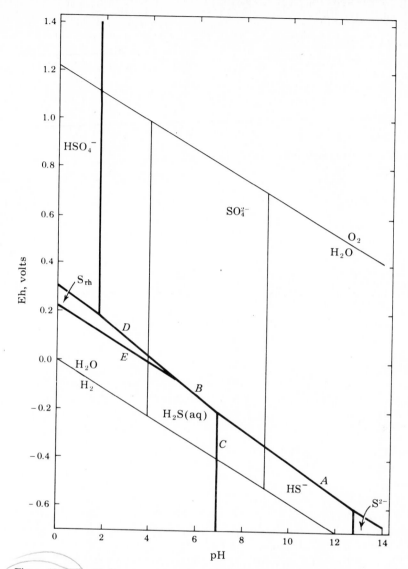

Figure 10-2 Eh-pH diagram for stable sulfur species at 25°C and 1 atm total pressure. Assumed total concentration of dissolved sulfur species = 0.001M.

is negligible. Although its concentration is very minute in the fields of HS⁻ and H_2S, it is capable nevertheless of precipitating many heavy-metal ions as insoluble sulfides.

In this diagram, as in any Eh-pH diagram, it should be noted once more that the boundaries are calculated for equilibrium conditions. Actually the reduction of sulfate to sulfide is very slow, so slow as to be undetectable unless bacteria are present, so that SO_4^{2-} may have at least a temporary existence in strongly reducing solutions. The oxidation of sulfides and native sulfur is also a slow reaction. Furthermore, a number of heavy metals form stable complex ions with various sulfur species, so that the diagram would be more complicated if such metals are present in large amounts.

The simple relations among the sulfur ions can be displayed in many ways. One alternative representation is shown in Fig. 10-3, where the oxidation state of a solution is shown by the partial pressure of oxygen rather than by electrode potentials. The absurdly low oxygen pressures given on the diagram, down to less than 10^{-80} atm, have no meaning as actual, measurable gas pressures; they are simply numerical measures of the oxidation state of the system, obtained by calculating free energies for reactions among the various sulfur species. For example, the line marked A, separating fields of SO_4^{2-} and HS^-, is obtained from the equation

$$\underset{+2.9}{HS^-} + \underset{0}{2O_2} \rightleftharpoons \underset{-178.0}{SO_4^{2-}} + \underset{0}{H^+} \qquad \Delta G° = -180.9 \text{ kcal}$$

The equilibrium constant is calculated from $\Delta G°$ and expressed as a quotient of activities:

$$\log K = -\frac{-180.9}{1.364} = +132.6$$

$$= \log a_{SO_4^{2-}} - pH - \log a_{HS^-} - 2 \log a_{O_2}$$

At the boundary, by definition, $a_{SO_4^{2-}} = a_{HS^-}$, so that the equation reduces to

$$\log a_{O_2} = \text{(approx.)} \log P_{O_2} = -\tfrac{1}{2} pH - 66.3$$

For comparison, the corresponding line on Fig. 10-2, also marked A, is drawn by calculating $\Delta G°$ and thence $E°$ and Eh for the reaction

$$HS^- + 4H_2O \rightleftharpoons SO_4^{2-} + 9H^+ + 8e^-$$

Thus each line on one diagram has a corresponding line on the other; the fields have different shapes, but the relations among them are identical. Note that the area outlining normal near-surface conditions, shown by light lines on both diagrams, is an oblique parallelogram on the Eh-pH diagram but a rectangle on the partial-pressure diagram.

Partial-pressure plots like Fig. 10-3 have the advantage that they can be more easily extended to higher temperatures. For example, the sulfur system at 227°C is shown in Fig. 10-4. Note that the field of native sulfur has vanished and that the field boundaries have shifted somewhat, but the general relations among the sulfur species in solution remain the same.

Similar diagrams with partial pressure (or activity) of oxygen plotted against pH may be substituted for the Eh-pH diagrams of the last chapter. To include other variables, plots can be made of P_{O_2} against P_{S_2}, or P_{O_2} against P_{CO_2}, or three variables can be handled in a three-dimensional diagram. The choice among these

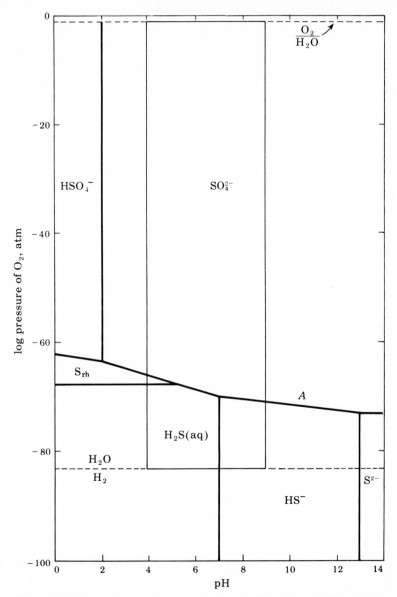

Figure 10-3 Stable sulfur species as a function of pH and P_{O_2} at 25° and 1 atm total pressure. Total concentration of dissolved sulfur species = 0.001M. The rectangle in light lines shows the usual limits of natural environments, corresponding to the parallelogram in Fig. 10-2.

many possibilities is dictated simply by the requirements of particular problems. For solutions at ordinary temperatures the standard Eh-pH diagrams are generally the most useful, but for reactions at higher temperatures, particularly reactions involving gases, partial-pressure diagrams are preferable. Details of

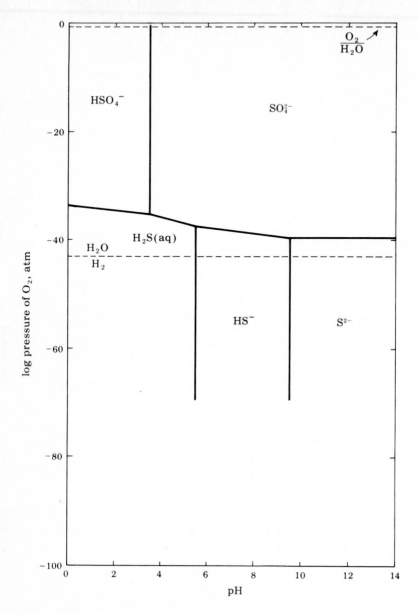

Figure 10-4 Stable sulfur species as a function of pH and P_{O_2} at 227°C and 1 atm total pressure. Total concentration of dissolved sulfur = 0.001M.

constructing many kinds of diagrams are explained in Garrels and Christ (1965), and a good example of the application of partial-pressure diagrams to geologic problems is given in a paper by Holland (1965).

From any of the sulfur diagrams it is evident that metal sulfides can be expected to form as sedimentary minerals only in very reducing environments, which means usually environments with abundant organic material. By far the most common sedimentary sulfide is the disulfide of iron, in the form of either pyrite or marcasite. Both of these minerals are reported from coal beds, locally in large concretions, and crystals of pyrite are often conspicuous in black shales and bituminous sandstones. As tiny disseminated grains, often appearing under the microscope as spherical aggregates of minute crystals (framboids), pyrite is a minor but widespread constituent of sedimentary rocks. In present-day black muds, iron sulfide most commonly occurs as a very-fine-grained black material, sometimes called hydrotroilite and assigned the formula $FeS \cdot nH_2O$, but recent evidence from x-ray examination indicates that it is mostly a cryptocrystalline variety of pyrite, marcasite, or pyrrhotite. This material alters to a more coarsely crystalline sulfide during consolidation of the sediment.

Marcasite and pyrite are dimorphs. The specific conditions that lead to formation of one or the other are not known, but marcasite is probably always metastable with respect to pyrite. Other sulfides are practically absent from present-day sediments, but some occurrences of sulfides of copper, silver, zinc, and lead in older sedimentary rocks can be plausibly explained as the result of sedimentary processes in the past.

In more oxidizing conditions sulfur goes into sulfate ion or one of the sulfate minerals. Commonest of the sulfates are the two compounds of calcium, gypsum and anhydrite, but many others can form under special conditions. Jarosite $[KFe_3(OH)_6(SO_4)_2]$ is occasionally reported as a sedimentary mineral; celestite $(SrSO_4)$ is a minor constituent of many limestones, although probably not as a primary sediment; and a variety of sulfates of potassium, magnesium, and sodium occur in evaporite beds. The sulfate minerals in general appear only in sedimentary rocks whose origin is demonstrably in waters well aerated by exposure to the atmosphere.

The change from sulfides to sulfates following an increase in the Eh of the environment is a commonplace in the zone of weathering. In humid regions the oxidation of sulfides leads to abundant sulfate ion in surface and subsurface waters; in arid regions the sulfate may show itself more tangibly in the form of gypsum crystals or in less common iron or copper sulfates. Locally the reaction is catalyzed by bacteria which make use of the energy released,

$$2FeS_2 + \tfrac{15}{2}O_2 + 4H_2O \rightarrow Fe_2O_3 + 4SO_4^{2-} + 8H^+ \qquad \Delta G^\circ = -582.8 \text{ kcal}$$

Under some circumstances, particularly in the very acid solutions formed during the weathering of sulfide ore deposits, the reaction may not go all the way to sulfate at first but may give powdery native sulfur instead (Fig. 10-2).

The opposite reaction, the reduction of sulfate to sulfide, takes place where sulfate comes in contact with organic matter. Experiments indicate that the reaction is too slow to be significant without the help of bacteria, a special kind of bacteria (anaerobic) which use this reduction process as both a source of energy and a source of oxygen. Taking the simplest organic compound, methane, as an example, we may write

$$2H^+ + SO_4^{2-} + CH_4 \rightarrow H_2S + CO_2 + 2H_2O \qquad \Delta G° = -25.6 \text{ kcal} \qquad (10\text{-}2)$$

The CO_2 and H_2O are the same as the products of respiration in aerobic organisms. The simple reaction of Eq. (10-2) should be regarded as symbolic only; it has not been demonstrated to take place in the laboratory with methane, but only with more complex organic materials. That reactions of this general sort go on in nature is indicated by the commonly observed production of abundant H_2S in the organic-rich sediments of stagnant lakes and isolated marine basins.

Native sulfur in sedimentary rocks generally occurs where gypsum or anhydrite is in contact, or has recently been in contact, with petroleum and natural gas. The largest deposits in this country are associated with the cap rock of salt domes along the Gulf Coast. Salt domes are subterranean plugs of salt, with diameters of a few hundred to a few thousand meters, evidently pushed up from below through thick beds of Cenozoic sedimentary rocks. Above the salt in a typical dome is "cap rock," consisting chiefly of gypsum or anhydrite plus calcite in an irregular layer some tens of meters thick. Within and under the cap rock of some domes are enormous masses of native sulfur, generally accompanied by solid bitumen or pockets of oil. Larger accumulations of petroleum—the accumulations responsible for much of the production from Gulf Coast oil fields—are trapped against the sides of the domes in the upturned and broken sedimentary layers.

The origin of salt-dome sulfur has been the subject of lively arguments. The immediate association with calcium sulfate minerals, the availability of abundant organic material to serve as a reducing agent, and the demonstrated presence of sulfur-reducing bacteria in the cap rock make it plausible to suppose that the sulfate is the ultimate source of the sulfur. The hypothesis has been strengthened by laboratory work showing that calcium sulfate is indeed reduced under conditions simulating those in a natural cap rock (Feely and Kulp, 1957). The experiments indicate that the reaction takes place in two steps, a bacteria-catalyzed reduction to sulfide, according to the general pattern of Eq. (10-2), followed by oxidation of the sulfide to sulfur, with or without the aid of other species of bacteria. The existence in the cap rock of sulfide compounds, both H_2S and small amounts of pyrite, helps to confirm this mechanism. The calcite of the cap rock may be reasonably interpreted as a by-product of the reaction, formed by a combination of Ca^{2+} liberated by reduction of the sulfate and CO_2 formed by oxidation of the carbon compounds. So neatly do the various observed constituents of the cap rock fit into this picture, and so convincing is the experimental corroboration, that alternative hypotheses ascribing the sulfur to supposed volcanic activity or deposition from hydrothermal solutions are largely discredited.

Additional support for the hypothesis of sulfate reduction hardly seems needed, but it is interesting to note that measurements of isotope ratios in the sulfur and carbon show convincing agreement with this mechanism. The two principal stable isotopes of sulfur, ^{32}S and ^{34}S, are slightly separated by low-temperature oxidation and reduction processes, in such a way that the heavy isotope is concentrated in sulfate minerals and the light isotope in sulfide minerals. The extent of separation depends on many factors, including temperature, completeness of reaction, and rate of reaction, so that quantitative prediction is not feasible. But if sulfide and native sulfur are formed by reduction of sulfate, they should have ratios of ^{32}S/^{34}S consistently higher than the ratios in associated calcium sulfate minerals, and this expectation is abundantly fulfilled. The ratios should also show more variation among themselves than is generally found in sulfur of volcanic origin, and this prediction is likewise borne out. The stable isotopes of carbon, ^{12}C and ^{13}C, are sufficiently separated in the process of photosynthesis so that carbon in organic matter is often distinguishable from inorganic carbon by a slight enrichment in ^{12}C; the fact that carbon in the calcite of salt-dome cap rocks has a fairly high ^{12}C/^{13}C ratio compared with most limestones is a good indication that this carbon came from organic compounds in petroleum rather than from atmospheric CO_2. The confirmation of the sulfate-reduction hypothesis by isotope work on two elements is a particularly good example of the usefulness of isotope ratios in helping to solve geologic problems (see also Sec. 21-7).

10-5 OTHER OXIDATION-REDUCTION PROCESSES

This discussion by no means exhausts the possibilities of oxidation-reduction reactions in sedimentary processes. The reactions of organic matter, both the reactions in living organisms and the processes of decay, can be looked upon as complex oxidations and reductions involving changes in oxidation state of the element carbon. We shall focus attention on these reactions in the next chapter. Nitrogen undergoes an interesting series of changes in oxidation state, from $+5$ in nitrates to -3 in ammonium salts, but these substances are so rare in sediments that they hardly warrant extended treatment here. The same would be true of changes from chlorides to perchlorates, and of iodides to iodates, changes that are significant only in evaporite deposits of a few localities.

More pertinent would be oxidation-reduction reactions for some of the rarer metals. Copper, for example, can exist in nature as the native metal and in compounds where it has oxidation numbers of $+1$ (for example, Cu_2S, CuCl) and $+2$ [for example, CuS, $Cu_2(OH)_2CO_3$]. Silver and mercury also show changes from native metal to compounds within the normal Eh range of sedimentary environments. The geochemistry of uranium and vanadium is largely concerned with changes from one oxidation state to another. Reactions of these metals, however, are seldom significant except near ore deposits; hence we shall defer discussion of their oxidation-reduction behavior to later chapters on metallic ores.

PROBLEMS

1 Of the following processes that may go on within sediments or during sedimentation, which ones involve oxidation, which reduction, and which neither?

(a) Precipitation of gypsum as seawater evaporates.

(b) Loss of organic matter from a sediment by decay.

(c) Formation of pyrite in marine sediments a few centimeters under the sediment-water interface.

(d) Precipitation of phosphate as fluorapatite from seawater.

(e) Flocculation of a ferric oxide sol brought to the sea by a river.

(f) Precipitation of manganese dioxide from groundwater percolating through sandstone.

(g) Precipitation of silica from the water of a hot spring.

(h) Growth of gypsum crystals in clays originally rich in organic matter.

(i) Formation of clay minerals by reaction between dissolved or colloidal silica and alumina in seawater.

2 What concentration of Fe^{2+} can exist in the presence of $Fe(OH)_3$ at a pH of 6.5 and an Eh of $+0.30$ volt? At a pH of 8.4 and an Eh of -0.30 volt?

3 Show that MnO_2 is a stronger oxidizing agent than $Fe(OH)_3$ at a pH of 8 and at a pH of 4.

4 According to Figs. 10-1 and 9-5, which of the following pairs of minerals could exist together at equilibrium at 25°C?

Pyrite and pyrolusite	Hausmannite and siderite
Hematite and pyrolusite	Rhodochrosite and pyrite
Magnetite and manganite	Rhodochrosite and hematite

5 Water which has drained through a podzol generally has a higher Mn/Fe ratio than the original water. Suggest a possible explanation.

6 Water emerges from a spring with a pH of 4 and an Fe^{2+} concentration of 1,000 ppm, and flows downhill over a limestone surface. Describe what you would see, and explain what is happening.

7 Under what conditions of Eh, pH, and relative concentrations of Fe^{2+} and Ca^{2+} would siderite precipitate with little or no accompanying calcite?

8 In well-aerated seawater the Eh is generally about $+0.40$ volt and the pH is about 8.2. Under such conditions, would manganese precipitate as MnO_2 if an amount of manganese is added sufficient to give a concentration of 1,000 ppm Mn^{2+}? 1 ppm Mn^{2+}? 0.001 ppm Mn^{2+}?

9 Write electrode reactions for the equilibria represented by lines B, C, D, and E on Fig. 10-2, and for each reaction show how an equation can be set up expressing Eh as a function of pH and concentrations of other ions. (The important thing here is the correct *form* of the equation; the standard potentials need not be calculated, but may be symbolized as $E°$.) Would any of these four lines shift their positions if total dissolved sulfur is increased from $0.001M$ to $0.1M$? In what direction would the shift be?

10 What lines on Fig. 10-3 are analogous to lines B, C, D, and E on Fig. 10-2? For these lines, write the equilibrium reactions and show how equations can be set up relating P_{O_2} to pH.

REFERENCES AND SUGGESTIONS FOR FURTHER READING

Berner, R. A., "Principles of Chemical Sedimentology," McGraw-Hill Book Company, New York, 1971. Chapter 10 is a brief summary of Berner's important work on the geochemistry of iron in sedimentary environments.

Carlson, L., T. Koljonen, P. Lahermo, and R. J. Rosenberg, Case study of a manganese and iron precipitate in a ground-water discharge in Somero, Finland, *Geol. Soc. Finland Bull.*, vol. 49, 159–173, 1977. Description of a deposit in which manganese oxide is precipitated before iron, because manganese-organic complexes are less stable than the corresponding complexes of iron.

Crerar, D. A., and Barnes, H. L., Deposition of deep-sea manganese nodules, *Geochim. et Cosmochim. Acta*, vol. 38, pp. 279–300, 1974. A good example of many recent papers on the chemistry of nodule formation.

Drever, J. I., Geochemical model for the origin of Precambrian banded iron formation, *Geological Society of America Bull.*, vol. 85, pp. 1099–1106, 1974. Chemical factors in the deposition of iron minerals at different depths and different distances from shore in a Precambrian sea, with postulated lower O_2 and higher CO_2 than the present ocean.

Holland, H. D., Some applications of thermodynamic data to problems of ore genesis. Part I, *Econ. Geology*, vol. 54, pp. 184–233, 1959; Part II, *Econ. Geology*, vol. 60, pp. 1101–1166, 1965.

Feely, H. W., and J. L. Kulp, Origin of Gulf Coast salt-dome sulfur deposits, *Am. Assoc. Petroleum Geologists Bull.*, vol. 41, pp. 1802–1853, 1957.

James, H. L., Chemistry of the iron-rich sedimentary rocks, Chap. W in Data of Geochemistry, 6th ed., *U.S. Geol. Survey Prof. Paper* 440-W, 1966. A comprehensive review of the long controversy over the origin of sedimentary iron ores, including many analyses, a theoretical discussion of facies based on Eh-pH diagrams, and a hypothesis of origin based on much field work and a critical evaluation of the literature.

Langmuir, D., Particle-size effect on the reaction goethite = hematite + water, *Am. Jour. Science*, vol. 271, pp. 147–156, 1971. Experimental study of goethite-hematite relations, with emphasis on the effect of particle size on stabilities.

Lynn, D. C., and E. Bonatti, Mobility of manganese in diagenesis of deep-sea sediments, *Marine Geology*, vol. 3, pp. 457–474, 1965. Oxidation-reduction behavior of manganese in the upper part of modern deep-sea sediments, separation of manganese from iron, and the origin of nodules.

Norton, S. A., Laterite and bauxite formation, *Econ. Geology*, vol. 68, pp. 353–361, 1973. Use of Eh-pH and pH-solubility diagrams to set limits on conditions under which laterite and bauxite can form.

Rickard, D. T., Kinetics and mechanism of pyrite formation at low temperatures, *American Journal of Science*, vol. 275, pp. 636–652, 1975.

Trudinger, P. A., I. B. Lambert, and G. W. Skyring, Biogenic sulfide ores: a feasibility study, *Econ. Geology*, vol. 67, pp. 1114–1127, 1972. Review of evidence regarding conditions and rates of reduction of sulfate to sulfide by bacteria, with special reference to stratiform sulfide ores.

ELEVEN

ORGANIC MATERIAL IN SEDIMENTS

The carbon compounds manufactured by organisms are not very stable. Exposed to oxygen, even the most resistant of them decay in a time brief by geologic standards, the carbon returning to the great reservoir of carbon dioxide in the atmosphere. Kept away from oxygen in stagnant water or by burial under accumulating sediments, organic compounds cannot decay completely but decompose into substances that are simpler and more stable than those in the original organisms. Such partly decomposed organic material is a constituent of most sediments, making up nearly 2% by weight of sedimentary rocks as a whole, and locally accumulates into predominantly organic sediments like coal, asphalt, and petroleum. In the language of the last chapter, organic material in sediments is a product of reducing environments and is itself an important reducing agent. We have had occasion to mention it often in connection with oxidation-reduction processes, and we now set out to examine it in more detail.

11-1 THE CHEMISTRY OF CARBON COMPOUNDS

General

Carbon is unique among the elements in the number and complexity of its compounds. Of known compounds, those containing carbon are at least ten times as numerous as compounds of all other elements combined. The study of these substances is a special branch of chemistry called *organic chemistry*, and the carbon compounds (except for a few simple ones like the oxides and carbonates) are called *organic compounds*. Organic chemistry is an intricate and difficult subject, but fortunately we need only the bare rudiments for an understanding of the general geologic behavior of carbon compounds in sediments.

The ability of carbon to form so many compounds may be correlated with two properties of the carbon atom: its small size and its possession of four valence electrons. With this number of electrons its bonds with other atoms are covalent rather than ionic, and the small size of the atom makes these bonds very strong. How strong they can be is attested by the hardness of diamond, which consists of carbon atoms linked together, each one to four others, by covalent bonds. Unlike the atoms of most elements, carbon atoms can join with each other in compounds as well as in the element itself, and this ability makes possible the formation of an almost indefinite number of molecular structures.

The existence of strong covalent bonds in carbon compounds not only helps to account for their enormous number, but explains also some of their general properties. Most carbon compounds, for example, are not very soluble in water, and those that do dissolve are not dissociated into ions, or dissociate only slightly; we shall meet a few exceptions in which special structures make possible the formation of partly ionic bonds. The strong bonds and the lack of dissociation mean also that reactions of carbon compounds are generally slow at ordinary temperatures.

In the next paragraphs we review briefly the chemistry of the kinds of carbon compounds important to the geology of sediments.

Hydrocarbons

Simplest of organic substances are the *hydrocarbons*, compounds that contain only the elements carbon and hydrogen. A familiar example is methane, CH_4, the chief constituent of natural gas and a product of partial decay of organic material in stagnant water ("marsh gas"). Chemically similar to methane are other hydrocarbons found in natural gas and petroleum; formulas and properties of some of these are given in Table 11-1. Mixtures of these hydrocarbons are familiar substances: the first four make up most natural gas; a mixture of those from hexane to decane is gasoline; those from $C_{17}H_{36}$ to $C_{22}H_{46}$ are constituents of lubricating oil; and those containing more than 22 atoms of carbon per molecule make up paraffin.

Note that the melting point, boiling point, and density of these substances increase with the molecular weight. Many other related organic compounds can be arranged in series of this sort, and as a general rule physical properties change regularly as the molecular weight increases.

In butane, C_4H_{10}, the carbon atoms are linked together in a chain. One way of showing this is by a diagrammatic representation of the molecule,

$$
\begin{array}{cccc}
\text{H} & \text{H} & \text{H} & \text{H} \\
| & | & | & | \\
\text{H--C--C--C--C--H} \\
| & | & | & | \\
\text{H} & \text{H} & \text{H} & \text{H}
\end{array}
$$

Table 11-1 The methane series of hydrocarbons

Formula	Name	Freezing point, °C	Boiling point, °C	Density, g/cm^3	
CH_4	Methane	-182	-161		⎫
C_2H_6	Ethane	-183	-89		⎪
C_3H_8	Propane	-190	-45		⎬ Fuel gases
C_4H_{10}	Butane†	-138	-1		⎭
C_5H_{12}	Pentane	-130	36	0.626	⎫
C_6H_{14}	Hexane	-95	68	0.659	⎬ Petroleum ether (naphtha)
C_7H_{16}	Heptane	-90	98	0.684	⎭
C_8H_{18}	Octane	-57	125	0.703	⎫ Gasoline
C_9H_{20}	Nonane	-51	151	0.718	⎪
$C_{10}H_{22}$	Decane	-30	174	0.747	⎭
$C_{11}H_{24}$	Undecane	-27	195	0.740	⎫ Kerosene
. . . .					⎬
$C_{16}H_{34}$	Hexadecane	$+18$	287	0.773	⎭

$C_{17}H_{36}$ to $C_{22}H_{46}$, semisolids, constituents of petroleum jelly and lubricating oil
$C_{23}H_{48}$ to $C_{29}H_{60}$, constituents of paraffin

† The data for butane and heavier hydrocarbons refer to the "normal," or straight-chain, compounds.

Here each dash indicates a covalent bond, in other words, a pair of electrons shared between two atoms. Carbon can form a total of four such bonds, so that each carbon atom must be the center of four dashes; hydrogen, with a single electron per atom, can form only one bond, so that every H is drawn with a single dash. A little juggling of atoms and dashes shows that these requirements can be met also with a different sort of diagram:

$$
\begin{array}{ccccc}
 & H & H & H & \\
 & | & | & | & \\
H - & C - & C - & C & - H \\
 & H & | & & H \\
 & H - & C - & H & \\
 & & | & & \\
 & & H & &
\end{array}
$$

Corresponding to these two representations, we find that two gases exist with the formula C_4H_{10}, and the different molecular structures give the two compounds slightly different chemical and physical properties. To distinguish them, we call the first gas normal butane and the second one iso-butane.

Now obviously the possibility of forming molecules with different structures that satisfy the rules of bonding will increase as the number of carbon atoms per molecule gets larger. The existence of more than one compound with the same simple molecular formula but different molecular structures is extremely common in organic chemistry, and accounts in large part for the enormous number of organic compounds. The phenomenon is called *isomerism*, the different substances with the same composition but different molecular architecture are called *isomers*, and spread-out formulas like those just used for the two butanes are called *structural formulas*. Such architectural formulas are the organic chemist's shorthand for describing the properties of organic compounds and for predicting their reactions. The necessity for distinguishing isomers is only one example of the manifold usefulness of these formulas.

To summarize, the hydrocarbons we have so far considered are characterized by chains of carbon atoms in their molecules. The chains are straight in the normal compounds, branched in the isomers. These compounds are also characterized by the fact that the link between each pair of atoms is a single electron-pair bond. Hydrocarbons with these characteristics are called *aliphatic* or *paraffin* hydrocarbons. They are the chief constituents of most petroleum and natural gas, and they are likewise obtainable in large amounts by distillation of coal.

Carbon atoms can be linked into rings as well as open chains. Two simple compounds with ring structures are cyclopentane, C_5H_{10}, and cyclohexane, C_6H_{12}:

These two are prominent ingredients of some petroleums. Collectively, compounds with this kind of ring are called *cycloparaffins* (or *naphthenes* or *polymethylenes*). They resemble the aliphatic hydrocarbons in that all bonds are formed by sharing single electron pairs, a characteristic we describe by saying that all compounds discussed so far are *saturated* compounds.

In contrast to saturated hydrocarbons are others with insufficient hydrogen atoms to satisfy the requirement of single electron-pair bonds. A simple example is the gas ethylene, C_2H_4:

In order that each carbon atom should have four bonds, two bonds must be drawn between the two carbon atoms; in other words, two electron pairs must be shared instead of one. Compounds of this sort are called *unsaturated*, and the linkage between the carbon atoms is called a *double bond*. Compared with saturated compounds, the unsaturated ones are in general much more reactive. Ethylene reacts with acids, for example,

$$C_2H_4 + HCl \rightarrow C_2H_5Cl$$

while the saturated compound ethane is indifferent to acids. Unsaturated compounds with straight chains and branching chains of carbon atoms (*olefin* hydrocarbons) form a series analogous to the methane series (C_2H_4, C_3H_6, C_4H_8, ...), and still other series are possible with compounds containing two or more double bonds per molecule. Also compounds exist with a higher degree of unsaturation, like acetylene, C_2H_2:

$$H-C\equiv C-H$$

Some of these unsaturated hydrocarbons are produced by the destructive distillation of coal, but the compounds themselves are practically nonexistent in nature. Presumably they are too reactive to persist in natural environments.

Certain unsaturated hydrocarbons with rings of carbon atoms, however, are more stable than those with carbon chains, sufficiently stable so that they are important constituents of some petroleums. The simplest ring hydrocarbon of this type is benzene, C_6H_6, for which the structure may be shown in two ways:

The second formula shows that the double bonds are not actually positioned between specific pairs of carbon atoms, as the first formula suggests; rather, the six extra electrons of the carbon atoms form a generalized bond (a so-called π-bond) that stabilizes the ring structure. Other ring hydrocarbons consist of a benzene nucleus with "side chains" of carbon atoms, and still others of two or more benzene rings joined together:

Toluene

Naphthalene

Collectively these compounds are spoken of as *aromatic* hydrocarbons, for the reason that many have pleasant odors.

This by no means exhausts the catalog of hydrocarbons. Such familiar substances as rubber and turpentine, for example, are hydrocarbons of still more complex structure than any we have considered. But for geologic purposes the four groups mentioned above—the paraffin, cycloparaffin, olefin, and aromatic hydrocarbons—are the only ones of general interest.

Alcohols

Structural formulas of alcohols are similar to those of hydrocarbons, but one or more of the H atoms is replaced by an OH group. The two simplest and most familiar alcohols are

$$
\begin{array}{cc}
\text{H} & \text{H H} \\
\text{HC}-\text{O}-\text{H} & \text{HC}-\text{C}-\text{O}-\text{H} \\
\text{H} & \text{H H}
\end{array}
$$

Methyl alcohol (Wood alcohol)   Ethyl alcohol (Grain alcohol)

These compounds appear in nature only ephemerally, as a result of some kinds of organic decay. Of more interest from a geologic standpoint is the trihydric alcohol glycerin (or glycerol):

$$
\begin{array}{ccc}
\text{H} & \text{H} & \text{H} \\
\text{HC} & -\text{C}- & \text{CH} \\
| & | & | \\
\text{O} & \text{O} & \text{O} \\
| & | & | \\
\text{H} & \text{H} & \text{H}
\end{array}
$$

This compound is one of the structural units in fats and oils, which are possible starting materials for the formation of petroleum hydrocarbons.

Organic Acids

Most organic compounds that behave as acids contain in their molecules the group

$$
-\text{C}\begin{array}{c} \nearrow \text{O} \\ \searrow \text{O}-\text{H} \end{array} \quad \text{or} \quad -\text{COOH}
$$

called the *carboxyl* group. Two simple examples are

$$
\begin{array}{c}
\text{H} \\
\text{HC}-\text{C} \\
\text{H}
\end{array}
\begin{array}{c}
\nearrow \text{O} \\
\searrow \text{O}-\text{H}
\end{array}
\quad \text{or} \quad \text{CH}_3\text{COOH}
\qquad
\begin{array}{ccc}
\text{H} & \text{H} & \text{H} \\
\text{HC}- & \text{C}- & \text{C}-\text{C}
\end{array}
\begin{array}{c}\nearrow\text{O}\\ \searrow\text{OH}\end{array}
\quad \text{or} \quad \text{C}_3\text{H}_7\text{COOH}
$$

Acetic acid   Butyric acid

The first is the acid of vinegar, the second the acid of rancid butter. The H atom of the carboxyl group is capable of dissociating slightly, the degree of dissociation (and thus the strength of the acid) depending on the length of the carbon chain and the kinds of atoms attached to it. Organic acids are found in soils and in decaying organic matter and, at least locally, are important agents of weathering.

Organic acids are of particular interest geologically because of their ability to decompose into carbon dioxide and a hydrocarbon:

$$
\text{C}_3\text{H}_7\text{COOH} \rightarrow \text{CO}_2 + \text{C}_3\text{H}_8 \tag{11-1}
$$

Butyric acid   Propane

This decomposition may be brought about by heating, by the action of certain bacteria, and by bombardment with the fast-moving particles of radioactive decay. The suggestion is obvious that this may be one of the steps in the formation of oil and natural gas.

Humic acids are a special type of organic acid formed by neutralizing a solution obtained when wood is treated with a strong alkali. Their molecules are large, and details of structure are uncertain. The humic acids themselves are only very feebly acidic; probably most of the materials called "humic acids" in geologic literature contain also molecules of smaller size and greater acidity (Sec. 4-4).

The group of acids called *fatty acids* are those whose molecules consist of a paraffin hydrocarbon with the carboxyl group substituting for one of its H atoms. Acetic acid and butyric acid, for example, are simple members of this group. The designation "fatty" is appropriate because some members of the group serve as structural units in the complex molecules of fats.

Fats and Oils

The reaction between an alcohol and an acid (the acid may be either organic or inorganic) is called *esterification*, and the products are water and a compound called an *ester*. For example,

$$\underset{\text{Ethyl alcohol}}{\overset{\text{H H}}{\underset{\text{H H}}{HC-C-OH}}} + \underset{\text{Acetic acid}}{\overset{O}{\underset{H-O}{C-CH}}}\ \longrightarrow\ \underset{\text{Ethyl acetate}}{\overset{\text{H H}\ \ O}{\underset{\text{H H}}{HC-C-O-C-CH}}} + \underset{\text{Water}}{H_2O} \quad (11\text{-}2)$$

Superficially the reaction resembles neutralization, in that OH from one molecule reacts with H from another to form water. Esterification, however, is a slow process, generally incomplete, and the resulting ester is commonly an undissociated, sweet-smelling liquid rather than a crystalline salt.

Geologically the most important esters are fats and oils, compounds whose molecules consist of glycerin combined with one or more long-chain fatty acids. A typical one is glyceryl palmitate:

$$\overset{\text{H}\quad O}{HC-O-C-(CH_2)_{14}-\overset{H}{\underset{H}{CH}}}$$
$$\overset{O}{HC-O-C-(CH_2)_{14}-\overset{H}{\underset{H}{CH}}}$$
$$\overset{O}{HC-O-C-(CH_2)_{14}-\overset{H}{\underset{H}{CH}}}$$

In solid fats the carbon chains are saturated, in liquid oils unsaturated. The chains are for the most part unbranched, and the more abundant ones contain even numbers of carbon atoms in the range 12 to 18. Fats and oils unite very slowly with water to set free glycerin and the fatty acids, a reaction that is the reverse of esterification (Eq. 11-2). This reaction is called *hydrolysis*. It resembles inorganic hydrolysis (Sec. 2-7) in that H_2O is broken up by the formation of weakly disso-ciated compounds, but little H^+ or OH^- is set free because both a weak acid and a weak base (alcohol) are formed. Organic hydrolysis reactions may be speeded up by heating or by the catalytic action of organic compounds called enzymes. The hydrolysis of fats is geologically important because it liberates organic acids which may in turn break up to form hydrocarbons (Eq. 11-1).

Certain other complex esters, especially the *waxes*, are included with fats under the more general term *lipids*.

Carbohydrates

Another important class of compounds produced by organisms is given the name carbohydrate, because their formulas look as if they consist of carbon united with water; in other words, they consist of carbon, hydrogen, and oxygen with the number of hydrogen atoms always twice the number of oxygen atoms. Simplest of the carbohydrates are sugars like glucose, $C_6H_{12}O_6$. More complex carbohy-drates may be regarded as chains of these 6-carbon units, each link in the chain after the first having lost a molecule of water: $C_6H_{12}O_6 \cdot C_6H_{10}O_5 \cdot C_6H_{10}O_5 \cdots$. Plants accomplish this linking together of carbohydrate units by reactions not yet fully duplicated in the laboratory; an example would be the formation of ordinary sugar (sucrose):

$$2C_6H_{12}O_6 \rightarrow C_{12}H_{22}O_{11} + H_2O \qquad (11\text{-}3)$$

Carbohydrates with only a few units per molecule (generally three or less) are called *sugars;* those with longer chains, represented by the general formula $(C_6H_{10}O_5)_x$, are *starches;* and those in which the subscript x attains values of the order of a few thousand are *celluloses*.

Most carbohydrates are readily decomposed, either by oxidation to CO_2 and H_2O or by anaerobic decay to a variety of carbon compounds simpler in composi-tion than the originals. The first step in decay is hydrolysis of the long chains [the reverse of Eq. (11-3)], setting free individual $C_6H_{12}O_6$ units. By dehydration and reduction it is conceivable that these units could be altered to hydrocarbons, but the process is not so simple and straightforward as it is for fatty acids. The simpler carbohydrates are not known in sediments of the past, but resistant celluloses have been reported from lignites of the early Tertiary.

Somewhat similar to cellulose, and occurring with cellulose in woody plants, are compounds called *lignins*. These resemble carbohydrates in gross composition, but the ratios of C to H and C to O are smaller, and that of H to O is larger. The structure is complex and difficult to characterize in simple terms. In general the lignins are more resistant than the carbohydrates to decay.

Proteins

Among the most complex of organic compounds are the proteins. These always contain nitrogen in addition to carbon, hydrogen, and oxygen, and some contain a little sulfur and phosphorus as well. The complex protein molecules can be broken down by hydrolysis (a reaction which in nature is catalyzed by enzymes, but which probably would take place slowly without their aid) into units called *amino acids*, of which two examples are glycine and alanine:

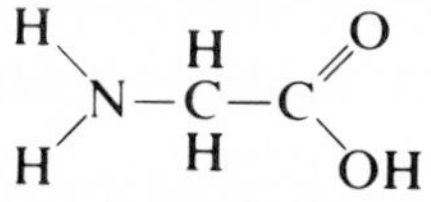

Glycine 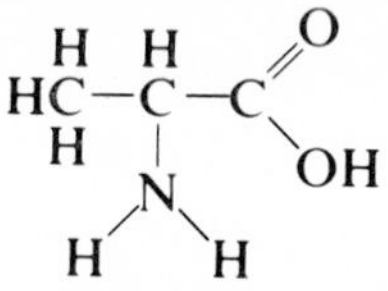   Alanine

The NH_2 group ("amino group") in one molecule can react with the carboxyl group of another molecule, giving a mechanism by which plants and animals can link together long chains of amino acids into the enormous molecules of proteins. A simple example of this linking mechanism is the reaction between two glycine molecules:

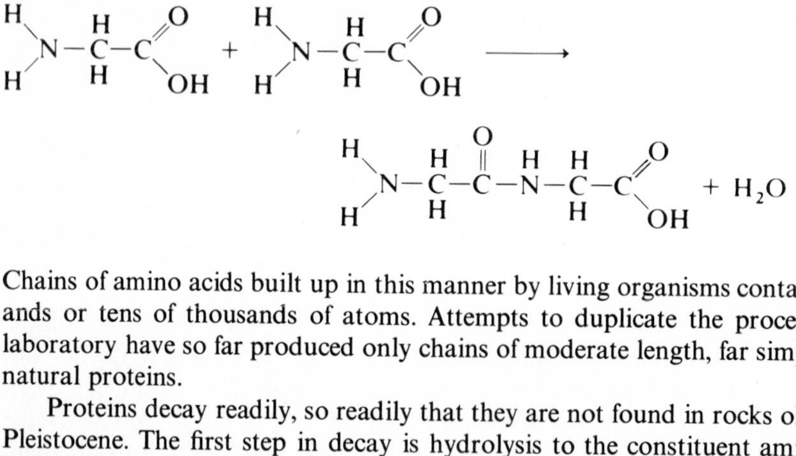

Chains of amino acids built up in this manner by living organisms contain thousands or tens of thousands of atoms. Attempts to duplicate the process in the laboratory have so far produced only chains of moderate length, far simpler than natural proteins.

Proteins decay readily, so readily that they are not found in rocks older than Pleistocene. The first step in decay is hydrolysis to the constituent amino acids [illustrated by the reverse of Eq. (11-4)]. The amino acids vary greatly in their resistance to decay; some disappear rapidly in either oxidizing or reducing environments, but a few are so stable that traces can be detected in rocks as old as Precambrian. During decay the nitrogen of an amino acid appears as N_2 or NH_3 (hence the odor of ammonia so common near dead animals). Under oxidizing conditions the carbon goes ultimately to CO_2, but in a reducing environment the carbon chain may remain intact. This means that amino acids, by loss of nitrogen, can be converted into simple organic acids, from which hydrocarbons may be formed just as they are from acids derived by hydrolysis of fats.

The substance called *chitin*, often described as a very resistant protein, is more properly regarded as a derivative of cellulose with an $NHCOCH_3$ group substituted for an H atom in each $C_6H_{10}O_5$ unit. Chitin serves as a structural material in many organisms and is so resistant to decay that it is preserved in the remains of invertebrates from rocks as old as the Cambrian.

Metals in Organic Compounds

Organic compounds containing metals have considerable geochemical importance because some of the rarer metals, particularly vanadium, molybdenum, and nickel, show remarkable enrichment in the organic matter of certain sediments. Three possible positions can be occupied by a metal atom in an organic molecule: (1) It may substitute for the hydrogen of a carboxyl group, forming a salt like sodium acetate. Many such salts are soluble in water, dissociating to form the metal cation and an organic anion. (2) It may be united directly to the carbon atom of an organic radical, forming a so-called *metallo-organic* compound; the most familiar example is lead tetraethyl, $Pb(C_2H_5)_4$, the antiknock ingredient in ethyl gasoline. (3) It may form the center of a complex ring structure, being attached to the carbon atoms of the ring through intermediate N, O, or S atoms. These are the *chelate* compounds, of which the chlorophyll of plants (Fig. 11-1) and the hemin in the hemoglobin of animals are familiar examples.

The relative importance of these three types of metal compounds in geologic materials is still unknown. The group of chelate compounds called *porphyrins* have received the most publicity, because representatives of this group have been found in petroleum and furnish one of the most convincing pieces of evidence that petroleum must be chiefly of organic origin.

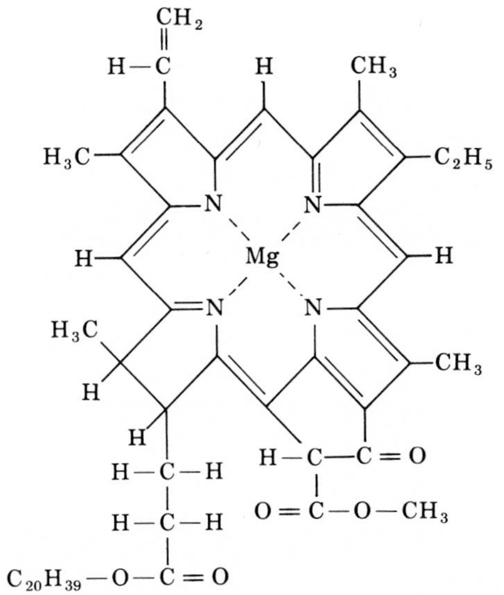

Figure 11-1 The structure of chlorophyll, a typical chelate compound. ·Each unlabeled corner of the complex ring structure is occupied by a carbon atom. The porphyrins of petroleum have similar basic structures, but (1) some of the side chains are modified or missing, and (2) the magnesium in the chelate ring is either absent or replaced by nickel or vanadium.

Optically Active Compounds

Some organic compounds, it has long been known, have the peculiar property of rotating the plane of polarization of a beam of polarized light which passes through them. One of Pasteur's great discoveries was that such "optical activity" is limited to compounds whose molecules contain a carbon atom attached to four *different* atom groups, for example,

$$
\begin{array}{c}
CH_3 \\
| \\
Cl-C-OH \\
| \\
NH_2
\end{array}
$$

The four groups may be arranged in space about the carbon atom in two different ways (see Fig. 11-2). Note that one arrangement is the mirror image of the other, so that the two cannot be superimposed.

Two isomers of this sort, called *optical isomers*, rotate the plane of polarized light by equal amounts in opposite directions. When such compounds are prepared in the laboratory, mixtures of equal amounts of the two isomers always result, so that no optical activity is apparent; but living organisms have the ability to manufacture one isomer to the exclusion of the other. Hence if a natural material containing carbon compounds can be shown to possess optical activity, the conclusion seems inescapable that living organisms played a role in its formation. The fact that petroleum contains optically active compounds is a second important piece of chemical evidence that it was formed by organic rather than inorganic processes.

11-2 ORGANIC REACTIONS

A few types of organic reactions are singled out for special mention, since they recur again and again in geologic environments.

Photosynthesis is a general term referring to any chemical combination aided by light, but if used without qualification, it means the reaction by which green plants manufacture carbohydrates from CO_2 and H_2O, using sunlight to supply

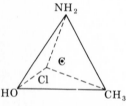 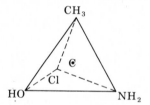

Figure 11-2 Optical isomers.

the necessary energy for this endothermic reaction, and using the green pigment chlorophyll as a catalyst. The overall process can be symbolized

$$6CO_2 + 6H_2O \rightarrow C_6H_{12}O_6 + 6O_2 \tag{11-5}$$

but it is actually much more complicated, involving a series of steps that are still imperfectly understood. This reaction is the ultimate source of all the organic material in plants and animals: not only carbohydrates, but fats and proteins as well, are manufactured from carbon fixed in this way. The reaction is essential also in continually renewing the supply of free oxygen in the atmosphere.

Aerobic decay, decay in the presence of ample free oxygen, can be symbolized as the reverse of photosynthesis. Like the forward reaction, it may take place in many complicated steps, but the end products are carbon dioxide and water. The same reaction takes place more rapidly when organic matter is burned, and it is likewise used by animals to furnish energy in the process of respiration. The three processes of decay, combustion, and respiration are largely responsible for maintaining the carbon dioxide supply in the atmosphere.

Anaerobic decay, decay in the absence or near absence of oxygen, is a more complicated process in which organic compounds undergo a sort of partial internal combustion. Much of the oxygen present in the original compounds combines with carbon to form carbon monoxide; the remainder of the carbon may go into hydrocarbons, or it may unite with hydrogen and a little of the oxygen to form compounds of complex structure. The direction taken by anaerobic decay depends on many factors: the nature of the original material, the temperature, the degree of exclusion of oxygen, and the types of bacteria present.

Reduction, in reference to organic reactions, means any process in which the amount of oxygen in a compound is decreased or the amount of hydrogen increased. If an organic compound containing oxygen is converted into a hydrocarbon, no matter by what method, the reaction involves reduction. If hydrogen is added to an unsaturated compound so as to form a saturated compound, the process is also reduction. Reduction reactions in which hydrogen is added are often referred to as *hydrogenation*.

Polymerization is a general name for any process by which large molecules are produced by causing small ones to join together. The reactions in plants by which 6-carbon sugar units are joined to form starches and cellulose, and by which amino acids are linked into the long chains of proteins, are typical polymerizations. One hypothesis about the origin of petroleum pictures the polymerization of simple hydrocarbons into complex ones, according to reactions like

$$C_2H_6 + C_2H_6 \rightarrow C_4H_{10} + H_2$$

Note that such reactions among saturated hydrocarbons must always result in the formation of hydrogen as a by-product.

The opposite sort of reaction, the splitting of big molecules into small units, is called *depolymerization;* the breakdown of starch into sugar and that of protein into amino acids come to mind as good examples. Depolymerization reactions of long-chain hydrocarbons, brought about by heat or catalysts or both, are often called *cracking;* a simple example is the cracking of octane:

$$C_8H_{18} \rightarrow 2C_2H_4 + C_4H_{10}$$

Note that the cracking of a saturated hydrocarbon must always give unsaturated compounds among the products, unless hydrogen is present while the cracking takes place.

Characteristics of Organic Reactions

As should be evident from this discussion, organic processes in nature differ in several important respects from the inorganic reactions we have discussed hitherto. Starting materials are generally complex mixtures rather than single compounds, and precise formulas for at least some constituents of the mixtures cannot be determined. Reactions are slow and often incomplete; the general way in which components of a mixture react can be described, but details of the reaction for each substance are often obscure. Furthermore, a complex organic molecule can generally react in many different ways, the different reactions being roughly equivalent energetically and one or another being favored by organic catalysts called enzymes.

These differences mean that concepts like equilibrium, free energy, and oxidation potentials, on which we have leaned heavily in previous discussions, are no longer so helpful. For organic reactions in the laboratory, where individual substances can be isolated and variables can be controlled, these quantitative concepts are still applicable, but for reactions in nature they give us little information that we could not guess from qualitative observations. In the study of organic chemistry the unifying, fruitful ideas that have geologic applications are concerned with the architecture of molecules: the way various carbon groups fit together to form molecular structures, and the properties that these groups impart to the molecules in which they occur. The ability to predict the behavior of organic materials in geologic environments depends chiefly on an understanding of how broad classes of molecules with similar structures behave when conditions of temperature, pressure, and compositions of solutions change.

Summary This very sketchy review of elementary organic chemistry has been limited strictly to the compounds and processes of interest in the study of geochemical phenomena. Hydrocarbons have occupied a prominent place, because they are the chief constituents of oil and natural gas and are likewise among the important substances given off when coal is heated. Of the more complex compounds, only those of possible interest as source materials of oil and coal have been mentioned: fats, carbohydrates, proteins, and a few less abundant but more resistant materials. The general processes of photosynthesis, decay, polymerization, and depolymerization are those to which geologically important substances are continually subjected.

The review has consisted entirely of statements of fact, linked together by dogmatic assertions about theoretical organic chemistry. This is a departure, of course, from our usual practice in this book of accepting no conclusions until their basis has been demonstrated. The conclusions about isomers, about structural formulas, about the makeup of complex compounds, all can be amply justified by experiment. But this is organic chemistry rather than geochemistry, and a full development would take us far afield from our main line of inquiry.

11-3 CARBON IN ROCKS

Igneous Rocks

Elemental carbon occurs in igneous rocks as the two crystal forms graphite and diamond. Under ordinary pressures graphite is the stable form at all temperatures; diamond is truly stable only at high pressures. Appropriately, the only igneous rocks with which diamond is found are ultramafic bodies whose pipelike form and brecciated structure suggest that they have moved rapidly upward from great depths. At the level where the diamond-bearing rocks presumably form, the pressure due simply to the weight of overlying rock would be sufficient to make diamond stable.

The source of the carbon in igneous rocks remains an open question. In rocks derived by differentiation from deep parts of the crust or upper mantle the carbon may well be an original constituent of the magma ("juvenile carbon"). In rocks formed from the partial melting of preexisting crustal rocks an obvious alternative source is the carbon of the original sediments, contained either in organic matter or in carbonate minerals.

Iron carbide [the mineral cohenite, $(Fe, Ni)_3C$] is a constituent of some meteorites and is reported as a minor mineral in some rare terrestrial occurrences of native iron. Its occurrence in meteorites has led to a speculation that it and other carbides may be an important form of carbon in the earth's deep interior, and that their reaction with water may have produced the methane and other hydrocarbons sometimes found in volcanic gases and in cavities of granitic rocks.

The principal carbon gases reported from fumaroles in volcanic areas and obtainable in traces by heating igneous rocks are CO_2, CO, and CH_4. The relative amounts of the three are probably determined by equilibrium in such reactions as

$$H_2 + CO_2 \rightleftharpoons CO + H_2O$$

$$CO + 3H_2 \rightleftharpoons CH_4 + H_2O$$

The position of equilibrium depends on the concentrations of H_2 and H_2O, hence on the oxidation state of the lava with which the carbon gases are associated. CO_2 should be the chief gas in oxidizing environments (high H_2O), and CH_4 in reducing environments (high H_2). These carbon gases found in and with igneous rocks are probably in part juvenile, but some of the methane observed escaping from lavas that have flowed over soils and vegetation may be explained more plausibly as a product of decomposition of the buried and heated organic matter.

In the form of carbonate, carbon occurs sparingly in igneous rocks as carbonate groups in the structure of apatite and scapolite crystals. Calcite is not unknown as a minor constituent of ordinary igneous rocks, but there is usually uncertainty as to whether it is a primary constituent, a remnant of partly assimilated limestone xenoliths, or a mineral introduced later during alteration of the rock by solutions. The peculiar rocks called carbonatites—masses up to several square kilometers in cross section consisting largely of simple carbonates and showing intrusive relations with their surroundings—almost surely represent igneous carbonate on a large scale, in the sense that they solidified from a magma consisting chiefly of carbonate material. This conclusion was for a long time in doubt, because carbonate magmas seemed impossible; carbonate minerals at magmatic temperatures, it was well known, decompose into CO_2 and refractory oxides. Recrystallization of big limestone inclusions, or later introduction of carbonate by hydrothermal solutions, were suggested as more likely hypotheses. But recent experimental work showing that water-rich carbonate magmas can indeed exist under geologically reasonable conditions at high pressures, together with observations of carbonate-rich lavas extruded from East African volcanoes, has given firm support to the idea that most carbonatites are truly magmatic bodies.

Black bituminous material is sometimes found in or near veins containing metallic sulfides. If an igneous source is postulated for the metals, it is certainly not unreasonable to assume a similar source for the bitumen (or for the hydrocarbons that polymerized to form the bitumen). Alternatively, the bitumen can be explained as organic material distilled out of sedimentary beds cut by the veins. A clear decision between the two hypotheses is seldom possible.

Thus, in general, when carbon and carbon compounds are found in or near igneous rocks, an argument arises as to whether the carbon has come as juvenile material from the earth's interior or is derived from the organic material and carbonate of adjacent or assimilated sedimentary rocks. It seems established beyond reasonable doubt that some carbon has escaped to the earth's surface from the interior during geologic time (Sec. 21-5), but locally field relations provide good evidence that sedimentary materials have made a substantial contribution to the carbon content of igneous rocks. For most occurrences, from geologic evidence alone, the relative amounts derived from the two possible sources can only be guessed. In some cases a decision as to the probable origin can be made by measuring the ratio of the stable carbon isotopes, ^{12}C and ^{13}C, since the lighter isotope shows a slight preferential concentration in organic materials.

Sedimentary Rocks

Elemental carbon in sedimentary rocks occurs only as detrital grains of diamond or graphite. The major part of sedimentary carbon is in the form of carbonates, but a large amount occurs also in various mixtures of organic compounds. It is this organically derived material that is of special interest here.

The nomenclature of sedimentary organic materials is ambiguous and redundant. This is almost inevitable in view of the wide range in characteristics of the different organic mixtures, the extreme local variability, and the difficulties in finding diagnostic properties either of a petrographic or chemical nature. The word *bitumen* is often used loosely for practically any organic material in rocks; more strictly it should refer to liquid or solid hydrocarbon mixtures largely soluble in carbon disulfide. Thus petroleum is liquid bitumen. *Asphalt* is a solid or semisolid bitumen with a boiling point in the range 65 to 98°C; strictly the term should be limited to solid material associated with or derived from oils with a high content of cycloparaffin hydrocarbons ("asphalt-base crudes"). The corresponding solid or semisolid bitumen associated with petroleum rich in paraffin hydrocarbons ("paraffin-base crudes") is called *ozokerite;* it consists chiefly of hydrocarbons whose molecules have 25 to 29 carbon atoms. Other less well defined solid bitumens are gilsonite and kerogen. In contrast to bitumen, *coal* consists mostly of organic material not appreciably soluble in organic solvents; it contains little or no hydrocarbon material as such, but gives hydrocarbons on destructive distillation. The hydrocarbons from coal include those belonging to both the aliphatic and aromatic series, and, unlike petroleum, those of the olefin series also. Another name often applied to solid organic material, especially to the organic constituents of black shale, is *sapropel;* it may comprise both coaly and bituminous substances, but is dominantly bitumen. The sapropel of oil shale, from which large amounts of petroleum hydrocarbons can be distilled, is called *kerogen.*

Metamorphic Rocks

The carbon of metamorphic rocks is chiefly in the form of graphite and carbonate minerals, both generally of ultimate sedimentary origin. Increasing metamorphic grade is often reflected in the degree of perfection of the crystal structure of graphite. *Anthracite* is a product of mild thermal or tectonic metamorphism of coal; increasing thermal metamorphism converts anthracite into graphite. The hard, black, carbon-rich materials called *anthraxolite* and *shungite* are probably metamorphic equivalents of bitumen.

11-4 ORIGIN OF PETROLEUM

Petroleum is a complex and variable mixture of hydrocarbons. Some idea of the extent of variation in composition is conveyed by Table 11-2, which shows the compositions of "straight-run" gasolines distilled from different petroleums at temperatures between 40 and 102°C. Besides the hydrocarbons, most petroleums contain small amounts of other substances, notably porphyrins and optically active compounds. Olefin hydrocarbons, free hydrogen, and carbon monoxide are nearly or completely lacking.

Table 11-2 Percentage composition of straight-run gasolines obtained by fractional distillation of petroleum from various sources. The compositions refer to the fractions obtained between 40 and 102°C†

Kind of hydrocarbon	Ponca City, Okla.	Wink-ler, Tex.	Con-roe, Tex.	Mid-way, Calif.	Green-dale, Mich.	Brad-ford, Pa.	East Texas
Normal paraffin	35.3	9.5	17.6	9.9	62.2	34.1	24.6
Branched paraffin	20.2	61.4	19.6	21.4	13.0	32.0	27.1
Alkylcyclopentane‡	23.1	20.4	16.6	40.7	7.8	13.4	25.9
Alkylcyclohexane‡	20.2	8.2	42.6	27.4	15.5	20.0	21.9
Aromatic	1.1	0.5	3.6	0.6	1.5	0.5	0.5

† *Source:* *American Chemical Society Monograph* 121, 1953.

‡ The term *alkyl* refers to a radical obtained by removing one H from a methane hydrocarbon. Examples of alkyl radicals are methyl (CH_3 —), ethyl (C_2H_5 —), and butyl (C_4H_9 —). An alkylcyclopentane is a hydrocarbon whose molecule consists of the cyclopentane ring with one or more of its hydrogens replaced by alkyl groups.

This particular kind of mixture of organic compounds has not been observed forming in nature and has not been produced in the laboratory under conditions that could conceivably exist in nature. Hence the field remains open for hypotheses of origin, and the number of hypotheses that have been proposed is very great. The problem is peculiarly difficult because petroleum is a fluid: this means that it may have migrated long distances from its place of formation and that it cannot retain fossils or structural features which might furnish hints about origin.

In the early years of this century an inorganic origin was widely advocated. The methane found in some volcanic emanations and in fluid inclusions of igneous rocks was supposed to indicate the slow and continuous sweating out of this gas from the earth's interior during geologic time; in the upper part of the crust the methane was then thought to polymerize, by a process never very precisely described, into hydrocarbons of higher molecular weight. Several weighty objections, drawn both from geology and from chemistry, have reduced proponents of this hypothesis to a very small minority. On the chemical side the presence of optically active compounds and of porphyrins is difficult to reconcile with an inorganic origin, and a plausible mechanism for the large-scale polymerization of methane is hard to devise. From a geologic standpoint the most crushing argument is the almost invariable occurrence of petroleum in sedimentary rocks. In the few places where oil is pumped from fractured igneous and metamorphic rocks, its source in adjacent sediments is clear. There is little argument today with the hypothesis that oil has an organic origin—that it forms by some sort of anaerobic change in the organic material of sediments and sedimentary rocks.

This is not to deny, of course, that methane does occur with some igneous rocks, in places accompanied by minute amounts of heavier hydrocarbon gases and solid bitumens. It is equally undeniable that the carbon and hydrogen of petroleum must have had an ultimate inorganic origin sometime in the earth's remote past. The point is simply that the constituents of petroleum as we find it today have been formed chiefly by the degradation of compounds produced by living organisms rather than directly from methane out of the earth's interior.

Clearly it would be desirable to have proof that petroleum hydrocarbons can actually form in nature from organic compounds in sediments. Definitive evidence of this sort seems peculiarly difficult to obtain. Modern analytical techniques have made it possible to demonstrate that hydrocarbons exist in many recent organic muds, in quantities that are very small but nevertheless sufficient, given a suitable method for concentration from large volumes of sediments, to account for the origin of petroleum. A difficulty is that these hydrocarbons are not in all respects similar to the hydrocarbons of petroleum. The aromatic constituents are less complex than in petroleum, and the proportion of hydrocarbons of intermediate molecular weight (C_3 to C_{14}) is far too small. Another notable difference is that the mixtures from recent muds often show higher concentrations of hydrocarbon molecules containing odd numbers of carbon atoms ($C_{13}H_{28}$, $C_{15}H_{32}$, and so on) than of those showing even numbers, while in petroleum there is no such preference. The emphasis on odd numbers is characteristic also of hydrocarbons that can be extracted in minute amounts from present-day land plants; hence a conceivable explanation for the hydrocarbons in recent muds is deposition of detrital plant material, with little or no chemical change after burial. On the other hand some muds, in which organic material of marine origin is abundant, show a more even distribution of molecules with odd and even numbers of carbon atoms, a distribution which probably requires formation of hydrocarbons in the sediment. The relative contribution of hydrocarbons from various sources to the formation of petroleum is still a matter of dispute, but there is general agreement that at least some of the hydrocarbons are generated from other kinds of organic matter as the sediment accumulates.

If the slow conversion of other organic substances to hydrocarbons is assumed, what kinds of compounds would be the most likely starting materials? This question is still a subject of active debate. Some workers think that almost any kind of organic material can serve as a source for petroleum; in support of this conclusion they point to the general similarity of petroleums produced in different geologic periods, and to the occurrence of petroleum with sediments formed in widely different environments. Others are impressed with the greater abundance of petroleum in sedimentary rocks of marine rather than freshwater origin, hence with its probable derivation from marine organisms. Since plants and animals of the sea contain less carbohydrate and lignin than terrestrial organisms, this restriction would suggest fats and proteins as the probable major precursors of petroleum.

From a chemical standpoint, fats have long seemed the most likely group of organic compounds to be the chief source of hydrocarbons. They have a molecular structure from which hydrocarbons can be derived by fairly simple chemical changes (Sec. 11-1); they are stable but only moderately so, hence should be able to survive initial decay and burial but should alter slowly in the anaerobic environment of a buried sediment. Carbohydrates, by contrast, require more extensive chemical change for conversion into hydrocarbons. Most of them are unstable, hence unlikely to endure through the early stages of decay; the more resistant carbohydrates and lignins go eventually into coaly material rather than hydrocarbons. Proteins, like fats, have structures from which hydrocarbons are derivable by simple degradational processes. Some amino acids from the hydrolysis of proteins may well contribute to the formation of oil, especially to the production of the lighter hydrocarbons, but in considerable part amino acids are so unstable that they are likely to disappear before or shortly after burial.

In the anaerobic breakdown of fats the first step is almost certainly hydrolysis, setting free the fatty acids from the glycerin of the original esters. Details of the further changes that produce hydrocarbons from the fatty acids are uncertain. In the laboratory, fatty acids break down with evolution of CO_2 [Eq. (11-1)] when heated to temperatures of 320 to 400°C, but such temperatures are unreasonable in sedimentary environments. The simple fact that petroleum is found in Pliocene and Pleistocene sediments that cannot have been buried to more than a few thousand feet is good evidence that the alteration to hydrocarbons must somehow take place at temperatures below a maximum of 120°, and probably well below 100°. Possibly the thermal decomposition that requires high temperatures in short-time laboratory experiments might be effective at much lower temperatures during geologic times; support for this idea comes from experiments by Hoering and Abelson (1963) showing that hydrocarbons are formed at a measurable rate in the laboratory by heating the kerogen of Green River shale to temperatures as low as 185°.

A number of suggestions have been made as to how the low-temperature conversion might be speeded up. Bacteria, for example, are known to produce methane from organic acids at ordinary temperatures, and conceivably might be responsible for other hydrocarbons; but the generation by bacterial activity of any hydrocarbon except methane has not been demonstrated experimentally. Bombardment of fatty-acid molecules with alpha particles from radioactive decay is another possible process. Such bombardment is known to split off carboxyl groups, leaving hydrocarbon residues; but the process results in formation of free hydrogen and helium, neither of which is generally present in petroleum or natural gas, and furthermore the amount of hydrocarbon that could be so produced is quantitatively insufficient to account for large accumulations of oil. Yet another possibility is the catalytic action of clay minerals; aluminosilicates are widely used as catalysts commercially in the high-temperature cracking of large hydrocarbon molecules, but they have never been proved effective at low temperatures. Thus a detailed mechanism for the production of petroleum hydrocarbons in nature still eludes discovery, but it seems a plausible guess that the slow degradation of fatty acids, perhaps aided by some kind of catalysis, is responsible for at least some part of the process.

Another aspect of the geochemistry of petroleum that is currently the subject of intensive research is the changes that may take place as oil migrates from one kind of rock to another and into environments with different temperatures and pressures. The hydrocarbons of petroleum are stable compounds in comparison with most organic substances and would not be expected to change rapidly or easily, but over long periods of time and in response to extreme environments they might well show some alteration. The differences in detail from one petroleum to another—differences in proportions of light and heavy hydrocarbons, in proportions of aliphatic, aromatic, and cycloparaffin compounds, in the content of porphyrins, sulfur, and traces of metals—can be attributed reasonably to chemical changes after original formation of the oil. The difficulty is that such differences can be explained equally well by variations in source material, in conditions of formation, or in geologic age. To sort out these numerous factors, especially when the mechanism of original hydrocarbon production remains uncertain, has so far proved impossible. Chemical trends in petroleums of different ages, from different kinds of reservoir rock, from different presumed source materials, from environments of different temperature and pressure, have often been pointed out; but the trends seem as often to be in one direction as in another, and their meaning is still a mystery.

In summary, petroleum and natural gas are in large part products of low-temperature anaerobic decay of organic matter buried with sediments. As the principal substances from which petroleum hydrocarbons arise, fats and some proteins seem the most likely choice, but this restriction is far from certain. Details of the conversion of organic matter into hydrocarbons, and of subsequent changes in the hydrocarbons during migration and aging, remain a matter of lively argument.

11-5 ORIGIN OF COAL

Coal is a mixture of compounds of high molecular weight and complex structure, containing a large percentage of carbon and small amounts of hydrogen, oxygen, and nitrogen; some coals also contain sulfur, phosphorus, and traces of other elements. Most of the compounds have not been identified. Free carbon is not present except in coals that have been metamorphosed. The various hydrocarbons and other simple organic compounds obtained from coal by heating are probably not present as such, but form at high temperatures by decomposition of complex original substances. In addition to organic material, most coal contains inorganic substances like quartz, feldspar, and clay minerals.

In contrast to petroleum, coal betrays its origin by fossils, by stratification, and by interbedding with sedimentary rocks. There seems no possible doubt that coal is a product of partial decomposition, under anaerobic conditions, of buried vegetation, largely terrestrial vegetation from a swamp environment. The chief geochemical questions about coal are matters of detail: how the transformation of plant material takes place, how the composition of coal is influenced by the nature of the vegetation and by the specific conditions of decay, what are the recognizable stages of the alteration, and so on. A further important question, but one that has no direct bearing on the formation of coal, is the reason for abnormal concentrations of rare metals like uranium and germanium in some coal beds.

Despite the difficulty of identifying particular compounds in coal, gross chemical composition is often useful in characterizing different coals and different megascopic constituents of coal. The ratio of carbon to the volatile constituents (H, O, and N) permits coals to be arranged in a general order of increasing carbon content; this order roughly parallels an arrangement according to disappearance of plant remains, or according to increase of hardness, or according to increase in ash content. Steps in this progression are lignite, subbituminous, and bituminous coals. The suggestion is obvious that these products represent stages in the progressive alteration of plant material, bituminous coals having been subjected to alteration over longer times and at higher temperatures and pressures than lignite. This assumption of a continuous sequence of changes is implicit in the term *rank*, lignite being a low-rank coal and bituminous coal relatively high-rank. It is by no means certain that *all* lignite would ultimately become bituminous coal, or that *all* bituminous coal has passed through a stage like present-day lignite, but the concept of progressive change is probably valid for most coals.

Many coals are conspicuously banded, the bands differing in such obvious characteristics as luster, hardness, and kind of fracture. On the basis of chemical composition and the nature of enclosed fossil remains, the bands can be grouped into two major kinds of material: the hard, shiny, obsidianlike coal derived principally from lignin and cellulose, and the dull, dirty coal derived chiefly from fats, waxes, resins, and resistant proteins. The former is called *humic* coal, the latter *sapropelic* coal (roughly equivalent terms are *attritus* and *anthraxylon*). Sapropelic coals are derived from material similar to that which forms bitumens and, in fact, grade into bitumens through hydrocarbon-rich varieties of coal called cannel coal and boghead coal.

The relative importance of lignin and cellulose in the origin of humic coal has occasioned much argument. Most cellulose decays so rapidly when a plant dies that its preservation by burial seems unlikely, but on the other hand plant structures are sometimes preserved which almost certainly were cellulose initially. Very likely both materials play a part in the formation of coal, and there seems no need to pursue the argument here.

If anaerobic decay of organic matter accounts for both coal and petroleum, what determines whether one or the other will form in a particular situation? An important factor is surely the kind of organic matter, since coal is derived from land vegetation consisting largely of lignin and cellulose whereas oil comes mainly from marine organisms, which contain principally lipids and proteins. The conditions of alteration may also play a role, but we know so little about details of the alteration to either coal or oil that speculation seems unprofitable.

11-6 ORGANIC MATTER IN BLACK SHALES

Nearly all sediments contain detectable organic matter, the general average being about 2%; in black shales the percentage is commonly a few percent but may rise much higher. Black shales, in fact, grade into pure organic materials, into coal on the one hand and into bitumen on the other. The organic matter of ordinary sediments may be either dominantly coaly or dominantly bituminous, or it may be a mixture of both. A particularly important variety of organic-rich sediment is the kind called *oil shale*, a shale containing large amounts of bituminous material from which hydrocarbons like those of petroleum can be distilled.

The origin of organic materials in black shales poses no problems we have not already considered in talking about coal and petroleum. Again we are dealing with partial anaerobic decay of buried organic matter, the only difference being that now an abundance of inorganic material has been deposited together with dead organisms.

In contrast to coal and oil, black shales can be observed in process of formation today. More accurately, places are known where organic matter in large amount is being deposited with fine-grained sediment that seems to have all the necessary qualifications to serve as the starting material for black shale. Such places are stagnant marine basins: deep spots in Norwegian fjords, the bottom of the Black Sea, the Cariaco trench off the coast of Venezuela. The necessary conditions are water that does not circulate, or that circulates only rarely, and a very slow deposition of clastic debris. The restricted circulation means that oxygen in the bottom water is used up in oxidizing organic matter and is not replenished, so that dead organisms falling to the bottom cannot decay completely to CO_2 and H_2O. Instead, the carbon compounds of their bodies only partially decompose, through reactions aided by anaerobic bacteria, and the products of these reactions accumulate as a black mud. The water column in such environments shows a decrease in dissolved oxygen from the aerated surface waters downward, then an increase in hydrogen sulfide toward the bottom. An example of these changes is given in Table 11-3, which lists analyses of water samples taken at increasing depths in a Norwegian fjord.

Table 11-3 Chemical characteristics of water in Bolstadsfjord, western Norway†

Measurements made June 17, 1932, at a spot where the depth of the fjord is 138 meters

Depth, meters	Tempera- ture, °C	Salinity, g/1,000 g	Oxygen, cm³/liter	Oxygen, % of saturation	H_2S, cm³/liter	pH
1	12.90	0.47	7.99	106.2	0	6.80
10	8.43	0.68	8.40	101.5	0	6.85
15	7.18	1.34	8.61	101.4	0	7.05
20	5.06	14.64	7.89	96.0	0	7.85
40	4.33	18.08	6.20	75.8	0	7.70
80	4.03	20.79	0	0	0.15	7.00
130	4.09	21.09	0	0	1.05	6.90

† *Source:* Strøm, 1936.

A few excellent detailed studies have been made of present-day black muds, but our chemical information is still limited largely to overall characteristics rather than to specific reactions. The pH of water in contact with black mud is generally between 5.5 and 7, somewhat more acidic than normal seawater. The redox potential varies, depending on whether oxygen is wholly or only partly excluded; it may reach extreme reducing values in the neighborhood of -0.5 volt. Below the surface of the mud, where the decomposition of organic material has gone on longer, the pH tends to rise and the Eh to become more reducing, but these changes may be reversed in some layers. The particular organic substances produced in the black mud are largely unknown. Methane is generally one recognizable product; heavier hydrocarbons, if formed at all, appear only in trace amounts. Sulfur from original protein is largely converted to hydrogen sulfide, which gives the mud its characteristic odor—hence the common expression for such areas, "foul bottoms." Some of the sulfur combines with iron to form iron sulfide, perhaps first as fine-grained, black hydrotroilite ($FeS \cdot nH_2O$), but this compound changes quickly to the more stable pyrite. Nitrogen appears chiefly in the form of ammonium ion, and phosphorus as one of the phosphate ions.

All gradations can be found, of course, between such extreme reducing environments and the oxidizing environment of shallow seawater on an open shelf. In the latter environment most organic material decays to carbon dioxide and water before burial can take place, regardless of how abundant organisms may be. A little of the more resistant material generally persists long enough to be buried, and local accidents of sedimentation may incorporate lenses and pockets of undecayed organic matter in the accumulating muds and sands. Once buried, the carbon compounds are largely protected from oxidation and undergo the same sort of anaerobic decay as in the black-mud environment. Probably the nature of the ultimate product depends in large part on the kind of material that was buried—humic material from original wood fragments, bitumen from marine organisms—but the particular environmental conditions and the degree of previous oxidation may also play a role.

The various kinds of organic matter found in sediments can often be correlated, at least tentatively, with the nature of the original environment. A marine black-mud environment is suggested by an abundance of bitumen containing scattered pyrite crystals; such material, by analogy with chemically similar material in coal, is often called *sapropel*, and the inferred environment is spoken of as *sapropelic*. A dark shale with less abundant organic matter and without pyrite may have formed in a less strongly reducing environment, perhaps where oxygen was nearly but not quite removed from bottom waters; a sediment and an environment of this sort are described by the term *gyttja*. Shiny coal-like material in terrestrial sediments is doubtless an alteration of wood fragments or other debris from land plants.

Such general correlations and inferences are pretty obvious, but many knotty chemical problems regarding organic material in sediments remain to be unraveled. We need, for example, more information about pH and Eh in the various kinds of environment where organic matter accumulates and about the relation of these quantities to the kinds of organic compounds formed. It would be useful also to have pH and Eh measurements in water standing in contact with organic materials of ancient sediments; perhaps with enough data of this sort the "reducing action of organic matter," so often referred to in geologic discussions, could be given a more precise meaning. Bacteria obviously play an important part in both aerobic and anaerobic decay. To what extent do the nature and number of bacteria in a particular environment determine the chemical changes that take place? Are the same bacteria always present in a given environment, so that they can be considered a constant factor, or is it possible for two environments, otherwise identical, to have different kinds of bacteria and therefore to produce radically different kinds of organic material? The problem of experimentally distinguishing different kinds of organic matter in sediments has been attacked with vigor in recent years, but still remains troublesome. And finally there is the cluster of problems mentioned in preceding sections regarding the formation of coal and oil: what are the specific reactions that take place, why does coaly material form from one kind of organic debris (or in one kind of environment) and bituminous material from another, what determines the extreme variations in composition of different coals and oils, and so on and on.

11-7 CARBON COMPOUNDS AS REDUCING AGENTS

We have referred frequently to the reducing nature of the environments in which carbon compounds accumulate. To some degree this manner of speaking puts the cart before the horse: carbon compounds do not accumulate *because* an environment is reducing, but more correctly an environment is reducing *because* carbon compounds accumulate there. The condition necessary for the organic matter to be preserved is the lack of access of atmospheric oxygen, but this in itself would not make an environment markedly reducing. The reducing quality is a result of the heaping up of undecayed and partly decayed plant and animal debris. Other reducing agents that may be present—methane, hydrogen sulfide, ferrous ion—are by-products of the alteration of the organic material which establishes the redox potential of the environment. It would simplify sedimentary geochemistry greatly if such a redox potential could be specified numerically, but to do this in any meaningful way seems all but impossible.

Ancient accumulations of carbon compounds—coal beds, oil pools, and the flakes and lenses of organic matter in ordinary sediments—likewise serve as reducing environments, often causing precipitation of metals from circulating waters, but again the assignment of numerical values to their reducing power is troublesome. From the point of view of a geochemist trying to make his science more quantitative, this is a discouraging circumstance, for organic matter is by far the most common and most widespread reducing agent in sediments and sedimentary rocks.

The reducing ability of organic compounds comes ultimately from photosynthesis,

$$6CO_2 + 6H_2O \rightarrow C_6H_{12}O_6 + 6O_2 \tag{11-5}$$

In this process carbon is evidently reduced and oxygen is oxidized. A strong reducing agent (the carbohydrate) and a strong oxidizing agent (O_2) are produced from a weak reducing agent (H_2O) and an equally weak oxidizing agent (CO_2); obviously such a reaction cannot take place spontaneously but must receive energy from an outside source, in this case from sunlight. The carbon compounds produced directly by photosynthesis undergo further changes in the plant and later in animals, many of these changes involving additional oxidation and reduction. The changes cannot be easily specified by changes in oxidation state, as they can for inorganic reactions. We could formally assign an oxidation number of 0 to the C in carbohydrates, -4 to the C in methane, $-2\frac{1}{2}$ to the C in butane, $+2$ to the C in acetic acid, but the numbers would correspond only roughly with the oxidation-reduction behavior of these compounds. In substances whose molecules are held together almost wholly by covalent bonds, the concept of change of oxidation number is less useful than in ionic compounds for measuring oxidizing and reducing strengths. We might have better luck by determining the oxidation-reduction behavior of organic compounds in reactions with substances of known strength as oxidizing and reducing agents, but here we encounter the difficulty that reactions of organic compounds are mostly very slow, so that equilibrium is not attained in a reasonable time.

For individual compounds the situation is not completely hopeless. Reactions can be tried at higher temperatures, where rates are faster; free energies can be obtained from heats of combustion and entropy changes; and from these numbers equilibria can be worked out for reactions at ordinary temperatures. For application to geologic environments, however, the effort is hardly worthwhile. Generally the particular organic compounds present are not all known; generally also a number of different reactions are possible, differing only slightly in amount of free-energy change, and the ones that predominate are determined more by rates than by equilibrium positions. In such circumstances attempts to derive a theoretical redox potential are foolish.

What recourse do we have? For present-day environments of deposition, we can get empirical results by direct measurements of redox potentials. Surprisingly few such measurements have been recorded in the literature, but what data we have indicate potentials of the order of -0.1 to -0.5 volt for environments in which organic matter is accumulating. These figures have the usual defect of potentials obtained for slow reactions: it is never certain to what extent they represent equilibrium potentials and to what extent they are merely effective potentials, changing slowly with time, determined by rates of particular reactions. But the numbers correlate fairly well with the kinds of metal ions and metal compounds found associated with the organic matter, and give at least a rough measure of reducing conditions in these environments.

Another possible measure of redox potential is the nature of inorganic compounds found associated with the organic matter. Pyrite, for example, can form only at Eh values below about $+0.2$ volt in acid solutions and -0.2 volt in basic solution, so that its presence in an organic-rich sediment means immediately that the redox potential is below these maxima (Fig. 9-5). This mineral, as mentioned in the last section, is the principal distinguishing feature of the "sapropelic" environment as opposed to the less reducing "gyttja" environment. The occasional presence of metallic copper in swamp muds gives a similar limiting value to the Eh of the environment.

The nature of associated minerals gives us about the only satisfactory empirical measure of the redox potential of ancient organic materials, as they affect substances dissolved in circulating waters. The precipitation of uranium compounds in groundwater by coal, for example, means that the coal must have a reducing ability at least sufficient to change the oxidation state of uranium from $+6$ to $+4$. The precipitation of metallic copper and metallic silver by coal or bitumen, and the reduction of sulfate in gypsum to free sulfur in the presence of petroleum, permit similar estimates of minimum Eh values. Some of these reduction reactions can be duplicated in the laboratory, for example by letting solutions of metal ions percolate through coal, but natural occurrences give a better idea of the reducing ability of the organic materials over long periods of time.

Inadequate as this sort of information is, it gives us a basis for speculating about the minimum redox potentials attainable with organic materials. Theoretically, one might suppose that such potentials would not be limited, as are those in aqueous solutions, by the fact that the hydrogen in water is liberated if redox potentials become very low (Sec. 9-4). In the interior of organic masses, completely out of contact with air and water, substances might conceivably form more strongly reducing than free hydrogen. There is no geologic evidence, however, for such extreme conditions; the substances found associated with organic materials are invariably compounds precipitable in the normal range of redox potentials. The lowest potential actually recorded in water associated with organic material is about -0.5 volt at pH 9, well above the limiting value of the hydrogen electrode.

The redox potential of an organic-rich environment should be sensitive to the pH of the environment, and this in turn should be determined at least in part by the substances produced as the organic matter decays. Here again the complexity of the possible reactions prevents a satisfactory theoretical analysis. Under aerobic conditions we should expect an abundance of CO_2, hence unusual acidity in adjacent solutions; the limiting amount of CO_2 would be determined by atmospheric pressure, and the limiting pH would then be that of a carbonic acid solution in equilibrium with CO_2 at 1 atm, or about 4.0. Locally the pH could go a little lower if organic acids are produced as intermediates. In anaerobic decay several possible reactions may influence the pH: a little CO_2 may appear if oxygen is not altogether excluded; organic acids may form; the sulfur of proteins may appear as H_2S; and nitrogen, in contrast to the other elements, may tend to increase the pH by forming ammonia. The net effect of these reactions is not readily predictable, but measurements of pH in swamps and stagnant marine basins give figures just under neutral, in the range 6 to 7.

In summary, basins where organic matter accumulates are the most reducing of sedimentary environments, but their reducing ability is difficult to express in quantitative terms. Redox potentials cannot be calculated from thermodynamic data because the formulas of all the substances present are generally not known, and experimentally determined redox potentials are open to question because of the slowness of organic reactions. Potentials can be roughly estimated from the kind of inorganic substances associated with the organic material, and these estimates corroborate the conclusion from measurements that Eh's are in the range -0.1 to -0.5 volt. Associated minerals also permit estimates of the effective reducing power of ancient organic matter with respect to circulating waters. The pH of solutions in contact with decaying organic matter is generally less than 7, reaching a minimum of about 4 where decay takes place under aerobic conditions.

PROBLEMS

1 In the inorganic hypothesis for the origin of petroleum, methane molecules are supposed to polymerize into heavier hydrocarbons. The organic hypothesis, on the other hand, pictures a breakdown of more complex compounds into relatively simple ones. A reaction like the following symbolizes the two points of view:

$$2C_2H_6 \rightleftharpoons C_4H_{10} + H_2$$

the inorganic hypothesis favoring the forward reaction and the organic hypothesis the reverse. By looking up free energies, determine which reaction is more probable from the standpoint of energy alone.

2 As an alternative, the buildup of heavy hydrocarbons may be thought to involve unsaturated compounds, for example:

$$C_2H_4 + C_2H_6 \rightleftharpoons C_4H_{10}$$

Would it be necessary to supply energy to make the forward reaction take place?

3 Write structural formulas for three isotopes of pentane.

4 Write structural formulas for normal hexane, cyclohexane, and benzene.

5 The sulfur associated with salt domes is often assumed to be the product of reduction of $CaSO_4$ (in the form of gypsum or anhydrite) by petroleum hydrocarbons, with bacteria performing the function of a catalyst. Write an equation for this reaction, using $CaSO_4$ as the formula of the sulfate, letting CH_4 represent the hydrocarbons, and assuming that accompanying solutions are acid.

6 Barium ion, Ba^{2+}, is often noted as a constituent of groundwater near oil fields but is practically never found in ground water elsewhere. Remembering that $BaSO_4$ is an extremely insoluble compound, suggest a reason for this peculiarity in the composition of groundwater.

7 To which group of hydrocarbons does each of the following belong: $(a) C_6H_{14}$, $(b) C_4H_8$, $(c) C_7H_8$? Of these three compounds, which might be prepared by distillation of petroleum? By destructive distillation of coal tar? From either?

8 What unsaturated hydrocarbon has the same molecular formula as cyclohexane?

9 To what group of organic compounds would each of the following belong: $(a) C_{12}H_{22}O_{11}$, $(b) C_4H_9OH$, $(c) C_6H_5CH_2COOH$, $(d) CH_3CHNH_2COOH$, $(e) C_6H_5OH$, $(f) CH_3COOC_2H_5$, $(g) C_3H_5(OOCC_{17}H_{35})_3$?

10 What hydrocarbon would be produced by the decomposition (driving off CO_2) of butyric acid, C_3H_7COOH?

11 During the aerobic decay of an animal body, what happens to the following elements originally contained in the organic compounds: C, H, N, S, P, Mg, Fe? Answer the same question for anaerobic decay.

12 Name three common inorganic compounds of carbon that occur in nature. Does carbon ever occur in nature as a constituent of compounds containing silicon?

13 Why are fats commonly regarded as an important source material for petroleum? What pieces of evidence are still needed to establish this hypothesis on a firm basis?

14 Explain and compare the various usages of the term sapropel.

15 From thermodynamic data, show that the concentration of CO is always low in volcanic gases relative to CO_2 and CH_4.

16 From the formulas of the amino acids glycine and alanine, would you expect either to show optical activity?

REFERENCES AND SUGGESTIONS FOR FURTHER READING

Barghoorn, E. S., W. G. Meinschein, and J. W. Schopf, Paleobiology of a Precambrian shale, *Science*, vol. 148, pp. 461–472, 1965. Porphyrins and optically active hydrocarbons found in a black shale about 1,000 million years old.

Hedberg, H. D., Geological aspects of the origin of petroleum, *Am. Assoc. Petroleum Geologists Bull.*, vol. 48, pp. 1755–1803, 1964. A review of the development of ideas about the origin of petroleum.

Hodgson, G. W., Vanadium, nickel, and iron trace metals in crude oils of western Canada, *Am. Assoc. Petroleum Geologists Bull.*, vol. 49, pp. 2537–2554, 1965. Details of the chemistry of the porphyrins in petroleum, and speculations about changes in the porphyrins during geologic time.

Hoering, T. C., and P. H. Abelson, Hydrocarbons from kerogen, *Geophys. Lab. Ann. Rept.* 1962–63, pp. 229–234, 1963.

Hunt, J. M., Distribution of hydrocarbons in sedimentary rocks, *Geochim. et Cosmochim. Acta*, vol. 22, pp. 37–49, 1961. Attempts to correlate the kinds of hydrocarbons with age and depth of sedimentary rocks containing them have given negative results.

Johns, W. D., and A. Shimoyama, Clay minerals and petroleum-forming reactions during burial and diagenesis, *Am. Assoc. Petroleum Geologists Bull.*, vol. 56, pp. 2160–2167, 1972. Experiments on clay minerals as catalysts for the formation of hydrocarbons from fatty acids and for the cracking of big hydrocarbon molecules.

Lind, C. J., and J. D. Hem, Effects of organic solutes on chemical reactions of aluminum, *U.S. Geol. Survey Water-Supply Paper* 1827-G, 1975. Review of the kinds of organic matter in natural waters, and effects of organic compounds in making aluminum more soluble by forming complexes.

Philippi, G. T., The influence of marine and terrestrial source material on the composition of petroleum, *Geochim. et Cosmochim. Acta*, vol. 38, pp. 947–966, 1974.

Strøm, K. M., Land-locked waters, *Norske Videnskaps Akad., Mat.-Naturv. Klasse*, 1936, No. 7. An old but often-quoted paper describing a detailed study of recent organic-rich sediments formed in the stagnant water of a Norwegian fjord.

TWELVE

EVAPORITES

The inorganic chemical sediments we have considered up to now have this in common: they all consist of materials that are only slightly soluble in water. They precipitate for the most part individually, as fairly simple substances, and we can learn a good deal about their chemical behavior simply from the solubility product and the Eh-pH relationships of each separate compound. We have now to discuss the more soluble substances in natural solutions, those which in special geologic environments form the salt deposits called *evaporites*, and we shall find that the chemistry of these substances must be treated from a different point of view. The solutions involved are concentrated, which means that solubility products are of little help because they are no longer even approximately constant, and activity coefficients are generally not predictable; two or more salts commonly crystallize simultaneously; and the situation is complicated by the existence of many double salts and many possible hydrates, which crystallize or recrystallize in response to minor changes of temperature and composition. There is little fundamental theory here to guide us. We shall be concerned chiefly with experimental observations on the kinds of salts that form under various laboratory conditions, which will serve as a guide to reconstructing the conditions under which salt deposits have formed in nature.

12-1 SOLUBLE MATERIAL IN STREAMS

The streams that carry salts to basins of deposition are in general very dilute solutions. The total salt content of stream water averages something like 100 ppm (0.01%), and very seldom exceeds 10,000 ppm. Despite the low concentration, the amount of erosion accomplished by solution alone is far from negligible. Accurate data are lacking for many drainage basins, but a reasonable average for dissolved material removed from the continents is about 80 tons per square mile per year. This amounts to an average lowering of the land surface by 1 ft in every 30,000 years.

The compositions of the dissolved material in several rivers, and an estimate of the world average, are shown in Table 12-1. The material is not entirely derived from erosion, for it includes a considerable quantity of salt carried inland from the ocean by winds, in the form of fine dust particles produced by evaporation of ocean spray. Such "cyclic salt" accounts for a large fraction of the Cl^- in Table 12-1, for a corresponding amount of Na^+, and for lesser quantities of the other substances.

The kind of salt carried by any one stream depends largely on the climate. Near the headwaters of a stream the composition may reflect adjacent rocks— abundant Ca^{2+} from limestone areas, Mg^{2+} from dolomite, alkali-metal ions and silica from granite rocks—but the effects of vegetation, of adsorption processes, and of mixing with tributaries very quickly alter the composition to a type largely responsive to the climate. In humid climates the abundance of CO_2 from decaying vegetation gives stream water a high carbonate content (using "carbonate" to refer to the total $CO_3^{2-} + HCO_3^- + H_2CO_3$) and makes it acid enough to dissolve

Table 12-1 Analyses of river waters, in parts per million†

	A	B	C	D	E	F	World average
HCO_3^-	93	101	183	108	149.2	17.9	58.4
SO_4^{2-}	25	41	289	19	0.44	0.8	11.2
F^-	0.0	0.1	0.2	0.5			
Cl^-	5.0	15	113	4.9	8	2.6	7.8
NO_3^-	1.2	1.9	1.0	0.3	0.44		1
Ca^{2+}	32	34	94	23	17.4	5.4	15
Mg^{2+}	4.9	7.6	30	6.2	5.2	0.5	4.1
Na^+	4.8	11	124	16	30.7	1.6	6.3
K^+	2.0	3.1	4.4	0.0	11.8	1.8	2.3
Fe	0.07	0.02	0.01	0.280		1.9‡	0.67
Al	0.304	1.01	0.012	0.238			
SiO_2	4.9	5.9	14	13	25.6	10.6	13.1
Total dissolved solids	173	221	853	191	249	43.1	120

A. Hudson River, Green Island, N.Y.
B. Mississippi River, Baton Rouge, La.
C. Colorado River, Yuma, Ariz.
D. Columbia River, 3 miles above The Dalles, Wash.
E. White Nile, near Khartoum, Sudan
F. Amazon River, near Obidos
  † Source: Livingstone, 1963.
  ‡ Computed from $Fe_2O_3 + Al_2O_3$, on assumption that only Fe_2O_3 is present.

limestone fairly rapidly; hence, in streams of humid regions, Ca^{2+} is generally the dominant cation and HCO_3^- the dominant anion. In arid regions, on the other hand, CO_2 is not added to the water in sufficient amounts to counterbalance alkalinity produced by hydrolysis; Ca^{2+} and CO_3^{2-} are largely precipitated out in the B horizons of soils, and stream water is characterized by Na^+ as the dominant cation and either SO_4^{2-} or Cl^- as the chief anion. Regardless of climate, K^+ is generally much lower than Na^+, a fact for which at least four circumstances can be held responsible: (1) the much higher Na^+ in cyclic salt from the ocean, (2) the more rapid weathering of plagioclase feldspar than of potassium feldspar, (3) the more extensive use of K^+ by vegetation, and (4) the greater adsorption of K^+ on clays and organic matter. All these statements about stream composition are generalizations of the broadest sort to which many exceptions will be found whenever the streams of any one area are examined in detail.

12-2 SALT DEPOSITS IN ARID REGIONS

The salts deposited in desert basins, and the salts obtainable from desert lakes, are for the most part a monotonous assemblage of sulfates, chlorides, and carbonates of sodium and calcium, with smaller amounts of compounds of potassium and magnesium (Tables 12-2 and 12-3). Two or three of these ions may dominate in a

Table 12-2 Analyses of water from salt lakes, in parts per million†

	A	B	C	D	E	F	G
HCO_3^-	232	187	1,390	17,400	4,946	240	180
SO_4^{2-}	4,139	4,960	264	6,020	2,368	540	13,590
F^-	1.6	4.9	0.8				
Cl^-	9,033	21,400	1,960	4,680	5,789	208,020	112,900
Br^-						5,920	
NO_3^-	1.2		1.6				
Ca^{2+}	505	1,230	10	3.9	36	15,800	330
Mg^{2+}	581	148	113	23	165	41,960	5,620
Na^+	6,249	14,100	1,630	12,500	7,707	34,940	67,500
K^+	112	594	134		435	7,560	3,380
Rb^+						60	
B	5.0	60					
SiO_2	20.8	49	1.4	101	70		
Total dissolved solids	20,900	42,700	5,510	40,700	22,000	315,040	203,490

A. Salton Sea, Calif.
B. Ponds at Bad Water, Death Valley, Calif.
C. Pyramid Lake, Nev.
D. Soap Lake, Wash.
E. Lake Van, Turkey
F. Dead Sea, Israel
G. Great Salt Lake, Utah
  † *Sources:* Livingstone, 1963, analyses A to F. Clarke, 1924, analysis G.

Table 12-3 Analyses of salt deposits, in weight percent†

‡	A	B	C	D	E	F
NaCl	96.49	0.74	24.51	18.47	20.88	27.55
Na_2SO_4	1.91	46.27	33.31	27.55	49.67	2.13
Na_2CO_3	0.96		25.95	52.10	7.02	
$NaHCO_3$			14.35		11.13	
$Na_2B_4O_7$					11.30	0.43
K_2SO_4			1.88			
$MgSO_4$		48.28				0.15
$CaSO_4$		4.45				0.41
$NaNO_3$						61.97
KNO_3						5.15
$KClO_4$						0.21
$NaIO_2$						0.94
Insoluble	0.12					0.39
H_2O (combined)	0.52					0.67

A. Osobb Valley, Nev.
B. Percy, Nev.
C. Antelope Valley, Nev.
D. Black Rock Desert, Nev.
E. Humboldt Lake, Nev.
F. Northern Chile

† *Source:* Clarke, 1924, pages 237–256.

‡ The formulas in this column are arbitrary combinations of the ions whose concentrations are determined by analysis. Actual salts with these formulas may or may not be present.

particular lake, depending on its immediate surroundings; for example, sodium sulfate may be the principal salt in a small basin among hills composed chiefly of granodiorite containing pyrite, and magnesium sulfate may be prominent in a similar basin surrounded by pyritic greenstones. Great Salt Lake has a composition similar to the ionic ratios in seawater; many lakes of western Nevada and central Oregon have a greater abundance of carbonate. When evaporation has progressed so far that most of the NaCl and $CaSO_4$ have crystallized out, the remaining water may be exceptionally rich in the more soluble salts of potassium, magnesium, and bromine; the Dead Sea is a good example of this type. In regions of volcanic activity and hot springs, desert basins may have salts of unusual composition, like the famous lithium salts of Searles Lake in California and the borate deposits of Death Valley. In detail, deposits of desert basins provide a complicated and rewarding geochemical study, but the subject is hardly of sufficiently general interest to detain us here.

One type of desert deposit merits more than passing mention because it is often cited as representing the extreme of oxidizing conditions in nature. This is the famous nitrate deposit, or group of deposits, in northern Chile. On the slopes of broad, barren valleys between the Coast Range and the Andes, in an area of prevailingly high temperatures and a rainfall as low as anywhere in the world, a unique accumulation of salts makes up a "caliche" layer on and just under the surface. The principal compound, the one for which these deposits are famous, is Chile saltpeter, $NaNO_3$. With it occurs the rare salt $Ca(IO_3)_2$, in amounts sufficient to be exploited as a source of iodine. Evidence of high oxidation potentials is furnished also by minor amounts of perchlorates [for example $Ca(ClO_4)_2$] and chromates (for example $CaCrO_4$). The fact that such soluble compounds and such strong oxidizing agents should be preserved here seems reasonable enough, in view of the climatic extremes; but the source of the various elements, especially of the nitrogen and iodine, has occasioned much argument. Former streams from the Andes have been called on, supposedly bringing nitrogen and iodine derived from decaying vegetation at higher and more humid elevations. Perhaps lightning discharges accompanying thunderstorms have produced nitrogen oxides in the atmosphere, and these have been dissolved in rainwater during a period of moister climate. Perhaps iodine and volatile nitrogen compounds have been set free by the decay of kelp along the seacoast and carried inland by winds. Hot springs and fumaroles associated with Andean volcanoes may have added to the supply of one or both elements. There is no reason, of course, why all these possible sources may not have contributed, but geologists are far from agreed as to which are the most important.

12-3 THE COMPOSITION OF SEAWATER

Most important and most studied of salt deposits are the ones formed by evaporation of seawater. To these we shall devote the rest of this chapter. We begin with a brief study of the composition of seawater and of some of the factors that serve to keep the composition approximately constant.

We recall first that seawater is a complex buffer system (Sec. 2-10), its pH maintained in the narrow range 8.0 to 8.4 by reactions involving H_2CO_3, HCO_3^-, CO_3^{2-}, and $CaCO_3$, by ion exchange on clay minerals, and to a lesser extent by reactions involving H_3BO_3 and $H_2BO_3^-$. The alkaline pH means that most of the Ca^{2+} and HCO_3^- brought to the sea by rivers must be precipitated as $CaCO_3$, and the high concentration of electrolytes leads to rapid coagulation of colloidal iron oxide and aluminum oxide. Silica also is quickly brought to a low figure, probably by the activity of organisms. This leaves Na^+, Cl^-, SO_4^{2-}, Mg^{2+}, and K^+, for which no control is immediately obvious. These five, together with the excess of Ca^{2+} not precipitated as carbonate, make up all but a tiny fraction of the salt in the sea (Table 12-4).

Can we conclude, then, that the five ions just mentioned have simply been accumulating in the sea all through geologic time? Do their present concentrations represent merely the total amounts so far dissolved by the weathering of rocks? This conclusion hardly seems likely, because the relative amounts of the ions are so different from their relative amounts in stream water (Table 12-1) and in ordinary rocks. In comparison with Na^+, the amounts of K^+ and Mg^{2+} seem much too small, and the amounts of Cl^- and SO_4^{2-} far too large. Are there

Table 12-4 Principal dissolved substances in seawater

Substance	Parts per million	Percent of total salt
Cl^-	18,800	55.05
Na^+	10,770	30.61
SO_4^{2-}	2,715	7.68
Mg^{2+}	1,290	3.69
Ca^{2+}	412	1.16
K^+	380	1.10
HCO_3^-	140	0.41
Br^-	67	0.19
H_3BO_3	26	0.07
Sr^{2+}	8	0.03

absorbed on clays

processes that might cause removal of K^+ and Mg^{2+} from seawater, and other processes besides weathering and erosion that might lead to accumulation of sulfur and chlorine?

An obvious source of extra sulfur and chlorine is volcanic activity, for volcanoes are known to expel huge amounts of these two elements in the form of volatile compounds. Two minor elements of seawater, boron and bromine, are also present in higher proportions than would be expected from weathering alone, and these also are elements often noted as constituents of compounds in volcanic gases and sublimates. Processes by which Cl^- and SO_4^{2-} are removed from the sea are not very effective. Both are precipitated locally in salt deposits, both are lost to a minor extent by adsorption on clays and by the incorporation of seawater in sediments, and both are removed temporarily when ocean spray is carried inland by wind. Sulfur is taken out of seawater temporarily by organisms and built into sulfur-bearing proteins; most of this organic sulfur is returned to the sea when the organisms die, but some may be fixed in organic sediments or by the conversion of organic sulfur into sulfide and its precipitation as pyrite. Very likely a balance is maintained between these processes of removal and the additions from weathering and volcanic activity, but it is possible that one or both elements are still increasing in concentration.

For controlling the concentration of Mg^{2+}, one possible suggestion is the conversion of calcite into dolomite. If this conversion takes place diagenetically (Sec. 3-10) in bottom sediments before they have been consolidated, magnesium may be continually in process of removal. The formation of glauconite must also remove a small amount of magnesium, and some potassium as well. An unsettled problem in oceanic geochemistry is the extent to which clay minerals other than glauconite can form out of materials dissolved or suspended in seawater; if such a reaction is possible, it also would be an effective process for tying up both K^+ and Mg^{2+}. Even if clays do not form as new minerals out of dissolved substances, detrital clays may be altered by reaction with the ions of seawater; the preponderance of illite in marine shales suggests, for example, that K^+ from seawater takes part in the slow alteration of other clay minerals. The simple adsorption of K^+ and Mg^{2+} on clays and other fine particles must also be an important mechanism for controlling the concentration of these ions. Both K^+ and Mg^{2+} are essential constituents of living organisms, so that minor amounts may be permanently removed in organic sediments.

The situation with respect to sodium is uncertain. Quite possibly its concentration in the sea is still growing. Processes of removal are not nearly so effective as for potassium, since sodium plays only a minor role in organisms and is not a major constituent of clay minerals. The only sodium minerals known to form in seawater (besides evaporites) are authigenic feldspars and zeolites, and the amounts of these are small. On the other hand a good deal of Na^+ is adsorbed on clay minerals, and removal by this means may be sufficient to maintain a balance with the amount supplied by weathering. Analytical data are not yet complete enough to settle the question.

To summarize this discussion, the ocean appears to have achieved a state of balance between supply and removal for at least three of its six most abundant ions: K^+, Mg^{2+}, and Ca^{2+}. The other three, Na^+, Cl^-, and SO_4^{2-}, probably have reached a state of balance also, but present data do not exclude the possibility that their concentrations are very slowly increasing.

Such questions of oceanic geochemistry are particularly interesting when projected into past time. Has the ocean always had approximately its present composition? If not, at what periods did important changes occur? Has the amount of water in the oceans been constant through most of geologic time, or has it increased, together with sulfur and chlorine, by the addition of volatile material from volcanoes? Questions like this are obviously important in a discussion of marine evaporite deposits, and study of the deposits themselves provides partial answers. Holland (1972), for example, has pointed out that the uniformity in the sequence of minerals in marine evaporites of different ages is good evidence for the approximate constancy in composition of seawater; by calculating the limits within which the composition can vary without changing the observed sequence, he concludes that concentrations of the major ions cannot have changed by more than a factor of two during the last 700 million years.

12-4 GEOLOGIC DATA ON MARINE EVAPORITE DEPOSITS

The most abundant substances in marine evaporites are calcium sulfate and sodium chloride. The calcium sulfate is mostly anhydrite, but this is probably not the mineral originally deposited; if precipitated initially as gypsum, the compound would have changed to anhydrite as a result of increased heat and pressure after burial. Rock salt and anhydrite are found in enormous deposits in the Silurian of New York and Michigan, in the Permian of Kansas, Texas, and New Mexico, and in deposits of comparable size on other continents. Smaller amounts appear frequently in strata of most other geologic periods from the Cambrian to the present. Chemically the origin of such deposits is scarcely a problem, except in details of the gypsum-anhydrite relationships, for calcium sulfate and sodium chloride are precisely the salts that would be expected to precipitate first in large amounts during the evaporation of seawater. Geologically the problem is simply one of reconstructing conditions under which an arm of the sea can be sufficiently isolated to permit long-continued evaporation without too much mixing with the rest of the ocean.

Of greater interest are the rare deposits of the more soluble salts obtainable from seawater—the chlorides and sulfates of magnesium and potassium (Table 12-5). The most famous of such deposits are those in the Permian Zechstein formation of northern Germany. Four distinct evaporite sequences are recognized in this formation, and some of the sequences extend beyond the borders of Germany into Poland, Holland, and northern England. Also of Permian age are the

Table 12-5 Important salts in marine evaporites†

Chlorides	
‡Halite	$NaCl$
‡Sylvite	KCl
Bischofite	$MgCl_2 \cdot 6H_2O$
‡Carnallite	$KMgCl_3 \cdot 6H_2O$
Chloride-sulfate	
‡Kainite	$KMgClSO_4 \cdot \frac{11}{4}H_2O$
Sulfates containing Ca	
‡Anhydrite	$CaSO_4$
‡Gypsum	$CaSO_4 \cdot 2H_2O$
Glauberite	$Na_2Ca(SO_4)_2$
‡Polyhalite	$K_2MgCa_2(SO_4)_4 \cdot 2H_2O$
Sulfates of Na and K	
Thenardite	Na_2SO_4
Glaserite, or aphthitalite	$K_3Na(SO_4)_2$
Simple sulfates of Mg	
‡Kieserite	$MgSO_4 \cdot H_2O$
Hexahydrite	$MgSO_4 \cdot 6H_2O$
Epsomite, or reichardtite	$MgSO_4 \cdot 7H_2O$
Sulfates of Mg and Na	
Bloedite, or astrakhanite	$Na_2Mg(SO_4)_2 \cdot 4H_2O$
Loeweite	$Na_{12}Mg_7(SO_4)_{13} \cdot 15H_2O$
Vanthoffite	$Na_6Mg(SO_4)_4$
Sulfates of Mg and K	
Langbeinite	$K_2Mg_2(SO_4)_3$
Leonite	$K_2Mg(SO_4)_2 \cdot 4H_2O$
Picromerite, or schoenite	$K_2Mg(SO_4)_2 \cdot 6H_2O$

† Formulas from Braitsch, 1962.
‡ Double daggers indicate major constituents.

potash deposits of Texas and New Mexico and those of northern Russia. Pennsylvanian evaporites occur in the Paradox Basin of Utah, and Devonian evaporites in the Williston Basin of North Dakota and Saskatchewan.

A good example for detailed study is the deposit near the German city of Stassfurt, a deposit which for a long time was the world's major source of potash. The evaporites of interest here are in the second of the four Zechstein sequences (Zechstein 2 or Stassfurt series). The geologic relations are apparently simple, and the generalized sequence shown in Table 12-6 can be recognized over a wide area. From bottom to top the dominant materials are rock salt and anhydrite, then a sulfate of Ca, Mg, and K, then chiefly magnesium sulfate, then chiefly chlorides of Mg and K. Halite is not restricted to the rock-salt layer but is a common mineral all through the deposit. The total thickness varies from place to place, locally exceeding 1,000 meters between the base of the anhydrite and the base of the salt clay. This kind of sequence is common in evaporite deposits, and in a general way it is the sort of thing we might expect from long-continued evaporation of seawater: first the precipitation of the less soluble salts NaCl and CaSO$_4$, then on extreme desiccation the crystallization of the more soluble potassium and magnesium sulfates, and finally of the very soluble chlorides.

Difficulties appear when the Stassfurt sequence is studied in detail. The first trouble is purely geological: in some places there is good evidence for a thrust fault at the top of the carnallite zone, so that the simplicity of the geologic picture may be more apparent than real. A second difficulty is the enormous thickness of the deposit, which seemingly would require evaporation of a layer of seawater about 100 km thick. To circumvent this difficulty, one can hypothesize that the basin of deposition was separated from the main ocean by a shallow bar, so that water could flow across the bar to make up for that being lost by evaporation. This hypothesis, now widely accepted for many of the larger evaporite deposits, was first elaborated by the German geologist Ochsenius. In the modern world

Table 12-6 Generalized section of salt deposits of the Stassfurt series (Zechstein 2), in the Stassfurt region, Germany†

The following sequence is representative of beds in the interior of the Zechstein basin. There is much local variation in thickness of the different units. Toward the margin of the basin the minerals containing MgSO$_4$ become less abundant and the chloride minerals more abundant. An important constituent of all the units, whether mentioned explicitly or not, is halite

	Thickness, meters
Gray salt clay	
Discontinuous zone of "Hartsalz" (mixture of sylvite, kieserite, halite).	
Locally also loeweite, picromerite, vanthoffite, langbeinite, kainite	0–20
Carnallite zone. Chiefly carnallite, some kieserite	15–40
Kieserite zone. Some carnallite and anhydrite	30–40
Polyhalite zone. Some anhydrite	5–10
Older rock salt and anhydrite	100–1,000

† *Source:* Stewart, 1963, p. 31.

**Table 12-7 Comparison of total thickness of salts in a represen-
tative section of the Stassfurt series with thicknesses obtained by
complete evaporation of seawater**

	Relative thickness of salts at Stassfurt	Relative thickness of salts from seawater
Anhydrite, $CaSO_4$	5.5	3.3
Halite, $NaCl$	100	100
Kieserite, $MgSO_4 \cdot H_2O$	2	8.0
Carnallite, $KMgCl_3 \cdot 6H_2O$	4.5	15.3
Bischofite, $MgCl_2 \cdot 6H_2O$		28.3

handwritten annotation: could react with $CaCO_3$

Source: Twenhofel, W. H., "Principles of Sedimentation," 2d ed.,
McGraw-Hill Book Company, New York, 1950.

Ochsenius could point to one place where the postulated mechanism is in operation today: the Gulf of Karabugaz east of the Caspian Sea is separated from the
sea by a sandbar through which a narrow channel carries water into the gulf, and
the greater evaporation in the extremely arid climate of the gulf causes a steady
loss of water and crystallization of abundant salts.

A further difficulty in correlating the Stassfurt sequence with salts that might
be expected from seawater arises when one calculates the relative amounts of
different salts that would form if present-day seawater were evaporated to dryness
(Table 12-7). Comparison of the figures with rough estimates of relative amounts
in the Stassfurt deposit shows obvious discrepancies, in that the $CaSO_4$ content of
the natural salts is too high and the proportion of potassium and magnesium salts
too low. These discrepancies can be reasonably explained away by supposing that
(1) at some stages deposits of gypsum were buried by fresh incursions of seawater across the bar before much halite or other salts had been deposited, and
(2) evaporation of the residual brines of the Zechstein sea was interrupted
before crystallization of the most soluble salts had gone to completion.

Another feature that is hard to explain is the great lateral variation in the
evaporite sequence of the Stassfurt series. Even more disturbing is the fact that
other salt deposits, which on geologic evidence are also of marine origin, show
very different sequences and assemblages of evaporite minerals. Most notable of
the differences from one place to another is the variation in amount of $MgSO_4$.
Some deposits, of which the Devonian beds of western Canada are a good example, have practically no magnesium sulfate minerals at all, so that the overall
composition is even farther from the theoretically expected composition than is
that of the Stassfurt deposit. To account for the commonly observed deficit in
$MgSO_4$, a reasonable guess is that seawater during evaporation may react slowly
with $CaCO_3$, either present as limestone beneath the deposit or brought in as
dissolved calcium bicarbonate by rivers. The reaction can be symbolized

$$Mg^{2+} + SO_4^{2-} + 2CaCO_3 \rightarrow CaMg(CO_3)_2 + CaSO_4$$

handwritten annotation: limestone (calcite) → dolomite + anhydrite

The dolomite (or magnesite) would be deposited near the base of a salt sequence, and the $CaSO_4$ would serve to augment the amount of gypsum or anhydrite. Support for this hypothesis comes from the fact that dolomite occurs very commonly below evaporite beds. The Stassfurt series provides a particularly good example: near the city of Stassfurt, which lies in the inner part of the Zechstein basin, magnesium sulfate minerals are abundant and dolomite is scarce, but toward the margins of the basin, the salts in this series change to facies poor in $MgSO_4$, and dolomite becomes prominent as the underlying rock. The hypothesis is clearly consistent with a great deal of apparently random variation in evaporite composition, since the extent of removal of $MgSO_4$ would depend on how effectively the evaporating brines come in contact with solid $CaCO_3$ and how much dissolved $CaCO_3$ is supplied by rivers.

Many of the apparent complexities of evaporite deposits can be handled by this kind of reasoning, without abandoning the fundamental postulate that the large salt beds owe their origin to evaporation of seawater similar in composition to that of the present ocean. To test the postulate rigorously, however, requires a closer look at the physical chemistry of salt deposition. The formation of evaporites takes place under very ordinary and easily definable conditions of temperature and pressure, so that comparison of laboratory results with field observation should provide considerable insight into details of salt-bed history.

12-5 USIGLIO'S EXPERIMENT

One possible method of attack is the direct one: let seawater evaporate in the laboratory, and keep track of the salts that precipitate at different stages of the desiccation. This procedure was tried in the middle of the last century by the Italian chemist Usiglio. Using a sample of Mediterranean water, Usiglio maintained the temperature at 40°C and let the water evaporate into an atmosphere kept dry with quicklime. At intervals some of the water was cooled quickly to 21°, and the precipitated salts were analyzed. In a general way Usiglio obtained the sequence of salts found at Stassfurt (Table 12-8): first a small amount of $CaCO_3$ (the Fe_2O_3 shown in the table probably represents contamination); then gypsum; then gypsum plus halite; then halite with $MgSO_4$ and $MgCl_2$. The later stages of evaporation gave variable results, depending on time of standing and on the magnitude of day-to-night temperature changes. Some of the crystals that Usiglio obtained could be recognized as substances found in salt deposits (epsomite, hexahydrite, picromerite, carnallite), but a number of the common Stassfurt minerals failed to appear (notably kieserite, kainite, polyhalite). These experiments, therefore, showed a good deal of correspondence with nature, but left some questions unanswered.

A second possible approach to the problem is to start with simple systems instead of the very complex mixture in seawater, systems with only two or three salts, whose behavior at different temperatures can be studied in detail. Then other salts may be added one by one to the system, so that the physical chemistry can be

Table 12-8 Order of precipitation of salts from seawater at 40°, according to the experiments of Usiglio.†

The numbers are weights in grams precipitated from an original volume of 1 liter

Volume, liters	Fe_2O_3	$CaCO_3$	$CaSO_4 \cdot 2H_2O$	NaCl	$MgSO_4$	$MgCl_2$	NaBr	KCl
1.000								
0.533	0.0030	0.0642						
0.316		Trace						
0.245		Trace						
0.190		0.0530	0.5600					
0.1445			0.5620					
0.131			0.1840					
0.112			0.1600					
0.095			0.0508	3.2614	0.0040	0.0078		
0.064			0.1476	9.6500	0.0130	0.0356		
0.039			0.0700	7.8960	0.0262	0.0434	0.0728	
0.0302			0.0144	2.6240	0.0174	0.0150	0.0358	
0.023				2.2720	0.0254	0.0240	0.0518	
0.0162				1.4040	0.5382	0.0274	0.0620	
Total	0.0030	0.1172	1.7488	27.1074	0.6242	0.1532	0.2224	
Salts in last bittern				2.5885	1.8545	3.1640	0.3300	0.5339
Total solids	0.0030	0.1172	1.7488	29.6959	2.4787	3.3172	0.5524	0.5339

† *Source:* Clarke, 1924, p. 220.

checked and rechecked at each step. In this manner our understanding of the basic principles of crystallization is much more firmly based; the only question is whether our artificially simple systems can be made to approximate nature closely enough for our generalizations to have geologic significance. This approach was tried by the Dutch chemist van't Hoff in the early 1900's, and it has proved the most fruitful means of unraveling the mysteries of salt deposition.

12-6 PHYSICAL CHEMISTRY OF SALT DEPOSITION

General

Some of the rules that hold for solubility in dilute solution are also valid in concentrated brines. For example,

1. Of a pair of salts in solution, the one that precipitates first during evaporation depends on relative solubilities and on concentrations. If A is 10 times more soluble than B, it will nevertheless precipitate first if its concentration is more than 10 times greater than that of B.
2. If two salts have a common ion, the solubility of each in a solution of the other is less than it would be in pure water. Thus NaCl is less soluble in KCl solution than in water.

3. If two salts do not have a common ion, the solubility of each is generally greater in a solution of the other than in pure water.
4. Solubility is markedly influenced by changes of temperature, but only slightly by changes of pressure. The temperature effect is different for different salts and in general is unpredictable. For most salts the solubility increases as the temperature rises; for many the solubility increases by a factor of 2 or 3 between 25 and 100°C, and for only a very few by a factor of more than 10.

Salt Pairs with a Common Ion

The simplest possible multiple-salt system is a pair like NaCl and KCl, which have an ion in common and which do not form double salts at ordinary temperatures. Their solubility relations can be represented conveniently by a diagram like Fig. 12-1, where O represents pure water and increasing concentrations of the two salts are plotted along the axes. Concentrations in this and succeeding diagrams are expressed as moles of the fictitious compounds K_2Cl_2 and Na_2Cl_2, rather than the simple salts KCl and NaCl. This is a matter of convenience, a device to avoid arithmetical difficulties in comparing amounts of salts of univalent and divalent ions. In effect, we treat the salt systems as if they included only the divalent ions Mg^{2+}, SO_4^{2-}, K_2^{2+}, Na_2^{2+}, Cl_2^{2-}. The last three have no actual existence, of course, but this is unimportant as long as it is only a means of expressing the concentrations at which particular compounds crystallize.

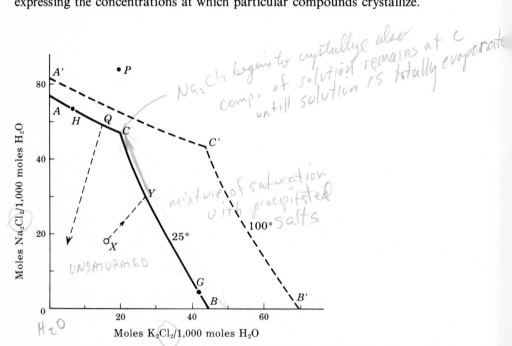

Figure 12-1 Crystallization of solutions containing NaCl and KCl at 25 and 100°C. (*Source: Braitsch, 1962, page 36.*)

Points *A* and *B* in Fig. 12-1 show the concentrations of NaCl and KCl, respectively, in saturated solutions of the pure salts. Point *G* shows the decreased solubility of KCl in a solution with a ratio of NaCl/KCl equal to $\frac{1}{10}$, and point *H* the solubility of NaCl when the ratio is $\frac{10}{1}$. A mixture with composition *C* is simultaneously saturated with both. Points between *O* and the line *ACB* represent unsaturated solutions, and points beyond *ACB* represent heterogeneous mixtures of the solid salts with saturated solutions. If a dilute solution with composition *X* is evaporated, its concentration changes along the line *XY* (directly away from *O*); at *Y*, KCl begins to crystallize; this makes the ratio NaCl/KCl in the solution greater, so that the composition changes along the line *YC* as KCl continues to crystallize; at *C*, NaCl begins to crystallize also, and the composition of the solution would remain at this point until evaporation is complete. Or we might consider the opposite kind of sequence, gradually dissolving a mixture of salts represented by point *P*: as water is added, the solution remains for a time saturated with both, hence has a concentration shown by *C*; when KCl has entirely dissolved, some solid NaCl remains; by further addition of water this would be dissolved and the concentration of the solution would change along *CQ*; at *Q* the last NaCl would disappear and further dilution would be shown by the line *QO*. Thus the diagram shows concisely and quantitatively how the solution changes and what salts dissolve or crystallize as the amount of water is increased or decreased.

The combination KCl-MgCl$_2$ behaves similarly, but the situation is complicated by the possibility of forming a double salt, KCl·MgCl$_2$·6H$_2$O (carnallite). In Fig. 12-2, point *A* represents saturation with MgCl$_2$·6H$_2$O alone; *B* represents saturation with KCl alone; and the dot-dash line *OL* shows the ratio of KCl to MgCl$_2$ in carnallite. When a dilute solution like *X* is evaporated, KCl is the first salt to crystallize; it continues to crystallize as the composition of the solution changes along the line *BD*, well past the composition of the double salt; only when the solution has reached the composition *D* does carnallite appear; at this point the composition of the solution remains fixed while carnallite crystallizes, both directly out of solution and by partial reaction of dissolved ions with some of the already precipitated KCl. The resulting solid salt is a mixture of carnallite and sylvite. If the original dilute solution has the composition *Y*, on the other hand, it contains more than enough MgCl$_2$ to react with the precipitated KCl; hence at point *D* a saturated solution will be in contact with carnallite alone. If evaporation continues, additional carnallite precipitates and the composition of the solution changes along *DC*; at *C*, bischofite (MgCl$_2$·6H$_2$O) begins to precipitate with carnallite, and the composition of the solution remains constant here until evaporation is complete. This time the resulting solid is a mixture of carnallite and bischofite.

For the opposite process, consider the dissolving of pure carnallite. The first saturated solution has a composition indicated by point *D*, much richer in MgCl$_2$ than in KCl, meaning that some of the carnallite is broken up and leaves a residue of solid sylvite. When the carnallite has entirely disappeared, some solid sylvite will remain, and the solution in the meantime has maintained the composition *D*.

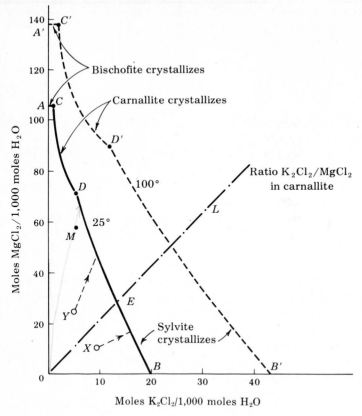

Figure 12-2 Crystallization of solutions containing KCl and MgCl$_2$ at 25 and 100°C. Solutions saturated with NaCl. (Point M refers to Problem 7.) *(Source: Braitsch, 1962, page 38.)*

Only as the residual KCl dissolves will the composition of the solution change along DB. When the last solid disappears, the solution reaches a composition E, and further dilution is shown by a change along EO.

We can describe the section of the saturation line BD as the "sylvite region," since here KCl alone is in equilibrium with saturated solutions; CD would be the carnallite region, and CA the bischofite region. Point C shows the composition of a saturated solution in equilibrium with both bischofite and carnallite, point D the composition in equilibrium with both carnallite and sylvite. Note that point D differs from point C in that during evaporation it can be approached from only one side, the KCl side, whereas C can be approached from either the carnallite or the bischofite side. In other words a salt sequence formed by simple evaporation might have sylvite followed by carnallite, carnallite followed by a bischofite-carnallite mixture, or bischofite followed by a bischofite-carnallite mixture, but never carnallite followed by sylvite.

Because carnallite apparently breaks up as it dissolves, giving solid sylvite and a solution rich in magnesium, it is said to dissolve *incongruently*. We have noted one other example of incongruent solution, the partial breakdown of dolomite at moderately high temperatures to give solid $CaCO_3$ and a solution with excess magnesium (Sec. 3-10). The phenomenon is very common among salts in evaporite deposits.

Reciprocal Salt Pairs

A salt pair without a common ion is much more complicated. We could start, say, with a solution containing KCl and $MgSO_4$; at once we have the apparent possibility of forming two other simple salts on evaporation, K_2SO_4 and $MgCl_2$, and a correspondingly greater possibility of forming double salts. Such a salt combination is called a *reciprocal salt pair*.

Several troublesome questions arise: (1) How many concentrations must be specified in order to describe all possible compositions of solutions of a reciprocal pair? A little reflection will show that the answer must be three: any solution containing the ions K^+, SO_4^{2-}, Mg^{2+}, and Cl^- can be described as a combination of any three of the possible four simple salts. Alternatively the composition can be specified by giving the concentrations of any three ions; the amount of the fourth ion is automatically fixed by the requirement of electrical neutrality. (2) How many salts can crystallize simultaneously from such a solution? Can all four simple salts be in equilibrium with the solution? Can several double salts be present also? The answer to these questions lies in a generalization called the phase rule, which we shall consider in the next chapter. For the present we shall not bother about a theoretical answer but simply accept the results of experiment. (3) How can the course of crystallization be represented diagrammatically? We are dealing now with three variables instead of two, so a complete diagram must be either a three-dimensional figure or a two-dimensional projection.

In answer to the third question, many devices have been tried. The representation most frequently used in modern work is a triangular diagram suggested by Jänecke, one of van't Hoff's followers in salt chemistry. Jänecke's triangle is an ingenious adaptation of a general kind of diagram often used in geology when it is necessary to show changing ratios among three quantities. Suppose, to take the simplest possible example, we wish to show compositions of various mixtures of three substances, A, B, C. On an equilateral triangle we let each corner represent 100% of one of the three pure substances, and the sides opposite represent 0% (Fig. 12-3). Then the percentage of A in a mixture is plotted as the distance from the side opposite toward the A corner, and the percentage of B as the distance from another side toward the B corner. Once the percentages of two substances are fixed, the third is automatically determined. Thus to represent a mixture of 65% A, 25% B, and 10% C, we draw one line through all points with 65% A and a second line through all points with 25% B; the point X, where the two lines intersect, is 10% of the distance toward C. The composition of point Y on Fig. 12-3 would be read as 25% A, 47% B, 28% C. Note that the diagram represents *three* substances, but only *two* independent variables, since any two of the compositional ratios determine the third.

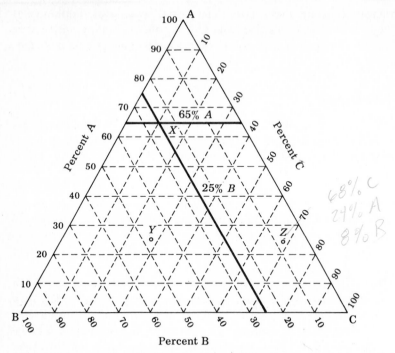

Figure 12-3 Interpretation of a triangular diagram. Point X represents a mixture of 65% A, 25% B, 10% C.

Plotting the composition of a solution with a reciprocal salt pair is somewhat more complicated. The composition is first expressed in molecular percentages of the four ions, and then the sum of the percentages for any three ions is recalculated to 100%. Thus for a salt pair AC and BD we may take the sum of three ions, say $(A^{2+}) + (B^{2+}) + (C^{2-})$, as equal to 100; if it becomes necessary to find the remaining ion for any given point, a little algebra shows that (D^{2-}) must be given by $2(A^{2+}) + 2(B^{2+}) - 100$. Then A^{2+}, B^{2+}, and C^{2-} are placed at the corners of a triangle, and ratios are plotted as usual. In effect, this procedure reduces three variables to two by focusing attention on ratios between ions rather than on actual concentrations. The Jänecke diagram shows relative amounts of different ions or salts, but not their actual concentrations in solution.

12-7 THE EVAPORATION OF SEAWATER

The Jänecke Diagram

The simple reciprocal-pair system $MgCl_2$-K_2SO_4 has only theoretical interest, since seawater always contains NaCl also. Details of the crystallization process have been worked out only for solutions saturated with NaCl, and the Jänecke diagram for this case (at 25°) is shown in Fig. 12-4. Corners of the diagram represent Mg^{2+}, K_2^{2+}, and SO_4^{2-}, corresponding to substances A^{2+}, B^{2+}, C^{2-} in the example of the last paragraph; the fourth ion, Cl_2^{2-}, does not appear on the diagram, but can be calculated (like D^{2-} in the last example) from the expression

$$(Cl_2^{2-}) = 2(Mg^{2+}) + 2(K_2^{2+}) - 100$$

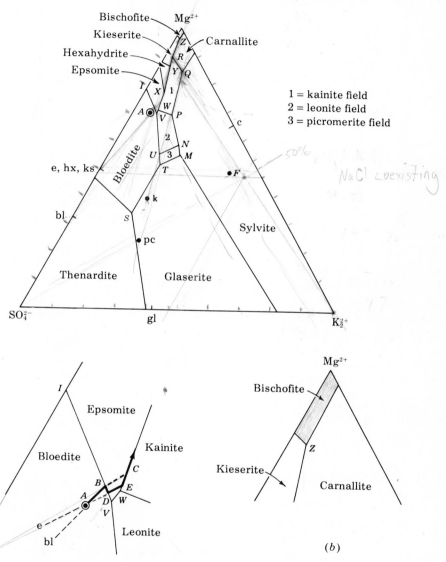

Figure 12-4 Jänecke diagram for system $NaCl$-KCl-$MgCl_2$-Na_2SO_4-H_2O at 25°C. Salt points: c = carnallite, bl = bloedite, gl = glaserite, e = epsomite, hx = hexahydrite, ks = kieserite, k = kainite, p = picromerite. A is composition of seawater. (a) is an enlargement to show part of the path of cystallization of seawater; (b) shows details at the Mg corner. (*Source: Braitsch, 1962, page 52.*)

for any combination of the other three ions. The Mg^{2+} corner of the diagram represents 100 % bischofite, $MgCl_2 \cdot 6H_2O$, and the K_2^{2+} corner 100 % sylvite. The SO_4^{2-} corner would have no meaning in the NaCl-free system, since it represents the impossible situation of 100 % $SO_4^{2-} + Cl^-$, but with Na^+ present this corner stands for pure Na_2SO_4 (thenardite). Compositions of pure salts that contain only two of the "end members" (K_2^{2+}, Mg^{2+}, SO_4^{2-}) are shown along the boundaries of the triangle (for example, c for carnallite, ks for kieserite), and pure salts that contain all three are represented by points in the interior (for example, pc for picromerite, k for kainite). Each labeled field within the triangle shows the salt that would precipitate first if a dilute solution of compositions within that field is allowed to evaporate until crystallization takes place. Thus sylvite would be the first salt to crystallize from a solution containing equal amounts of K_2^{2+} and Mg^{2+} and only one-fifth as much SO_4^{2-} (point F on Fig. 12-4).

Note that for most of the salts represented, the point showing the composition of a salt (the "salt point") lies outside the field where the salt would precipitate. This is analogous to the situation in Fig. 12-2, where the point representing carnallite on the crystallization line (point E) lies far outside the "carnallite region" (line CD) where carnallite would be the first precipitate. This means that most of the salts in Fig. 12-4, like carnallite, show incongruent solution. In using Fig. 12-4, it will be helpful to remember that this diagram is analogous to Fig. 12-2, but that the regions of crystallization are now two-dimensional fields rather than lines. To show a point of infinite dilution on Fig. 12-4, corresponding to point O on Fig. 12-2, we would need to use another dimension, putting this point somewhere above the diagram. Then Fig. 12-4 can be visualized as a warped surface, and evaporation of a dilute solution would mean movement down from the O point until the surface is intersected. We are concerned, however, not with dilute solutions but only with the course of crystallization after one of the salts has started to precipitate. By analogy, on Fig. 12-2 we would be concerned only with crystallization along the warped line $ACDB$, not with dilute solutions in the region $OACDB$.

Crystallization of Major Components

The path of crystallization of most interest is that of normal seawater, whose composition is shown on the diagram by the point A. This point lies in the bloedite field, meaning that bloedite [$Na_2Mg(SO_4)_2 \cdot 4H_2O$] is the first mineral to form when seawater has evaporated far enough for magnesium salts to start crystallizing. (Remember that the solution is continuously saturated with NaCl, so that halite will precipitate before and all during the crystallization of the magnesium and potassium minerals.) The Mg^{2+}/SO_4^{2-} ratio in bloedite is shown by the bloedite salt point (bl) on the left-hand side of the triangle. As bloedite crystallizes, the two ions are removed from the solution in this ratio, and the composition of the solution must then change along a line directed away from the salt point. Before much bloedite has crystallized, therefore, the solution will reach a composition shown by a point on the boundary between the bloedite and epsomite fields. Here bloedite becomes unstable and as crystallization progresses, epsomite will precipitate in its place.

We can imagine two extreme cases for the transition at this boundary. The bloedite crystals as they form may separate out of the liquid as a layer on the bottom, which thereafter has little contact with the remainder of the solution; in this case the precipitation of bloedite will simply stop and the crystallization of epsomite will begin, the composition of the solution now changing along a new path away from the epsomite salt point (e). Alternatively the solution as it evaporates may be stirred sufficiently so that bloedite crystals remain in effective contact with the liquid; in this case the bloedite crystals will react with the liquid when its composition reaches the boundary, changing over into the now more stable epsomite. This changeover means a corresponding change in the composition of the solution, since the Mg^{2+}/SO_4^{2-} ratio is different in the two minerals. The two cases can be visualized with the aid of Fig. 12-4a, which is a portion of Fig. 12-4 near the seawater point magnified and somewhat distorted to make the relations clearer. In the first case, with complete separation of bloedite crystals, the composition of the solution follows path ABC, merely changing direction at the boundary so that it now moves away from the Mg^{2+}/SO_4^{2-} ratio for epsomite. In the second case, where bloedite disappears by reaction to form epsomite, the composition of the solution shifts along the boundary to a point D, on an imaginary line drawn from the epsomite salt point through the seawater point. At D the last bloedite would vanish, and the composition of the solution thereafter changes along DE as new epsomite crystallizes.

Note that the first case would give ultimately a salt deposit with a bloedite layer overlain by epsomite, while the second case would give only a layer of epsomite. All gradations between the extreme cases can be imagined. If bloedite reacts only partially, the end result would be a thin layer of bloedite, or perhaps a layer of mixed bloedite and epsomite, overlain by a layer containing epsomite alone.

As epsomite continues to crystallize, the changing composition of the solution must bring it soon to the epsomite-kainite boundary. This is a different kind of boundary, along which the two salts crystallize together (as could be predicted from the fact that the kainite and epsomite salt points are on opposite sides of an extension of the line). Simultaneous crystallization causes the composition of the solution to change along the boundary toward point X. Here epsomite becomes unstable with respect to hexahydrite, and whether any is preserved or not depends on how effectively the epsomite crystals are brought into contact with the solution at this point. With further evaporation hexahydrite and kainite crystallize together, the composition of the solution changing along XY. At Y hexahydrite in its turn becomes unstable with respect to kieserite, and at R kainite becomes unstable with respect to carnallite. In the last stages of evaporation carnallite and kieserite crystallize together, the brine changing its composition toward point Z. Here bischofite begins to precipitate along with the other two salts, and the composition of the brine remains fixed until the last liquid disappears. The bischofite field is so small that a magnified sketch of the upper corner of the triangle (Fig. 12-4b) is necessary to show the final stages clearly.

Complete evaporation of seawater at 25°C, then, if equilibrium is preserved in the sense that each salt reacts completely with the solution at the point where it becomes unstable, should lead to a mixture of three salts only—kieserite, carnallite, and bischofite (plus, of course, halite and small amounts of calcium salts). Lack of complete reaction would permit other salts to be preserved in an evaporite sequence, notably bloedite, epsomite, hexahydrite, and kainite. In either case the minerals sylvite, leonite, and picromerite should be absent.

The relations shown in Fig. 12-4 are the result of many years of work by van't Hoff and his colleagues, plus later refinements by other workers. They are not, however, the results that one obtains immediately by evaporation of solutions containing the ions K^+, Mg^{2+}, SO_4^{2-}, Cl^-, and Na^+, as Usiglio discovered long ago. During rapid evaporation kieserite and kainite are not formed; other salts appear (notably $MgSO_4 \cdot 5H_2O$ and $MgSO_4 \cdot 4H_2O$) that are not represented on Fig. 12-4; and the fields of sylvite and carnallite are considerably larger. The difficulty is the old bugaboo of experimental geochemistry, the slowness of reaction rates. The two new hydrates of $MgSO_4$ are metastable compounds that crystallize early but on standing change slowly to the stable hydrates kieserite and hexahydrite. Kainite forms only when the solution and its precipitated salts are allowed to stand for long periods, so that stable equilibrium rather than metastable equilibrium can be attained. The early formation of metastable minerals is one of the reasons why salt chemistry is so complicated, and why an equilibrium diagram like Fig. 12-4 takes so long to work out. Presumably the equilibrium diagram is the appropriate one for comparison with natural occurrences, since evaporation of seawater in nature takes place slowly and salts remain in contact with brines for very long times.

Other Components

Figure 12-4 is essentially a diagram for the reciprocal salt pair $MgSO_4$-KCl. It is true that we have included NaCl in the system, by making the provision that the solution is at all times saturated with this salt; but in the upper part of the diagram sodium actually enters only one of the minerals (bloedite), so that we can focus attention on the sulfates and chlorides of K and Mg alone. In other words, during the evaporation of seawater halite crystallizes practically independently of the other salts. The presence of Na^+ in solution modifies the fields of the K and Mg minerals, but the fields can be described wholly in terms of the ratios K^+/SO_4^{2-} and Mg^{2+}/SO_4^{2-}. This independent behavior of NaCl is a fortunate circumstance, because if Na^+ were an important constituent of the K and Mg salts we should have to regard this ion also as an independent variable, and a two-dimensional diagram like Fig. 12-4 would not suffice to portray the crystallization process.

Of the six major ions of seawater (Table 12-4), the Jänecke triangle thus summarizes the important variations in all but Ca^{2+}. Again we are favored by a lucky circumstance: most of the Ca^{2+} originally present is precipitated early in the crystallization sequence as gypsum or anhydrite. Gypsum begins to form when seawater has evaporated to about one-fifth of its original volume, halite when the volume has shrunk to about one-tenth of the original, bloedite not until more than

98% of the water has evaporated. By this time so much of the Ca^{2+} has precipitated that only about 0.0001 mole/mole of H_2O remains in solution, an amount so small that it can have little effect on the crystallization of the K and Mg salts. To a good approximation, then, the later stages in evaporation of seawater can be treated as a problem with only three variables (concentrations of any three of the ions K^+, Mg^{2+}, SO_4^{2-}, Cl^-), and can be accurately represented by a triangular diagram like Fig. 12-4, despite the presence of other ions in the solution.

The relations of the two principal calcium sulfate minerals, gypsum and anhydrite, are still not altogether clear. The one that precipitates first, in salt lakes or salt lagoons or in the laboratory, is gypsum, but the common mineral of evaporite deposits is anhydrite. Textural relations suggest that one can change to the other fairly rapidly and that such changes have occurred on a large scale in nature. High temperature and low humidity favor anhydrite, as might be predicted from the formulas, so that the predominance of anhydrite in buried salt deposits is plausibly accounted for as a diagenetic alteration of primary gypsum. A major question remains, however, as to whether anhydrite can ever be a primary precipitate from seawater. Thermodynamic data for the system $CaSO_4$-H_2O indicate that, in the presence of pure water, gypsum should be the more stable precipitate below 56°C, and anhydrite above this temperature (Sec. 1-5). The transition temperature is lowered by the presence of other salts in solution: in seawater it is about 20°, and as seawater becomes more concentrated during evaporation, the transition point drops steadily. According to these data the evaporation of seawater at 25° should give anhydrite as the first precipitate, so that theoretically the anhydrite of salt deposits could be primary. Some geologists have considered that textural evidence supports this assumption. On the other hand, laboratory attempts to precipitate anhydrite under simulated natural conditions have uniformly failed. Even at fairly high temperatures and high salt concentrations, where anhydrite is unquestionably the stable form, metastable gypsum always appears as the first precipitate. This fact, together with the absence or near absence of primary anhydrite in present-day evaporites, has led most geologists to conclude that the anhydrite of marine evaporites is entirely secondary.

Despite the fact that much of the Ca^{2+} in seawater is removed as gypsum early in the course of normal evaporation, so that Ca^{2+} should theoretically play only a minor role in crystallization of the later K and Mg compounds, salts containing Ca together with K, Mg, or Na are not at all uncommon in marine evaporites. One of these salts, polyhalite $[K_2MgCa_2(SO_4)_4 \cdot 2H_2O]$ is often a major constituent. To get a complete picture, therefore, we should make an effort to fit this salt into the sequence shown by Fig. 12-4. This proves to be a difficult undertaking experimentally, because polyhalite, like anhydrite, forms only extremely slowly under laboratory conditions. The crystallization is so slow that polyhalite can hardly be expected to form in nature as a primary precipitate at temperatures near 25°. Its field of stability can be represented approximately on Fig. 12-4 as a small area covering parts of the bloedite, kainite, sylvite, and leonite fields. Ideally, therefore, polyhalite should crystallize together with bloedite, epsomite, and kainite during the early stages of formation of the Mg-K salts, but (like kainite and kieserite) it does not appear experimentally unless the salts and brine remain in contact for months or years.

The less abundant constituents of seawater play only a minor role in salt deposition. HCO_3^- is removed early by precipitation as $CaCO_3$, or possibly as dolomite or magnesite. Boron forms rare secondary minerals. Br^- enters the crystal structure of chlorides, particularly sylvite and carnallite, as an isomorphous replacement of Cl^-. Most of the Sr^{2+} substitutes for Ca^{2+} in gypsum or anhydrite. The distribution of the rare constituents is an interesting study, especially for the information their presence gives about conditions during the evaporation process, but the present discussion must be limited to the major ions.

12-8 THE EFFECT OF TEMPERATURE

Theoretically, according to Fig. 12-4, complete evaporation of seawater at 25°C should lead to one of two results: (1) If equilibrium is maintained, so that crystals can react with the remaining brine, the ultimate product should be a mixture of anhydrite, halite, kieserite, carnallite, and bischofite (point z on Fig. 12-4). (2) If crystals accumulate so that reaction with the brine is restricted, successive layers should be formed in which the following minerals are prominent: polyhalite, bloedite, epsomite, epsomite-kainite, hexahydrite-kainite, kieserite-kainite, kieserite-carnallite, kieserite-carnallite-bischofite, with anhydrite and halite present in all. A comparison of these predictions with the salt sequence at Stassfurt (Tables 12-6 and 12-7) shows at once that agreement is poor, whichever alternative is assumed. Polyhalite, if formed at all, should be very minor, but at Stassfurt it is a major constituent. Kainite should be absent (case 1) or should be present in large amount early in the sequence (case 2); actually it appears only sporadically in the extreme upper part. Salts like loeweite, picromerite, vanthoffite, langbeinite should not be present at all. The combination kieserite-sylvite, common in the "Hartsalz" of the upper layer, is strictly forbidden by the equilibrium diagram. The absence of bischofite at Stassfurt is also conspicuous, but this may mean simply that evaporation did not go to completion.

Of all the marine salt beds that have been studied in detail, the Stassfurt deposit comes closest to matching predictions from experiment, and even here the fit is evidently unsatisfactory. What do the discrepancies mean? Has the ocean changed its composition from the Permian to the present? Were atmospheric temperatures and pressures very different in the Permian? Have the salt beds been modified since their formation?

Van't Hoff, the first to pinpoint the major discrepancies between experiment and observation, suggested that a difference in the temperature of evaporation might be responsible. To explore this possibility, he and his colleagues investigated the K^+-Mg^{2+}-SO_4^{2-}-Cl^--Na^+ system at higher temperatures as well as at 25° (Fig. 12-5). At 83° he found that two major difficulties were removed: the

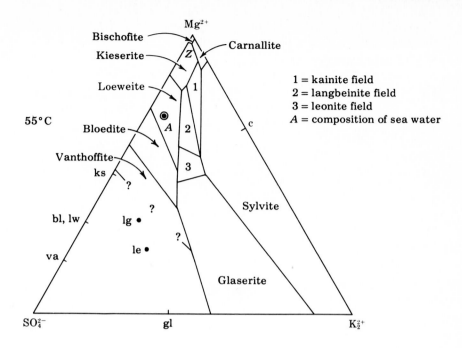

1 = kainite field
2 = langbeinite field
3 = leonite field
A = composition of sea water

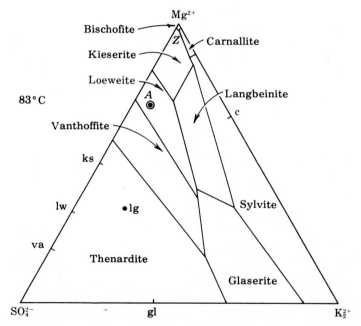

Figure 12-5 Jänecke diagrams at 55 and 83°C. *(Source: Stewart, 1963.)*

field of kainite has disappeared, and the fields of kieserite, carnallite, and sylvite have changed their shape so that the combination kieserite-sylvite is stable. On this basis van't Hoff concluded that the basin in which an arm of the Zechstein sea evaporated must have had a temperature within 17° of the boiling point of water.

12-9 POSTDEPOSITIONAL CHANGES IN SALT BEDS

Van't Hoff was a physical chemist, and his conclusion about the evaporation temperature was a straightforward chemical deduction. It initiated a long controversy among geologists, some maintaining that chemistry had "proved" extraordinarily high temperatures in the Permian, others pointing to the complete lack of evidence for such temperatures in the sediments and particularly in the fossils of adjacent beds. As a possible means of reconciliation, Jänecke and some others at length suggested that the indicated high temperature might not be the temperature of original deposition, but a temperature attained by the salt beds later on, after their burial under other sediments. The salts should be very sensitive to temperature changes, as Figs. 12-4 and 12-5 show, and a temperature of 83°C would be reached, with an average geothermal gradient, at depths of only a few thousand meters. In other words the salt deposits can be regarded as metamorphic rocks, whose mineral composition has adjusted itself to a new equilibrium at higher temperature and pressure. The temperature of this transformation is so low, however, that it becomes a matter of dispute whether the term "metamorphism" or "diagenesis" is more appropriate.

Thermal metamorphism is a plausible explanation for some changes in the salt beds, notably for the change from kieserite-carnallite to kieserite-sylvite, but later workers (especially D'Ans and Kühn) have pointed out that it is far from adequate to account for all the differences between theoretical and actual salt sequences. Much more important is metamorphism, or diagenetic alteration, by solutions percolating through the salt beds. Since the salts are so very soluble, they are easy prey to solutions of all kinds. Dilute solutions can remove great masses of salts bodily, and more concentrated solutions of various compositions can bring about profound changes in mineral assemblages. Large bodies of polyhalite, for example, are probably always secondary, a likely reaction being the alteration of kainite by solutions rich in Ca^{2+}; a process of this sort would account for the absence of kainite and the presence of polyhalite in the early part of the sequence at Stassfurt (Table 12-6). The incongruent dissolving of carnallite may be responsible for the formation of large deposits of sylvite—a mineral which is the most important commercial source of potassium, although it would not be formed at all in the normal low-temperature evaporation of seawater. Very probably the entire discontinuous upper layer at Stassfurt, containing "Hartsalz," loeweite, picromerite, vanthoffite, and other minerals not in the normal low-temperature sequence, has been formed by the action of later solutions; probably also this upper portion is separated from the carnallite zone by a thrust fault and is not, as van't Hoff thought, part of a continuous sequence. The action of solutions is the only reasonable explanation also for layers of sylvite and carnallite that cut across other salt beds, as in the deposits of New Mexico.

It remains an open question whether alteration by solutions, coupled with equilibrium during primary deposition and possibly with the effects of thermal metamorphism, can satisfactorily account for all details of salt-mineral assemblages. Uneven topography in the basin of deposition, crystallization of salts in a thermal gradient, retrogressive metamorphism as the temperature drops, plastic movement of salts during orogenic disturbances, are among recent suggestions for additional factors that may play a role in the history of salt deposits. An increasing number of puzzling details that turn up as salt beds are examined ever more closely seems to ensure that a complete explanation of these curious sediments is a possibility only for the distant future.

Summary The crystallization of salts from seawater has warranted discussion in considerable detail, since this is a classical example of the successful application of physical chemistry to a geologic problem. It illustrates beautifully how useful the chemical approach can be, how difficult it is to allow for all geological variables in laboratory experiments, and how misleading a purely chemical conclusion can be when unrestrained by geological judgment. It shows also, as do many other subjects in recent geochemistry, how discouragingly complicated an apparently simple phenomenon can be when it is examined in great detail.

To begin with, van't Hoff's problem seemed so ideally simple: the salt sequence at Stassfurt, at that time the only sequence which had been studied with care, appeared to be the result of long-continued evaporation of seawater; he needed only a few chemical data to validate this postulate. Two possible methods of attack suggested themselves: he could try to duplicate the complexities of nature, starting with seawater under conditions as close as possible to natural ones, or he could use the more rigorous approach of assembling precise data on simple systems and trying from them to discover the principles that govern the behavior of complex natural systems. For any geochemical problem a similar choice must be made. Sometimes the first approach proves more useful, sometimes the second. The second is obviously more basic, and ideally every problem should ultimately be handled in this way, but complexities are often so formidable that the number of necessary simple experiments is prohibitively large. For the salt problem, Usiglio had tried the first approach, and van't Hoff decided on the second.

To reconstruct in imagination the reasoning of van't Hoff and his followers, we limit ourselves at first to the four ions most abundant in the later stages of salt formation, K^+, Mg^{2+}, SO_4^{2-}, Cl^-. Taking these in pairs, then in threes, then all four, we study the compositions of saturated solutions and the salts in equilibrium with them for various initial ratios. Provisional comparison with the natural deposits at this point shows little agreement. We bring our system closer to actual seawater by adding NaCl, and now the experimental difficulties multiply, because we get different salt assemblages depending on how long the salts and brine remain in contact. With perseverance we determine the true equilibrium assemblages here, and with even more perseverance we determine, at least roughly, the effect of adding the remaining principal ion of seawater, Ca^{2+}. Comparison with natural salts is now more rewarding, but discrepancies persist. So we try yet another variable and, with further tedious experiments, determine the effect of temperature on the equilibrium assemblages. Now at last we find fair correspondence between the experiments and the salt sequence at Stassfurt, provided we assume that evaporation in the Zechstein sea took place at a temperature over 83°C. Our experiments have been successful, for we have proved that the Stassfurt salts were indeed deposited from seawater, and we have turned up the hitherto unsuspected "fact" that northern Germany was very warm during the Permian.

But now we find that further observational evidence does not support our conclusions. Stratigraphers tell us that sediments and fossils of the Zechstein adjacent to the salt deposits are completely normal, giving no indication of high temperatures. The Stassfurt sequence, when looked at more closely, shows many details that are not explained by the quiet evaporation of seawater. There is even a suggestion that the deposit is not a simple sequence at all, but is broken by a fault. Still more damaging is an accumulation of evidence from other parts of the Stassfurt series, from other salt series within the Zechstein, and from marine potash deposits elsewhere in the world: salt sequences seem notable for their variability rather than for the uniformity we might expect, and almost the only one that shows much resemblance to our experimental results is the sequence at Stassfurt.

So we ask how our beautifully simple chemical conclusions might be modified by the geological variables we have hitherto neglected. A little reflection shows a host of possible modifying factors. There is the increase of temperature due to burial, which might account for the apparently high-temperature assemblages. There is the possible slow reaction of seawater in early stages of evaporation with $CaCO_3$, leading to an impoverishment in Mg^{2+} and SO_4^{2-}, which might explain the common presence of dolomite and magnesite in evaporite sequences and might, at least in part, explain the often-noted high $CaSO_4/NaCl$ ratio. There is a probability of extensive alteration of the salts by circulating solutions. There are possible complications in the shape of the evaporation basin, possible changes in the relation of land and sea during evaporation, possible plastic movement of the salt after deposition. With all these possibilities the lack of uniformity among salt deposits and the failure of most deposits to follow experimentally based predictions closely no longer seem so strange.

The consequences of some of the geologic variables can be tested by modifications of the basic experiments, so that details of the history of particular deposits can often be guessed. Early predictions from the experiments were obviously far too sweeping, but the experiments themselves remain the basis for exploring many aspects of salt behavior. Van't Hoff failed in his original objective, to prove that the salts at Stassfurt formed by simple evaporation of seawater, but the work he initiated remains a monument to the power of physical chemistry to clarify geologic problems.

PROBLEMS

1 Salt lakes from which nearly pure epsomite ($MgSO_4 \cdot 7H_2O$) precipitates are sometimes found in undrained basins in areas underlain by chlorite schist containing pyrite. Explain.

2 Explain why lakes in arid regions generally have a higher pH than lakes in humid regions.

3 Explain why each of the following is *not* a common constituent of evaporite deposits in desert basins:

$$FeSO_4 \quad BaSO_4 \quad Na_2S \quad MgBr_2$$
$$Al_2O_3 \quad CaO \quad CaCl_2 \quad MnCO_3$$

4 Sodium and potassium are about equally abundant in the rocks of the earth's crust, but dissolved sodium is much more abundant than potassium in seawater. Explain.

5 Ba^{2+} and Fe^{3+} have concentrations in seawater far lower than would be expected from the amounts of these ions supplied to the sea by weathering and erosion. Suggest possible mechanisms by which these ions may be continuously removed from seawater.

6 In which of the following would you expect $MgSO_4$ to be most soluble? In which would it be least soluble? Explain. (a) $1M\ K_2SO_4$; (b) $1M\ NaCl$; (c) distilled water.

7 Using Fig. 12-2, describe what would happen if a solution of composition M is evaporated to dryness (a) at 25°, (b) at 100°C.

8 (a) On Fig. 12-3, what proportions of A, B, and C are shown by point Z?

(b) On Fig. 12-3, show the composition of chalcopyrite, $CuFeS_2$, letting A stand for Cu, B for Fe, and C for S.

9 Using Fig. 12-4, describe the course of crystallization of a solution with a ratio of $Mg^{2+}/K_2^{2+} = 1/1$ and a ratio $Mg^{2+}/SO_4^{2-} = 10/1$.

10 Which of the salts that appear in Fig. 12-4 show incongruent solution?

11 Suggest an explanation of the fact that bischofite is not found in the Stassfurt deposits.

12 In the analyses of Table 12-1, suggest possible explanations why, in comparison with the world average,

(a) Total dissolved solids in the Amazon are low.

(b) Na^+ and K^+ in the Hudson River are low.

(c) The ratios SO_4^{2-}/HCO_3^- and Cl^-/HCO_3^- in the Colorado River are high.

(d) The ratio Cl^-/SO_4^{2-} is high in both the Amazon and the White Nile.

REFERENCES AND SUGGESTIONS FOR FURTHER READING

Bentor, Y. K., Some geochemical aspects of the Dead Sea, *Geochim. et Cosmochim. Acta*, vol. 25, pp. 239–260, 1961. A description of the unusual combination of dissolved materials in the Dead Sea, and speculations about their origin.

Blount, C. W., and F. W. Dickson, Gypsum-anhydrite equilibria in the systems $CaSO_4$-H_2O and $CaSO_4$-NaCl-H_2O, *Am. Mineralogist*, vol. 58, pp. 323–331, 1973.

Braitsch, O., "Entstehung und Stoffbestand der Salzlagerstätten," vol. 3 of von Engelhardt and Zemann (eds.), "Mineralogie und Petrographie in Einzeldarstellungen," Springer-Verlag OHG, Berlin, 1962. History of investigations and review of current hypotheses on the origin of marine salt deposits. Particularly good for its treatment of quantitative relations, much of it original with the author.

Butler, G. P., Modern evaporite deposition and geochemistry of coexisting brines, the Sabkha, Trucial Coast, *Jour. Sedimentary Petrology*, vol. 39, pp. 70–89, 1969. Describes the precipitation of aragonite, gypsum, and halite on a salt flat, and the alteration of gypsum to anhydrite and of aragonite to dolomite.

Clarke, F. W., The data of geochemistry, 5th ed., *U.S. Geol. Survey Bull.* 770, 1924. This famous volume is now seriously out of date, but for some geologic materials it still provides the best and most convenient compilations of representative analyses.

Friedman, G. M., Significance of the Red Sea in problems of evaporites and basinal limestones, *Am. Assoc. Petroleum Geologists Bull.*, vol. 56, pp. 1072–1086, 1972. Argues for accumulation of evaporite deposits on the flats of a shifting sea margin, rather than in a partly enclosed basin.

Holland, H. D., The geologic history of sea water—an attempt to solve the problem, *Geochim. et Cosmochim. Acta*, vol. 36, pp. 637–652, 1972. The composition of evaporites shows that concentrations of the major constituents of seawater have remained within a factor of 2 of their present concentrations for at least 700 million years.

Jones, B. F., H. P. Eugster, and S. L. Rettig, Hydrochemistry of the Lake Magadi basin, Kenya, *Geol. Soc. Am. Bull.*, vol. 41, pp. 53–72, 1977. One in a series of papers by Eugster and his colleagues on an unusual evaporite basin in East Africa containing a thick bed of trona and deposits of the rare sodium-silicate mineral magadiite.

Livingstone, D. A., Chemical composition of rivers and lakes, chap. G in Data of geochemistry, 6th ed., *U.S. Geol. Survey Prof. Paper* 440-G, 1963. This article and others in Professional Paper 440 are being prepared seriatim as a new edition of F. W. Clarke's 1924 bulletin (above).

Stewart, F. H., Marine evaporites, chap. Y in Data of geochemistry, 6th ed., *U.S. Geol. Survey Prof. Paper* 440-Y, 1963. The best brief account in English of the current status of work on salt deposits of marine origin.

THIRTEEN

THE PHASE RULE

Abruptly now we leave the geochemistry of sediments and turn to the more obscure processes in which igneous rocks have their origin. For these processes we cannot use as direct an approach as we have hitherto, because many igneous rocks are formed in environments inaccessible to observation and because laboratory work is made difficult by the high temperatures, high pressures, and corrosive substances involved. We must depend much more on inferences from geologic and geophysical observations, on extrapolation from simple experiments, and on weighing hypotheses which cannot be rigorously established or disproved.

Some igneous rocks, of course, we can watch forming before our eyes. In the lava flows at active volcanoes we see a viscous silicate liquid, exuding gases as it cools, gradually losing its mobility and freezing to a glass, to a mass of crystals, or to a mixture of both. We can analyze the cooled rock and so determine the composition of the original liquid; we can measure the temperature, the rate of temperature decrease, the thickness and viscosity of the flow—all the important geologic variables except the precise nature and amount of the original gas content. Armed with these data, we can then proceed to study the chemistry of the freezing process, to see just how a silicate melt converts itself into a crystalline or glassy solid. And by generous extrapolation we can extend our findings to those igneous rocks whose birth we cannot witness, those that crystallize deep within the crust of the earth.

In studying the chemistry of crystallization we have the same choice of approaches as in the study of evaporites. We can try the frontal attack, simply melting an igneous rock in the laboratory and letting it cool under controlled conditions; or we can take the more laborious synthetic route, working out the physical chemistry of simple silicate systems first and then adding components one at a time until we reach a system complex enough to show some resemblance to natural silicate liquids. The first approach was tried long ago but has proved disappointing. Letting lavas cool under laboratory conditions established a few general rules, for example, that glasses form when cooling is fast and crystals when cooling is slow, and that silica-rich melts are more viscous and less easily crystallized than melts with abundant FeO and MgO; but in most such experiments the cooled product did not greatly resemble the original rock, and the mixture was so complex that the course of crystallization could not be followed easily. The second method of attack, starting with simple artificial silicate mixtures, has been far more successful. The work of Bowen and his collaborators at the Geophysical Laboratory in Washington during the early years of this century, together with more recent work at this and other laboratories, has given us an understanding of all but minor details in the crystallization of common igneous rocks.

The freezing of a melt has this much in common with crystallization of salts from seawater, that both involve the successive separation of various solids from a liquid. An important question here, as in the formation of evaporites, is the limitation on the number of solids that can separate from a given liquid. The guiding principle in answering such questions is the phase rule, which we did not state explicitly in the last chapter but which is expressed implicitly in diagrams like Figs. 12-2 and 12-3. It is time now to bring the phase rule out in the open, to see how it systematizes the phenomena of crystallization, either from a solution or from a melt.

13-1 FORMULATION OF THE PHASE RULE

General

The phase rule is one of many thermodynamic principles we owe to the work of the American chemist Willard Gibbs. Gibbs' derivation of the phase rule is given in modified form in Appendix XI. Here we shall adopt a simpler inductive approach, showing that the phase rule is a plausible inference from specific examples.

A One-component System: Water

Water vapor, we say, condenses to liquid water on cooling. If we start with pure water vapor and maintain a constant pressure, liquid water appears abruptly at a particular temperature. The pressure is then called the *vapor pressure* for this temperature. At greater pressures the temperature of condensation is higher, and at smaller pressures the temperature is lower, as is shown graphically by the curve *TC* (the "vapor-pressure curve") in Fig. 13-1.

If, during condensation, we maintain a fixed pressure and continue to abstract heat, the temperature stays constant as long as any vapor is left to condense. When all the water is in liquid form, the temperature begins to drop, and falls

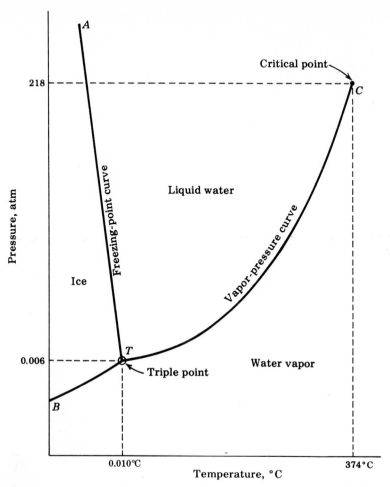

Figure 13-1 Pressure-temperature diagram for water. Not to scale. Dense forms of ice stable at high pressures are not shown.

steadily until the point is reached where freezing starts. Here the temperature once more becomes constant and remains fixed until the liquid disappears. Raising the pressure lowers the freezing point, a relation expressed by the line TA on Fig. 13-1. This line and the vapor-pressure curve intersect at T, which evidently is a point at which vapor, liquid, and ice can all exist together. At temperatures lower than that represented by point T liquid water cannot be present; vapor on cooling condenses directly into ice, the vapor-pressure curve for ice being shown as TB.

In phase-rule language, we are dealing here with a *system of one component*, meaning that we have set up an experiment, either actual or imaginary, involving only a single chemical substance, here described by the formula H_2O. We are concerned with two variables, pressure and temperature. As these variables change, the substance H_2O exhibits three *phases* (vapor, liquid, solid), which may be described as the homogeneous regions, or parts, of the system.

Consider now the various lines and areas in Fig. 13-1. At the far right, beyond the line TC, only the single phase water vapor is present. Either temperature or pressure or both may be varied, and the single phase remains, unless, of course, the change is sufficient to bring the system to the line TC. Because two variables can change, within limits, without altering the number of phases, we say that the system has two *degrees of freedom*, or is *divariant*. By contrast, at a pressure-temperature combination represented by a point on TC, two phases (liquid and vapor) can exist together and the number of degrees of freedom is smaller. If temperature increases, the liquid phase will disappear unless the pressure increases also; if pressure decreases, temperature must decrease in order to preserve the two phases. In other words, we may alter *one* variable at will but this change automatically fixes the other variable if we want our two-phase system to persist. We say the system has only one degree of freedom, or is *univariant*, meaning that only one variable can be assigned arbitrarily. Finally, point T (the *triple point*) represents coexistence of three phases, solid, liquid, and vapor; if either pressure or temperature changes, no matter in what direction, at least one of the phases must disappear. Thus, when three phases are present, the system has zero degrees of freedom, or is *invariant*.

Two Components: A Salt Solution

Now we add a second component to the system by dissolving salt in the water. The lines in Fig. 13-1 will obviously be displaced, because both the vapor pressure and the freezing point of water are lowered by dissolved solids. The amount of displacement depends on the concentration of the salt. In other words, we now need a three-dimensional graph to represent the system, the third axis showing increase in concentration. The line TC of Fig. 13-1 becomes a curved surface in the three-dimensional figure, expressing the fact that equilibrium between liquid and vapor for a solution depends on pressure, temperature, and composition. Note that a two-phase system containing liquid and vapor now has two degrees of freedom instead of one, for we may vary arbitrarily any two of the three variables, pressure, temperature, concentration. The triple point T of Fig. 13-1 becomes a line in the three-dimensional figure, since the pressure and temperature where ice, liquid, and vapor coexist change with the composition of the solution; and such a three-phase system now has one degree of freedom instead of none. To attain a condition of no degrees of freedom, we would have to cool the solution until salt begins to crystallize in addition to ice, in other words until four phases (ice, salt, solution, vapor) are present.

A complication is introduced if the dissolved salt forms hydrates. Consider, for example, the system $CaSO_4$-H_2O. At some temperature-pressure combinations the solid salt anhydrite is stable, at other combinations the hydrated salt gypsum. (For simplicity we disregard the hemihydrate $CaSO_4 \cdot \frac{1}{2}H_2O$, which also may appear in this system.) Thus five phases are now possible rather than four:

anhydrite, gypsum, solution, ice, and vapor. The number of components is still two, inasmuch as one of the substances may be derived by reaction among the others $(CaSO_4 + 2H_2O \rightarrow CaSO_4 \cdot 2H_2O)$; the number of components is taken as the *minimum* number of chemical substances necessary to describe a system. For this case we would find that the pressure-temperature-composition diagram is similar to that for ordinary salt, with the one exception that the surface representing equilibrium between solid and solution would have an abrupt break along a line showing the transition from anhydrite to gypsum. Any surface would again represent two-phase equilibrium, the possibilities being

Solution-vapor	Anhydrite-solution
Gypsum-vapor	Ice-solution
Gypsum-anhydrite	Ice-vapor
Gypsum-solution	Ice-gypsum

A line in the three-dimensional diagram would represent three-phase equilibrium; the possibilities are

Solution-vapor-gypsum	Ice-solution-gypsum
Solution-vapor-anhydrite	Ice-vapor-gypsum
Vapor-gypsum-anhydrite	Solution-gypsum-anhydrite
Ice-solution-vapor	

The diagram would have one point, as before, representing equilibrium between four phases, ice, gypsum, solution, and vapor. There could be no corresponding point for ice, anhydrite, solution, and vapor, because anhydrite is not stable with respect to gypsum at so low a temperature. Nor could all five possible phases exist together at any temperature-pressure combination, again because only gypsum is stable at temperatures where ice exists. Thus the addition of a hydrate to the system adds a surface to the pressure-temperature-composition diagram, but does not change the relation between number of phases and number of degrees of freedom.

Summary This discussion may be summarized in tabular form as shown in Table 13-1. The regularity shown by these figures may be condensed into the relation

$$f = c - p + 2 \qquad (13\text{-}1)$$

where f is number of degrees of freedom, c is number of components, and p is number of phases. This is Gibbs' phase rule, here derived from a few very simple cases but applicable to systems with any number of components and phases.

Table 13-1

	Number of components	Number of phases	Degrees of freedom
Water alone	1	1	2
	1	2	1
	1	3	0
Water + salt	2	1	3
	2	2	2
	2	3	1
	2	4	0
Water + gypsum + anhydrite	2	1	3
	2	2	2
	2	3	1
	2	4	0
	2	5	Impossible

Discussion

The simple definitions given above for phase, component, and degree of freedom would need further qualification for general use, but for most geologic applications they are sufficiently restricted. A few additional comments may be helpful.

A phase we have defined as one of the homogeneous parts of a system. The requirement of homogeneity, it should be noted, does not imply that a phase must be continuous. For example, separate crystals of ice in water at 0° would all be part of the same phase, whether or not the crystals are in contact.

A component, according to the previous discussion, is one of the minimum number of chemical substances necessary to describe the composition of a system. This is the most difficult of the three concepts to grasp. We say, for example, that the system $CaSO_4$-H_2O has two components; why is it not equally correct to speak of three components (Ca^{2+}, SO_4^{2-}, H_2O) or four components (Ca, S, O_2, H_2)? The point to be remembered is that the phase rule is concerned with *variations* in conditions, and the resulting *variations* in phases. In the example we have used, no change of conditions would lead to a variation in the ratio of Ca^{2+}/SO_4^{2-}, or Ca/S, or H_2/O_2. According to the specifications of the experiment these ratios are fixed, and therefore the separate substances cannot be independent components. It would be easy to imagine different experiments in which one or all of these ratios might be variable (the reaction of $CaSO_4$ with $BaCl_2$ solution, for example, or the high-temperature reduction of $CaSO_4$ by H_2) and in which the number of components would be correspondingly greater. The number of components is not fixed by the overall chemical composition of a system, but by the kinds of variation within the system which are being considered.

The number of degrees of freedom has been defined as the number of conditions that may be varied arbitrarily. The kinds of conditions we have considered are temperature, pressure, and concentrations. One could easily add other conditions to this list, for example, the strength of the electric field in which a system finds itself. If variation of the field produces variations in the number of phases present, the phase rule would have to be changed to read

$$f = c - p + 3$$

In most geological and chemical problems additional variables are not important, so that Eq. (13-1) is the most generally useful form of the phase rule. For some geological applications, in systems where pressure is constant or nearly so, the number of variables is reduced by 1, and the phase rule would read

$$f = c - p + 1 \qquad (13\text{-}2)$$

In other words, the expression for the number of degrees of freedom depends on the number of possible kinds of variation in a given experiment.

13-2 THE PHASE RULE APPLIED TO SALT CRYSTALLIZATION

A few examples from the salt combinations in the last chapter will illustrate ways of using the phase rule.

A mixture of NaCl, KCl, and H_2O (Fig. 12-1) has three components; hence for constant pressure $f = 3 - p + 1$, or $f = 4 - p$. This means that the maximum number of phases that can exist together is four; with this number the system would have zero degrees of freedom, and with any more the value of f would be negative. At a particular temperature, then, we might have solid NaCl and solid KCl both in contact with solution and ice, but any change in temperature would cause one of the four to disappear. A mixture of NaCl, ice, and solution (the solution containing KCl as well as NaCl) would have $f = 3 - 3 + 1 = 1$; hence all three could be present at different temperatures if the concentration of KCl is suitably varied. If ice is not present, both solid salts could be in contact with the solution, but of course their concentration in the solution is fixed for any one temperature (point C on Fig. 12-1).

A mixture of KCl, $MgCl_2$, and water (Fig. 12-2) also has three components. The possible compound carnallite ($KCl \cdot MgCl_2 \cdot \cdot 6H_2O$) is not an additional component, because it may be formed from the others. Hence the phase rule makes the same predictions as before, and we could interpret them to mean that KCl and carnallite, or $MgCl_2$ and carnallite, may be present at some fixed temperature along with ice and solution; but it would be impossible for all three salts to be present with ice and solution at any temperature, because this would give $f = -1$. For an arbitrary temperature ($f = 1$), two of the salts but not all three could be present simultaneously in contact with the solution (point C or D on Fig. 12-2).

A reciprocal salt pair is completely specified in terms of three salts or three ions (Sec. 12-6), hence furnishes only three independent components. Together with water this gives four components, so at constant pressure $f = 4 - p + 1 = 5 - p$. If temperature is regarded as variable ($f = 1$), then at any selected temperature as many as three salts can be at equilibrium together with the solution (since when f is 1, p may be as high as 4, and 3 salts + 1 solution gives 4 phases). This situation would be represented by points like S and T on Fig. 12-4, where the fields of three salts meet at a point. If only two solid salts are present, the composition of the solution may vary as well as temperature (a line on Fig. 12-4). At some particular temperature, as many as four salts could coexist ($f = 0$); this would not be shown on Fig. 12-4 but would be represented as a transition point in the solid figure constructed by adding temperature as a third dimension to the diagram.

Thus the phase rule sets limits on the maximum number of salts that can exist at equilibrium with any given solution. These limits are expressed graphically, but not stated explicitly, by diagrams like Fig. 12-4. In making use of the phase rule, we shall most frequently embody it in diagrams of this sort ("phase-rule diagrams"), rather than try to calculate the number of degrees of freedom numerically.

13-3 EUTECTICS AND SOLID SOLUTIONS

General

We turn now to problems of silicate melts. Such a melt in nature (a *magma*) consists during crystallization of a single liquid phase and one or more solid phases. Over much of the crystallization period we may consider pressure approximately constant, so that $f = c - p + 1$. The number of major components is generally nine or ten, and this is likewise the maximum number of degrees of freedom (since there must be at least one phase present, in which case $f = 9 - 1 + 1$). To represent such a system diagrammatically, would require a model in nine- or ten-dimensional space.

We start with much simpler systems, where representation is not so formidable a problem. Fortunately it turns out that the actual crystallization of a magma takes place in large measure as a series of steps, or in two or three independent series of steps, each one involving only a few of the many components. Hence the results of experimental work on simple binary and ternary combinations of silicates have a greater immediate relevance to problems of crystallizing lavas than we might guess from the bare statistics.

Artificial melts consisting of only two components show two extreme types of behavior: such a melt may freeze to form a mixture of separate crystals of the two component substances, or it may form a single type of crystal representing a solid solution of the two components. Most melts show behavior between these extremes, but intermediate cases are best understood as modifications of the extremes. For examples of the two types, we shall consider systems with a direct bearing on crystallization processes in magmas.

Eutectics

If a mixture of anorthite $(CaAl_2Si_2O_8)$ and diopside $(CaMgSi_2O_6)$ is melted in a platinum capsule and allowed to cool slowly, the solid produced is a mixture of separate crystals of the two minerals. Details of the process can be followed by using various proportions, by noting the temperature at which crystals first appear, and by quenching (cooling rapidly by plunging into water) from various temperatures to preserve the remaining liquid as a glass. From melts rich in diopside the first crystals to form are diopside; if more and more anorthite is included in the melt, the temperature of first crystallization drops steadily. Melts rich in anorthite give anorthite as the earliest-formed crystals, and the melting point becomes progressively lower as diopside is added. These experimental facts may be represented by a temperature-composition diagram like Fig. 13-2. The line DE shows the melting points of diopside-rich mixtures, the line AE the melting points of anorthite-rich mixtures. The curves intersect at E, which is the lowest temperature at which a liquid phase can exist in this system.

The curves show not only melting points but also the course of crystallization of any mixture. A melt with 20% anorthite, for example, cools to give diopside crystals at 1350°C (line MN). As diopside is removed from the liquid, the composition of the liquid changes and its melting point decreases, the melting point for any composition being represented by a point on DE. When the liquid composition has reached 58% diopside and 42% anorthite (corresponding to E), anorthite begins to crystallize together with the remainder of the diopside, and the temperature remains constant until crystallization is complete. The lowest melting temperature E is called the *eutectic temperature*, and the composition corresponding to E is a *eutectic mixture*.

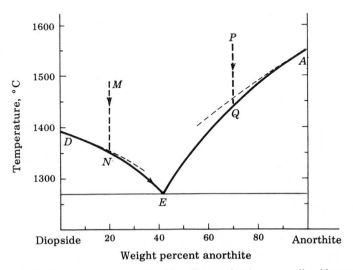

Figure 13-2 Temperature-composition diagram for the system diopside-anorthite at 1 atm. *(Source: Bowen, 1928, page 26.)*

Figuratively, we say that a mixture with 20% anorthite "follows the line *MN* until it intersects *DE* at *N*, then follows *NE* to *E*," just as in the last chapter we spoke of salt mixtures as following certain lines on the diagrams during evaporation. Similarly a mixture with 30% diopside follows *PQ* to *Q*, then *QE* to *E*. The liquid remaining at *E* in the last stages of crystallization would be the same for the two mixtures, or for any other mixture.

The melting of a solid anorthite-diopside mixture on heating follows the same course in reverse. Liquid appears first at the eutectic temperature, and the first liquid has the eutectic composition. There is no change in temperature or composition until all of one mineral has melted; then the temperature rises as the residue of the other mineral melts, the melting point changing along *ED* or *EA* as the composition of the liquid changes.

Solid Solution

A mixture of albite ($NaAlSi_3O_8$) and anorthite melted and allowed to cool very slowly gives crystals of only a single kind, a "solid solution" of the two minerals. (Actually solid solution in plagioclase is not quite complete, but for present purposes this complication will be ignored.) Details of the crystallization, somewhat more complicated than for systems with a eutectic, are shown in Fig. 13-3. Pure albite melts at 1118°C, pure anorthite at 1552°, and mixtures have intermediate melting ranges. A melt with 40% albite, for example, cools to about 1470° before

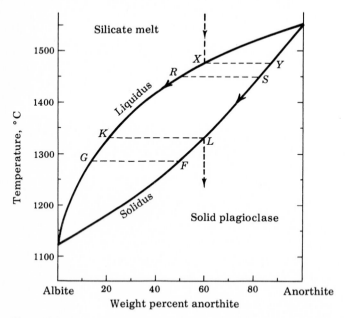

Figure 13-3 Temperature-composition diagram for the plagioclase feldspars at 1 atm. (*After Bowen, 1928, page 34.*)

crystals appear (point X). The composition of the first-formed crystals is not the composition of the melt nor the composition of pure anorthite, but somewhere between; the experimentally determined composition is shown by point Y. As crystals of composition Y are formed, the melt is impoverished in anorthite and enriched in albite, so that its melting point must steadily decrease (line XK). If the first-formed crystals remain suspended in the melt, they are out of equilibrium as the temperature falls and must react with the melt to form solid solutions of higher albite content. Thus at any temperature along the line XK the remaining melt is in equilibrium with a solid rich in anorthite, and the composition of both liquid and solid change progressively. At any one temperature the composition of the liquid is given by a point on XK, and that of the solid by a point on YL; thus at 1450° a liquid of composition R is in equilibrium with crystals of composition S. Crystallization ends when the last remaining melt reaches the composition K and the crystals have the composition L—identical, of course, with the composition of the original melt.

When plagioclase crystals are heated, the first liquid to form has a composition more soda-rich than the crystals. For example, if the crystals contain 50% albite, the first melt contains 87% albite and 13% anorthite (points F and G, respectively). As the temperature rises, both liquid and solid change their composition (along FLY and GKX, respectively), until the last crystals melt to give a liquid of the composition 50% albite, the same as the original solid.

The upper line in Fig. 13-3, showing compositions of the liquid phase at different temperatures, is called the *liquidus* of the system. The lower line, showing compositions of the solid phase, is called the *solidus*. (By way of contrast, note that the "solidus" in Fig. 13-2 consists of the vertical lines on either side of the diagram.)

Nonequilibrium Systems

All the preceding has been based on the assumption that equilibrium is maintained during cooling. This means that crystals must remain in contact with the melt until crystallization is complete, a condition not necessarily fulfilled in geologic environments. Let us see what effect lack of equilibrium would have.

Consider first the albite-anorthite system under conditions in which crystals are removed as fast as they are formed, say, by gravitative settling of crystals out of the liquid. In this case the enrichment of albite in the liquid is greater than before, since the anorthite-rich crystals cannot react with it. For an original composition with 40% albite, the last liquid to crystallize may have a composition far more albite-rich than point K; in fact, if removal of crystals as they form is complete and continuous, the last liquid may be practically pure albite. The net result of such failure of equilibrium would be a pile of plagioclase crystals of nonuniform composition, rich in anorthite toward the bottom and in albite toward the top. If removal of crystals is less rapid, they may have an opportunity for partial reaction with the liquid, giving rise to the zoned crystals which are so prominent in many natural plagioclases.

For mineral pairs which form eutectics, the early removal of crystals would have less influence on the course of crystallization. Cooling progresses to the eutectic temperature as before, and the last liquid to crystallize would have the eutectic composition. In the resulting solid aggregate, the mineral whose crystals form first would be more abundant toward the bottom; but the bulk composition near the top would show no enrichment (beyond the eutectic) toward the other mineral, and the composition of the individual minerals would of course show no change.

The difference between the two extreme cases of continuous equilibrium and complete removal of early-formed crystals is exactly analogous to the similar two cases in salt deposition (Sec. 12-7).

13-4 THEORETICAL CRYSTALLIZATION CURVES

In this discussion, as in the treatment of salt crystallization in the last chapter, we are proceeding on a purely empirical basis. To obtain a phase diagram for a crystallization process, the technique is tedious but simple: various mixtures of the end members are heated until they form homogeneous melts, and then, as the melts cool, the sequence of solid phases crystallizing out is recorded. In actual laboratory work this seemingly straightforward procedure is often beset with difficulties, arising from trouble in identifying solid phases and in preserving them down to low temperatures where they can be examined carefully, and from sluggish crystallization or crystallization of metastable phases rather than stable ones. But basically the operation is not complicated. The lines on phase diagrams merely summarize the results of large numbers of such experiments. It is reasonable to ask whether the procedure might be shortened, or the results systemized, by handling the problem theoretically.

At first glance the problem seems simple. Take, for example, the right-hand side of the anorthite-diopside diagram (Fig. 13-2): the line is drawn through points showing compositions of a liquid in equilibrium with pure anorthite over a range of temperature. To express the equilibrium more formally,

Solid anorthite $\rightleftharpoons$ anorthite in melt

The equilibrium constant would be the quotient of the two activities. The activity of the solid, by convention, is taken as 1, so the expression for the constant is merely

$$K_a = a_{an(melt)}$$

The variation of the constant with temperature is given by the van't Hoff equation [Eq. (8-3)]:

$$\log K = -\frac{\Delta H}{2.303R}\frac{1}{T} + C$$

The integration constant C can be evaluated by noting that $K = 1$ and $\log K = 0$ at the melting point of pure anorthite, T_m. Solving for C in this manner gives

$$\log K = \frac{\Delta H}{2.303R}\left(\frac{1}{T_m} - \frac{1}{T}\right)$$

This equation has the form

$$\log K = A - B\frac{1}{T}$$

where A and B are constants if ΔH is constant. Since $K = a_{an}$,

$$\log a = A - B\frac{1}{T}$$

Thus $\log a$ plotted against $1/T$ should give a straight line, and a simple plot of a against T should be a curve. To see how closely the curve would fit the experimental curve, a value of ΔH can be assumed (or the heat of fusion of anorthite may be obtained from tables) which will make the curve fit the upper part of AE in Fig. 13-2. Then if a is equated to the concentration of anorthite in the melt, a curve is obtained shown by the light dashed line on the drawing. The fit of the two curves is reasonably good only for mixtures with more than 80% anorthite.

Why is the agreement so poor for all but anorthite-rich liquids? Note that three assumptions are hidden in the last paragraph: (1) ΔH, the heat of melting of anorthite, is assumed constant over a temperature range of 100 to 200°C; (2) ΔH is assumed not to be affected by the presence of diopside in the melt; (3) the activity of anorthite in the melt is assumed to be measured by its concentration. The first assumption is probably not a serious source of error. The other two are equivalent to assuming that diopside-anorthite mixtures behave as ideal solutions, in other words, that each component is completely independent of the other and is merely diluted by it. The assumption of ideal behavior is generally far from true for silicate mixtures; ΔH is modified by heats of solution, and activity deviates markedly from concentration except in mixtures near the pure components. Enough is known about some mixtures so that corrections can be applied to ΔH and a, and the fit of the theoretical and empirical curves can be greatly improved. But in general the necessary information is lacking, and for geological purposes the empirical diagrams are more informative than theoretical analysis.

13-5 COMPLEX BINARY SYSTEMS

Combinations of Eutectics and Solid Solutions

The simplest combination of a eutectic and solid-solution diagram would be that shown in Fig. 13-4, for which there is no example among common rock-forming minerals. Each substance lowers the melting point of the other, as in Fig. 13-2, but the solids that crystallize out are not the pure end members. Crystallization of a

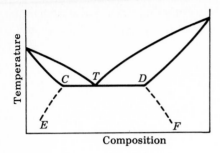

Figure 13-4 Hypothetical diagram for a binary system showing both a eutectic and partial solid solution.

melt of any composition would end at the temperature T, and the last crystals to form would be a mixture of composition C and composition D. The lines CE and DF represent hypothetical changes in the compositions of the solids at temperatures below the melting point.

Double Eutectic

In the system silica-nepheline, solids of three different compositions may form: SiO_2 (as tridymite in laboratory experiments with dry melts, as quartz in nature); nepheline, $NaAlSiO_4$ (nepheline in nature has a somewhat more complex formula); and albite, $NaAlSi_3O_8$. In phase-rule language, however, only two components are present, since one of these substances is derivable from the other two:

$$2SiO_2 + NaAlSiO_4 \rightarrow NaAlSi_3O_8$$

According to the phase rule the three solids cannot exist together, because there would then be a total of four phases (3 solids + 1 liquid), and this would lead to a negative value for f:

$$f = c - p + 1 = 2 - 4 + 1$$

Either of the pairs silica-albite or nepheline-albite may exist together, and both pairs form simple eutectics, as shown in Fig. 13-5. The interpretation of this diagram is no different from that for Fig. 13-2: any melt on cooling approaches one of the eutectic mixtures, the choice depending on whether the original composition is more rich or less rich in silica than pure albite. Removal of early-formed crystals cannot affect the course of crystallization and can at best lead only to a solid having a eutectic mixture in one part and a slight concentration of one of the minerals in another. The diagram illustrates, as of course it must, the impossibility of the three possible solids existing together, or of silica and nepheline crystallizing together from the same melt. (We ignore in this discussion the minor complications resulting from solid solutions at the nepheline end of the diagram.)

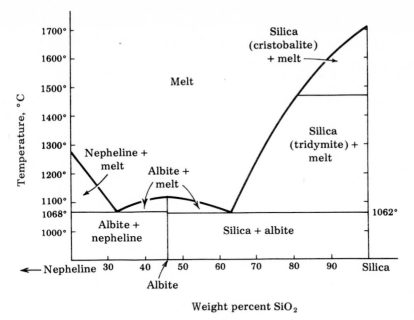

Figure 13-5 Simplified diagram of the nepheline-silica system at 1 atm. *(After Barth, 1962.)*

Eutectic with Incongruent Melting

The silica-leucite system is another example of a two-component system with three possible solids (silica, K-feldspar, leucite), but the relation between the solids is different. Potassium feldspar is unstable at temperatures above 1150°, decomposing into leucite and a silica-rich liquid:

$$KAlSi_3O_8 \rightarrow KAlSi_2O_6 + SiO_2$$

The K-feldspar is said to *melt incongruently*. Liquids with compositions near that of K-feldspar, then, cannot yield this mineral as a direct product of crystallization from a melt; leucite must crystallize first and then must change to K-feldspar on cooling by reaction with the liquid. This behavior is diagramed in Fig. 13-6.

The right-hand part of the diagram (compositions 0 to 58% $KAlSi_2O_6$) is obviously a simple eutectic, either silica or K-feldspar precipitating first and the liquid thereafter changing in composition toward the eutectic point. Melts with more than 58% $KAlSi_2O_6$, however, have a more complicated cooling history. Consider first a liquid with only 15% SiO_2 (point *A*). As it cools, leucite begins to crystallize at about 1590° (point *B*). Removal of leucite means that the liquid becomes richer in silica, its changes in composition being shown by points on the line *BI*. When the temperature falls to 1150°, leucite is no longer stable in the presence of the melt; it reacts with the melt to form K-feldspar, and more K-feldspar precipitates directly, so that the composition stays at point *I* until

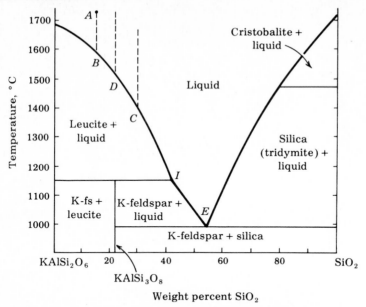

Figure 13-6 Temperature-composition diagram of the system leucite-silica at 1 atm. (*Source: Schairer, F., and N. L. Bowen, Geological Society of Finland Bull., vol. 20, p. 74, 1948.*)

the liquid is used up. The final solid, then, consists of K-feldspar plus the remaining leucite which did not react. If the original melt had 30% SiO_2 instead of 15% (point *C*), the cooling history would be similar except that now all the leucite would react at *I* and some liquid would be left over. As cooling continues, K-feldspar would precipitate from this liquid, and the composition of the liquid would change along the line *IE* until the eutectic composition is reached. The final solid would consist of K-feldspar plus silica, all trace of the early-formed leucite having disappeared. If, finally, the original melt has precisely the composition of K-feldspar (point *D*), leucite again crystallizes first; at point *I* it all reacts with the liquid, but this time there would be no liquid left over. The resulting solid would be, of course, pure feldspar. Thus the net product of crystallization in this system is similar to that for the silica-nepheline system: silica and K-feldspar if the original composition is more silica-rich than feldspar, leucite and K-feldspar if the original composition is less silica-rich.

The important difference between the two kinds of crystallization lies in the greater possibility of separation of minerals in the leucite system *when equilibrium is not maintained*. The discussion above assumes equilibrium; early-formed leucite is available to react with the melt, and cooling is so slow that all possible reactions can go to completion. But now suppose that, in a melt containing initially 15% SiO_2, leucite crystals are removed as fast as they form. When the temperature falls to 1150° (point *I*), leucite is not available; K-feldspar crystallizes from the melt, and the composition changes toward the eutectic. Ultimately point *E* is reached, and the last liquid to crystallize is a eutectic mixture. The resulting solid would have leucite concentrated in one part, and a mixture of silica and feldspar in another. In other words, two "rocks" of very different composition may be formed from the same melt. Silica may crystallize from a liquid initially rich in leucite, and leucite may be preserved from initial crystallization in a melt more silica-rich than K-feldspar—possibilities that have no counterpart in the silica-nepheline system.

A similar, and petrologically more important, system is the combination silica-forsterite:

$$\underset{\text{Forsterite}}{Mg_2SiO_4} + SiO_2 \rightarrow \underset{\text{Enstatite}}{2MgSiO_3}$$

(Recall that forsterite is an end member of the olivine series and that enstatite is a pyroxene. Hence this system is a simple analog of reactions that are important in the crystallization of mafic igneous rocks like gabbro and basalt.) The artificial system (Fig. 13-7) differs from natural rocks in that clinoenstatite rather than enstatite is the intermediate substance formed (because of the higher temperatures required in a dry melt), but the phase relations are similar to those observed in nature. The clinoenstatite melts incongruently at 1557°, so that forsterite is the first substance to crystallize from all melts containing initially less than 51 mole percent SiO_2. If equilibrium is maintained, the early-formed forsterite would react with the liquid at 1557° to form clinoenstatite; whether or not the forsterite is all

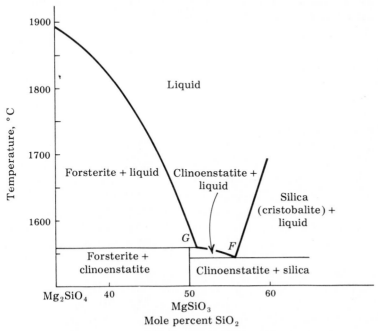

Figure 13-7 Part of the system MgO-SiO_2 at 1 atm, showing the incongruent melting of enstatite. (*Source: Barth, 1962, page 97.*)

changed to clinoenstatite at this point depends on the original composition of the melt. But if equilibrium is not maintained, early-crystallized forsterite may settle out and the remaining liquid may reach the clinoenstatite-silica eutectic, even if the initial composition is very low in silica. The analogy to crystallization in mafic rocks is obvious: early-formed olivine may fail to react completely with the magma to form pyroxene, so that olivine crystals may be preserved in a rock apparently saturated with silica, and quartz may form from a magma initially deficient in silica. Thus thick basalt flows and dolerite sills may have accumulations of olivine near their base and interstitial quartz near their top; and olivine grains in a basalt may be surrounded by tiny pyroxene crystals ("reaction rims"), as if they were caught in the process of reacting with the liquid.

The phenomenon of incongruent melting is obviously analogous to incongruent solution in the last chapter (Sec. 12-6). One refers to the breakup of a complex compound by partial melting, the other to a breakup by partial dissolving in water.

Solid Solution with a Minimum Melting Point

Albite and K-feldspar form a continuous series of solid solutions, just as do albite and anorthite, but mixtures of intermediate composition have lower melting points than either of the end members (Fig. 13-8). The diagram for this system, unlike the preceding ones, is drawn for a water-vapor pressure of 2,000 kg/cm² rather than for a dry melt, because in the dry-melt diagram the minimum melting-

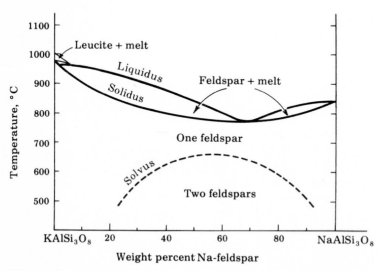

Figure 13-8 The alkali-feldspar system at a water-vapor pressure of 2,000 kg/cm². (*After Bowen and Tuttle, 1950.*)

point relation is obscured by complications due to formation of leucite in K-feldspar-rich mixtures. The lower curve in the diagram (*solvus*) shows "unmixing" of the two feldspars in the solid state: the solid solution is stable only at high temperatures, breaking up on cooling into separate crystals which approach the compositions of the end members more and more closely as the temperature falls. Unmixing may lead to intimate, very fine-grained intergrowths like those that make up anorthoclase, or to the coarser intergrowths called perthite and antiperthite, or possibly even to large separate crystals. An alkali feldspar that has cooled very rapidly may remain as a metastable solid solution; sanidine crystals in lavas are a good example.

Summary The freezing of a binary, or two-component, system may be represented by a temperature-composition diagram on which combinations with one degree of freedom are represented by lines and combinations with zero degrees of freedom by points. These diagrams show either eutectic mixtures, or solid solutions, or combinations of the two, and may be complicated by incongruent melting and by solid solutions with minimum or maximum melting points. Although binary systems are far simpler than natural lavas, they help to explain some of the petrographic relations found in igneous rocks.

13-6 TERNARY SYSTEMS

General

A closer approach to natural silicate liquids can be obtained with artificial melts having three components. Here we face a greater difficulty in representation, for ternary systems have a maximum of three degrees of freedom (temperature and two ratios expressing composition) and hence require three-dimensional diagrams for complete display. For many purposes a two-dimensional plot showing only compositions is sufficient; temperature may be added in the form of contour lines if needed. The representation most commonly used is a triangular diagram like Fig. 13-9. Compositions of mixtures of the three components are plotted on the diagram just as compositions of salt solutions were plotted on the triangular diagrams of the last chapter.

Figure 13-9 is a hypothetical diagram for the simplest kind of ternary system, one in which the three components crystallize from a melt without forming solid solutions. Each side of the triangle shows a binary eutectic, of the kind we have discussed before; we are now, so to speak, looking down vertically on three such diagrams along the temperature axis. From a melt containing all three components, one of the three, in general, begins to crystallize first. As cooling proceeds, the crystallizing component is joined by a second, and in the last stages all three crystallize together. This sequence may be illustrated by the mixture X. From it the first material to crystallize is B; as B is removed, the composition of the liquid changes along a line directed away from the B apex; when the composition reaches point P, component C starts to crystallize also; as both C and B are removed, the composition of the remaining liquid changes along PE; when the composition reaches E, the third component begins to crystallize, and the composition remains at this point until crystallization is complete. Lines such as DPE, FE, and GE, representing simultaneous crystallization of the two components, are called *cotectic lines;* point E is a *ternary eutectic*.

If temperature is regarded as a third dimension, at right angles to the plane of the paper, it may be shown conveniently by means of contour lines (the dashed lines in Fig. 13-9). Note that the course of crystallization of mixture X is down the temperature gradient; the ternary eutectic is the low point of the "temperature surface," and the cotectic lines are the bottoms of "valleys."

Crystallization in a system of this sort, like crystallization of a binary eutectic mixture, is little affected by removal of early-formed crystals. The resulting solid may have local concentrations of the early-crystallizing component or components, but the final eutectic composition will be unchanged.

A ternary system which forms a series of solid solutions among all its components would be more difficult to portray on a triangular diagram. There could be no cotectic lines outlining fields of crystallization, so that the triangle would be featureless except for contour lines suggesting the shape of liquidus and solidus surfaces.

Rather than pursue a discussion of other hypothetical systems, we shall consider two specific cases of interest in chemical petrology.

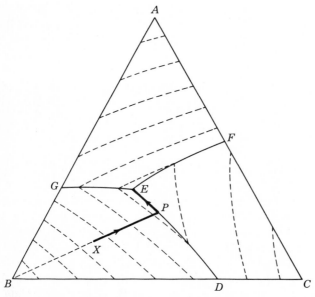

Figure 13-9 Hypothetical diagram for a system with a ternary eutectic and three binary eutectics. Dashed lines are temperature contours.

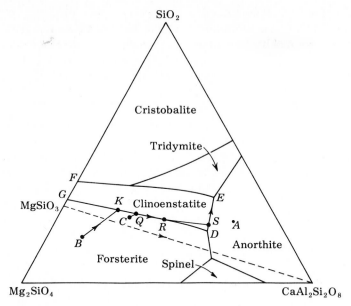

Figure 13-10 Phase diagram for the system silica-forsterite-anorthite at 1 atm. *(After Bowen, 1928.)*

Forsterite-Silica-Anorthite

This system is a ternary eutectic modified by the incongruent melting of clinoenstatite and by the formation of spinel ($MgAl_2O_4$) from mixtures rich in forsterite and anorthite. In Fig. 13-10 the eutectic is shown as point E. The silica field above the eutectic is complicated only by the tridymite-cristobalite transition; as in preceding diagrams, the high-temperature forms of silica appear rather than quartz because dry melts crystallize at higher temperatures than water-rich natural magmas. Mixtures with compositions near the eutectic crystallize as described in the last section. For example, in mixture A the first crystals to form would be anorthite; the next would be clinoenstatite; and these two would be joined by tridymite when the composition of the remaining liquid reaches E.

The spinel field at the bottom of the diagram is, strictly speaking, part of a four-component system, and cannot be accurately shown on this triangle. But no ordinary rocks have compositions in this part of the diagram, so for present purposes the complication may be disregarded.

The incongruent melting of clinoenstatite is shown by the left-hand edge of the triangle. This edge, in effect, is Fig. 13-7 as seen looking down from above; points G and F are the same on the two diagrams. The incongruent melting persists even when large amounts of anorthite are present, as is shown by the fact that line GD remains on the silica side of the composition of clinoenstatite (indicated by the dashed line). Details of crystallization may be illustrated by two examples:

1. From a mixture with composition B, forsterite would crystallize first, and the composition of the liquid would move directly away from the forsterite corner. At K on GD, forsterite would become unstable and would react with the liquid to form clinoenstatite, while more clinoenstatite precipitates directly; these processes would make the composition of the liquid change along KD. At D, anorthite would start to crystallize, and the composition of the liquid would remain at this point until the supply is exhausted. The resulting solid would consist of anorthite, clinoenstatite, and the remainder of the forsterite that had failed to react.
2. The mixture C would also give forsterite as the first-formed crystals. Here, however, the amount of forsterite would be small when the liquid composition reaches Q on GD, and reaction with the liquid would quickly convert all of it to clinoenstatite. From this point (R) on, clinoenstatite would crystallize directly and the composition of the liquid would change along RS, directly away from the $MgSiO_3$ point. At S, clinoenstatite would be joined by anorthite, and final crystallization together with silica would take place at the eutectic.

This outline assumes that equilibrium is maintained during crystallization. If, on the contrary, early-formed forsterite is removed as fast as it forms, the system may crystallize to give one solid rich in forsterite and another containing silica; for if the reaction between forsterite and liquid is prevented, there is nothing to keep the liquid from ultimately reaching the eutectic, no matter how undersaturated with silica it may have been initially. Like the two-component system shown in Fig. 13-7, this gives a convenient explanation for reaction rims around olivine grains and for the gravitational segregation of olivine from basaltic magmas.

Diopside-Albite-Anorthite

The deceptively simple diagram for this system (Fig. 13-11) shows only a single cotectic line connecting the binary eutectics diopside-anorthite (E) and diopside-albite (X). The bottom line of the triangle represents the solid solution series albite-anorthite; it is a top view, so to speak, of Fig. 13-3. Melts with compositions above the cotectic line give diopside as the first crystals; melts with compositions below the line give plagioclase. If diopside crystallizes first, the composition of the melt changes along a line directly away from the diopside corner. If plagioclase appears first, the cooling history is more complicated.

Consider, for example, the cooling of a mixture represented by point A. The feldspar that appears first will not have a composition 45% albite and 55% anorthite, as one might expect (point D), but a composition considerably richer in anorthite (point F). Formation of these first crystals containing about 80% anorthite causes the liquid to change in composition toward albite; but as cooling progresses and as the crystals take up more and more albite, the direction of change is altered. In other words the composition of the liquid follows a curved

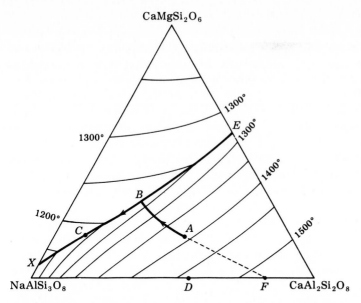

Figure 13-11 Phase diagram for the system diopside-albite-anorthite at 1 atm. The light lines are temperature contours. *(After Bowen, 1928.)*

path *AB*, intersecting the cotectic at *B*. Here diopside begins to crystallize also, and the composition of the melt changes along the cotectic toward *X*. The last liquid (*C*) will be used up when the solid plagioclase attains the composition 55% anorthite (*D*). The precise path of crystallization cannot be predicted from this diagram, as it could be from preceding diagrams, but the general outline is well defined. (See Prob. 13-9 at the end of this chapter.)

If equilibrium is not maintained in this system, if in particular early-formed plagioclase is removed, the ultimate composition of the melt is not restricted by the redissolving of the anorthite-rich crystals. The resulting solid might then contain anorthite-rich feldspar in one part and albite-rich feldspar in another part. Early separation of diopside from diopside-rich melts, on the other hand, would not change the character of the crystallizing material.

Compositions in this system are close enough to ordinary basalt to make possible predictions about crystallization of saturated mafic magmas. Note, for example, that no definite "order of crystallization" is to be expected. The first mineral to crystallize in a magma may be either pyroxene or plagioclase, depending on the initial composition, and during most of the cooling the two principal minerals will crystallize simultaneously. The composition of the resulting rock will be uniform if equilibrium is approximately maintained, but segregation of early-formed calcic plagioclase gives the possibility of formation of different rocks from a single original magma. Minor and temporary deviations from equilibrium during cooling may lead to formation of conspicuously zoned plagioclase crystals.

13-7 ADDITIONAL COMPONENTS

In Fig. 13-11 we have built up from artificial silicate systems a mixture whose behavior shows considerable resemblance to that of basaltic lava. The actual chemical resemblance is not very impressive: our mixture has only five oxides (SiO_2, Al_2O_3, MgO, CaO, Na_2O), whereas basalt has another five important components (FeO, Fe_2O_3, K_2O, H_2O, TiO_2) and a host of minor ones. Even among the five oxides shown on Fig. 13-11, all possible variations are not indicated, for no independent variation of silica with the other oxides is allowed for. The diagram shows only how the course of crystallization is influenced by changes in the two ratios Na_2O/CaO and MgO/Al_2O_3. It is surprising that variations in these two alone can determine in large measure how a basaltic magma will crystallize.

Adding more components to the mixture, or allowing for greater variation among the original five, means additional difficulty both in securing experimental data and in representing the data on diagrams. We shall not try to follow details of the quantitative experimental work but shall look briefly at some qualitative interpretations.

If silica is permitted to vary independently, the system would be in effect a combination of those shown by Figs. 13-11 and 13-10. One could represent such a system by erecting a pyramid on Fig. 13-11 as a base, using SiO_2 as the upper vertex, or alternatively by erecting a pyramid on Fig. 13-10 as a base with $NaAlSi_3O_8$ as the vertex; neither would be entirely satisfactory because they do not permit enough independent variation among the oxides CaO, MgO, and Al_2O_3. The representation would demand a four-dimensional figure, and the only possible way to handle it practically would be to construct a series of pyramids with different values for the ratios between these oxides. Even without actual diagrams, however, some aspects of the system are clear: moderate variation of SiO_2 from the amounts implicit in Fig. 13-11 would not markedly affect the precipitation of plagioclase but would make possible precipitation of either early olivine or late quartz in addition to pyroxene on the CaO-MgO side of the diagram. Extreme deficiency of silica would lead to formation of nepheline (Fig. 13-5).

Independent variation of alumina, within limits, would be accommodated by changes in the composition of the pyroxene and changes in the albite-anorthite ratio of the plagioclase. High alumina would permit the formation of aluminum-rich pyroxene and calcic plagioclase; low alumina would lead to lime-rich pyroxene and sodic plagioclase. The essential nature of the minerals crystallizing would not be changed unless the variation of alumina were extreme.

Of the five oxides mentioned above which have so far been omitted from consideration, TiO_2 and Fe_2O_3 are perhaps least important because they appear largely in accessory minerals (especially magnetite and ilmenite) which do not greatly affect crystallization of the silicates. Some titanium and ferric iron find their way into pyroxenes, but the field of pyroxene crystallization is not greatly altered. The oxide FeO is interchangeable with MgO in pyroxene and olivine, the melting points of these minerals depending on the Fe/Mg ratio just as the melting point of plagioclase depends on the Ca/Na ratio. Some of the possible mixtures with FeO can be represented conveniently by using hypersthene ($MgFeSi_2O_6$) as the vertex of a pyramid with Fig. 13-11 as a base (Fig. 13-12). The cotectic line of Fig. 13-11 now becomes a cotectic surface, separating the pyroxene "volume" of the pyramid from the plagioclase volume. Crystallization of mixtures rich in MgO and FeO would start with the appearance of pyroxene, its composition depending on the Ca/Mg/Fe ratio, and the composition of the melt would change along a curving line toward the cotectic surface; when the composition reaches the surface, plagioclase would precipitate also, and thereafter the change in composition would be shown by a line on the surface.

The oxide H_2O, even in small quantities, has a huge effect on the crystallization process. For one thing, water lowers the temperature of crystallization markedly. The experiments we have considered so far involve dry melts at temperatures ranging up to 1700 or 1800°C; basaltic lavas in nature, because of their content of water and other volatile substances, crystallize at temperatures in the range 850 to 1100°, and more siliceous lavas at still lower temperatures. Under these conditions some of the solid phases exhibit different crystal forms: the magnesian pyroxene appears as ordinary orthorhombic enstatite (or hypersthene) rather than clinoenstatite, and silica appears as quartz rather than tridymite or cristobalite. Then water itself enters the structure of some of the crystallizing solids, giving minerals like hornblende or biotite in place of or in addition to pyroxene. If much water is present, more hydrous minerals like chlorite, antigorite, and zeolites may form at lower temperatures.

The oxide K_2O goes mostly into the formation of potassium feldspar (sanidine, orthoclase, or microcline). Fortunately this feldspar appears chiefly during late stages of the crystallization process, so that the presence of K_2O does not greatly influence the course of crystallization shown by Fig. 13-10 or Fig. 13-11. If water is present, potassium may also enter into the composition of biotite, again fairly late in the crystallization process.

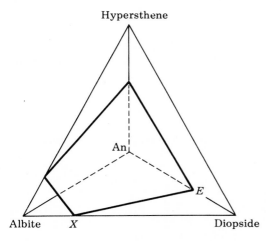

Figure 13-12 Approximate diagram of the system albite-anorthite-hypersthene-diopside. The surface separating the plagioclase volume from the pyroxene volume is shown as a plane, but it should be a curved surface. Line XE corresponds to line XE on Fig. 13-11. *(Source: Barth, 1962, page 114.)*

Summary The chemistry of crystallization of lavas can best be approached through a study of artificial silicate melts with relatively simple compositions. Equilibrium between crystals and liquid in such melts requires that the number of crystalline solids be limited by the chemical composition of the melt, the relation between them being expressed in the phase rule. On the basis of the phase rule, diagrams can be constructed which summarize concisely the experimental results for simple systems. Many of the laboratory experiments differ from natural conditions not only in the small number of components but also in the much higher temperatures needed because of the absence of water and other volatile constituents, but, nevertheless, results of some of the experiments show striking resemblances to the behavior of lavas in nature. The resemblance is close enough so that we can undertake in the next chapter a reconstruction of natural crystallization processes based on the experimental work.

The phase rule is a broad generalization of physical chemistry, with applications to many fields besides the crystallization of igneous rocks. We have seen already its usefulness in working out the details of salt crystallization, and we shall meet it again in discussing metamorphism. Outside of geology it has many other applications, in both theoretical chemistry and the chemistry of industrial processes.

PROBLEMS

1 For a system in which the effect of pressure is small, the phase rule becomes

$$c - p + 1 = f$$

where c is number of components, p is number of phases, and f is number of degrees of freedom. Using this rule, and remembering that MgO and SiO_2 form the two compounds $MgSiO_3$ (enstatite) and Mg_2SiO_4 (forsterite), answer these three questions about each of the following combinations: (a) Can the combination exist at equilibrium? (b) If it can exist, how many degrees of freedom does it have? (c) If it can exist at some one temperature, what would happen if the temperature were lowered a few degrees?

    A. Forsterite and liquid containing MgO and SiO_2.
    B. Forsterite, enstatite, solid silica, and liquid containing MgO and SiO_2.
    C. Enstatite, solid silica, and liquid containing MgO and SiO_2.
    D. Liquid containing MgO + SiO_2, with no solid phases present.

2 Using the diagram of Fig. 13-2, describe what would happen if (a) a melt containing 10% anorthite and 90% diopside is cooled from 1500° to 1000°, and (b) a solid mixture containing 50% of each is heated from 1000° to 1500°. Assume that equilibrium is maintained in both cases. If equilibrium is not maintained, what are the possibilities of forming fractions of different compositions by crystal separation?

3 Forsterite and fayalite form a continuous solid-solution series; pure forsterite melts at 1890° and pure fayalite at 1205°. Draw an equilibrium diagram for this system and describe what happens when (a) a melt containing 10% fayalite and 90% forsterite is cooled from 1900° to 1200° under equilibrium conditions, (b) the same mixture is cooled through the same range, but crystals are removed as fast as they form, (c) a mixture of 50% fayalite and 50% forsterite is heated from 1000° to 1800°.

4 The following questions refer to Fig. 13-5:

    (a) If a mixture containing 80% silica and 20% nepheline is heated to 1300°, calculate the composition of each phase present.

(*b*) Describe the course of crystallization under equilibrium conditions of a mixture containing 40% SiO_2, and of a mixture containing 70% SiO_2.

(*c*) If equilibrium is not maintained, is it possible for nepheline to crystallize from any mixture containing more than 50% SiO_2? Explain.

5 Using Fig. 13-6, describe how the following mixtures would crystallize, first assuming maintenance of equilibrium and second assuming removal of early-formed crystals: (*a*) 10% SiO_2, 90% leucite; (*b*) 30% SiO_2, 70% leucite; (*c*) 50% SiO_2, 50% leucite. Would it be possible to obtain a rock with leucite from a melt containing a greater proportion of silica than is present in orthoclase? If so, under what conditions? If not, why not? What would be the composition of the first liquid obtained from the melting of pure orthoclase? From the melting of an aplite containing 50% orthoclase and 50% quartz?

6 Using Fig. 13-11, describe qualitatively the crystallization of the following mixtures, first assuming maintenance of equilibrium and then assuming removal of early-formed crystals: (*a*) 10% diopside, 40% anorthite, 50% albite; (*b*) 70% diopside, 15% anorthite, 15% albite.

7 Sketch a triangular diagram for a ternary system having one ternary compound and one binary compound, assuming that all compounds melt congruently and that all possible pairs of substances form eutectics. Indicate directions of crystallization by arrows on all cotectic lines. Show how your diagram would be altered if the binary and ternary compounds form a solid-solution series, the binary compound having a higher melting point than the ternary.

8 Using Fig. 13-10, describe the crystallization of mixtures whose compositions lie (*a*) in the anorthite field, (*b*) in the cristobalite field, (*c*) in the clinoenstatite field, (*d*) in the forsterite field near the forsterite corner, (*e*) in the forsterite field near the clinoenstatite border. In each case, assume that equilibrium is maintained, and then describe the changes that would follow if the early-formed crystals were removed.

9 On most of the phase diagrams in this chapter the path of crystallization of a given mixture can be plotted quantitatively. On Fig. 13-11, however, the path of crystallization and the location of the end point were described only qualitatively. The diagram can be made quantitative by considering the triangle as the top surface of a triangular prism, with the three binary systems as sides. If the albite-anorthite face of the prism is then drawn below the bottom edge of the triangle (as if folded up into the plane of the triangle), plagioclase compositions from the triangle can be projected onto the binary diagram. The compositions from the triangle are points on the liquidus; the corresponding solidus points can be located on the binary diagram and projected back to the bottom edge of the triangle. Using this construction, plot the crystallization of the two mixtures in Prob. 6 quantitatively. (The construction is explained in detail in Smith, 1963, page 166.)

REFERENCES AND SUGGESTIONS FOR FURTHER READING

Barth, T. F. W., "Theoretical Petrology," 2d ed., John Wiley & Sons, Inc., New York, 1962. A standard petrology textbook. Part III has many examples of phase diagrams for silicate systems.

Bowen, N. L., "The Evolution of the Igneous Rocks," Princeton University Press, Princeton, N.J., 1928. This is the classical work on the application of the phase rule to the crystallization of silicate melts. Chapter 4 gives a detailed description of the experimental systems investigated up to 1928.

Findlay, A., "The Phase Rule and Its Applications," 9th ed. (revised by D. N. Campbell and N. O. Smith), Dover Publications, Inc., New York, 1951. A classical exposition of the phase rule, including the theoretical derivation of the rule and its application to many kinds of systems.

Ricci, J. E., "The Phase Rule and Heterogeneous Equilibria," D. Van Nostrand Company, Inc., Princeton, N.J., 1951. Reprinted by Dover Publications, Inc., New York, 1956. Another excellent general treatment of the phase rule.

Smith, F. G., "Physical Geochemistry," Addison-Wesley Publishing Company, Inc., Reading, Mass., 1963. Chapter 9 gives an exhaustive discussion, with many diagrams, of various possible two- and three-component systems.

FOURTEEN

CRYSTALLIZATION OF MAGMAS

The phase relations in artificial silicate melts considered in the last chapter provide a basis for discussing the formation of igneous rocks from magmas. We have already seen, in a general way, how the experimental results can be applied to the crystallization of olivine, pyroxene, and plagioclase in magmas of basaltic composition. Although natural silicate melts contain many more components than were used in the experiments, the similarity of the artificially produced crystals to natural minerals indicates that the additional components do not greatly modify the general sequence of events during crystallization. The most serious lack in the examples we have chosen is the absence in most of them of water vapor and other volatiles that are normally present in natural magmas. Water dissolved in a magma causes crystallization to take place at lower temperatures than in dry melts, helps to overcome the slowness of reactions in a viscous liquid, and permits the formation of hydrous mineral phases like the amphiboles and micas. Despite the lack of water and other components in the artificial systems so far considered, they serve as a valuable guide in working out the probable events in the geologic history of igneous rocks.

In this chapter we focus attention on the geologic problems of magmatic crystallization, using chemical reasoning wherever possible to supplement geologic data. We begin with the mafic rocks, for which the application of chemistry is somewhat simpler than for felsic rocks.

14-1 MAFIC AND INTERMEDIATE MAGMAS

Most rocks whose chemical compositions are relatively high in iron and magnesium and low in silica consist dominantly of plagioclase and one or more of the ferromagnesian minerals (olivine, pyroxene, hornblende, sometimes biotite). These are the common rocks gabbro and diorite, and their extrusive equivalents basalt and andesite (analyses IV, V, VI of Table 14-1). The plagioclase may show considerable variation in composition (zoning), suggesting that its constituents were in process of reaction with the melt at the time consolidation took place. The mafic minerals too may give evidence of reactions which were incomplete when freezing occurred: not only zoning, but reaction rims of pyroxene around olivine, or of hornblende around pyroxene. The reactions involved are chiefly within the systems $CaAl_2Si_2O_8$-$NaAlSi_3O_8$ and CaO-MgO-FeO-SiO_2, which on the basis of this geologic evidence appear to be in large measure independent of one another. Pyroxene crystals may grow around plagioclase grains, and plagioclase may include grains of pyroxene, but minerals of intermediate composition are absent. The same sort of relationship is suggested by artificial melts (Figs. 13-2, 13-10, and 13-11): plagioclase and pyroxene may crystallize together as a eutectic, or along a cotectic curve or surface, but do not form solid solutions or intermediate compounds. This independence of the two mineral series makes the crystallization of mafic magmas much simpler than it might otherwise be, and permits the direct application of conclusions from relatively simple phase diagrams.

Table 14-1 Average compositions of representative igneous rocks†

	I	II	III	IV	V	VI	VII	VIII
	Alkali granite	Grano-diorite	Quartz diorite	An-desite	Basalt (tho-leite)	Olivine basalt	Nephe-line syenite	Peri-dotite
SiO_2	73.86	66.88	66.15	58.17	50.83	47.90	55.38	43.54
TiO_2	0.20	0.57	0.62	0.80	2.03	1.65	0.66	0.81
Al_2O_3	13.75	15.66	15.56	17.26	14.07	11.84	21.30	3.99
Fe_2O_3	0.78	1.33	1.36	3.07	2.88	2.32	2.42	2.51
FeO	1.13	2.59	3.42	4.17	9.06	9.80	2.00	9.84
MnO	0.05	0.07	0.08		0.18	0.15	0.19	0.21
MgO	0.26	1.57	1.94	3.23	6.34	14.07	0.57	34.02
CaO	0.72	3.56	4.65	6.93	10.42	9.29	1.98	3.46
Na_2O	3.51	3.84	3.90	3.21	2.23	1.66	8.84	0.56
K_2O	5.13	3.07	1.42	1.61	0.82	0.54	5.34	0.25
H_2O	0.47	0.65	0.69	1.24	0.91	0.59	0.96	0.76
P_2O_5	0.14	0.21	0.21	0.20	0.23	0.19	0.19	0.05

† *Sources:* Analysis IV from Chayes, 1969, p. 2; all others from Nockolds, 1954, pp. 1012–1025.

Bowen (1928) summarized the freezing of mafic magmas by a generalization called the *reaction principle*. This is best illustrated with a diagram:

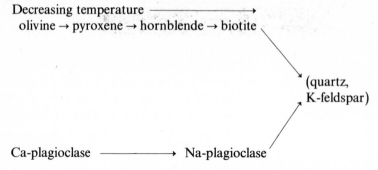

Decreasing temperature ————————→
olivine → pyroxene → hornblende → biotite

(quartz, K-feldspar)

Ca-plagioclase ————————→ Na-plagioclase

The mafic minerals in the upper row make up, in Bowen's terminology, a *discontinuous reaction series*. This means that each substance, as the magma cools, reacts with the melt to form the next mineral in line and that the reaction takes place at a definite temperature or over a restricted temperature interval. The first step in the sequence, from olivine to pyroxene, is easily studied experimentally and can be represented by a simple diagram (Fig. 13-7); in nature the temperatures would be lower (because water is present) and the two minerals would contain iron as well as magnesium, but the relationship is the same. The later steps pyroxene-hornblende and hornblende-biotite are more complicated, because these reactions involve water and thus depend on partial pressures of gases as well as silicate compositions, but petrographic and experimental evidence leaves no doubt that the changes are discontinuous, as Bowen pictured them. In contrast to this discontinuous sequence, the plagioclase feldspars of the lower line form a *continuous reaction series*, in which crystals react with the liquid continuously until freezing is complete (Fig. 13-3).

The reaction-principle diagram shows temperature decreasing from left to right, but specific numbers are not given because in any particular case the temperature of appearance of a mineral depends on the precise composition of the magma, the pressure, and the volatile content. The diagram indicates, however, that crystallization in the two series proceeds simultaneously as the temperature falls, so that lime-rich plagioclase appears together with olivine or pyroxene and soda-rich plagioclase together with hornblende or biotite. Crystallization may start in one series or the other, but through most of the freezing process two kinds of crystals are being formed at the same time. The point in the series at which crystallization begins depends on the original composition; the point where crystallization ceases depends on the composition and also on whether equilibrium is maintained as the magma cools.

This representation of the reaction principle, as Bowen and others have pointed out, is too restricted for application to most actual rocks. In many basalts textural relations indicate that calcium-rich pyroxene (diopsidic augite) crystallized simultaneously with olivine, rather than by reaction of olivine with the liquid, and that the pyroxene produced at a later stage by conversion of olivine is a different, calcium-poor variety. In other words a series of calcium-rich pyroxenes should be added as a third branch to the simple diagram above, a branch which probably converges with the olivine-pyroxene branch at later stages of cooling. For lavas rich in potassium, still another branch should be included, representing the crystallization of potash feldspars simultaneously with some of the plagioclase.

The progression of minerals in both continuous and discontinuous series shows a general decrease in the ratio of O to Si (thus 16 : 4 for olivine, 12 : 4 for pyroxene, 11 : 4 for amphibole, 10 : 4 for mica; see Sec. 5-6) and a general increase in complexity of the silicate structures (isolated SiO_4 tetrahedra in olivine, chain structures in pyroxene and amphibole, sheet structures in micas). These changes reflect changes in the structure of the silicate melt as temperature falls: at high temperatures the melt contains many free SiO_4 groups, but these polymerize into more and more complex aggregates as cooling proceeds. One may picture the freezing of a magma as a progressive growth and linking together of silicon-oxygen structures, first in the liquid and then in the minerals that crystallize from it.

The minerals in parenthesis at the right-hand side of the reaction-principle diagram are not related as are the others, in that these minerals do not necessarily form by reaction of previously crystallized material with the melt. Quartz and alkali feldspar are rather the minerals that form from the last remaining melt, the "residual liquid" of the mafic magma. How much of this sort of liquid will be present depends on the initial composition and on the effectiveness with which early-formed crystals are removed from the system. Early separation of olivine crystals is particularly effective in concentrating silica in the remaining melt (Sec. 13-5). Much experimental work has shown that almost any melt containing the alkali-metal oxides along with other common oxides of igneous rocks will show an enrichment of soda, potash, alumina, and silica in the last liquid to crystallize. Bowen called a final liquid containing only these four substances "petrogeny's residua system." Its possible compositions can be represented by a triangle whose corners are SiO_2, $KAlSiO_4$, and $NaAlSiO_4$ (Fig. 14-1). This is, so to speak, the ultimate goal of the crystallization process, the liquid that will tend to form even from magmas of quite different initial compositions and histories, provided that the removal of crystals from contact with the melt is reasonably effective during consolidation.

Bowen's reaction principle is so much a part of current petrologic theory that we find it difficult to appreciate the profound influence of this generalization on geologic thinking when it was introduced early in this century. Previous speculations had followed one of two assumptions: either igneous rocks were thought to represent eutectics, by analogy with the crystallization of metals, or the minerals of igneous rocks were supposed to show a definite "order of crystallization," an order allegedly revealed under the microscope by greater or less perfection of crystal form and by the growing of one mineral completely around or into embayments of another. These ideas encountered serious difficulties. If, on the one hand, igneous rocks are eutectics, it is odd that the groundmass, or last-crystallizing material, is so different from one rock to the next. If, on the other hand, there is a

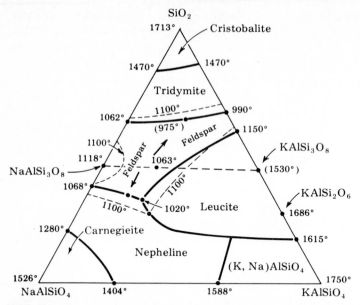

Figure 14-1 Approximate phase diagram for the system SiO_2-$NaAlSiO_4$-$KAlSiO_4$ at 1 atm pressure with dry melts. The dashed line connecting $NaAlSi_3O_8$ and $KAlSi_3O_8$ is not a phase boundary, but is drawn through points having the composition of feldspar. The point of lowest temperature on this line (the minimum melting point of a solid-solution series) is 1063°C. This is a shallow saddle on the temperature surface; from here "valleys" slope down to a low point on the feldspar-nepheline cotectic, and to the temperature minimum on the silica-feldspar cotectic at about 975°. The general area of lowest temperatures on the diagram is outlined by the 1100° contour (light dashed lines). Possible directions of crystallization within this area are shown by the heavy arrows, for silica-rich mixtures (upper arrow) and silica-deficient mixtures (lower arrow.) *(After Schairer, 1950.)*

universal order of crystallization, microscopic evidence that the order looks different in different rocks is hard to explain away.

The reaction principle disposed of both difficulties at a stroke. Most igneous rocks cannot be simple eutectics because of the existence of continuous reaction series; the point where crystallization stops in such a series depends on the cooling history and thus may be different from one rock to another even if initial magmatic compositions are similar. The final consolidation of a rock will in general be represented by a point on a cotectic line or a cotectic surface, not by a eutectic point. Order of crystallization depends on initial composition: the first mineral to crystallize may be either plagioclase or some member of the discontinuous series, so that one rock may show plagioclase enclosing pyroxene and another rock pyroxene enclosing plagioclase. A certain amount of regular order would be expected, in that hornblende should follow pyroxene, pyroxene should follow olivine, and Na-plagioclase should follow Ca-plagioclase; this corresponds with observation, and of course it was such observations that led to the original concept of a definite and invariable order. The groundmass of a mafic rock is commonly an intergrowth of plagioclase and one or more mafic minerals, with the order of development ambiguous—as might be expected, inasmuch as minerals of the two series crystallized simultaneously.

Bowen's work not only cleared up basic concepts about the freezing of magmas but provided explanations for many petrographic details. The zoning of plagioclase, for example, and reaction rims of pyroxene around olivine, represent failure to maintain complete equilibrium between crystals and melt during cooling. The differentiation sometimes observed in thick sills and flows of mafic rock—the development of an olivine-rich layer just above the basal chilled border, and of interstitial granophyre (a mixture of quartz and K-feldspar) near the top of the body—is convincingly explained by the settling out of heavy, early-formed olivine crystals, with a resulting enrichment of the remaining magma in silica. The behavior of inclusions of foreign material in magmas, particularly the partial melting, fragmentation, and disappearance of felsic xenoliths in mafic rocks and the changes in composition of mafic xenoliths in granitic bodies, can be accounted for by the partial adjustment of material in the xenoliths to the equilibrium conditions of their new environment. In all these ways and many more, the reaction principle serves to explain common field and laboratory observations of igneous rocks.

14-2 MAGMATIC DIFFERENTIATION

A persistent question about the geochemistry of igneous processes is the extent to which a magma can change its composition before or during crystallization. For each separate variety of igneous rock, must we assume a different original magma? Or can we suppose that there is only a single kind of "original" magma, and that the multiplicity of igneous rocks results from changes in its composition? These are the extreme views; we could equally well take an intermediate position, and ask if there might be more than a single kind of original magma, but not many kinds, and if the great observed variety of igneous rocks could be explained by alterations in the few fundamental types. Field observations seem to show clearly that some magmas do undergo changes of composition in the course of their history, for it is common to find within a single igneous mass different varieties of rock grading imperceptibly into one another. Furthermore, the composition of lava in successive flows from a volcano often shows gradual change during the volcano's lifetime, and it is certainly more reasonable to think that the material in a single large magma reservoir beneath is changing than to suppose that the volcano taps a new reservoir with a different kind of magma each month or each year. On the other hand, to suppose that such different and very abundant rocks as granite and basalt are produced by changes in a single original magma puts a strain on the imagination. Can we get any help from the reaction principle in answering questions of this sort?

To begin with, the reaction principle provides a reasonable mechanism by which a magma might change its composition during crystallization. The minerals crystallizing in each reaction series undergo changes in composition as the temperature drops; therefore, if crystals are separated from the magma as they form—in other words, if equilibrium between solid and liquid is not maintained—the composition of the remaining magma must gradually alter. The rocks resulting from the freezing of the last residue of the magma may have compositions quite different from rocks made up of the early crystals that have separated out.

If olivine or leucite should separate during the cooling, the change in composition may be very great, giving one rock containing a mineral deficient in silica and another containing free quartz, both derived from the same magma. To express this difference in composition, the rock containing olivine or leucite is said to be *undersaturated with silica*, and the rock containing free quartz is described as *oversaturated*. The concept of *saturation* here refers to the amount of silica in a mineral or rock relative to the amount that ideally could combine with the metallic elements present. Saturated minerals are those that can crystallize from a magma together with quartz, in other words those that contain all the silica their metallic constituents can accommodate; examples are feldspars, pyroxenes, amphiboles, and micas. Undersaturated minerals are those that cannot crystallize at equilibrium with quartz; examples are olivine, nepheline, leucite, and melilite. One can use the same terms in a similar sense for magmas (or for rocks): a saturated magma is one from which only saturated minerals can crystallize, an undersaturated magma would give at least some undersaturated minerals, and an oversaturated magma would give quartz together with saturated minerals. Thus if olivine or leucite should separate from a saturated magma in the early stages of crystallization, the rock eventually formed from these separated crystals would be undersaturated, while the rock formed from the last part of the magma to crystallize would contain free quartz and hence be oversaturated (Sec. 13-5).

The principal process by which separation of crystals takes place is commonly thought to be a simple settling out due to gravity, since crystalline silicates in general have densities different from that of the melt in which they form. The wide applicability of this mechanism has been seriously questioned recently, on the basis of experiments demonstrating viscosities of magma too high for crystal settling to be effective. Some examples of separation, for example the olivine-rich layers found near the base of thick sills and flows, are certainly best explained by gravitational settling, but other mechanisms involving movement of partly crystallized magma or progressive crystallization inward from the cool walls of a magma chamber could equally well account for many observed cases of separation. Whatever mechanical process may be involved, segregation of early-formed crystals is evidently an effective means of producing at least some of the observed variability of igneous rocks.

It is tempting to speculate about the possible wider applicability of differentiation by crystal fractionation to problems of igneous geochemistry. Consider the facts that basalt is the most abundant rock of the ocean basins, and one of the more abundant on the continents; that basaltic magma has come to the surface in huge quantities throughout geologic history, from earliest times to the present; and that a great variety of different kinds of rock can demonstrably be produced by the simple segregation of crystals during the slow cooling of basaltic magma. Recall too that the familiar sequence of common igneous rocks (gabbro, diorite, tonalite, granodiorite, granite, and their extrusive equivalents) is often recognized in the succession of plutons that make up a batholith and also in the kinds of lavas emitted during the lifetime of a volcano, and that this is precisely the sequence that would be expected to form by differentiation of basaltic magma. Can we draw from these facts an inference that basalt is the primordial magma, available in large amounts at some level at or below the earth's crust, and that most other igneous rocks have formed from it by differentiation? This was an early idea of Bowen's, and it is easy to see why the success of his experimental work would incline him toward such a speculation.

Like so many attractively simple hypotheses in geology, this one does not stand up under close scrutiny. For one thing, basalt is not a single uniform kind of rock but shows many subtle variations. Then there is abundant evidence that some basalts have not differentiated in the expected direction, toward a final product containing soda plagioclase, potash feldspar, and quartz. Some islands in the Pacific, for example, have basalts accompanied by trachyte or by nepheline-bearing rocks as differentiates. In the Skaergaard intrusive of eastern Greenland, basaltic magma has differentiated to give rocks containing abundant iron-rich pyroxene and olivine, with only a little quartz-bearing material formed at the very end. The sequence of lavas erupted from a volcano, while often showing agreement with the hypothesis, may show the reverse order or may be completely irregular. Such anomalies can be explained away by adding assumptions—that there are several kinds of "primordial" magma, that differentiation is partly controlled by factors other than simple cooling, that orogenic episodes interrupted the cooling—but a hypothesis is weakened by multiplying assumptions.

More serious are the objections that differentiation of basaltic magma can produce only a small amount of the residual liquid that would crystallize to form granite, and that no evidence is found in large granite bodies of a downward gradation into more mafic material. It may well be that in a gross, worldwide sense, granitic rocks as well as the different varieties of basalt have been generated by some sort of differentiation process in the mantle. But the idea of a single layer of uniform liquid or potentially liquid basalt quietly crystallizing to form huge masses of granite and granodiorite as residual products is too simple to fit the geologic evidence.

The objections that have been raised to Bowen's more extreme ideas should not detract from the magnificence of his achievement in working out the regularity of pattern behind the apparent complexities of igneous rock formation. It is one of the finest examples of the insight that physicochemical principles can give into some kinds of geologic processes.

14-3 VARIATION DIAGRAMS

In mapping an area of igneous rocks, one often finds a considerable variety of rock types in close proximity. The question then arises: Are all these types genetically related? In other words, do they all represent products of differentiation of a single kind of original magma, or do they belong to two or more different series or groups of rocks, perhaps intruded at different times or from different depths? An answer to this question must be sought in large part from geologic and petrographic observation, but chemical relationships may also provide clues. If one or two or several differentiation sequences are represented, compositions of the rocks in each sequence ought to show predictable relationships.

The chemistry of differentiation is too complicated to permit detailed predictions, but at least one might guess that compositions in a given sequence would change smoothly and regularly. This can be tested, provided that enough chemical analyses are available, by means of *variation diagrams*, which are graphs showing each important element or oxide plotted against an appropriate variable. In a very simple case, for example parallel layers in a compound sill or dike, the amount of each oxide may be plotted against distance from one of the contacts. For the recent lavas of an active volcano, an appropriate variable for the graph would be time of eruption. More commonly, where neither geometric nor temporal relationships are accurately known, some one oxide is selected to plot the others against. Since silica is generally the most abundant oxide and since differentiation often gives rocks of steadily increasing silica content, most variation diagrams have weight percent of silica as the independent variable.

As an example, Fig. 14-2 is a variation diagram for the lavas of Crater Lake. In interpreting a diagram of this sort, one must keep in mind that the increase in silica from left to right necessarily means a decrease in other oxides, regardless of differentiation. In Fig. 14-2 silica is shown increasing from about 50 to 73%; the sum of other oxides must then decrease from 50 to 27%, and if no differentiation occurs (except for the increase in silica), each oxide should diminish in this same ratio. In other words, just from the method of plotting, we would expect each oxide to show a decrease from left to right to about half of its initial concentration; the reactions that lead to differentiation will then be significant only if they cause a markedly smaller or larger change. Inspection of the diagram shows that CaO, MgO, and Fe_2O_3 decline to much less than half their initial values, as would be expected from differentiation, since these oxides are concentrated in the low-silica minerals (olivine, pyroxene, calcic plagioclase) that make up the earlier rocks of the sequence. The alkali metals show no decrease at all but actually increase across the diagram, as is reflected in the abundance of alkali feldspars and micas in the later stages of differentiation. Alumina shows a slight initial rise due to early separation of nonaluminous olivine and rhombic pyroxene, but thereafter its apparent falling off is only about what would be expected from the method of plotting. Thus the trends of the curves are consistent with the hypothesis that all these rocks belong to a single differentiation sequence, and the smoothness of the curves is an indication that the rocks are genetically related. To illustrate how the diagram distinguishes rocks belonging to this sequence from other igneous rocks, two analyses (small circles) are plotted for rocks from different areas; although a few points lie near the Crater Lake curves, most are so far away that an origin by differentiation from a single magma would seem unlikely.

Several other kinds of variation diagram have been proposed. Combinations

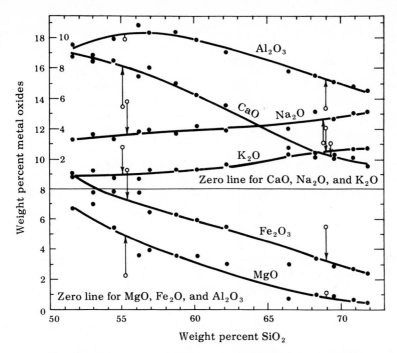

Figure 14-2 Variation diagram for lavas of Crater Lake. Each vertical row of dots represent a chemical analysis, with total iron calculated as Fe_2O_3. Only a few of the 30 analyses from which the diagram was constructed are shown. The small circles show analyses of two rocks that do not belong to this series. (*Source: Williams, 1942, page 154.*)

of oxides (for example, $\frac{1}{3}SiO_2 + K_2O - FeO - MgO - CaO$) are often substituted for SiO_2 as the independent variable. Another suggestion is the calculation of a "differentiation index" for the independent variable; the index is designed to show how far the differentiation of an igneous rock has progressed toward "petrogeny's residua system" (Sec. 14-1), and is defined as the sum of quartz + orthoclase + albite + nepheline + leucite + kalsilite (idealized minerals, calculated from the chemical analysis). Triangular plots are often useful, since they can show simultaneous variation in three important elements or combinations of elements (Fig. 14-6). The choice of variation diagram in any particular case depends on the relationships which need emphasis.

A variation diagram by itself clearly does not constitute unambiguous proof of differentiation from a single magma. If a group of rocks is related by differentiation, then the curves on the diagram should be smooth and should show a qualitatively predictable pattern; if, on the contrary, the rocks of the group have formed from separate intrusives, points on the diagram may or may not approximate smooth curves. Despite this ambiguity, variation diagrams are frequently useful to bring out relationships among rock groups that would not be obvious from the analyses alone.

14-4 FELSIC MAGMAS

Igneous Origin of Granites versus Granitization

Although silica-rich liquids can be produced by differentiation of mafic magmas, it seems doubtful, as noted above, that this mechanism can account for large bodies of granitic rocks. What alternative process can be suggested for the origin of granites? This is a question that has been debated since geology became a science, and the end is not yet in sight.

Field evidence is not even altogether convincing as to whether granite is, in fact, an igneous rock. The common intrusive felsic rocks (granite, granodiorite, tonalite, and several others) have chemical compositions similar to those of volcanic rocks like rhyolite and dacite (Table 14-1) that are demonstrably igneous in origin, but relations between corresponding intrusive and extrusive types are much less close than for mafic rocks. Volcanic rocks of felsic composition are in large part pyroclastics, fragmental rocks produced by explosions in gas-rich, viscous magma; both the pyroclastics and the small flows and domes are generally made up chiefly of glass, and gradations from the glass into completely crystalline intrusive rocks are not often found. Bodies of granitic intrusive rocks commonly lack the fine-grained chilled borders and the segregations of early-formed crystals that so clearly demonstrate the igneous history of many mafic intrusives. Without these evidences of igneous origin for granites we must fall back on field observations like sharp contacts, tongues of granite invading adjacent rocks, homogeneity of composition of granitic bodies, and patterns of oriented mineral grains that suggest flow in a viscous liquid. Observations of this sort are sufficient to persuade most geologists that at least *some* granite masses have moved as liquid magma for considerable distances.

Field evidence of this kind, however, is by no means universally found. Some large granite masses show gradational and intertonguing contacts with adjacent metamorphic rocks, a concordance of structure between layering in the metamorphic rocks and banding in the granite, and an abundance of metamorphic xenoliths in the granite showing various stages of breakup and assimilation. Observations like this have convinced some geologists, especially those who have worked extensively in areas of high-grade metamorphic rocks, that such granites should not be considered igneous rocks at all, but rather a product of extreme metamorphism of material that was originally sedimentary and volcanic rock. According to this view, granite showing such intimate relations to metamorphic rocks has formed practically in place, remaining in large part solid during the entire transformation. Various mechanisms have been proposed to explain details of this ultra-high-grade metamorphism, all of the mechanisms falling under the general descriptive term *granitization*.

Details of the long argument between "granitizers" and "magmatists" over the origin of granite need not detain us here, since the pertinent evidence is largely textural and structural rather than chemical. Most geologists at present adopt an intermediate position, agreeing that some granites have formed approximately in place by extreme metamorphism, while others have clearly moved into their present positions as masses of fluid or partly fluid material. For some granite bodies the evidence as to origin is clear, for others ambiguous. Regardless of possible arguments in specific cases, enough granite has surely formed by freezing of a melt so that the chemical problems of crystallization in a silica-rich magma merit attention.

On the laboratory side, granites pose a more difficult experimental problem than basalts because dry silica-rich melts are extremely viscous and freeze to a glass rather than a crystalline solid even when cooled very slowly. In the laboratory study of basaltic melts the addition of water vapor would be desirable for a closer simulation of nature, but it is not essential for an understanding of gross features of the crystallization process; in the study of granitic melts, on the other hand, addition of water is indispensable for any headway to be made at all. The development of experimental techniques making it possible to follow the course of crystallization under controlled pressures of water and other volatiles is one of the real breakthroughs of the last few decades. Some of the advances in understanding the behavior of silica-rich melts made possible by these techniques are described in the following paragraphs. In a later section we will return to the question of the origin of such silica-rich melts in nature.

The System Quartz-plus-Alkali-Feldspar

A rough approximation to the composition of many granites, especially those low in mafic minerals, can be given in terms of the three components SiO_2, $KAlSi_3O_8$, and $NaAlSi_3O_8$. From Fig. 13-5, 13-6, and 13-8 in the last chapter one could guess at the general form of the phase diagram for this system: on a triangular plot the side quartz-albite should have a simple binary eutectic; the side quartz-orthoclase should also have a eutectic (complicated at low water-vapor pressures by the incongruent melting of orthoclase); and the side orthoclase-albite should represent a solid-solution series with a minimum melting point. Inside the triangle the simplest possible relationship among the three components would be shown by a cotectic line connecting the two eutectics. These relationships have been demonstrated experimentally for water-vapor pressures of a few thousand atmospheres (Fig. 14-3).

The diagram resembles the one for diopside-plagioclase (Fig. 13-11), except that the point of lowest temperature is near the middle of the cotectic line rather than at one end. In other words, the minimum melting point between the two alkali feldspars persists even when quartz is added, so that the lowest-melting mixture of all three consists of quartz plus a roughly equimolal mixture of the feldspars. The situation is not far different from a simple ternary eutectic (Fig. 13-9), except that the "cotectic" represented by the minimum melting points is a shallow trough instead of a sharply defined valley. Paths of crystallization, shown by arrows in Fig. 14-3 for various mixtures, all lead into the temperature "basin" near the middle of the cotectic line.

The relationships in Fig. 14-3 mean that any magma in this restricted range of

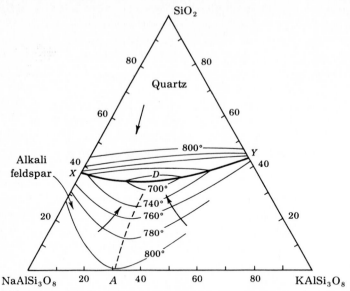

Figure 14-3 Phase diagram for the system SiO_2-$NaAlSi_3O_8$-$KAlSi_3O_8$ at a water-vapor pressure of 2,000 kg/cm². This diagram is similar to the upper half of Fig. 14-1, but temperatures are lower because of the water vapor. The bottom edge of the triangle is a "top view" of Fig. 13-8, and A is the minimum melting point of the binary system. The dashed line marks the approximate bottom of the temperature trough extending to the low-temperature basin on the quartz-feldspar boundary at D. Arrows show approximate directions of change in composition of various melts as crystallization takes place. *(Source: Tuttle and Bowen, 1958, page 55.)*

compositions should give, on cooling, a residual liquid consisting of roughly equal amounts of SiO_2, $KAlSi_3O_8$, and $NaAlSi_3O_8$ and that this liquid should freeze to form a rock of the same composition. If granites are formed by crystallization-differentiation in a molten magma, therefore, we might expect to find that the composition of many granites would approximate equal quantities of quartz, potash feldspar, and sodic plagioclase. The two feldspars may form a single phase of homogeneous mix crystals if the cooling was rapid (unlikely in a granite), or a perthite showing partial unmixing of the solid solution, or separate crystals if the cooling was very slow. Hence alkali granites may show various textural features, but the chemical compositions should be similar. By plotting analyses on the same ternary diagram, Tuttle and Bowen (1958) have demonstrated that a large number of granites do in fact have compositions close to the predicted temperature minimum (an example is analysis I, Table 14-1). This is convincing experimental evidence that alkali-rich granites are a product of crystallization from a melt.

Granitization and the Quartz-Feldspar Diagram

At first sight the work of Tuttle and Bowen seems to have undermined the granitization hypothesis, for they have shown beyond any doubt that liquid magma is involved in the origin of granite, at least of granite rich in sodium and potassium. But granitization is a concept of many facets, and if suitably interpreted, it may still be reconciled with the experimental data. The essential idea of granitization is that an igneous-appearing rock should form from preexisting material with little overall movement away from its original location, and that at least in part the transformation should involve reactions of solids. This kind of alteration would be permitted by the experimental results.

For it follows from the experiments, not only that a quartz-plus-alkali-feldspar mixture should be the *last* material to crystallize from a siliceous melt, but that this same mixture should be the *first* liquid to form when a silicate rock material is heated in the presence of water vapor. A shale, for example, on being heated gradually during metamorphism will show the usual changes in mineral composition (clay minerals to muscovite, biotite, cordierite, garnet, sillimanite, and so on) in the solid state; but if it becomes hot enough for melting to start, the first liquid will have a composition represented by points near the ternary minimum in Fig. 14-3. We may imagine this first liquid being squeezed into pockets and cracks in the metamorphic rock, and there solidifying to form the pods and stringers of quartz plus feldspar (pegmatite) that are often found in areas of high-grade metamorphism. As the liquid increases in amount, we may think of it as circulating with greater ease, so that the mixture of liquid plus remaining solid crystals becomes increasingly homogeneous. At any stage the temperature change may be reversed, and the material may freeze to a rock resembling granite if the melting has gone far enough, or to a migmatite ("mix rock") if partly altered fragments and bands of recognizably metamorphic rock remain. Thus a granite might be formed practically in place, with part of its constituents never having become molten.

Some advocates of granitization would say that we have stretched the meaning of the term too far. "Granitization," in their eyes, should be restricted to the making over of a solid, without formation of appreciable amounts of liquid. The alteration of the solid minerals should be a strictly metamorphic process, brought about by recrystallization, solid diffusion, and exchange of ions with a small quantity of mobile liquid or vapor moving through the rock (metasomatism). Against a hypothesis of granitization in this extreme form—at least against its widespread operation in nature—Tuttle and Bowen raise almost insuperable objections. For if granites were to originate in such a manner, the clustering of compositions in the low-temperature trough of the quartz-feldspar diagram would be a sheer accident, unexplainable except by fabricating additional ad hoc hypotheses. A liquid stage, or at least partly liquid stage, seems a necessary part of the history of any granite mass.

Thus the experimental work shows that granitization by metasomatism is untenable, but provides a plausible mechanism for a kind of granitization that involves partial melting. Definitions become fuzzy here: what fraction of original rock material must melt before the formation of granite from it should be described as an "igneous" rather than a "metamorphic" process? The definitions are perhaps less important than an understanding of the process. For present purposes, "granitization" will be used to describe the formation of granitic rocks by partial melting, without much overall movement of the partly melted material.

But clearly the melt need not always remain at its initial site, especially in an area that is undergoing tectonic disturbance during metamorphism. The viscous silicate liquid, perhaps accompanied by solid crystals and larger solid blocks, can be squeezed upward into regions of lower pressure, or possibly can rise upward simply because of its low density. Hence the hypothesis of partial melting may be used to account not only for alkali granites formed in place, but for granite masses that have moved far from their place of origin and show the usual structural evidences of igneous intrusion. Can we jump to the conclusion that *all* granites have their origin in such partial melting of preexisting rocks, the apparent differences in behavior of different granite bodies being explainable by the extent to which they have moved while in a molten or semimolten condition?

So sweeping a conclusion is hardly warranted. Granitic material in small amount has demonstrably formed by the differentiation of more mafic magmas, and locally rocks of granitic appearance may well have been formed by metamorphism without appreciable melting. The relative importance of these processes is still a matter of legitimate debate. But Tuttle and Bowen have shown beyond question, at least for alkali-rich granites, that differential melting in itself provides a *possible* explanation for most of the observed phenomena of large granitic bodies.

Other Granitic Magmas

A serious limitation of Tuttle and Bowen's argument is the restriction of their experiments to magmas rich in sodium and potassium and low in calcium, magnesium, and iron. What happens if these latter three elements are present in amounts that are not negligible? Can the experimental conclusions be extended from alkali granites to other abundant felsic intrusive rocks, like granodiorite and tonalite? Possible answers to these questions are suggested by several recent experimental studies, of which a good example is one by von Platen (1965).

Von Platen started with an obsidian very low in calcium, iron, and magnesium, resembling in composition the simple quartz-plus-alkali-feldspar mixtures used by Tuttle and Bowen. Various compositions were simulated by adding to the obsidian measured quantities of anorthite and biotite. The mixtures were melted completely under a pressure of 2,000 bars of water vapor, allowed to cool slowly, and quenched at various stages of crystallization. In this way von Platen could follow, with careful control of all important variables, the sequence in which minerals separate from melts with compositions covering nearly the entire range of common felsic magmas.

One important conclusion is that the crystallization of quartz and feldspar is practically unaffected by small amounts of ferromagnesian material. In melts to which biotite had been added, the iron and magnesium crystallized during cooling as biotite or magnetite or both, but details of appearance of quartz and feldspar were the same whether biotite was present or not. This conclusion is perhaps not unexpected, but the experimental confirmation puts speculation about granitic magmas on firmer ground.

The effect of adding calcium to the quartz-alkali-feldspar system can be displayed by constructing a tetrahedral model with SiO_2, $KAlSi_3O_8$, $NaAlSi_3O_8$, and $CaAl_2Si_2O_8$ as the corners, and then projecting lines and surfaces from the tetrahedron onto the qz-or-ab face. In this way the effect of calcium is shown as a series of modifications of Tuttle and Bowen's qz-or-ab diagrams. As an illustration of von Platen's results, Fig. 14-4 pictures the phase relations for two albite-anorthite ratios, 1.8 and 7.8. The figure shows several effects of increasing anorthite. For one thing, the range of compositions in which feldspar is the first mineral to crystallize expands at the expense of quartz (line XY moves to WY and then to VY as anorthite increases from zero to an ab/an ratio of 1.8). Even more striking is the fact that the single feldspar field of Fig. 14-3 (representing $KAlSi_8O_8$-$NaAlSi_3O_8$ solid solutions) is broken into two parts, one showing initial crystallization of plagioclase and the other initial crystallization of alkali feldspar. In other words the entire field of the diagram now has three subfields, corresponding to the three major mineral constituents of practically all granitic rocks. The point at which the subfields join is not strictly a ternary eutectic, as it appears to be, because it is on a projection of a line from the tetrahedral model; it

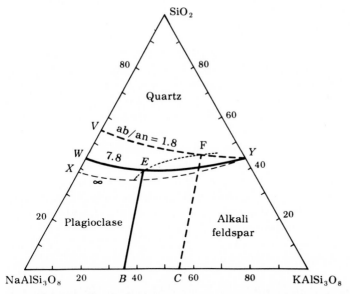

Figure 14-4 Phase diagram for the system qz-or-ab-an, projected onto the qz-or-ab plane. Water-vapor pressure 2,000 kg/cm². The light dashed line XY, the same as XY on Fig. 14-3, is the quartz-feldspar boundary for the simple system qz-or-ab, for which the ratio ab/an = ∞. The solid line WY is the same boundary when the ratio ab/an = 7.8. Addition of anorthite to the qz-or-ab system causes the alkali-feldspar solid solution to break up into two phases, plagioclase and Na-K-feldspar, whose fields are separated by the boundary BE. Hence E represents a liquid composition from which three phases crystallize together, rather than the two phases (quartz + solid solution) represented by point D on Fig. 14-3. Increasing anorthite causes the plagioclase field to expand, so that the three-phase point moves along the light dotted line, reaching F when the ratio ab/an = 1.8. (*Source: Von Platen, 1965, page 366.*)

nevertheless represents roughly compositions from which the three minerals will crystallize simultaneously or nearly so. Note that this point shifts toward increasing silica and potash feldspar as the proportion of anorthite in the melt becomes larger (point E shifts to F as the ab/an ratio falls from 7.8 to 1.8). In a melt with high calcium, therefore, the first mineral to crystallize should be plagioclase over a wide range of compositions (in other words, the plagioclase field on the diagram becomes large), and the ultimate residual liquid should be rich in the constituents of quartz and orthoclase. These experimental conclusions fit nicely the common field observations that, in a granodiorite, plagioclase generally has more nearly euhedral crystals than either quartz or orthoclase, and that the felsic dikes (aplites and pegmatites) accompanying a granodiorite usually consist very largely of quartz and K-feldspar.

Von Platen's experimental results are clearly consistent with Bowen's hypothesis of origin of granitic rocks by crystallization-differentiation of a basaltic magma. They are by no means restricted to this hypothesis, however, since the experiments could equally well be used to predict the compositions of liquids formed by progressive melting of any kind of silicate material. If a random assemblage of sedimentary and volcanic rocks is slowly heated to the point of incipient melting, the first liquid should have a composition represented by a point on the light dotted line (EF) of Fig. 14-4, the precise composition being determined by the ab/an ratio of the starting material and by such additional factors as pressure and amounts of other volatiles besides water. Changes in the composition of the melt as temperature continues to rise are predictable from von Platen's diagrams, or better from the tetrahedral model: the initial liquid would be relatively rich in silica and the components of alkali feldspar, and the amount of calcium would increase progressively with the temperature rise. The sequence of changes, provided that approximate equilibrium is maintained, is the same as that for a residual liquid during crystallization except that the order is reversed.

Thus experiments like von Platen's demonstrate that Tuttle and Bowen's conclusion about the origin of simple granitic melts can be extended to materials of more complex composition. Silica-rich magmas of widely varying compositions can form either as liquids left when crystals settle out of a mafic magma or by partial melting of preexisting silicate rocks.

14-5 EFFECT OF PRESSURE ON DIFFERENTIATION

The discussion so far has concerned the application of laboratory experiments at fairly low pressures to field observations of igneous rocks. The pressures have been limited to a few thousand atmospheres, corresponding to processes that may reasonably be expected to occur in the upper few kilometers of the earth's crust. How far can our conclusions be projected downward? Can we expect that the same processes of differentiation, partial melting, and crystallization will operate under the enormous pressures near the base of the crust and in the upper mantle?

Experiments on simple silicate systems show clearly that high pressures lead to important modifications of the familiar low-pressure reactions. Increase of pressure on most silicate minerals in the dry state leads to an increase in melting point, because crystalline silicates in general are more dense than the corresponding liquids (Sec. 8-12), but if the pressure is exerted by water vapor the melting point is generally lowered. The same relations hold for mixtures of silicates. Because the change in melting point varies greatly from one mineral to another, the composition of eutectic mixtures is often markedly affected by pressure. As an example, Fig. 14-5 shows the effect of high pressures, with and without water vapor, on the simple binary system anorthite-diopside previously illustrated in Fig. 13-2.

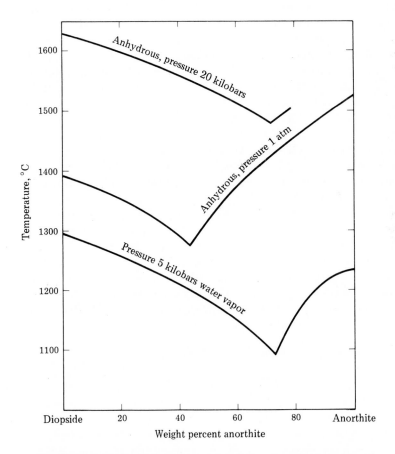

Figure 14-5 Effect of pressure on melting in the anorthite-diopside system. The middle curve, for a dry melt under ordinary atmospheric pressure, is the same as Fig. 13-2. The upper curve is for a dry melt under a pressure of 20 kilobars, and the lower curve is for melting under a water-vapor pressure of 5 kilobars. The upper curve is not complete because anorthite at high pressures melts incongruently.

For interpreting high pressures in diagrams like Fig. 14-5, it is useful to keep in mind the rough correspondence between pressure and depth within the earth. The downward increase in pressure is far from uniform because rock densities are variable, but on the average in the crust and upper mantle a change in pressure of 1 kilobar can be equated to a change in depth of about 3.5 km. (This assumes an average rock density of about 3.5 g/cm^3.) Thus the lithostatic pressure at the base of the continental crust is roughly 10 kilobars, and at the base of the oceanic crust about 2 kilobars.

Pressure affects not only melting points and eutectic compositions, but the nature of stable mineral phases. If molten basalt, for example, is cooled slowly under a pressure of a few thousand atmospheres of water vapor, the crystalline product is not basalt but a rock rich in amphibole. Depending on the composition of the melt and the pressure of water vapor, the amphibole may be accompanied chiefly by plagioclase or chiefly by pyroxene, or it may be practically the only mineral formed. The pressure at which amphibole becomes a dominant mineral is surprisingly low; basalt of almost any composition apparently becomes unstable under a water-vapor pressure of 1,500 atm. If plagioclase forms at all, it appears late in the crystallization process, not near the beginning as it commonly does at lower pressures. These results mean that basaltic magma may be generated by the melting of rocks of widely variable mineral composition: amphibolites, hornblendites, pyroxene hornblendites. Whether basalt and gabbro in the form we know them, as plagioclase-pyroxene-olivine rocks, are present at lower levels in the crust depends on the amount of water present; material with the chemical composition of basalt may well exist, at least locally, in the form of widespread amphibole-rich rocks.

Even more extraordinary are the results of laboratory work on the crystallization of basaltic magma at very high pressures in the absence of water vapor. An uncommon, unusually dense rock called *eclogite*, with the chemical composition of basalt but consisting of garnet and a soda-rich pyroxene (omphacite), has long figured in petrologic speculations as a possible high-pressure equivalent of basalt; Yoder and Tilley (1962) have demonstrated experimentally that a basalt melt under sufficiently high pressure will indeed crystallize as eclogite. The transition from basalt to eclogite may be roughly symbolized by the equation

$$\underbrace{NaAlSi_3O_8 + CaAl_2Si_2O_8}_{\text{Plagioclase}} + \underset{\text{Diopside}}{CaMgSi_2O_6} + \underset{\text{Forsterite}}{Mg_2SiO_4} \rightarrow$$

$$\underbrace{NaAlSi_2O_6 + CaMgSi_2O_6}_{\text{Omphacite}} + \underset{\text{Garnet}}{CaMg_2Al_2Si_3O_{12}} + \underset{\text{Quartz}}{SiO_2} \quad (14\text{-}1)$$

The equation is simplified by omitting minor constituents and by leaving iron out of the formulas of the pyroxenes, olivine, and garnet, but it shows the essential nature of the change: soda enters the heavy pyroxene omphacite rather than feldspar, and garnet replaces the usual plagioclase and olivine. The transition depends on both temperature and pressure, occurring at pressures of 10 to 12 kilobars if the temperature is near 500°, and at 14 to 16 kilobars for temperatures near 1000°. Such pressures correspond to depths below the surface of 35 to 55 km, so that the transition should be possible near the base of the crust and in the upper mantle.

If, therefore, a body of basalt magma is generated in the upper mantle, and if the pressure of water vapor is low, its crystallization at that depth would follow a very different pattern from the one embodied in Bowen's reaction principle. Either omphacite or garnet would crystallize first, depending on the initial composition; eventually it would be joined by the other, and the two would crystallize together during most of the cooling. Since both minerals are complex solid solutions, there would doubtless be considerable change in the composition of the solids by reaction with the liquid during cooling. Details are not known from experiment, but clearly extensive differentiation should be possible, and the course of differentiation would be quite different from that at usual pressures.

14-6 ORIGIN OF BASALTIC AND GRANITIC MAGMAS

In preceding paragraphs we have mentioned older speculations about basalt as a possible primordial magma, and have noted the possible origin of more felsic magmas either by fractional crystallization of molten basalt or by partial melting (anatexis) of many kinds of crustal rock materials. We have not, however, faced squarely the question of where the major kinds of igneous rock found in nature have their source. Like many of the deeper questions in geology, this one cannot be given an unequivocal answer. Because the principal evidence bearing on the question comes from geophysics rather than geochemistry, we will limit the discussion here to a brief review of current ideas.

The existence of any large body of basaltic magma within the earth can be dismissed at once by the observation that transverse seismic waves travel through much of the earth's interior, proving that the entire crust and mantle must be largely solid. Study of the speeds of seismic waves at different levels in the interior, however, has shown that over much of the earth a layer in the mantle between depths of approximately 70 to 150 km (the asthenosphere) transmits wave motion more slowly than layers above and below, hence consists of less rigid material. It is very likely material near its melting point, and probably contains pockets and stringers of liquid. A further suggestion that liquid may be locally generated by melting at this level is the observation that basaltic eruptions in Hawaii and Japan are often preceded and accompanied by earthquakes with foci 60 to 100 km beneath the surface. The simple fact that basalt is widespread over the earth's surface, forming the major part of the crust under all the oceans and occurring also in huge accumulations on the continents, is additional evidence that this kind of lava has a source deep within the earth, at a level where rock material is less heterogeneous than in the continental crust.

What kind of rock would we expect to find in this part of the mantle? A first guess might be solid basalt, which would simply melt to form the lava we see after its ascent to the earth's surface. This guess is impossible, however, because at depths not much greater than the thickness of the continental crust the high pressures would convert basalt to eclogite. Some of the mantle material may be eclogite, as is indicated by the occasional finding of eclogite xenoliths in basaltic lavas. More common as xenoliths, however, are fragments of peridotite, and geophysical evidence also makes it probable that peridotite is the dominant material of the upper mantle. It is probably a peridotite containing an aluminum mineral, garnet or spinel, in addition to the usual olivine, orthopyroxene, and clinopyroxene. Experiment has shown that basaltic liquid in considerable amounts can be generated by heating such material under conditions simulating those in the upper mantle. The proof is perhaps not quite complete, but it seems likely that most basalt magma is formed by local heating of garnet peridotite, or a mixture of eclogite and garnet peridotite, at depths of the order of 100 km under the earth's surface.

This does not end the problems of basalt. Superficially basalt from different places looks like a very uniform sort of rock, but analyses show differences in composition that can be correlated with manner of occurrence—for example, the silica-saturated or nearly saturated basalt of the ocean floor and the Columbia River plateau (tholeiitic basalt), and the undersaturated, alkali-metal-rich basalt of many oceanic islands (alkali olivine basalt). Attempts to link the observed compositional differences with depth of origin, kinds of peridotite at the source, and possible differentiation or contamination during ascent of the magma to the surface constitute a presently active field of geochemical research.

Granitic rocks, in contrast to basalt, are largely restricted in distribution to the cores of mountain ranges and to areas which from geologic evidence appear to have been the cores of mountain ranges in the past. They are commonly accompanied by layered rocks that are metamorphosed and highly deformed. Although the rock of a batholith may appear to be uniform over large areas, detailed study generally shows that it is made up of units that differ somewhat in texture and composition. Such differences, and the more considerable differences between batholiths in different areas, contrast sharply with the relatively minor variations in the composition of basalt. The variability of granites and their association with regions of mountain building and metamorphism strongly suggest that many granite bodies originate by the partial melting of other rocks within the crust during orogenic disturbances. Some of these bodies clearly have moved away from the place where they were formed, as coherent masses of viscous fluid; others have remained intimately associated with remnants of the metamorphic rocks from which their material was derived. On this hypothesis, the broad similarity of granites the world over does not imply a common source but results simply from the chemical fact that the product of partial melting of practically any heterogeneous silicate mixture has a composition within a restricted range.

It cannot be concluded, however, that all granites have formed by this process of differential anatexis of crustal materials. Some may well be products of differentiation of more mafic magma formed in the upper mantle. The differentiation would necessarily be limited to the uppermost part of the mantle, where early separation of calcic plagioclase and olivine could concentrate silica and the alkali metals in the remaining liquid; it could not happen under the high pressures at deeper levels, where the solids separating would be garnet or sodic pyroxene [Eq. (14-1)]. One can imagine that a body of basaltic liquid formed deep in the mantle might rise to a position near the crust-mantle boundary, and there undergo fractionation that would produce at least a small amount of silica-rich liquid. Good evidence that some granites have formed in this way has come from studies of strontium isotope distribution, which indicate that these granites consist in large part of material that has not undergone the processes of weathering, sedimentation, and metamorphism characteristic of crustal rocks. In other words, the substance of these granites is "juvenile," in the sense that it has not previously been part of the upper crust. The kinds of evidence available about the origin of a given granite body often seem directly contradictory, and there is much current argument about the relative importance of crustal anatexis versus mantle differentiation in the formation of granite batholiths.

A similar lively argument concerns the origin of andesite, a lava with intermediate amounts of silica (Table 14-1). From a strictly chemical standpoint, andesite could form readily as a stage in the differentiation of basaltic magma, or equally readily as an advanced stage in the partial melting of crustal rocks (although the required temperature would be somewhat higher than temperatures normally found in the crust). Andesite is similar in composition to the less siliceous varieties of granitic rocks (tonalite and diorite), and like them is characteristically abundant in regions that have undergone orogenesis and mountain building—for example, the chains of volcanoes along the margins of the Pacific basin. The origin of andesite is doubtless intimately associated with the origin of granitic rocks.

The concepts of plate tectonics provide a basis for speculation about mechanisms by which the large observed volumes of granitic intrusive rocks and andesitic lavas have formed. If orogenic zones are visualized as parts of a continental margin under which a subducting plate is moving, several ways of forming abundant silica-rich melt can be imagined. The downward-moving plate consists of uppermost mantle, oceanic crust (basalt), and overlying sediments that are made up in part of material eroded from the continent. This heterogeneous assemblage is carried down to depths of a few hundred kilometers beneath the continent, hence must be gradually heated until some of its materials are metamorphosed and ultimately begin to melt. Part of the molten material would then rise into the fractured crustal and upper mantle rocks above the moving plate, some of it feeding volcanoes at the surface and some collecting beneath the surface to form intrusive masses. In this sequence of events magmas might form in a variety of ways: by partial melting of the sediments, basalt, or peridotite in the subducting plate, or by nearly complete melting followed by fractional crystallization; and the magmas so formed could be modified by reaction with the rocks above through which they move. Guesses about the kinds of material from which different magmas are derived, based on observed compositions of igneous rocks from orogenic zones, provide a fertile field for geochemical controversy.

Abundant silica-rich melt, seemingly, could be produced most easily by a generous contribution of material from the continental crust, either from partial melting of the continent-derived sediment on top of the subducting plate or from assimilation of crustal rocks as a more mafic magma makes its way upward toward the surface. Against this attractive hypothesis two objections can be urged: (1) andesite is abundant in some islands of the southwest Pacific above active subduction zones far from any major source of continental material, and (2) the low ratio of radiogenic to nonradiogenic strontium isotopes (Sec. 21-3) in andesites precludes any important contribution from the continental crust. Strontium isotope ratios, in fact, suggest that most of the sediment carried by a subducting plate is scraped off and piled against the continent as the plate begins its descent, so that little crustal material is available for magma generation. If andesites, then, are derived largely from fractional melting or fractional crystallization of more mafic material, a further question can be raised as to what particular variety of basalt or mantle peridotite is the major source. On this question, trace-element concentrations provide useful information but still not definitive answers. Granitic rocks vary greatly in their strontium isotope ratios; based on this evidence, some granitic magmas may be formed, like andesites, largely by differentiation of mafic material, and others by partial melting of crustal rocks. Thus plate tectonics, isotope ratios, and trace-element concentrations narrow the possibilities for generating andesitic and granitic magmas, but the number of variables is so great that specific mechanisms remain elusive.

14-7 UNUSUAL IGNEOUS ROCKS

The rocks we have so far discussed are the abundant ones, those belonging to the series gabbro-diorite-granodiorite-granite and their volcanic equivalents (analyses I to VI, Table 14-1). It is possible, as we have seen, to derive rocks of these compositions by crystallization-differentiation from a basaltic magma, although the more siliceous members of the series may originate in other ways. Because these rocks are so abundant, because they are at least formally related by Bowen's reaction series, and because the sequence of compositions can be demonstrated within many igneous bodies and in many areas of igneous rocks, these varieties have come to be regarded as "normal" igneous rocks and the above sequence as the "normal" succession.

A good many igneous rocks, however, do not fit this pattern at all. For the most part these are rocks that occur only locally and in small masses and hence come to be regarded as "abnormal." Because of their compositional peculiarities, they have received an amount of attention from petrographers out of all proportion to their abundance, and they are responsible for much of the profusion of names that makes petrography so forbidding a subject to the beginner. It would take us much too far afield to consider the chemical problems of all these unusual rocks. We can get a good idea of the kinds of questions that arise by discussing briefly three varieties: iron-rich rocks, alkalic rocks, and ultramafic rocks.

Iron-Rich Rocks

As mentioned earlier, differentiation in some gabbro and diabase intrusives appears to give as a final or near-final product a rock whose outstanding characteristic is an abundance of iron-rich minerals rather than the expected quartz-plus-alkali-feldspar mixture. A little quartz may indeed be present, but the low melting end product (ferrogabbro) consists largely of plagioclase plus an iron-rich pyroxene, and in some cases iron-rich olivine (fayalite) as well. A famous example of such a differentiation trend in an apparently normal mafic magma is the Skaergaard intrusive of eastern Greenland.

In the "normal" differentiation process, the pyroxenes and olivines that crystallize first are rich in magnesium, and the Fe/Mg ratio increases as differentiation progresses. The enrichment in iron with falling temperature is shown, for the olivine series, in the diagram for Prob. 13-2. A similar magnesium-iron relationship can be demonstrated for the pyroxenes, but the diagram is more complicated because of the greater variation in composition of these minerals. In the normal sequence the FeO content of pyroxene and olivine does not climb much above 20%, while in the ferrogabbros at Skaergaard it may be as high as 37% in the pyroxene and 67% in the olivine. What circumstance could bring about this trend of crystallization?

The enrichment of silica toward the end of the usual differentiation series depends on a number of factors. Much of the silica originally present in the magma is taken up in the formation of pyroxene and plagioclase, the amount so used depending on the ratio of silica to magnesium, iron, calcium, and sodium. In some magmas the amount may be just sufficient to satisfy these metals, and none will be left over; in others the original amount may be slightly greater, and the proportional excess in the residual liquid becomes steadily larger as crystallization proceeds, so that free quartz is an ultimate product. If olivine separates early and does not react with the melt, the amount of silica left over during crystallization becomes greater. Also, if any process removes iron during crystallization, less silica can go into pyroxene and olivine, and more will remain in the residual fluid. Now one possible process for removing iron is oxidation; if conditions are sufficiently oxidizing for magnetite to form early, some iron will be eliminated from the crystallization sequence and less will be available to form pyroxenes. Hence a high concentration of silica in the residual liquid, and an abundance of quartz in the latest differentiates, is favored by oxidizing conditions. One could guess, then, that the normal sequence of differentiation requires partial oxidation of iron to magnetite, while extremely reducing conditions would permit all the iron to remain in the melt and hence to continue reacting with silica down to the

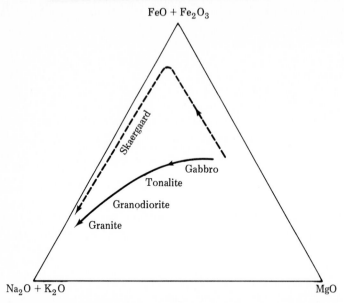

Figure 14-6 Contrasting trends of differentiation for the "normal" igneous rock series (gabbro to granite) and for the Skaergaard intrusive. The initial magmas for both series have similar proportions of magnesium, iron, and the alkali metals. Under relatively reducing conditions (Skaergaard), iron accumulates in the residual melt until a late stage of differentiation. Under more oxidizing conditions (normal sequence), much iron is removed as magnetite, so that Mg/Fe in the melt remains roughly constant and the alkali metals steadily increase. (*After Carmichael, Turner, and Verhoogen, "Igneous Petrology," McGraw-Hill Book Company, New York, 1974, pp. 49 and 568.*)

lowest temperatures of crystallization. We conjecture, therefore, that differentiation at Skaergaard took place under abnormally low partial pressures of oxygen (Fig. 14-6).

This explanation is considerably oversimplified. It should include a discussion of the effect of the oxidation state of iron on the kind of pyroxene that crystallizes, hence on the Mg/Ca ratio in the pyroxene and the amount of calcium left to form plagioclase. But even such a rudimentary account makes it clear that an important factor determining the trend of differentiation in a magma is its state of oxidation.

Alkalic Rocks

The alkali metals sodium and potassium are concentrated during normal differentiation in the residual fluid, crystallizing from it to form the alkali feldspars which, along with quartz, make up the granitic end product of the differentiation process. In some igneous rocks the amount of sodium or potassium or both is too large to be accommodated in the feldspars, and the excess forms unusual alkali-rich minerals. In rocks with a high Si/Al ratio an excess of alkali metal is commonly expressed by the formation of soda pyroxenes (for example, aegirite, $NaFeSi_2O_6$)

and soda amphiboles; in those with a low Si/Al ratio the excess goes into feld-spathoid minerals, of which the two commonest are nepheline (ideally $NaAlSiO_4$, but generally containing some potassium, and more silicon than is shown by this formula) and leucite ($KAlSi_2O_6$). The origin of these alkali-rich rocks has excited much petrographic speculation.

One might inquire first why the question arises at all, why we cannot simply assume that these rocks have crystallized from magmas of the same composition. In the context of the last section, for example, we might suppose that partial melting during the extreme metamorphism of alkali-rich sediments could produce liquids of these compositions. It is possible that some alkalic rocks have indeed been formed in this manner. For others, however, their geologic relationships, both among themselves and with more ordinary igneous rocks, suggest that some sort of differentiation process has played a role in their history.

Going back to the normal differentiation of basaltic magma, we recall that silica and the alkali metals are concentrated in the residual liquid because the early-crystallizing minerals are relatively deficient in these substances. Differentiation commonly yields enough silica to the residual fluid for the alkali metals to be completely tied up in feldspar and for at least a little silica to be left over as quartz. How much silica gets into the residual fluid depends on (1) the details of the differentiation history, especially the early separation of olivine and the amount of oxidation of iron, and (2) the composition of the original magma. If the magma contains any excess silica, no matter how slight, over the amount necessary to satisfy the cations of the pyroxenes, amphiboles, and feldspars, this excess must become increasingly concentrated in the liquid as differentiation proceeds. If, on the contrary, the amount of silica is somewhat less, the final material to crystallize would be largely feldspar without quartz, and the deficiency of silica would be expressed by the persistence of olivine. And if, finally, the amount of silica is still less, the concentration of alkali metals and aluminum in the residual liquid will be great enough so that some silica must go into feldspathoid minerals rather than feldspars. Thus the trend of differentiation depends rather delicately on the relative amounts of silica and alkali metals in the original magma. Any slight excess of original silica leads to a high concentration in the final liquid, hence to the crystallization of quartz; a deficiency leads to a concentration of the constituents of the feldspathoids, with the proviso that a slight deficiency may be accommodated by adjustments in the olivine-pyroxene ratio.

This description of possible variation in the differentiation process is far from complete. It should take account also of possible slight variations in the calcium, iron, and aluminum content of the original magma, which determine the nature of the pyroxene and the plagioclase forming at various stages of cooling. But even without these details, the dependency of differentiation on silica and the alkali metals should be clear. While differentiation most commonly gives an end product containing quartz, with a slight difference in original composition it can just as well lead to a residual product consisting largely of alkali feldspar or a product with feldspathoid minerals. In this way we find a reasonable explanation for the fact that the low-silica basaltic lavas of the Pacific islands are often accompanied by minor amounts of trachyte (a lava consisting largely of alkali feldspar plus minor mafic minerals), phonolite (like trachyte but with nepheline in addition), and nepheline basalt, but rarely by the quartz-bearing lavas dacite and rhyolite.

A melt that crystallizes to give largely alkali feldspar or alkali feldspar plus nepheline is part of "petrogeny's residua system," just as are melts that crystallize to give alkali feldspar plus quartz. This is apparent from Fig. 14-1: alkali-rich melts fall below the dashed line joining $KAlSi_3O_8$ and $NaAlSi_3O_8$ and silica-rich melts above the line, but all are in the low-melting "basin" of the diagram. Thus we broaden our concept of "normal" differentiation: on the basis of both field observation and experimental work, silicate melts of basaltic composition differentiate toward residual liquids rich in the components of alkali feldspars and either silica or feldspathoids, the difference depending chiefly on minor variations in original composition (see analyses in Prob. 6 at the end of this chapter).

Just as differentiation of a slightly silica-rich basalt magma is unsatisfactory in explaining large granite masses, so is differentiation of basalt a poor explanation for the larger bodies of intrusive alkalic rocks. The chief examples of these are nepheline syenites (analysis VII, Table 14-1); famous localities are central Arkansas, southern Ontario, southern Norway, and Kola Peninsula in northern Russia. Such bodies are small compared with granite batholiths, but the larger ones have surface areas measured in hundreds of square miles. Hypotheses suggested to account for the existence of these rocks are very numerous.

One idea involves a mechanism of differentiation different from the crystallization processes we have been considering. If a body of granitic magma is particularly rich in volatiles, the volatile constituents (water, carbon dioxide, hydrogen chloride, and so on) can be imagined streaming upward through the liquid, perhaps carrying alkali-metal ions with them. The top of the magma body would thus become enriched in the alkali metals, and if this part of the melt should move away from the remainder, we should have a reasonable precursor for a nepheline syenite. One argument in favor of this hypothesis is the common presence of minerals containing volatiles in and near nepheline-syenite masses. Another is the experimental fact that the simple alkali silicates (K_2SiO_3 and Na_2SiO_3) and alkali chlorides are very soluble in hot water vapor under high pressure, so that the association of abundant alkalies with moving volatiles is reasonable. An argument against the hypothesis is the field observation that many nepheline syenites show no sign of the association with or gradation into granitic rocks that one might expect.

Another suggestion is that alkali-rich magmas may result from contamination of more normal magmas by other rocks. If, for example, a moving magma should come in contact with limestone, or should engulf large blocks of limestone, the calcium of the limestone would presumably react with silica from the magma to form pyroxene, so that the remaining liquid would be depleted in silica. Chemically this hypothesis is sound enough, but the geologic evidence is ambiguous: a few bodies of alkalic rocks show the expected relation to limestones, but many do not. The contamination mechanism is capable of almost infinite variation. For example, instead of imagining the addition of rock material to remove silica, one can suppose that excess alkali metal is added directly by reaction of the magma with appropriate sedimentary or metamorphic rocks. For such speculations there is commonly little direct evidence one way or the other.

Perhaps, like so many geologic questions, this one about the origin of alkalic rocks has multiple answers: almost certainly some alkalic lavas are produced by simple differentiation, probably some intrusive bodies result from contamination, perhaps alkalic magma can form by differential melting of appropriate material during metamorphism, and perhaps the streaming-volatile hypothesis is responsible for some concentration of alkalis. A decision in particular cases must usually rest on geologic rather than chemical evidence.

Ultramafic Rocks

The ultramafic rocks pose a different set of problems. These are igneous rocks consisting of a single mafic mineral (for example, pyroxene in pyroxenite, olivine in dunite) or a combination of mafic minerals (for example, pyroxene and olivine in peridotite), with little or no feldspar (analysis VIII, Table 14-1). At first glance such compositions seem easy to explain within the framework of the differentiation hypothesis, for we may suppose that olivine and pyroxene crystallize early, settle out, and collect (to form "cumulate" rocks) in the lower part of a cooling magma. Such a process is clearly responsible for the "olivine layer" near the base of many thick sills and flows (for example, the Palisades sill in northern New Jersey) and provides a reasonable mechanism for the origin of some larger ultramafic bodies. As a general explanation it seems dubious, however, because many ultramafic rocks show clear textural and intrusive relationships which can hardly be accounted for by simple crystal settling. From the geologic relationships alone, it seems necessary to postulate an ultramafic magma with a composition not far different from that of the rocks we observe. Confirmation of the existence of such magmas has come from the recent discovery of rare rocks in Precambrian terrains having ultramafic compositions and the textures of lava flows.

Magmas of this sort pose problems, however, because of the high temperatures required for such material to melt. Experiment shows that most ultramafic mixtures, in the absence of water, could not become liquid at temperatures below 1400 or 1500°C. With a few thousand atmospheres of water-vapor pressure the melting point can be lowered to about 1300°, but even this is higher than practically all recorded lava temperatures. Nevertheless, rocks adjacent to ultramafic intrusives commonly show at most only mild metamorphism, rather than the drastic changes that one would expect from such extreme heating. This is the major conundrum in the study of ultramafic rocks: the conflicting evidence for the existence of ultramafic liquids, on the one hand, and for low-temperature intrusion on the other.

To make matters more complicated, many ultramafic bodies show a partial or complete alteration of the original pyroxene and olivine to serpentine. From experimental work it is known that serpentine cannot exist at temperatures much above 500°, no matter how high the water pressure may be (Sec. 19-5). Above this temperature, material with the composition of serpentine can exist only in the form of solid olivine and/or pyroxene crystals together with water vapor.

Two ways out of this dilemma have been suggested. Perhaps some ultramafic material is intruded as a "crystal mush," an aggregate of solid crystals lubricated by compressed water vapor, with the temperature never rising higher than a few hundred degrees. Serpentinization would result from reaction between crystals and vapor as the mixture cooled. On this hypothesis one might expect that intrusion of the mush into cracks and pockets would result from deformation produced by orogenic forces and that the solid crystals might be broken and sheared during the intrusion, as they often appear to be when ultramafic rocks are studied in thin section.

A second possibility is that some ultramafic rocks result from the crystallization of true magma, but that the extreme temperature produces little effect on surrounding rocks because the magma is accompanied by almost no volatile material. The magma is perhaps the result of local heating of crystals that have settled out from a larger mass of more normal silicate liquid, or possibly it is a sample of the upper part of the mantle, trapped during the downward movement of a subducting plate. One can suppose further that the magma is so dry that it is capable of taking up water from its surroundings, and that this added water is responsible for serpentinization. Movement of water vapor toward the magma would tend to keep the adjacent rocks cool and so prevent the formation of metamorphic minerals.

Both hypotheses may well be correct for particular cases. Whether the two are sufficient to account for all the baffling characteristics of ultramafic rocks remains an open question.

This survey of three unusual kinds of igneous rocks gives an idea of the possible complications of the differentiation process, but it is no more than a superficial introduction to the chemical problems that such rocks present.

PROBLEMS

1 A typical section through the Palisades sill in northern New Jersey shows layers as indicated in Table 14-2. Suggest an explanation for this layering. To establish the correctness of your explanation, what sort of chemical and geologic data would you need?

Table 14-2

	Thickness, ft
Fine-grained diabase	40
Coarse-grained diabase with interstitial material consisting of quartz plus alkali feldspar	150
Medium-grained diabase	800
Layer with abundant olivine	15
Fine-grained diabase	50

2 If a small xenolith of granite is engulfed by a partly crystallized basalt flow, what would you expect to happen to the granite?

3 Explain how the engulfing of large amounts of dolomite might change the composition of a granite magma so that nepheline would crystallize from it.

4 A pegmatite dike consists typically of coarse-grained quartz and alkali feldspar, and presumably forms by crystallization of a water-rich liquid with this composition. Where a pegmatite dike is observed to cut both a sandstone and an ultramafic rock, it is often found that within the ultramafic body the amount of quartz in the dike is greatly reduced and may even be replaced by corundum. Suggest an explanation.

5 The formation of crystals in a basaltic magma can give rise to a residual liquid enriched in silica, from which quartz can crystallize, or to a liquid deficient in silica, from which feldspathoid minerals may crystallize. The differences in composition which cause differentiation to go one way or the other may be very slight. This can be illustrated by a simple calculation based on the analyses in the first two columns of Table 14-3. The figures in the first two columns are divided by the molecular weights of the oxides (third column) to get relative numbers of millimoles (fourth and fifth columns). Assume that TiO_2 unites with FeO to form ilmenite, $FeO \cdot TiO_2$, and that Fe_2O_3 unites with FeO to form magnetite, $FeO \cdot Fe_2O_3$; this removes 63 and 62 millimoles of FeO, respectively. Similarly 6 and 11 millimoles of CaO are used up in forming apatite (roughly, $3CaO \cdot P_2O_5$). These subtractions leave the numbers in the sixth and seventh columns. Now assume that the remaining oxides combine with the maximum amount of SiO_2 as follows: MgO and FeO to form pyroxene, $(Mg,Fe)SiO_3$; the alkalies to form feldspar, $(K,Na)AlSi_3O_8$; the remaining Al_2O_3 with CaO and SiO_2 to form anorthite, $CaAl_2Si_2O_8$; and then the remaining CaO to form pyroxene, $CaSiO_3$. This is not the way differentiation occurs, of course, but it indicates whether or not any silica will be left over after the various cations have reacted with the maximum amount. Go through the calculation to see how much silica is in excess, or how much is lacking to supply the cations, in the two analyses. In the one where silica is deficient, could the deficiency be accommodated if half of the MgO and FeO react to form olivine

Table 14-3†

	Weight percent		Molecular weights	Millimoles/ 100 g		Millimoles to form silicates	
	I	II		I	II	I	II
SiO_2	49.7	50.7	60	828	839	828	839
TiO_2	2.9	3.0	64	45	47		
Al_2O_3	13.7	14.0	102	134	137	134	137
Fe_2O_3	2.9	2.4	160	18	15		
FeO	8.7	11.7	72	121	163	58	101
MgO	8.3	4.8	40	208	120	208	120
CaO	9.1	8.3	56	162	148	156	137
Na_2O	3.2	3.0	62	52	48	52	48
K_2O	1.0	1.3	94	11	14	11	14
P_2O_5	0.5	0.8	111	4	7		
	100.0	100.0					

† I: Hawaiian basalt, average of 56 analyses (Washington), quoted by Barth in "Die Entstehung der Gesteine," Springer-Verlag OHG, Berlin, 1939, p. 66.

II: Oregon basalt, average of 6 analyses (Washington), quoted by Barth, *ibid.*, p. 71. Water and MnO omitted; analyses recalculated to 100%.

rather than pyroxene, and if the olivine settles out of the magma at an early stage? (Note that such a calculation is idealized and can be only approximate at best. Some Ti and Fe^{3+} will go into pyroxene rather than oxides; some Al will go into pyroxene; no allowance is made for the formation of sphene, hornblende, or biotite.)

REFERENCES AND SUGGESTIONS FOR FURTHER READING

Bowen, N. L., "The Evolution of the Igneous Rocks," Princeton University Press, Princeton, N.J., 1928. Chapter 5 is Bowen's statement of his reaction principle, and Chapter 6 is an application of the principle to the crystallization of basaltic magma. Succeeding chapters discuss variation diagrams, the mechanism of crystallization differentiation, assimilation, and various unusual igneous rocks.

Chayes, F., The chemical composition of Cenozoic andesites, *Oregon Dept. of Geology and Mineral Resources Bull.*, vol. 65, pp. 1–11, 1969.

Eggler, D. H., CO_2 as a volatile component of the mantle, *Physics and Chemistry of the Earth*, vol. 9, pp. 869–881, 1975. The effect of various CO_2-H_2O mixtures on melting points and silica saturation in olivine-pyroxene melts.

Green, D. H., Petrogenesis of the high-temperature peridotite intrusion in the Lizard area, Cornwall, *Jour. Petrology*, vol. 5, pp. 134–188, 1964. Unlike most ultramafic intrusives, the one described in this paper has high-grade metamorphic rocks along its borders.

Joesten, R., Mineralogical and chemical evolution of contaminated igneous rocks at a gabbro-limestone contact, *Geological Soc. America Bull.*, vol. 88, pp. 1515–1529, 1977. Field and laboratory evidence for formation of pyroxenite and nepheline syenite in small amounts by assimilation of limestone in gabbro.

Kushiro, I., Effect of water on the composition of magmas formed at high pressures, *Jour. Petrology*, vol. 13, pp. 311–334, 1972. Experimental demonstration that silica-rich magmas like andesite and dacite, as well as basalt, can form by partial melting of peridotite if water is present.

Moorbath, S., and J. D. Bell, Strontium isotope abundance studies on Tertiary igneous rocks from Skye, *Jour. Petrology*, vol. 6, pp. 37–66, 1965. Moorbath and Bell show that the granites of Skye have a relatively high ratio Sr^{87}/Sr^{86}, hence are probably formed at shallow depths, in contrast to the more mafic rocks of the same area whose lower ratio suggests formation at deeper levels.

Nockolds, S. R., Average chemical composition of some igneous rocks. *Geol. Soc. America Bull.*, vol. 65, pp. 1007–1032, 1954. A standard compilation of analyses.

Rose, W. T., N. K. Grant, G. A. Hahn, I. M. Lange, J. T. Powell, J. Easter, and J. M. Degraff, Evolution of Santa María volcano, Guatemala, *Jour. Geology*, vol. 85, pp. 63–87, 1977. A good example of recent studies of lava sequences. Several kinds of variation diagrams are used to interpret major-element chemistry, and inferences about magma sources are drawn from trace-element concentrations and strontium isotope ratios.

Tuttle, O. F., and N. L. Bowen, Origin of granite in the light of experimental studies in the system $NaAlSi_3O_8$-$KAlSi_3O_8$-SiO_2-H_2O, *Geol. Soc. America Memoir* 74, 1958. Experimental work on the silica-alkali-feldspar system at various pressures of water vapor applied to the crystallization of granitic magmas.

Von Platen, H., Kristallisation granitischer Schmelzen, *Beitr. Mineralogie und Petrographie*, vol. 11, pp. 334–381, 1965. An extension of the experimental work of Tuttle and Bowen to systems containing anorthite and biotite, with applications to the origin of granodiorite and quartz diorite and to the granitization of sedimentary rocks.

Williams, H., The geology of Crater Lake National Park, *Carnegie Inst. Washington Publ.* 540, 1942. The last part of this memoir contains a discussion of the chemical evolution of the lavas at Crater Lake.

Wyllie, P. J., "The Dynamic Earth," John Wiley & Sons, Inc., New York, 1971. Chapters 5 and 6 cover the chemical relations of crust and mantle, chap. 8 is an excellent treatment of magma generation based on experimental work, and chap. 14 relates magma formation to plate tectonics. The book is particularly good for its historical account of the development of modern ideas about the chemistry of igneous rocks.

Yoder, H. S., "Generation of Basaltic Magma," National Academy of Sciences, Washington, 1976. The book is an up-dating and amplification of the famous memoir by H. S. Yoder and C. E. Tilley, Origin of basaltic magmas: an experimental study of natural and synthetic rock systems, *Jour. Petrology*, vol. 3, pp. 342–532, 1962. A detailed inquiry into the formation of basaltic magma from mantle material: conditions of melting, processes of melting, and differentiation trends, based largely on the experimental work of the author and his colleagues at the Geophysical Laboratory of the Carnegie Institution.

FIFTEEN

END STAGES OF MAGMATIC CRYSTALLIZATION

Most silicate magmas, as we have seen, solidify over a considerable temperature range. The last material to crystallize consists of a fluid made up chiefly of water plus the constituents of alkali feldspar: sodium, potassium, aluminum, and silica. Commonly, but not universally, silica is in excess of the amount necessary to form feldspar, so that crystallization of the residual fluid gives a solid material with the composition of granite. This kind of water-rich fluid may be produced in small amount by the crystallization of basaltic magma, and in larger amount by crystallization of a granitic or granodioritic magma; it may form also by initial melting of heterogeneous rock material during extreme metamorphism. It is a common geologic fluid and a most interesting one, for it often contains, in addition to water plus silica and the constituents of feldspar, a variety of minor constituents that form curious kinds of rock and some types of ore deposits. To the details of formation and crystallization of such residual fluids we turn our attention in the following pages.

15-1 KINDS OF EVIDENCE AVAILABLE

Here we proceed on a much shakier basis than in our discussion of normal igneous rocks. Laboratory data are less extensive, for the good reason that handling corrosive gases at high temperatures and pressures is experimentally difficult. Theoretical studies are handicapped by the number of variables that must be considered and by scarcity of data from which reasonable ranges of the variables can be guessed. On the observational side, geologic data are far from satisfactory because the materials we find in the field are not solid equivalents of the residual fluids as such, but only a residue left after much of the fluid has escaped into the atmosphere, ground water, or ocean. A good deal of the discussion in this chapter must necessarily be speculative, based in part on general physicochemical principles, in part on experimental data, and in part on the hints we can pick up from geologic observations.

We might expect, to begin with, that the residual fluid from a crystallizing magma would show a concentration of substances of low melting points, the so-called *volatiles*, which were originally dissolved in the silicate melt. The general nature of these volatiles can be inferred from the composition of volcanic gases (Chap. 16), from tiny fluid inclusions in the minerals of igneous rocks, and from the kind of alteration found in wall rocks around an intrusive body. Such evidence suggests that the volatiles may vary a good deal from one body of magma to another, but that they consist predominantly of water with lesser amounts of carbon dioxide, hydrogen chloride, hydrogen fluoride, sulfur compounds, nitrogen, and compounds of boron. In addition to the volatiles, geologic evidence indicates the presence of compounds of metals: chiefly NaCl, with lesser amounts of compounds of potassium, calcium, magnesium, iron, and traces of many others. Some of the minor metals may show strong local concentrations; good examples are copper, lead, and zinc in volcanic sublimates, and beryllium, lithium, uranium, and thorium in some pegmatite dikes. As a generalization, we may conclude that the residual fluid contains material of at least three kinds: substances with particularly stable molecules that are gaseous at magmatic temperatures, compounds that are very soluble in water, and elements whose ions do not fit readily into the structures of growing silicate minerals. These materials, so to speak, are the misfits of the magma, the constituents that for one reason or another cannot readily become a part of normal igneous rocks.

To give the discussion a concrete geologic framework, we shall use the conventional model of a body of magma cooling in a well-defined " chamber " a few kilometers or tens of kilometers below the earth's surface. The residual fluid will be thought of as forming in the interstices between the growing crystals, as collecting into pockets and fissures in the nearly solid mass, and as escaping in part into the rocks enclosing the magma body. At best such a model is tremendously oversimplified. It is not realistic at all for incipient melting during metamorphism, where the fluid we are considering would be more " initial " than " residual." But to avoid tedious repetition, we shall find it convenient to use a single model, a model that has the merits of mechanical simplicity and demonstrable correspondence with at least one kind of geologic situation where residual fluids have been active.

15-2 PHASE RELATIONS

The " Second Boiling Point "

In imagination, then, we consider a body of magma cooling somewhere below the surface. At first, if temperature and pressure are sufficiently high, the magma consists of a single liquid phase. The potentially volatile materials are dissolved in the liquid, some of them chemically attached to silicon-oxygen groups and some as molecules simply wandering in the spaces between the groups. Crystallization starts, and the system now consists of liquid plus one or more solid phases. (A second liquid phase consisting of sulfides is possible, or at extreme compositions even a second liquid silicate phase, but for present purposes these possibilities will be neglected.) As crystallization progresses, the volume of remaining liquid becomes smaller and smaller, and in this decreasing volume the volatiles are concentrated. The question then arises: Will the volatiles, confined in a smaller and smaller volume of liquid, eventually separate as a gas phase? In other words, will the liquid boil?

This is a much trickier question than at first appears. Two factors are operating to change the vapor pressure of the cooling liquid: the vapor pressure tends to decrease because of the cooling, and at the same time it tends to increase because the most volatile constituents are becoming steadily more concentrated. Which factor is more important? Obviously the answer must depend on the amount and nature of the volatiles present. Even if this query could be answered in favor of a net increase in vapor pressure, separation of a vapor phase would not be assured, because boiling occurs only when the vapor pressure exceeds the external pressure. What is the external pressure? Here we run head on into the inadequacies of our model. Are the walls of the magma chamber impermeable and infinitely strong, so that pressure can build up indefinitely? In this case boiling could be prevented; but such a picture, carried to extremes, is clearly absurd. Are the walls of the magma chamber impermeable, but limited in their ability to confine gas by the pressure of overlying rock? In this case boiling would occur if the amount of volatile material is large enough to build up a vapor pressure in excess of the lithostatic pressure (for example, roughly 2,500 atm at a depth of 10 km), and we may have here a possible mechanism for explosive volcanic eruptions. On the other hand boiling would not occur if the amount of volatiles is small, so that the maximum vapor pressure reached before crystallization is complete does not exceed the lithostatic pressure. Are the walls of the magma chamber impermeable to a viscous silicate liquid, but freely permeable to gases? With this condition there would be nothing to prevent separation of a gas phase all during the crystallization, for the only pressure to be overcome is atmospheric pressure at the depth of the magma chamber, or at most the hydrostatic pressure of water in fissures above the magma chamber. Thus boiling in the cooling liquid may or may not occur, depending on a variety of factors which cannot be evaluated even semiquantitatively.

If boiling does happen, note that it is a different sort of phenomenon from ordinary boiling. This boiling is in response to *cooling*, and results from an increase in vapor pressure due to confinement of dissolved gases in a smaller and smaller body of liquid. Ordinary boiling follows an increase in vapor pressure due to simple heating. The temperature at which a hypothetical magmatic liquid might boil on cooling is often called its *second boiling point*, and the phenomenon is spoken of as *retrograde* or *resurgent* boiling.

Solubility of Water in Silicate Melts

Experimental data on silicate-water mixtures give a roughly quantitative idea of the temperatures, pressures, and compositions involved in the separation of a gas phase from a magma. Such experiments are difficult, and many older data are probably not reliable. Work by Burnham and Jahns (1962), summarized in Fig. 15-1, indicates that the amount of water dissolved in a granitic liquid increases indefinitely as the pressure is raised, but that it would not exceed 15% by weight at temperatures and pressures to be expected in a cooling magma. The silicate mixture used in this investigation, obtained from a pegmatite, contains a little more silica and a little less CaO and K_2O than average granite, but exploratory experiments by others show that the solubility of water is not greatly affected by a change in composition even from granite to basalt. If we accept the data of Fig. 15-1 as typical of granitic melts, we can estimate that the residual liquid of an intrusive body at a depth of 10 km would boil when crystallization during cooling has increased its water content to about 7%.

The highest temperature at which retrograde boiling might occur would be the temperature at which crystals begin to appear in a cooling melt, in other words along the liquidus of the silicate-water system. The temperature of the liquidus for the pegmatite of Burnham and Jahns's study is plotted as a function of water-

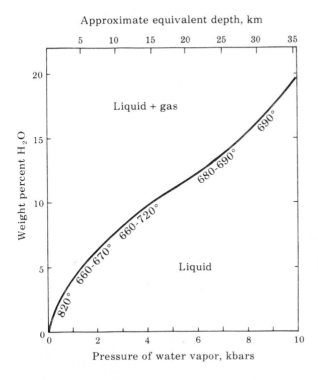

Figure 15-1 Solubility of water in a granitic melt at temperatures slightly above those of the water-saturated liquidus. Numbers along the curve show approximate temperatures at which the experimental points were obtained. The melt used was from a pegmatite at Harding, N. Mex. (*Source: Burnham and Jahns, 1962, page 735.*)

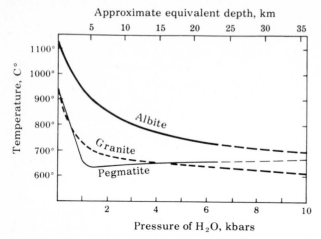

Figure 15-2 Effect of water vapor on the liquidus of Harding pegmatite, the melting point of albite, and the incipient melting temperature of granite. The temperatures for the pegmatite liquidus and the start of melting of granite are close to the temperatures of the ternary minimum in the silica-alkali-feldspar system (Fig. 14-3); the temperature of melting of the single mineral albite is considerably higher. *(Sources: For albite and pegmatite, Burnham and Jahns, 1962, page 732. For granite, Tuttle and Bowen, 1958, page 83.)*

vapor pressure in Fig. 15-2. Besides the pegmatite, the diagram includes a melting-point curve for albite and an incipient melting curve for granite. The effect of water on the melting behavior of these substances is very striking. Note, for example, that dry granite does not begin to melt until the temperature rises above 900°, but in the presence of 1 kilobar (approximately 1,000 atm) of water vapor, incipient melting occurs near 700°.

Separation of a water-rich gas phase from a magma can be brought about by a change in pressure as well as by concentration during crystallization. Again a rough quantitative estimate can be made from Fig. 15-1. If, for example, a granitic magma at 700°C containing 10% dissolved water moves upward from a depth of 20 km to a depth of 4 km, it must lose more than half of its water, since at 1,000 atm the solubility is reduced to about 4%.

Water Vapor as a Solvent

We are describing the vapor that separates as pure water, but actually this is too simple. At the temperatures and pressures of Burnham and Jahns' experiments the water is well above its critical point (see next section) and is so greatly compressed that its density is comparable with that of ordinary liquid water. Under such conditions gaseous (or supercritical) water has considerable solvent ability, so that the "gas" which appears in the experiments must contain appreciable silica, sodium, and potassium. At first sight this seems odd, because we are accustomed to thinking of water vapor as having no solvent power whatsoever; in fact, we often separate nonvolatile material from water by heating to drive off water vapor, never giving a moment's thought to the possibility that some of the material might dissolve in the vapor and escape. But conditions of the experiments we are considering here differ radically from ordinary laboratory conditions, and a measurable solubility of silica and silicates in gaseous water at high temperatures and pressures is confirmed by a variety of observations. An accurate description of the phases in the experimental system, then, would include one or more crystalline solid phases, a silicate liquid phase containing a few percent of dissolved water, and a fluid phase consisting of a supercritical water-rich solution.

In nature the situation would be further complicated because the fluid phase that separates during cooling or crystallization would contain not only minor amounts of dissolved silica and silicates but much larger quantities of the substances mentioned in the introductory paragraphs as constituents of volcanic volatiles. The composition of such a high-temperature gaseous solution can only be guessed, but evidence from volcanic emanations and from fluid trapped in tiny cavities in igneous minerals suggests that the solution may be fairly concentrated, particularly in sodium chloride. With such a solution we can no longer be certain that it is above its critical point, since dissolved salts are known to raise the critical point of water. If it happens, either initially or in the course of later cooling, that the temperature of the gaseous solution drops under its critical point, a further separation of phases becomes possible. In addition to the solid phases we can imagine now three fluid phases: the remaining viscous silicate melt, a water-rich liquid containing abundant dissolved salts, and a water-rich vapor. To explore the possibilities more fully, we digress a little to consider the general phenomenon of the critical point in multicomponent systems.

15-3 THE CRITICAL POINT

Definitions

Carbon dioxide under pressure in a stout-walled glass tube shows a clear separation of liquid and vapor. If the tube is warmed, the meniscus separating the two phases becomes thinner and less curved; at 35°, quite suddenly, it vanishes altogether. If the tube is then cooled, the meniscus reappears at the same temperature. We say that carbon dioxide has a *critical temperature* of 35°, meaning that above this temperature two fluid phases cannot exist. The minimum pressure necessary to maintain a liquid at or just below the critical temperature is the *critical pressure*. Above the critical temperature no amount of pressure can cause the appearance of a liquid. The *critical volume* is the specific volume at the critical temperature and the critical pressure.

Nomenclature is confusing about the phases just below and just above the critical point. A substance immediately above its critical point cannot properly be described as either a "gas phase" or a "liquid phase," since it has the ability of the former to expand indefinitely and the close molecular spacing of the latter. The term "fluid" is applicable, but this term also refers to ordinary liquids and vapors—to any material capable of flow under infinitely small forces. Some authors restrict the word "gas" to substances above their critical temperatures, and "vapor" to low-density phases below their critical points; more commonly the two terms below the critical point are used interchangeably. For the present discussion a substance above its critical temperature will be referred to loosely as a "gas" or a "fluid," or specifically as a "supercritical fluid" when the meaning is not clear from the context.

Note that a supercritical fluid may have any density, from that of a rarefied gas to one comparable with the density of the low-temperature liquid. The ability of hot, compressed gases to dissolve nonvolatile substances, referred to in the last section, applies only to supercritical fluids of high density. This, of course, requires high confining pressure. Nothing about the supercritical state itself causes greater solubility; a tenuous supercritical gas has just as little solvent power as the vapor of the substance near its boiling point. But high-density supercritical fluids have molecular spacings equivalent to those in a liquid, and they would be expected to show similar dissolving ability. To repeat: it is not primarily high temperature, but high pressure, that makes non-volatile materials soluble in water vapor escaping from a magma.

Effect of Solutes

The critical point of a substance is changed by the addition of solutes, just as its boiling point and freezing point are changed. Experimentally, solutes of greater volatility are found to lower the critical point, whereas less volatile solutes, with a few exceptions, are found to raise it. For example, a 2% solution of NaCl in water (approximately $0.4M$) has a critical point of 399°, and a 5% solution one of 424°, in contrast to 374° for pure water. These points are plotted on Fig. 15-3, which is an extension of the upper right-hand corner of Fig. 13-1. In both diagrams point C, at the upper end of the vapor-pressure curve, is the critical point for pure water (374°C, 224 kg/cm^2). The vapor-pressure curve for water does not continue beyond this point, because at higher temperatures and pressures there is no distinction between liquid and gas. The three lighter solid lines on Fig. 15-3 are similar vapor-pressure curves for NaCl solutions of different concentrations; these solutions have lower vapor pressures and higher critical points than pure water. The upper light dashed line extending beyond C is drawn through the critical points of increasingly concentrated NaCl solutions. The field above this line represents a single fluid phase, and the field below the line shows equilibrium between unsaturated solutions and vapor. The lower dashed line is the vapor-pressure curve for saturated solutions, in other words the line representing three-phase equilibrium between solid salt, solution, and vapor. To show the relations in full detail would require a three-dimensional diagram, with NaCl/H$_2$O ratios as the third variable.

In a partly frozen magma the various solids in equilibrium with residual fluids would be present in large amounts. Hence a pertinent question to ask about the

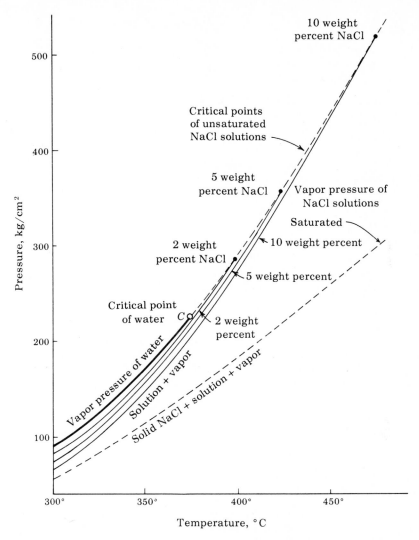

Figure 15-3 Vapor-pressure curves and critical-point curve of NaCl solutions. *(Source: Ölander and Liander, 1950.)*

experimental data for NaCl is this: Suppose that solid NaCl is present in great excess as the temperature and pressure are raised beyond the limits shown in Fig. 15-3, so that the solution becomes indefinitely more concentrated; would this saturated solution, in continuous contact with the solid, ever reach a critical point? In other words, with solid NaCl present, would we ever find a point where liquid and vapor approach the same composition? From the fact that the critical-point curve and the saturation curve (the two dashed lines) on Fig. 15-3 diverge so sharply, we might guess that the answer would be negative. Fortunately, for this system we have enough experimental data to check our guessing (Ölander and Liander, 1950; Sourirajan and Kennedy, 1962).

The data for the temperature range 350 to 700°C and for pressures up to 1,300 bars are shown by the heavy solid lines on Fig. 15-4, which are extensions of the two dashed lines on Fig. 15-3. The two curves are extrapolated beyond the experimental data by light lines, which are not drawn to scale. On the left-hand side of the drawing is shown the one-component diagram for pure water, identical with Fig. 13-1; on the right-hand side is the corresponding (hypothetical) diagram for pure NaCl. Actually these should be in different planes: the drawing is a projection of a three-dimensional figure, with NaCl-H_2O compositions as its third axis; the one-component diagram for H_2O may be considered to lie on the forward face of the figure, and the NaCl diagram on the back face; the two solid lines (CZ and TEU) are lines in space between the two faces. The upper solid line CZ connects critical points of unsaturated solutions, with solutions rich in water near the left-hand end at C and those rich in NaCl near the critical point of pure NaCl at Z. The lower solid line TEU connects the triple points of the two pure substances and, like the triple points, represents equilibrium between two fluid phases and one solid phase. (This line is the edgewise view of a curved surface in the three-dimensional figure, one side of the surface representing compositions of a liquid solution and the other side compositions of a gaseous solution.) Although experimental data are lacking for the extreme NaCl side of the diagram, the existing data show clearly that the vapor-pressure curve for saturated solutions goes through a maximum and hence does not intersect the critical-point curve. This means that at *any* temperature two fluid phases can coexist. In other words, a critical point is not approached so long as solid NaCl is present.

Substances much less soluble than NaCl behave differently. Silica, for example, raises the critical point only slightly, even when excess solid is present, simply because the solubility is very low at temperatures around 400°. This situation is shown diagrammatically on the left-hand side of Fig. 15-5. The drawing is not to scale, because the vapor-pressure curve for saturated silica solutions (EQ) is so close to the vapor-pressure curve for pure water that the two would be indistinguishable on any reasonable scale, and the critical points of all silica solutions up to saturation are within a fraction of a degree of the critical point of water. In this system the critical-point curve CQ intersects the vapor-pressure curve at Q, the "critical end point"; in other words, at Q the vapor pressure becomes equal to the critical pressure (or the composition of the "solution" becomes identical with the composition of the "vapor"), and the distinction between liquid and gas vanishes. Hence above the temperature represented by Q only a single fluid phase can exist in contact with solid silica.

The right-hand side of Fig. 15-5 is also of interest. This is the high-temperature end, the end showing mixtures with much silica and only a little water. Again we use a two-dimensional projection of a three-dimensional figure; the one-component curves for pure silica (hypothetical) may be considered to lie on the back face, and the solid lines (ZR and UR) extend out into space toward the H_2O (front) face. The heavy solid line shows experimental data (Kennedy et

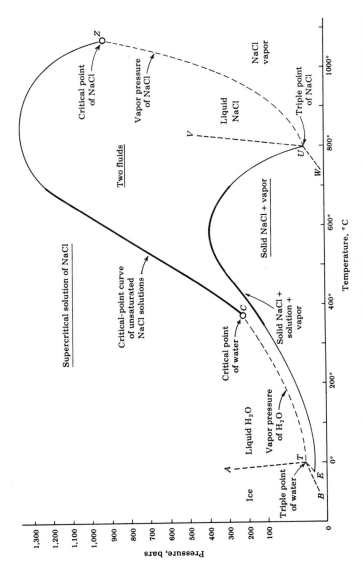

Figure 15-4 Pressure-temperature diagram for the system NaCl-H₂O. This is a projection of a three-dimensional diagram, with NaCl/H₂O ratios as the third axis. Dashed lines show one-component plots for H₂O (front face of the three-dimensional figure) and NaCl (back face); these are not drawn to scale. Solid lines (heavy lines drawn to scale from experimental data, light lines hypothetical and not to scale) are curves in space connecting the end faces. Underlined designations of phases refer to regions in space; designations without underlining show phases of pure water and salt on the end and planes of the three-dimensional figure. (*Source: Experimental data from Sourirajan and Kennedy, 1962; hypothetical reconstruction of ends of diagram after Morey, 1957.*)

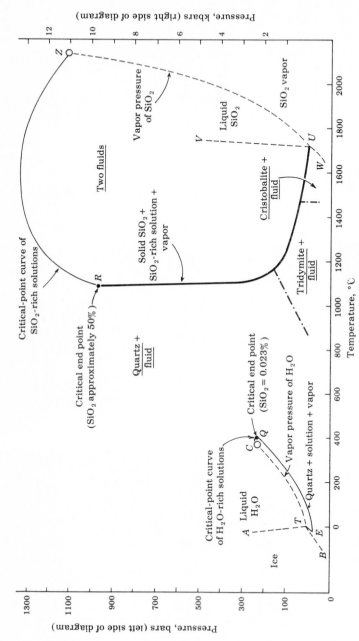

Figure 15-5 Pressure-temperature diagram for the system SiO_2-H_2O. This is a projection of a three-dimensional diagram, with SiO_2/H_2O ratios as the third axis. Dashed lines show one-component plots for H_2O (front face of the three-dimensional figure) and SiO_2 (back face); these are not drawn to scale. Solid lines are curves in space connecting the end faces. The heavy solid line and the dash-dot lines are drawn approximately to scale from experimental data; the light solid line on the right-hand side is hypothetical and not to scale; the position of the light solid line on the left-hand side is known from experimental data but is displaced on the drawing for clarity. Underlined designations of phases refer to regions in space; designations without underlining show phases of pure water and silica on the end planes of the three-dimensional figure. (*Sources: Experimental data from Kennedy et al., 1962; hypothetical reconstruction of the diagram as a whole after Morey, 1957.*)

356

al., 1962), while the light lines are hypothetical and not to scale. Both the critical point and the melting point of silica are lowered by the addition of water; the sharp reduction of the melting point, from 1720° to less than 1200° by a water-vapor pressure of less than 2 kilobars, is particularly striking. The experimental work demonstrates the existence of a critical end point (point *R*), where the compositions of liquid and vapor in equilibrium with solid silica become identical. The critical-point curve for silica-rich solutions must extend from this point to the (unknown) critical point of pure SiO_2, shown diagrammatically at *Z*. In contrast to the $NaCl-H_2O$ system, neither the critical-point curve nor the vapor-pressure curve for saturated solutions is here continuous from one end member to the other. In other words, saturated solutions near both ends of the series have critical points, and over a large range of temperature and pressure only a single fluid phase can exist in contact with solid quartz.

In summary, then, critical-point relations in a *saturated* solution of a nonvolatile substance may follow one of two patterns. A solute may become indefinitely more soluble as the temperature is raised, so that a continuous series of solutions exists from pure water to pure solute (for example NaCl); in this case no critical point can be reached as long as excess solute is present, and two fluid phases can coexist at any temperature. Or a solute may show only limited solubility, in which case the solution has a critical point slightly higher than the critical point of pure water (for example, SiO_2); at temperatures above this point, and below the corresponding point for limited solubility of water in the molten solute, only one fluid phase can be present in contact with excess solute. Presumably most substances with high solubility, especially substances whose solubility increases rapidly with rising temperature, would show the first type of behavior, but quantitative predictions are not possible from present experimental data.

The Critical Point in Magmas

If the effect of simple salts on the critical point is impossible to predict, the effect of the complex mixture in a magma would seem completely hopeless. Here we must deal not only with many nonvolatile substances, but with a variety of volatile solutes as well. Will such a mixture be above its critical point at some stage of the cooling process (second case above), or will it stay below the critical point at all temperatures (first case)? No answer is possible on theoretical grounds, and the experimental data are not conclusive.

Burnham and Jahns' experiments, on the one hand, indicate that water and the major constituents of granite have only limited miscibility. In other words, a granite-water mixture behaves like a silica-water mixture, fitting the second case described above—as we might perhaps expect, since the principal constituents of granite are the very insoluble quartz and aluminosilicates. There is the added complication here that the greater number of components makes it possible for a liquid silicate phase to exist in addition to one or more solid phases, but between the gaseous water-rich phase and either a solid-solid or solid-liquid mixture, there is a clean separation. The gas phase can contain only small amounts of silica and the components of feldspar, and in an ideally simple cooling melt it would necessarily be above its critical point.

On the other hand an abundance of NaCl or other soluble salts in the original melt might lead to a separation of a concentrated gaseous solution that would be below its critical point (first case above) even in the temperature range of a crystallizing magma, 600 to 700°C. The presence of volatiles like HCl and CO_2 would have an opposite influence on the critical point, so that predictions become mere speculation.

An additional complication is the experimental fact that the simple alkali silicates K_2SiO_3 and Na_2SiO_3 are very soluble at high temperatures and affect the critical point in the same way as does NaCl. It seems possible that these substances might be present in a melt, over and above the amounts needed to combine with Al_2O_3 in the formation of feldspar; if such an excess were present originally, it would leave no record in the rocks ultimately formed, since the excess would simply be dissolved in ground water and removed. A sufficient excess of Na_2SiO_3 and K_2SiO_3, as has been experimentally demonstrated by Tuttle and Bowen (1958), would make water far more soluble in the melt and would permit the melt to remain liquid down at least to temperatures near 400°. In this case a water-rich phase might not separate at all, the composition of the melt simply changing continuously during crystallization from a viscous, silica-rich liquid at the beginning to a mobile, water-rich liquid at the end, so that only a single fluid phase would be present during the entire process. Such an extreme is hardly likely, but even a small excess of alkali silicates would complicate the phase relations.

Thus a gamut of possibilities can be envisioned: (1) A water-rich phase may separate from the silicate melt during crystallization, containing so much dissolved salt that it is below its critical point to start with or becomes so as the temperature falls; in this case a maximum of three fluid phases can coexist with the crystallizing silicates. (2) A water-rich phase may separate, either so dilute or mixed with so much volatile material that it remains above its critical point until long after crystallization is complete; in this case the maximum number of fluid phases is two. (3) A water-rich phase may fail to appear, either because the amount of water dissolved originally was so small that it could be accommodated in the crystallization of micas and amphiboles or because excess alkali silicates increased its solubility in the melt; in this case no more than one fluid phase can be present. Which of the three possibilities is most likely depends on the amounts of water, salts, and volatiles in the original melt, and these are generally not determinable. Quite conceivably all three possibilities may be represented in different magmatic bodies.

Significance of the Critical Point

Why is the existence or nonexistence of the critical state in residual magmatic fluids important? Two reasons may be suggested. The first, now largely of historical interest, is the idea that supercritical fluids may have unique properties, particularly unique ability to dissolve and replace minerals. In the older literature of geology, processes supposedly dependent on supercritical fluids were often given a special name, "pneumatolytic," in contrast with "hydrothermal" processes, which were thought to be limited to subcritical conditions. Modern experimental work has shown that such a distinction is illusory, since there is no discontinuity in the solvent action of a substance on passing through its critical region. Supercritical fluids may indeed be powerful solvents, but as we have noted earlier, their solvent ability depends simply on their pressure and temperature and not on their supercritical condition as such.

The second and more important reason for concern with critical phenomena lies in the number of separate fluids that may be imagined to form during the late stages of crystallization of a granitic magma. From observations near granite contacts, four fairly distinct geologic effects are commonly ascribed to residual fluids: formation of pegmatite dikes (coarse-grained quartz-feldspar rocks), formation of aplite dikes (fine- to medium-grained, sugary-textured quartz-feldspar rocks), formation of quartz-sulfide veins, and alteration of wall rocks. The alteration effects may perhaps be due to the same solutions that form either pegmatites or veins, but to explain pegmatites, aplites, and sulfide veins as products of the same solution is difficult. An explanation for the radically different compositions and textures becomes much easier if two or more different fluids can be called on, and the most obvious way to obtain different fluids is by a phase separation. The separation of fluid phases, either by simple vaporization or by retrograde boiling, is possible only if the initial fluid is below its critical temperature; the "initial fluid" in this sense can refer either to the original silicate melt (case 2 of a preceding paragraph), or to a separated fluid rich in water and salts (case 1). Thus the possible kinds of explanation we can use for late magmatic phenomena are severely limited by the information we can get, or the guesses we can make, about the presence or absence of supercritical fluids in complex crystallizing melts.

In the following sections we shall use this discussion of the critical point as a background in considering the origin of pegmatites, aplites, and veins.

15-4 PEGMATITE AND APLITE DIKES

Typical pegmatites and aplites have a similar mineral composition: chiefly quartz and alkali feldspar, often a little biotite or muscovite, less commonly hornblende. The difference between them is textural, pegmatites having crystals much larger than those in ordinary granite and typically showing much variation in texture and composition, aplites having a sugary texture superficially resembling sandstone and remaining monotonously uniform over long distances. Pegmatites are by far the more spectacular of the two, their crystals sometimes attaining enormous size and showing conspicuous segregation into quartz-rich and feldspar-rich zones. Pegmatites are notable for the common occurrence in them of minor black tourmaline and pink garnet; a few contain large amounts of much less common minerals, minerals of a characteristic group of elements which includes lithium, beryllium, niobium, tantalum, tin, rare earths, uranium, thorium, tungsten, and zirconium. The word "pegmatite," although generally referring to dikes with quartz and alkali feldspar as principal constituents, is not restricted to such compositions but can be used for similarly coarse-grained varieties of any igneous rocks. Nepheline-syenite pegmatites, for example, contain chiefly nepheline and feldspar, and gabbro pegmatites consist of coarse-grained plagioclase plus hornblende. In general, pegmatites have a lesser content of mafic minerals and a more sodic plagioclase than the igneous rocks with which they occur.

Pegmatites and aplites are commonly found as dikes cutting intrusive igneous rocks, often extending beyond the contacts into adjacent metamorphic rocks. In addition to dikes with parallel walls, they may form pockets and irregular masses within an intrusive. Pegmatites also are common as pockets and stringers in metamorphic rocks far from intrusive contacts. Aplite and pegmatite are often found together in the same dike, sometimes separated by a sharp contact and sometimes grading into one another.

From their composition and mode of occurrence, these two rocks have long been regarded as material crystallized from the residual liquid of a silicate magma. The coarse texture of pegmatites has commonly been ascribed to crystallization from a residual melt particularly rich in water. This sounds like a reasonable explanation, but it seems to run counter to the experimental results described above showing that the amount of dissolved water in granitic melts is limited to 10 to 15%. How can the postulated water-rich residual fluid evolve, if water necessarily boils out of the melt before crystallization has progressed very far? Two possible answers have been suggested by experimental work.

One answer depends on the demonstration by Tuttle and Bowen, as mentioned above, that the amount of water miscible with a silicate melt is much greater if the melt contains alkalis and silica in excess of the amounts needed to form feldspar. By this mechanism the residual melt can be imagined to become increasingly fluid and water-rich as it cools, and to remain liquid down to temperatures below 600°C. Hence pegmatites would form very late in the cooling history, from a liquid so mobile that large crystals could grow easily. A difficulty is the lack of evidence of excess alkali in and around pegmatites, but this can be plausibly explained by the great solubility of the simple alkali silicates and their consequent removal in solution.

A second answer, probably the more likely one, is provided by some experiments of Jahns and Burnham (1969). In a typical experiment, a mixture of quartz, K-feldspar, and Na-feldspar is heated to melting with a measured amount of water slightly less than that necessary for saturation, and allowed to cool slowly. Crystallization from the homogeneous liquid begins with the separation of a fine-grained quartz-feldspar aggregate on the walls of the container. The crystallization causes increasing concentration of water in the remaining liquid, and eventually conditions for retrograde boiling are reached. A water-rich phase separates in bubbles, and the character of the crystallization changes abruptly. Relatively large, well-formed crystals of quartz and feldspar grow out into the bubbles, becoming larger as the proportion of the water-rich phase increases. Crystallization takes place chiefly from the viscous, silicate-rich phase, but the existence of a water-rich fraction provides space for the crystals to grow. Ultimately a crystalline solid is produced showing a gradation from sugary, aplitic-looking material on the outside to coarse-grained material with the texture of pegmatite at the center. The experiment can be modified by changing the original amount of water and by maintaining different pressures during crystallization. If the pressure is high enough at all times to prevent the formation of an aqueous phase, the crystalline product has the fine-grained texture of aplite throughout. If the proportion of water is large and if the pressure permits boiling to occur early in the experiment, the relative amount of "aplite" with respect to "pegmatite" is small. By varying pressure and temperature during crystallization, Jahns and Burnham were able to produce zoning and replacement relations like those in natural pegmatites. The resemblance of some of the experimental products to natural rocks is very striking.

The mixtures used by Jahns and Burnham differ from the melts that form natural pegmatites in at least one important respect, that they do not contain the other solutes—for example, HCl, HF, CO_2, NaCl, and alkali silicates—which probably accompany a residual magma in nature. Nevertheless, the similarity of the mixtures to natural residual fluids is sufficiently close to make a generalization reasonably safe, that aplites represent a residual melt which crystallized without separation of an aqueous phase, and pegmatites a melt in which two phases were present while crystals formed. The two kinds of material may eventually form completely independent dikes, depending on volatile content and pressure-temperature relations at various stages of cooling, or boiling may occur irregularly during cooling so that pegmatite and aplite form immediately adjacent to one another. Pegmatites that occur in metamorphic rocks at a distance from any intrusive may represent the water-rich fluid that forms by incipient melting of the original rock. Thus Jahns and Burnham's work has provided a mechanism, based solidly on experiment, which explains satisfactorily the principal features of aplites and pegmatites observed in nature.

15-5 QUARTZ-SULFIDE VEIN AND REPLACEMENT DEPOSITS

More difficult to account for is the separation from a magma of the material that ultimately leads to the crystallization of quartz-sulfide veins. The composition of such deposits shows a wide range, from pure quartz to nearly pure mixtures of metallic sulfides. The sulfides may be simple ones like pyrite and galena, or complex sulfosalts like enargite and jamesonite. Vein deposits may contain carbonates, together with quartz or practically to the exclusion of quartz, and metals may occur as native elements (for example, gold), oxides (for example, pyrolusite, cassiterite), or carbonates (for example, rhodochrosite) as well as sulfides. For brevity the deposits will be referred to as "veins," but the same associations of minerals occur also as widespread disseminations and replacement deposits. Such vein and replacement deposits make up one of the most important classes of metallic ores.

Many quartz-sulfide veins and areas of replacement occur in and adjacent to igneous intrusives, so that a common origin seems clear. On the other hand a considerable number of such deposits are found at long distances from intrusives, some even in areas where no sign of igneous activity is apparent; the lead-zinc deposits of the Mississippi Valley are a good example. It is by no means certain, therefore, that all quartz-sulfide veins have an origin connected with igneous activity, but enough of them do show an intimate association with intrusives to suggest that the cooling of a silicate melt is at least one possible method of formation.

Because vein and replacement deposits have compositions very different from those of igneous rocks, because their minerals generally show a clear order of crystallization, with the latest ones to form often projecting out as euhedral crystals into open cavities, and because the temperatures of formation, as determined in a variety of ways, are far below the temperatures at which their constituents could be molten, these deposits are almost certainly formed by crystallization out of a hot aqueous solution rather than by freezing of a residual silicate melt. This is a distinction that cannot be pressed too far, because pegmatites also are deposited from solution in the sense that their material was once mixed with water and other volatiles which have disappeared; but the *amount* of water present, relative to the amount of material deposited, must have been far greater in the quartz-sulfide veins. This difference is the reason that we commonly speak of pegmatite "dikes," as opposed to quartz-sulfide "veins," although this distinction in nomenclature is not universally followed.

If we assume that quartz-sulfide veins, like aplites and pegmatites, can be a product of the later stages of magmatic crystallization, what is the relation between the three kinds of material? In the hypothetical sequence of events during the transformation of a homogeneous silicate melt into solid igneous rock, at what point do the solutions form that will ultimately deposit quartz plus sulfides of the common ore metals?

The simplest assumption, based on the line of thought in the last section, would be to suppose that a fluid analogous to the water-rich fluid of Jahns and Burnham's experiments might be the precursor of vein deposits. Perhaps the residual liquid of differentiation separates into two fluids after part of its quartz and feldspar have crystallized to form aplite; when the two fluids are present together, further crystallization of the silicate-rich part gives pegmatite; and the remaining water-rich phase, on further cooling, might deposit quartz and sulfides. Ore deposits are often widely separated from pegmatites, so the fluid would have to be capable of traveling far from its point of origin; but this is not unreasonable, since the ore-forming fluid would be much richer in water and hence presumably more mobile than the pegmatite-forming fluid which it leaves behind.

Attractive as this hypothesis looks, it is not very successful in explaining geologic relationships. One could reasonably expect, for example, that sometimes the water-rich fluid would not escape from the silicate-rich fluid, and that its content of sulfides would be deposited with the pegmatite minerals. As a matter of observation, however, sulfides do not commonly occur with pegmatites in more than minor amounts. Pegmatites may contain ore minerals, but they are largely nonsulfide minerals of a special group of metals—lithium, beryllium, tin, uranium, thorium, niobium, tantalum—quite different from the usual associations in vein and replacement deposits. Even more damaging evidence against the hypothesis is the lack, or at least the great scarcity, of observed transitions between pegmatites and quartz-sulfide veins. Surely, if pegmatites and veins are formed from different fractions of the same original fluid, gradations between the two kinds of material should be reasonably common; yet one of the conspicuous observational facts about ore deposits is that pegmatites and quartz-sulfide veins, where they do occur together, show sharp contacts. Gradational contacts have been looked for, and have occasionally been reported in the literature; but in nearly every case a later, more careful examination has shown that the supposed gradation does not exist. The field evidence makes it clear that the fluids responsible for pegmatites and quartz-sulfide veins must be sufficiently separated, either in origin or in timing, so that the two kinds of deposits cannot intergrade.

Do phase separations offer any further possibilities, which might take care of this difficulty? In the previous discussion of the critical point, one conceivable set of circumstances was described (alternative 1 on page 358) in which three fluid phases rather than two might be present during part of the cooling history. This requires that the water-rich phase from the original separation contain so much nonvolatile dissolved material that its critical point is a hundred degrees or more above the critical point of pure water. Such a fluid, when cooled through its critical point, could separate into two phases, both of them water-rich but one of them now a fairly tenuous vapor with far less dissolved salt than the other. Could such a vapor be the material which, after drastic cooling and condensation, might constitute the fluid responsible for quartz-sulfide veins?

This hypothesis has the advantage that the vapor at high temperatures would presumably be much less viscous than either the aplitic or pegmatitic fluid, hence might separate from them completely. Furthermore, most of its dissolved material would probably remain in solution until the temperature had dropped several tens or hundreds of degrees, so that transitions to pegmatites would be unlikely. One can expand the hypothesis by supposing that a subcritical vapor of this sort might be capable of separating from the original magma at any time, perhaps locally where pressure is drastically reduced, and would not be limited to places where aplite and pegmatite had formed previously; this would account for ore deposits adjacent to intrusives where pegmatites and aplites are scarce or absent. But the hypothesis is little more than free speculation, since there is no good field evidence for the supposed sequence of separations, and no proof from experiments that the third fluid would be able to transport the common heavy metals in sufficient amounts to form ore deposits.

Thus the ultimate origin of ore-forming fluids from a cooling magma remains obscure. If, as suggested in the above speculative reconstruction, they separate first as high-temperature, high-density gases, their later history must involve drastic cooling and partial condensation, since evidence seems convincing that actual deposition of most quartz-sulfide ores is from a liquid rather than from a gas. Liquid aqueous solutions with this history, at temperatures in the range 100 to 300°C, are difficult to distinguish from hot underground waters of other origins: heated ground water containing metals leached from the rocks it has traversed, water derived from volcanic activity, or water set free during progressive metamorphism of sedimentary and igneous rocks. The solutions responsible for ore formation may come from any of these sources and very likely are often mixtures from more than one source.

The problem of distinguishing minerals formed by water with a source in cooling magma from those formed by heated ground water has been attacked most successfully by study of the heavy isotopes of oxygen (Sec. 21-4). Ground water is almost entirely meteoric water, meaning that it has its source in rain, and in rainfall the ratio of heavy to light oxygen isotopes (chiefly $^{18}O/^{16}O$) has a narrow range of values distinctly smaller than the ratio in the primary minerals of igneous rocks. This ratio measured in minerals of veins or of rocks that have been altered by hot solutions thus serves as a "fingerprint" for identifying the source of the water, or for estimating how much of the water was of meteoric origin. In many ore deposits so studied, the ratio has turned out to be low, meaning that the water responsible for the deposits was in large part meteoric. This suggests that much postulated "magmatic" water may be ground water that has been incorporated into the magma at depth, has mixed with whatever dissolved water the magma contained, and then has risen through the magma to the relatively shallow levels where veins are formed. In other words, a magma body may be pictured as a source of heat, setting up a convectional system in its surroundings in which ground water moves toward the magma at depth, is heated and rises through (or with) the magma into the upper part of the crust, and then moves out of the magma as a part of the solutions that form veins and alter the adjacent rocks. Whether water that has had this history should be described as "magmatic" water, since it has been for some time part of the magmatic fluid, is a nice semantic question; it is certainly not a residual fluid left over from magmatic crystallization. A more substantial question, to which there is still no satisfactory answer, is whether the metal content of quartz-sulfide veins and replacement deposits is derived largely from the sedimentary rocks through which the ground water moved before entering the magma, or from the cooling magma itself.

This discussion has been limited to generalities about the possible relations between processes of magmatic crystallization and ore deposits. More specific questions about the behavior of individual metals in high-temperature gaseous and liquid solutions will be touched on in Chap. 17.

15-6 RECAPITULATION

The formation of igneous rocks from molten silicate mixtures is a part of geology to which physical chemistry has been broadly and successfully applied. The sequence of crystallization of minerals and the variety of rocks to which this sequence can give rise are strictly chemical problems—conditioned, of course, by the setting under which the crystallization occurs and by mechanical accidents that may disturb the course of cooling. Laboratory experiments with grossly simplified silicate systems show enough resemblance to natural systems that the process of crystallization can be described in terms of a few physicochemical principles. Chief of these principles is the phase rule, which specifies a limitation on the number of solids that can coexist at equilibrium in a system of given composition. Add to the phase rule a few generalities about the melting behavior of mixtures and solutions (which, if desired, can be expressed in the formal language of free energies and entropies of melting), add a few specific data on the melting points of individual compounds, and the main features in the behavior of igneous melts can be summarized neatly and succinctly. We say that we can *explain* the formation of igneous rocks, with an elegance not even approached in most parts of geology.

But many peripheral problems regarding igneous rocks do not yield so readily to a chemical explanation. First there is the baffling similarity between rocks that have moved and intruded other rocks as molten masses, and rocks that have been altered in texture and composition, without much bodily movement, by the processes of metamorphism. Then there are the odd varieties of igneous rock—the alkalic rocks, the ultramafic rocks, and many others less easily classified—which do not fit into the "normal" scheme of crystallization differentiation without additional assumptions. Finally there is the tremendous problem of the volatiles and their role in the final stages of crystallization.

In attacking the problem of the volatiles, we have faced some geochemical difficulties of an all-too-common sort. In the first place the problem is poorly defined: we have only vague and general ideas about the nature and amounts of the volatiles which constitute our problem. Perhaps it would be more accurate to say that we only infer the existence of the volatiles from geologic evidence which is hard to decipher because most of the materials that formed the evidence have long since vanished. In the second place we have no more than shrewd guesses about the conditions of temperature, pressure, and mechanical nature of the environment in which our volatiles are supposed to be produced and to operate. Thirdly, basic thermodynamic data for the probable range of conditions with which we must deal are far from adequate. We are confronted with a large number of uncontrolled variables and with ranges of variations over a very wide field— conditions that are the antithesis of the carefully controlled environments attainable in a chemical laboratory. Small wonder that here the rules of physical chemistry can help us only a little. All that we can do, all that we have done in this chapter, is to imagine possible hypotheses within the broad limits of the geologic evidence, and then discard hypotheses that lead to patent discrepancies with chemical data. This still leaves us with many alternatives. We cannot claim to have anything like a single concise description based on general principles, to explain how a magma behaves in the final stages of cooling.

PROBLEMS

1 What conditions are necessary for a liquid to exhibit a "second boiling point"? Could this phenomenon be demonstrated with a single pure substance? With a mixture of only two substances? With a mixture of three substances?

2 Consider a hypothetical body of granitic magma, with horizontal dimensions 100 by 100 km and a thickness of 10 km, situated in the crust so that its top is 10 km below the surface. (*a*) What would be the pressure at the top of the body due to the weight of overlying rock? (*b*) What would be the pressure at the base of the magma? (*c*) At the level of the top of the body, what would be the temperature be if the hot magma were not present and if the temperature were determined only by the normal geothermal gradient (measured values lie between 10 and 50°C/km)? (*d*) At approximately what temperature would the last liquid of the magma solidify, assuming that the granite is made up of approximately equal parts of quartz, Na-feldspar, and K-feldspar and that it is under water-vapor pressure equal to the hydrostatic pressure? (*e*) If the magma contains 1% of dissolved water by weight, what weight of water is present? (*f*) Calculate the volume of this weight of water at 700°C and 1,000 atm, using the simple gas laws. (*g*) Calculate the volume under the same conditions using the measured value of the density, 0.28 g/cm^3.

3 The critical pressure of water is approximately 220 atm. At what depth in the crust is all water in a supercritical state (i.e., at a supercritical pressure, but not necessarily at a supercritical temperature)?

4 If water is confined in a cylinder under a movable piston, so that no air space is left, it is possible to convert the water from a liquid at 25° and 1 atm to water vapor at 25° and 0.01 atm without at any time having a separation of two phases. Using the pressure-temperature diagram for water (Fig. 13-1), show the sequence of operations by which this could be accomplished.

5 Which of the following mixtures would show critical behavior (a critical point or critical end points)?

    (*a*) 1 g of KCl and 1,000 g of H_2O.
    (*b*) A mixture of KCl and water in which solid KCl is always present.
    (*c*) 0.1 g of SiO_2 and 1,000 g of H_2O.
    (*d*) A mixture of SiO_2 and water in which solid SiO_2 is always present.

6 What is the geological evidence for the statement: "Many pegmatites are formed by the residual solutions from the crystallization of granitic magmas"? What is the geological evidence that not all pegmatites have formed in this manner? How does chemical evidence support the hypothesis that pegmatites may form in two different ways?

7 Is the following statement correct? Why or why not? "A pegmatite dike forms by the crystallization of a melt of the same composition."

8 How could you prove that the material of a quartz-sulfide vein is not an igneous rock, i.e., was not formed by crystallization of a melt of roughly the same composition?

REFERENCES AND SUGGESTIONS FOR FURTHER READING

Burnham, C. W., Water and magmas: a mixing model, *Geochim. et Cosmochim. Acta*, vol. 39, pp. 1077–1084, 1975. A simple model for the effect of H_2O on bonding in silicate melts, from which the solubility of water in melts can be predicted.

Burnham, C. W., and R. H. Jahns, A method for determining the solubility of water in silicate melts, *Am. Jour. Sci.*, vol. 260, pp. 721–745, 1962. Solubility of water in melts prepared from a pegmatite and from pure albite, at pressures up to 10 kilobars.

Jahns, R. H., The study of pegmatites, *Econ. Geology*, 50th Anniversary vol., pp. 1025–1130, 1955. A critical discussion of geologic data and inferences about the origin of pegmatites.

Jahns, R. H., and C. W. Burnham, Experimental studies of pegmatite genesis: I. A model for the derivation and crystallization of granitic pegmatites, *Econ. Geology*, vol. 64, pp. 843–864, 1969.

Kennedy, G. C., G. J. Wasserburg, H. C. Heard, and R. C. Newton, The upper 3-phase region in the system SiO_2-H_2O, Am. Jour. Sci., vol. 260, pp. 501–521, 1962. Experimental data on phase relations among quartz, tridymite, and two fluids (one silica-rich, one water-rich) at temperatures over 1000°C and pressures up to 10 kilobars.

Luth, W. C., and Tuttle, O. F., Hydrous vapor phase in equilibrium with granite and granite magmas, Geol. Soc. America Memoir 115, pp. 513–548, 1969. Composition of the water-rich phase in equilibrium with a granitic melt.

Morey, G. W., The solubility of solids in gases, Econ. Geology, vol. 52, pp. 225–251, 1957. A theoretical discussion of critical points of solutions and a review of experimental data.

Ölander, A., and H. Liander, The phase diagram of sodium chloride and steam above the critical point, Acta Chem. Scand., vol. 4, pp. 1437–1445, 1950. An early study of the NaCl-H_2O system, notable for the clarity of its diagrams.

Sourirajan, S., and G. C. Kennedy, The system H_2O-NaCl at elevated temperatures and pressures, Am. Jour. Sci., vol. 260, pp. 115–141, 1962. Experimental data on the solubility of salt in liquid water, water vapor, and supercritical water.

Taylor, H. P., $^{18}O/^{16}O$ evidence for meteoric-hydrothermal alteration and ore deposition in the Tonopah, Comstock Lode, and Goldfield mining districts, Econ. Geology, vol. 68, pp. 747–764, 1973. A good example of many recent papers by Taylor and his coworkers on the use of oxygen isotopes in determining the origin of water in vein-forming solutions.

SIXTEEN

VOLCANIC GASES

In the last chapter we speculated about the volatile substances associated with igneous intrusives, volatiles which we never see in action and which we can study only from the effects they produce. To become better acquainted with the volatiles, we look now at the one place where these substances can actually be observed, the area around an active or recently extinct volcano. During a volcanic eruption we see gases bubbling out of liquid lava and forming sublimates on the walls of fissures; here there can be no doubt that we are dealing with substances escaping from solution in a silicate melt. These gaseous substances, we may confidently assume, are similar to the volatile materials given off by intrusive bodies at depth. Since compounds of the common ore metals are almost universal, if minor, constituents of volcanic emanations, we may also expect that a study of these gases will be useful as a basis for inquiring into the origin of ore deposits.

16-1 RELATION OF VOLATILES TO VOLCANIC ACTIVITY

During volcanic eruptions gases appear at many places in addition to the bubbles in moving lava. Enormous volumes billow upward from the central vent, and smaller amounts issue from fissures in the cone and from local spots on the partly cooled lava. It is reasonable to suppose that all this gas, or at least the major part, was originally dissolved in the lava; that it escapes because of reduction in pressure as the lava nears the surface; and that the rapid liberation of gas is responsible for some of the more violent and spectacular aspects of volcanic activity. In a few places, where rough estimates have been possible of the amount of gas given off during an eruption, it turns out that the weight of gas is of the order of a few percent of the total weight of fluid lava emitted by the same eruption. This agrees nicely with experimental determinations of the solubility of water in silicate melts (Sec. 15-2).

In our usual reconstruction of events accompanying volcanic activity, we imagine a "magma chamber" at shallow depth (a few hundred to a few thousand meters) beneath the volcano, into which magma has moved from an unknown source below. Such a chamber is not entirely a matter of imagination, because small intrusive bodies are often mapped in areas where old volcanoes have been deeply dissected. The magma, having ascended into a region of relatively low pressure, is unable to retain its volatile constituents in solution, and these collect at the top of the magma chamber. The initial eruption takes place when the pressure of the released gases becomes high enough to force a way through the overlying rocks. The outrushing gases may carry with them frothy bits of molten lava and chunks of rock torn from the fissures through which they move, this material collecting at the surface into a "cinder cone." Fluid lava commonly follows the original outburst of gas, welling up along fissures opened by the pressure from beneath. Gas continues to rise from deeper levels in the magma chamber, leading to a long succession of explosive eruptions as the conduits to the surface become clogged and reopened. Fluid lava continues to pour out on the surface, becoming less frothy as the eruption goes on. Finally eruptive activity ceases when the supply of rising gas is no longer sufficient to force its way through the lava and broken rock blocking the main conduit. Gases may continue to escape quietly through small cracks for long periods, up to hundreds or even thousands of years.

Not all eruptions by any means follow this simple sequence of events, but the pattern is common enough to make plausible the general picture of a silicate melt releasing its dissolved volatiles at shallow depth. Possible differences in the depth of the magma chamber, the composition of the lava, and the structure of the overlying rock give ample latitude for modifying the hypothesis as necessary to fit particular eruptions.

It should be noted, however, that this "explanation" of volcanic activity is not at all profound. It takes us only a little way beyond bare description of volcanic phenomena and leaves untouched the deeper questions about the causes of vulcanism. What makes the magma rise high in the crust? Where and how is the magma formed? What is the source of the heat dissipated in a volcanic eruption? Are dissolved gases responsible for pushing lava to the surface, or does the lava move in response to orogenic forces? Questions of this sort are the fundamental ones which must be answered really to "explain" volcanic activity, but they are more the province of geophysics than of geochemistry.

16-2 GASES FROM FUMAROLES

General

The mechanical problem of obtaining samples of volcanic gas for analysis is a peculiarly difficult one. Ideally the best place for sampling would be the throat of a volcano in eruption, but the obvious practical difficulties make it necessary to find a place where activity is less violent. The smaller gas vents, or *fumaroles*, are a good second choice, and it is from fumaroles that nearly all of our data about the nature of volcanic gases have come. Even at fumaroles the collection of samples uncontaminated by air is all but impossible. And the best fumarole samples, as representatives of the original gases, are suspect on at least two counts: (1) they may represent only the residual gases, the gases given off at a late stage in cooling after most of the volatiles have already escaped, and (2) if the fumaroles are at a distance from the central conduit of the volcano, the gases may be contaminated by volatile products distilled from vegetation and sedimentary rocks buried by the lava. Despite these objections, fumarole samples are the most direct means available for studying gases associated with magmas.

The scientific study of fumaroles has a long history, going back to the early 1800s. Much qualitative and semiquantitative information on gases from volcanoes in the Mediterranean, the West Indies, and Iceland was obtained during the nineteenth century by French, German, and Italian geochemists. The use of modern techniques for the collection and analysis of gases dates from the well-known work of Day, Jaggar, and Shepherd at Kilauea between 1910 and 1920 (summarized by Shepherd, 1938). In recent years good samples of volcanic gases have been obtained in many parts of the world, especially Japan, Kamchatka, New Zealand, Iceland, and East Africa. Available data on gas compositions have been summarized in thoughtful discussions by White and Waring (1963), Gerlach and Nordlie (1975), and Giggenbach and Le Guern (1976). Before considering the general conclusions which these authors have drawn, let us look at a few specific examples of fumarole investigations.

Kilauea and Erta'Ale

The collection of satisfactory gas samples is possible only under exceptional circumstances, when an active fumarole near a lava source can be approached with the necessary equipment. Such circumstances are found most commonly at basaltic volcanoes, and accordingly our best information about gas compositions relates to volatiles that have come from mafic lavas. Two basaltic centers that have offered particularly good opportunities for collecting are Kilauea on the island of Hawaii and Erta'Ale in Ethiopia.

The caldera of Kilauea for a long time in the early 1900s was partly filled with a lake of molten basalt. At times the lake crusted over, and gases from beneath the crust were concentrated at a few places where cracks extended down to the molten material below. The escaping gases brought up fragments of molten rock through the cracks, so that small mounds ("spatter cones") were built up around the gas orifices. On rare occasions it was possible to approach a mound closely, insert a pipe deep into a gas vent, and collect a sample of hot gas with little contamination from air. A similar lava lake, with similar spatter cones, existed in the caldera of Erta'Ale at various times in the 1960s and 1970s. At both volcanoes the temperature of the samples at the time of collection was in the range 1100 to 1200°C.

Examples of analyses of gases from Kilauea and Erta'Ale are given in Table 16-1. The analyses are corrected for admixed air by subtracting free oxygen and corresponding amounts of N_2, Ar, and CO_2, on the assumption that appreciable quantities of free O_2 are not likely as an original constituent of volcanic gas. In addition to the gases listed, Shepherd notes that boron compounds were present in the Kilauea samples, but he had no means of determining their amount. He notes also that hydrocarbon gases and helium were looked for but not found; in samples collected more recently at Kilauea, traces of methane have been detected.

The general similarity of gases from these two widely separated volcanoes is striking. In most samples water is the chief constituent, CO_2 next, and SO_2 third; the preponderance of these three substances has been noted in high-temperature gases from many other volcanoes. Gases containing chlorine (expressed as HCl) are minor and variable in amount; the amounts in the Kilauea samples are especially low, and a similar scarcity of halogens has been noted at other oceanic volcanoes. Amounts of CO, COS, H_2, and S_2 are small. The SO_3 in the Kilauea samples probably represents partial oxidation of SO_2 by admixed air.

Table 16-1 Analyses of gases from Kilauea and Erta'Ale (mol-%)

	Kilauea			Erta'Ale		
H_2O	36.18	61.56	67.52	84.8	69.9	79.4
CO_2	47.68	20.93	16.96	7.0	15.8	10.4
CO	1.46	0.59	0.58	0.27	0.68	0.46
COS				0.001	0.01	0.009
SO_2	11.15	11.42	7.91	5.1	10.2	6.5
SO_3	0.42	0.55	2.46			
S_2	0.04	0.25	0.09	0.4†	0.0	0.5†
HCl	0.08	0.00	0.20	1.28	1.22	0.42
H_2	0.48	0.32	0.96	0.85	2.11	1.49
N_2	2.41	4.13	3.35	0.10	0.25	0.18
Ar	0.14	0.31	0.66	0.001	0.001	0.001

Sources: Shepherd, 1938, p. 321 (Kilauea); Giggenbach and Le Guern, 1976, p. 26 (Erta'Ale). The three Kilauea samples are Shepherd's Nos. J8, J11, J13; the first two from Erta'Ale are representative individual samples, and the third is an average of 18 samples. Shepherd calculated all chlorine as Cl_2; his numbers are expressed here as HCl, for ease of comparison with the Erta'Ale analyses.

† May include some H_2S.

Despite the general similarity from sample to sample, the analyses show much variability in detail. Some of the variability, especially at Kilauea, can be ascribed to different temperatures, different collecting procedures, and different amounts of admixed air. The 18 samples from Erta'Ale, however, were collected with the same apparatus, at a single fumarole, within less than a half hour, during which time the temperature varied by no more than 10 degrees, so that much of the observed fluctuation must represent actual short-term changes in composition of the emitted gas.

Showa-shinzan

Fumarole gases from a more felsic lava were studied by Oana and his colleagues at the volcano Showa-shinzan in northern Japan. Their work was summarized by White and Waring (1963), from whose paper Table 16-2 is taken. The eruption of this volcano in 1944–1945 consisted in part of the extrusion of a dacite dome, the interior of which remained hot for many years and continued to give off gases in large quantities. Representative analyses of gases from fumaroles at various temperatures are shown in the table. Conditions for collecting were not as favorable as at Kilauea and Erta'Ale; samples were taken long after the original eruption, and the great excess of H_2O probably represents dilution by water vapor from heated ground water. Carbon dioxide is the most abundant gas after water vapor; in contrast to the gases from basalt, sulfur compounds are relatively minor and the

Table 16-2 Analyses of gases from Showa-shinzan†

The "active" gases include all gases other than H_2O, air, excess N_2, and the inert gases. The figures for the active gases are in volume percent, recomputed to total 100%. The last three lines of the table give the total of the active gases, excess N_2, and H_2O, also in volume percent and totaled to 100%

		Temperature, °C					
		750	700	645	464	328	194
"Active" gases	CO_2	65.0	61.1	64.3	91.1	89.5	76.4
	CH_4	0.08	0.14	0.14	0.14	0.15	0.16
	NH_3	0.06	0.007	0.01	0.10	0.007	0.01
	H_2	25.0	24.5	21.3	5.12	6.96	13.6
	HCl	5.39	8.61	8.61	1.51	1.48	4.66
	HF	2.76	3.54	3.51	0.88	0.65	0.43
	H_2S	0.10	0.62	0.53	1.07	1.05	4.27
	SO_2	1.66	1.50	1.60	0.12	0.14	0.50
Total active gases		0.723	0.592	0.569	0.859	0.948	0.258
N_2		0.026	0.019	0.021	0.042	0.052	0.026
H_2O		99.25	99.39	99.41	99.10	99.00	99.72

† *Source:* White and Waring, 1963, p. 24.

halogen gases are more prominent. In a general way the amounts of CO_2 and H_2S appear to increase as the temperature falls, while the amounts of H_2, HCl, HF, and SO_2 decrease, but the changes are far from regular.

Kamchatka

Observations of volcanoes on the Kamchatka Peninsula and the adjacent Kurile Islands by Naboko (1959) and others have brought to light interesting relationships between gas compositions, kinds of lava, and fumarole temperatures. As a general rule gases from all these volcanoes show water as the most abundant constituent, the carbon gases next, and the sulfur gases third. The principal form of sulfur at high temperatures is SO_2, at low temperatures H_2S. Fumarole gases near a basaltic volcano (Klyuchevskaya) are unusually rich in halogen compounds, those from an andesitic volcano (Sheveluch) in sulfur compounds, and those from another andesitic volcano (Avacha) in boron compounds. The greater abundance of halogen gases with more mafic lavas is just the reverse of the apparent relationship shown by the gases at Showa-shinzan, Kilauea, and Erta'Ale. At Klyuchevskaya gases collected near the volcano's crater were relatively rich in sulfur gases, while those from moving lava far from the crater were rich in halogens.

Parícutin

During the eruptive activity of the Mexican volcano Parícutin between 1943 and 1952, fumarolic activity was too feeble to permit the collection of satisfactory samples. Crude analyses, however, showed a striking difference between the gases coming from cracks in the crater near the volcano's throat and those in fumaroles on the lava that flowed from the base of the volcanic structure: the former were rich in sulfur vapor and SO_2, the latter in the halogen compounds HCl, HF, and NH_4Cl. This is similar to the observation at Klyuchevskaya mentioned above, and is confirmed by reports of the same sort of differences elsewhere. Evidently sulfur gases escape in quantity when the lava is freshest and hottest, while the halogen gases are released only slowly after the lava has started to flow. A possible explanation is a difference in solubility, the sulfur compounds being less soluble in the molten rock and therefore escaping more quickly.

Summary This brief sampling of fumarole studies suggests some general conclusions that are amply borne out by more complete reviews like that of White and Waring. Gases from fumaroles associated with many kinds of volcanoes the world over show much general similarity in composition, but also a great deal of perplexing variation. Water vapor is nearly always the dominant gas, often amounting to 90% or more of the total. The carbon gases (CO_2, CO, COS, CH_4) are usually next in order of abundance; of these, CO_2 is generally dominant, COS and CH_4 are always low, and CO is only rarely reported to exceed CO_2. The sulfur gases (SO_2, H_2S, S_2, SO_3) commonly follow the carbon gases in abundance; SO_2

is usually the most important sulfur gas at high temperatures and H_2S at low temperatures, but White and Waring caution that this rule has many exceptions. Of the halogen gases HCl is by far the commonest. Its amount is highly variable from one fumarole to another; it is often thought to be especially characteristic of high-temperature fumaroles, but White and Waring note that this generalization is not borne out by their compilation. HF is usually not determined; where it is reported, the amount is generally less than one-tenth that of HCl. The heavier halogens are minor constituents and are seldom determined. Nitrogen is present chiefly as the free element, but a little NH_3 is sometimes reported; NH_4Cl is a common sublimate, and in tiny particles it is often responsible for the bluish color of smoke rising from fumaroles. Free hydrogen is highly erratic, but is commonly not abundant. Sublimates at the mouths of fumaroles show that the hot gases contain small amounts of many metal compounds.

In reviewing their work at Erta'Ale, and comparing their analyses with gas compositions reported by others, Giggenbach and Le Guern reach the conclusion that uncontaminated gases from high-temperature fumaroles at a wide variety of volcanoes have a fairly uniform composition that can be approximated as:

H_2O	CO_2	SO_2	H_2	HCl	CO	
80	10	7	1–2	0.5	0.5	mole-%

They are aware, of course, that the composition changes somewhat with temperature, with oxidation state, and with the age and extent of degassing of a given body of lava, and that contamination by other gases or by reaction with rocks and organic matter can cause much local variation. There are also minor unexplained fluctuations, as the analyses from Erta'Ale illustrate. Evidence is not convincing as to whether gas composition shows any clear relation to kind of lava or kind of eruptive activity. It may well be that volcanic gas in general has the approximately uniform makeup that these authors suggest, perhaps determined in large part by the solubilities of different gases in silicate melts.

16-3 EQUILIBRIUM RELATIONS IN VOLCANIC GASES

Some of the variation in gas compositions may be due to shifting of equilibria as the temperature changes. This possibility can be explored by postulating a gas of given composition, in which all possible reactions among its constituents have attained equilibrium under given conditions of temperature and pressure, and calculating the effects on various equilibrium reactions that temperature-pressure changes would produce. If the calculated compositions approximate compositions observed in nature, the existence of equilibrium is demonstrated. And if the temperature and pressure at which a sample was taken are not known, the calculations should permit these variables to be estimated from the gas composition.

Of the many possible reactions that must be considered, the following are a few examples:

$$H_2 + CO_2 \rightleftharpoons CO + H_2O$$

$$4H_2 + 2SO_2 \rightleftharpoons S_2 + 4H_2O$$

$$N_2 + 3H_2 \rightleftharpoons 2NH_3$$

$$2H_2 + S_2 \rightleftharpoons 2H_2S$$

$$CO_2 + 4H_2 \rightleftharpoons CH_4 + 2H_2O$$

All these reactions, it should be noted, involve the breaking down and rearrangement of gas molecules. This means that some or all of them may be slow, even at the high temperatures of a volcanic environment, so that equilibrium may not always be reached. The application of equilibrium calculations, therefore, is limited not only by the possibility of contamination of gas samples but also by the fact that attainment of equilibrium in rapidly moving fumarole gases is far from assured.

The calculations follow a pattern we have often used before. Standard free energies for the various substances at different temperatures must first be obtained, then from these the equilibrium constants, and from these in turn the proportions of gases present for various assumed overall compositions. Since each substance takes part in several reactions, all occurring at once, the calculation involves handling a considerable number of simultaneous equations, and is best made by using a computer. Calculations of this sort have been carried out by a number of authors, and typical results are illustrated in Figs. 16-1 and 16-2. As is

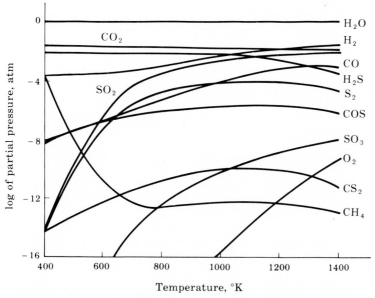

Figure 16-1 Calculated change in equilibrium composition of volcanic gas with temperature. Assumed atomic ratios: H/O/C/S = 275.5/142.2/2.680/1.000, from analyses of a typical Kilauea gas. Assumed total pressure = 1 atm. (*Source: Heald, Naughton, and Barnes, 1963, page 547.*)

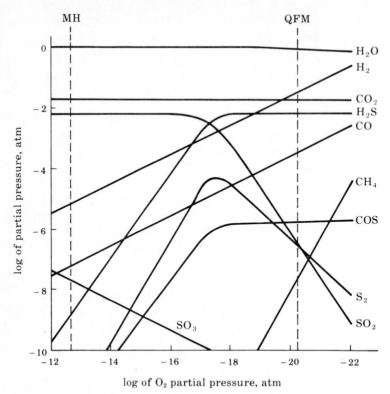

Figure 16-2 Change in equilibrium composition of volcanic gas with change in oxidation state. Assumed atomic ratios same as in Fig. 16-1. Temperature 627°C, total pressure 1 atm. The dashed lines are O_2 pressures that might be found in lavas, as indicated by mineral assemblages: MH for the combination magnetite-hematite (sometimes found in rhyolite), and QFM for quartz-fayalite (olivine)-magnetite (possible in basalt). The oxygen fugacities (or partial pressures) of most lavas lie between these extremes. (*Heald et al., 1963, page 549.*)

shown by Fig. 16-1, the often noted prominence of SO_2 and H_2 at high temperatures, and of H_2S and CO_2 at low temperatures, can be nicely accounted for by displacements of equilibrium. Figure 16-2 shows variations in composition produced by change in oxidation state, notably the increase in SO_2 and SO_3 at the expense of H_2S, and the decrease in H_2 and CO, as the partial pressure of oxygen increases. Oxygen fugacities shown in this figure are in the range that might be encountered in lavas under the surface, and extrapolation of the diagram toward the left gives a good picture of the changes that occur when fumarole gas mixes with oxygen from the air. Both figures illustrate clearly the relatively minor role of S_2, CO, COS, CH_4, and SO_3 in uncontaminated gases, regardless of temperature or oxidation state.

Predicted effects of changing temperature on gas compositions show reasonable correspondence with analytical results, indicating that reactions among fumarole gas constituents are fast enough to maintain a fairly close approach to equilibrium. Some authors have noted discrepancies between measured temperatures in fumaroles and the temperatures indicated by gas compositions, as if the gases had attained equilibrium under conditions somewhat different from those under which they were collected. The most recent and most elaborate of such studies (Gerlach and Nordlie 1975), however, indicates strongly that the apparent discrepancies have reasonable explanations and that equilibrium in fumarole gases responds quickly to changes in temperature and pressure. Gerlach and Nordlie show also that compositions of fumarole gases are generally compatible with compositions of the lavas with which they are associated, although the sensitivity to lava composition is not very great.

In summary, analytical data and calculations from them are consistent with a hypothesis that the original gases dissolved in lava are fairly similar from one volcano to another. Some initial differences may exist, but in large part the observed variation is accounted for by adjustment of equilibria to changing temperature and pressure and by admixture of air and meteoric water. The hypothesis is far from proved, but it is the simplest conclusion we can draw at present from a mass of complex and confusing data.

16-4 SUBLIMATES

In addition to substances gaseous at ordinary temperatures, volcanic emanations contain compounds which are volatile only at magmatic temperatures and which therefore collect as sublimates around the mouths of fumaroles. A great variety of substances has been reported in volcanic sublimates, some of them minerals that are found nowhere except in this environment. The sublimates are of particular interest because they include compounds of the common ore metals and hence suggest a relation between volcanic emanations and ore deposits.

The most abundant compounds in the sublimates are generally chlorides and sulfates of Na, K, Ca, Mg, Al, and Fe, oxides of Si and Fe, ammonium chloride, and native sulfur. Of the elements reported in minor amounts or traces, the more abundant are usually F, Br, B, P, As, Zn, Cu, Pb, Mn, and Sn. Others, generally very minor, include Li, Be, Ag, Ni, Co, V, Mo, Ga, Ge, Ti, Zr, Cr, Cd, Sb, Bi, Sr, Ba, Se, Te. Note that the list includes nearly all the metals commonly found in sulfide ores.

As a rule, the compounds found as sublimates are those that might be expected on the basis of volatilities. The common prevalence of chlorides is a reflection of the fact that for most metals the chlorides are the most volatile naturally occurring compounds. Even when the actual compound observed is not a chloride, one can reasonably suppose that the metal was present in the gas as a chloride (or a fluoride) and was deposited in another form by reaction with water. For example,

$$2FeCl_3(g) + 3H_2O \rightarrow Fe_2O_3 + 6HCl$$

$$SiF_4(g) + 2H_2O \rightarrow SiO_2 + 4HF$$

$$2AlCl_3(g) + 3H_2O + 3SO_3 \rightarrow Al_2(SO_4)_3 + 6HCl$$

A good example of recent studies of sublimates is the exhaustive survey of fumarolic incrustations at Central American volcanoes by Stoiber and Rose (1974). These authors list a bewildering number of sulfate, chloride, and oxide minerals that have been identified, plus several that are not yet named. They note also the presence in some of the sublimates of traces of heavy metals, particularly copper, vanadium, and molybdenum.

More informative with regard to possible relations between fumarole deposits and metallic ores are some older observations by Zies (1929) at the Alaskan volcano Katmai. In 1919, seven years after an eruption that produced a deposit of rhyolitic pumiceous ash at least 30 meters thick, the principal sublimates around fumaroles in the ash were magnetite, hematite, amorphous silica, gypsum, halite, alum, and ammonium chloride. Trace amounts of many metals were found in the magnetite, principally Zn, Pb, and Cu, with lesser Ti, Mn, Ni, and Co. Four years later the sublimates were found to have completely changed character, probably in response to a general lowering in the temperature of the fumaroles. The previously abundant magnetite was nearly gone, and the chief substances now were ZnS, CuS, PbS, and $PbCl_2$. Evidently the original deposits had been worked over by acid solutions, derived in part by condensation of steam from the fumaroles and in part from meteoric water, and the fumaroles had supplied enough H_2S to fix some of the metals as insoluble sulfides.

The fact that the sulfides were observed in process of formation from volcanic emanations invites the question: How far can these observations lead us toward an explanation for the origin of ore deposits? The situation is obviously different in many respects from the formation of ores in veins beneath the surface, so a good deal of caution is necessary. We can best proceed by dividing the question into parts.

What is the source of the metals? The geologic relations at Katmai are not entirely certain, but recent observations indicate strongly that the gases rose from the cooling ash itself. Analysis of the pumiceous ash shows that zinc, lead, and copper are its principal trace elements, just as they are in the sublimates, and this has led to a suggestion that possibly the metals were not contained in the original gas but were added to the gas as it moved through the ash. This hardly seems more than a quibble, however, for both metals and gas would still be volatile constituents of the magma, and that is the chief point we are trying to establish.

How are the metals transported? Obviously in the gas, and presumably in the form of chlorides. In this respect conditions may be very different from conditions during the formation of a vein. Indirect evidence suggests that most vein deposits were formed by deposition from a liquid rather than a gas, so that the metals would presumably exist as ions in solution rather than as volatile compounds. Early stages of vein formation could conceivably involve gaseous transfer, but this gets us into a realm of speculation that is best put off until the next chapter.

How can the metals be carried as chlorides in the presence of H_2S? This involves the position of equilibrium in reactions like

$$PbCl_2(g) + H_2S \rightleftharpoons PbS(s) + 2HCl$$

Whether the metal will move as the volatile chloride or precipitate as the sulfide must depend on the ratio of HCl to H_2S, on the volatility of the chloride, on the relative stability of the sulfide and chloride, and on how the equilibrium shifts with temperature. Experimental measurements combined with reasonable assumptions indicate that movement of some of the metals in a gas phase is possible as long as the amount of H_2S is not too great. This generalization helps very little with an understanding of liquid ore solutions, however, because most of the metal sulfides are so insoluble in water that transportation of metal ions together with H_2S is hard to understand. Again we defer discussion to the next chapter.

Is the transporting medium acid or alkaline? With its content of HCl, CO_2, and H_2S, the fumarole gas could hardly be anything but acid. This is not sufficient justification, however, for supposing that ore solutions at depth must also be acid. At higher pressures, in a liquid in contact with silicates and perhaps with carbonates, alkaline conditions seem at least equally likely.

In summary, then, sublimates at fumaroles provide evidence that the common ore metals can indeed be a part of volatile emanations from magmas. Beyond this it is questionable whether the observations help us much in an understanding of ore formation. If we choose to be optimistic, we may suppose that separation of metals from a magma in the form of volatile chlorides might be possible underground as well as near the surface, and that these chlorides are ultimately taken into solution as the emanations cool off and mix with ground water. In a more skeptical mood we could maintain that volcanic phenomena are a special case, at pressures so low that they can have no relevance to processes beneath the surface, and that therefore the separation of ore solutions from magmas at depth may follow an entirely different pattern. Evidence is insufficient to permit a final choice between these viewpoints, but fragmentary data have given economic geologists ammunition to argue their merits back and forth for many years.

16-5 HOT SPRINGS

In many volcanic areas, particularly areas in which actual eruptions have long since ceased, fumaroles are accompanied by or replaced by hot springs. At first glance it seems that hot-spring water might resemble more closely a possible ore solution than does gas from a fumarole, but a moment's reflection shows serious difficulties. The water in most hot springs is largely of meteoric origin: rainwater that has penetrated deeply enough into the ground to become heated by the still-hot rocks beneath. This is demonstrated, for example, by the topographic position of hot springs and by the common dependence of flow on rainfall. Now a hypothetical ore solution may contain some meteoric water, but we expect at least a small part of many ore solutions to be the product of condensation of volatile material from a magma. If the water of a hot spring does contain some material of magmatic origin, how can we determine what proportion of the water this represents? And how drastically would such material have been altered by reaction with wall rocks on its way to the surface?

The best method for estimating the proportion of magmatic fluids in hot-spring waters is the measurement of isotope ratios, the same method that is used in finding the contribution of magmatic material to the fluids responsible for zones of hydrothermal alteration (Sec. 15-5). The oxygen of meteoric waters has less of the heavy isotopes (a lower $^{18}O/^{16}O$ ratio) than oxygen from water in equilibrium with a magma. Unfortunately, hot meteoric ground water may acquire a slightly abnormal content of the heavy isotopes by reaction with igneous rocks in contact with it, so that the measurement is not unambiguous. In some hot springs the ratio of hydrogen isotopes ($^2H/^1H$) is also different from that in magmatic water, and use of both oxygen and hydrogen ratios permits a better guess as to the magmatic contribution (Sec. 21-4). Study of isotopes has shown that most hot-spring waters are predominantly meteoric; in many springs the measurements indicate that not more than 10%, and probably not more than 5%, of the water can have an igneous source. Nevertheless it is worthwhile to examine briefly the dissolved material in hot springs, because some of the constituents, particularly the more unusual ones present in minor amounts, may provide clues about additions to the circulating meteoric water from feeble emanations still rising out of the nearly frozen magma beneath.

A compilation of analyses of subsurface waters by White, Hem, and Waring (1963) shows great diversity in the compositions of water from hot springs. The familiar anions and cations are universally present, together with many minor constituents, but amounts and ratios of the different substances are so variable that generalizations are difficult. The authors of the compilation nevertheless try to separate waters with possible magmatic contributions from thermal waters of other origins. The very hot waters of springs in areas of recent or active volcanism, where some addition from a magma is most likely, are characterized by high ratios of Li to Na and of B to Cl, and by low Ca/Na, Mg/Ca, I/Cl, and usually Br/Cl ratios. Heavy metals are generally present only in traces. Water from springs associated with mineral deposits is less certainly of volcanic origin, especially where the springs are far from evidences of recent volcanism and where the major-ion ratios are different from those in volcanic areas; but the water near mineral deposits, as might be expected, shows larger heavy-metal concentrations. Nearly all the metals of sulfide deposits have been reported, generally in amounts from a few to several hundred parts per billion, and in widely varying ratios from one spring to another. In a few hot springs there is good evidence for deposition of metals at the present time; notable examples are Sb, As, and a little Ag and Au at Steamboat Springs, Nevada, and Hg at Boiling Springs, Idaho. In analyses that include the gas content of hot-spring waters, the gases found are similar to those from fumaroles: generally CO_2 is the most abundant, accompanied by minor CO, H_2, N_2, NH_3, H_2S, and Ar in varying amounts. The pH of hot springs is so obviously influenced by the nature of adjacent rocks and often by oxidation of H_2S that it is not a reliable indicator of the original character of the water, but perhaps there is significance in the fact that the water of very many hot springs is nearly neutral, with pH's in the range 6 to 8.

Especially interesting as a possible contemporary ore solution in a volcanic area is the water from hot brine wells near the Salton Sea in southern California (Skinner et al., 1967). This is an extraordinary water, containing far more copper (about 8 ppm), silver (2 ppm), zinc (500 ppm), and lead (80 ppm) than has been reported in any other naturally occurring water with appreciable flow. Evidence from isotopes indicates that this water, like other investigated thermal waters, is at least in large part of meteoric origin. The heavy-metal content may be derived by the leaching of sedimentary rocks through which the water travels, but since the wells are near some small rhyolite and obsidian domes of Quaternary age, there is a good presumption that some of the metals may be supplied by a magma reservoir beneath.

Thus hot springs, despite their gross contamination by meteoric water and by substances dissolved from their walls, give us much the same information about the composition of magmatic volatiles as do fumaroles. We find gases containing carbon, sulfur, nitrogen, and the halogens, plus trace amounts of boron and compounds of many metals—the repetition of this list from spring to spring and volcano to volcano becomes monotonous.

It should be noted that in this discussion of hot springs we have tacitly assumed a pretty narrow definition of an "ore-forming solution" as a liquid consisting largely of condensed volatile material from a magma. This was convenient to bring out the contrast between juvenile and surficial components in hot-spring waters, but actually it assumes far more knowledge than we actually possess about the nature of ore solutions. Ore-forming liquids may, for example, be much closer to hot-spring waters than this discussion would indicate; quite possibly ore minerals could be deposited in large amounts from meteoric waters that have derived little but heat from a magmatic source, and that have picked up their content of metals by circulating widely through ordinary rocks. And there is no reason, of course, why ore solutions should all be alike: perhaps some are largely juvenile and others largely meteoric. For the present we shall not pursue this argument, but continue to concentrate attention on the inferences we can make about possible magmatic contributions to ore solutions and other volatile materials.

16-6 GASES FROM HEATED ROCKS

One other possible method of attack on the problem of volatiles is to heat igneous rocks in the laboratory, in the hope that small amounts of the gases present at the time of consolidation have been trapped in or on the mineral grains and that the gases may be released by heating. Volatiles may be trapped in at least three ways: (1) as constituents of minerals, like the water and fluorine in micas and hornblende; (2) as "fluid inclusions," tiny cavities within mineral grains filled with liquid or gas or both, sometimes very numerous; and (3) by simple adsorption on surfaces and in cracks. Gases held in all these ways should be liberated if the rock is made hot enough.

Here again, just as in studies of fumarole gases and hot springs, contamination is a serious problem. If a rock has been long exposed to air, the gases originally adsorbed would be partly replaced by atmospheric gases, and weathering may change the nature of chemically held volatiles. If the rock has been heated, subjected to pressure, or permeated by hot solutions since its formation, even the volatiles sealed in cavities may have been altered and new cavities may have formed along zones of incipient fracture in mineral grains. These possibilities may be minimized by using extreme care to select specimens for heating which show no sign of weathering or hydrothermal alteration, but it is doubtful that they can be eliminated completely.

Shepherd (1938), following up his work at Kilauea, conducted experiments of this sort with basaltic lavas from Hawaii and dacitic lavas from Mount Pelée in Martinique. The gases he obtained were the same as those from the lava lake, and the variability in proportions from specimen to specimen was similar to the variability from one fumarole to another. Other experiments on a considerable variety of lavas have yielded similar results.

Experimental work designed more specifically to analyze the constituents of fluid inclusions has given results consistent with those of most other work on volatiles. The principal gas reported is CO_2, along with minor amounts of N_2, NH_3, H_2, H_2S, and Ar. The chief dissolved ions are generally Na^+, Ca^{2+}, and Cl^-, sometimes in amounts equivalent to $2M$ NaCl or more, and minor constituents are K^+, F^-, HCO_3^-, SO_4^{2-}, Fe^{2+}, and Fe^{3+}. The solutions are generally near the neutral point. No consistent differences in composition have been reported for fluids from different minerals, from different igneous rocks, or from minerals of sulfide veins.

Summary and speculations We possess no really satisfactory way of analyzing volatile material from cooling magmas, but indirect evidence from three sources—fumarole gases and sublimates, hot springs, and gases from heated rocks—gives a consistent picture of what these volatiles must be. Nearly always water is far and away the most abundant substance. Other substances volatile at usual temperatures include nitrogen, argon, the two oxides of carbon, the sulfur gases H_2S and SO_2, the halogen acids HCl and HF, and sometimes hydrogen. Methane and ammonia are sometimes noted as minor constituents. Less volatile substances include sulfur, ammonium chloride, and compounds of many metals; probably most of the metals are present in the form of chlorides, but some would more probably be native (mercury, arsenic) and a few may exist as fluorides or oxyhalides. This list of materials is not only uniform from one volcanic area to another but is also consistent with inferences about the nature of the volatiles responsible for contact metamorphic zones and for ore deposition.

The consistency, however, is purely qualitative. Quantitative relations among the different volatile substances vary bewilderingly not only from one volcanic center to another but even between adjacent fumaroles and hot springs. Some of the variability can be ascribed to different degrees of approach to equilibrium in a complex mixture, some to differences in temperature, some to different amounts of contamination, but a part of it must go back to the magma itself. Do pockets of different kinds of volatiles exist within the same magma chamber? Do magmas show regional differences in the character of their volatiles? Does the nature of the volatiles depend on the composition of the magma? These are questions about which we have a little indirect evidence, but not enough quantitative data to supply definitive answers.

The demonstrated presence of many metallic compounds in volcanic emanations has been the basis for a great deal of speculation. Because metals like copper, zinc, and lead are found in sublimates and in hot-spring deposits, the inference is obvious that ore deposits of these same metals could be formed by similar emanations escaping from magmas at depth. Furthermore, since emanations at the surface are notoriously variable in composition, it is certainly legitimate to suppose that volatiles at depth may have special concentrations of some one metal or a group of metals. Hence to "explain" an ore deposit, one need only postulate as a source of the metals a solution derived from magmatic emanations having the appropriate composition. One can extend such theorizing also to ore deposits formed in sedimentary environments: if a sedimentary source for the metals is not immediately apparent, one can always call on "volcanic springs" or solutions that have percolated through volcanic ash to supply the metals—and of course one may ascribe to such solutions any composition that is convenient to bolster the hypothesis. If a manganese deposit is in question, one assumes a solution containing manganese and little else; if a uranium deposit needs explanation, one assumes a solution or a tuff peculiarly rich in this metal.

Such a postulate, correct though it may be, is not a very satisfying explanation. It merely transfers the area of ignorance from the ore deposit itself to a magma (often a magma whose existence is itself a hypothesis), and our information about magmas is so very limited that we can neither check the hypothesis nor use it to make predictions. Building hypotheses in this manner is a harmless pastime, but it contributes little to our knowledge of ore deposition.

Another subject worthy of speculation is the relation between the composition of volcanic emanations and the gases one might obtain by heating an ordinary sedimentary rock, say a shale or an arkose. A measure of similarity would be expected: heating a sedimentary rock would generally give water as the chief volatile, carbon dioxide as a second constituent if the rock contained any carbonate minerals, and nitrogen from air adsorbed on the sedimentary particles. If the rock contained organic matter, other gases might be expected, notably methane, carbon monoxide, and ammonia. Sulfur and hydrogen sulfide might come either from organic matter or from the decomposition of pyrite. Thus many of the typically volcanic gases could be duplicated in this way. A notable exception would be hydrogen chloride, which could not be produced in more than traces from ordinary sedimentary rocks; but if one supposes that ground water or connate water is heated at the same time, some of this deficiency could be supplied. Even the minor metallic constituents of volcanic gases could probably be duplicated in vapors from heated sediments, for sedimentary rocks contain trace amounts of many metals, and these could be volatilized as chlorides.

This brings us back to the question of how much a volcanic gas may be contaminated by, or even formed from, the rocks through which it travels. To take an extreme view, one could suppose that magma itself contains very little gas originally, but simply distills gases out of the rocks around it; or alternatively one could guess that the magma consists of material that was originally sedimentary but that has been heated until most of it melted and its volatile constituents were released. Certainly for volcanoes in island arcs or on continental margins above active subduction zones, an origin of the magma by partial melting of sedimentary material carried down by the moving plate seems plausible. Volatile material contained in the magma rising above the subducting plate could well be largely water originally present in the sediments, perhaps augmented by material from the cool crust through which the magma moves on its way to the surface. Volcanoes along mid-ocean ridges, on the other hand, or fissure-volcanoes like those that fed the basalt flows of the Columbia River plateau, bring to the surface magma generated within the mantle, at depths of 60 km or more, and their volatile constituents may be largely water and other gases originally contained in the mantle peridotite.

Note that the magma of volcanoes above subduction zones, according to this speculation, would be in part recycled material, material that had been at the earth's surface, was carried down by plate movements, and now reappears at the surface in the form of lava and volcanic gases. The lava and volatiles from the upper mantle, however, have no obvious connection with surface phenomena. Can we assume that they are samples of juvenile material, liquid rock and gases that have never before been at the surface? Or, in the long evolution of the earth, have rocks from the surface been incorporated in the upper mantle, so that mid-ocean volcanoes also bring up largely recycled material? To questions of this sort we shall return in a later chapter.

PROBLEMS

1 The data on enthalpies and entropies in Appendix VIII make it possible to obtain rough values for free-energy changes at magmatic temperatures. The results are not accurate because they depend on the assumption that ΔH and ΔS remain constant over a temperature range of several hundred degrees, but for geologic problems the rough values often serve a useful purpose. The procedure has been explained in Sec. 8-11; for another example, the free-energy change in the reaction

$$H_2 + Cl_2 \rightleftharpoons 2HCl$$

at 1000°C is found as follows: Write down first the figures for ΔH° and S° from Appendix VIII for the three substances:

$$
\begin{array}{cccc}
 & H_2 + & Cl_2 \rightleftharpoons & 2HCl \\
\Delta H^\circ: & 0 & 0 & -44.12 \text{ kcal} \\
S^\circ: & 31.21 & 53.29 & 89.28 \text{ cal/deg}
\end{array}
$$

Addition and subtraction give, for the reaction, $\Delta H° = -44.12$ kcal and $\Delta S° = +4.78$ cal/deg. Then at 1000°C or 1273K,

$$\Delta G° = \Delta H° - T \Delta S° = -44.12 - 1,273 \frac{4.78}{1,000} = -50.2 \text{ kcal}$$

Make a similar calculation for the reaction $2H_2 + O_2 \rightleftharpoons 2H_2O$, and from the two free-energy values calculate the equilibrium constants. If a sample of volcanic gas at 1000°C contains 1 atm of H_2O vapor and 0.1 atm of HCl, calculate the partial pressures of H_2, Cl_2, and O_2 in equilibrium with the compounds.

2 Would you expect to find H_2S and SO_3 in the same sample of volcanic gas? H_2S and HCl? H_2S and Cl_2? HCl and SO_3? Discuss these combinations from the point of view of chemical equilibrium. (In calculating high-temperature free energies for sulfur compounds, note that $\Delta H°$ for S_2 gas is *not* zero.)

3 Look up equilibrium data for the reaction

$$SiCl_4(g) + 2H_2O \rightleftharpoons 4HCl + SiO_2(s)$$

Under what conditions (of total pressure, temperature, and HCl/H_2O ratio) would you expect SiO_2 to be deposited from volcanic gas? Under what conditions would the silicon of SiO_2 be volatilized as $SiCl_4$?

4 The free-energy change in the reaction

$$PbS(s) + 2HCl \rightleftharpoons PbCl_2(g) + H_2S$$

is about $+8.6$ kcal at 627°C and $+7.5$ kcal at 927°C. Discuss the effect of temperature, total pressure, and HCl/H_2S ratio on the deposition of PbS from volcanic gas.

5 Ammonia is often reported as a minor constituent of volcanic gases. Some geologists have assumed that it is a "juvenile" gas, originally dissolved in the magma; others have thought that ammonia can arise only by the decomposition of organic material in sedimentary rocks with which the hot lava has come in contact. If you were faced with this question in relation to the gases from a particular fumarole, what geologic evidence would you look for in order to decide it? Is there any chemical objection to the assumption that ammonia may be a juvenile magmatic gas (provided nitrogen is assumed to be present)? Using the data in Appendix VIII, discuss the pressure-temperature conditions under which appreciable amounts of ammonia could exist in volcanic gas.

REFERENCES AND SUGGESTIONS FOR FURTHER READING

Gerlach, T. M., and B. E. Nordlie, The C-O-H-S gaseous system, *Amer. Jour. Sci.*, vol. 275, pp. 353–410, 1975. Computer-based calculations of high-temperature equilibria, plus analyses of gases from Iceland, Etna, and Kilauea, are used to determine the ranges and trends of composition of volcanic gas in equilibrium with basalt.

Giggenbach, W. F., and Le Guern, F., The chemistry of magmatic gases from Erta'Ale, Ethiopia, *Geochim. et Cosmochim. Acta*, vol. 40, pp. 25–30, 1976.

Heald, E. F., J. J. Naughton, and I. L. Barnes, The chemistry of volcanic gases, *Jour. Geophys. Research*, vol. 68, pp. 539–557, 1963. Description of methods of collecting and analyzing volcanic gases, and diagrams showing results of equilibrium calculations based on the analyses.

Naboko, S. L., Volcanic exhalations and products of their reactions as exemplified by Kamchatka-Kurile volcanoes, *Bull. Vulcanol.*, Ser. II, vol. 20, pp. 121–136, 1959. Summary of many years of study of fumarole gases, including discussion of volatile metal compounds.

Shepherd, E. S., Gases in rocks and some related problems, *Am. Jour. Sci.*, vol. 235a, pp. 311–351, 1938. Analyses of gases obtained by heating specimens of igneous rocks; for comparison, includes a summary of Shepherd's earlier analytical work on gases collected at Kilauea.

Skinner, B. J., D. E. White, H. J. Rose, and R. E. Mays, Sulfides associated with the Salton Sea geothermal brine, *Econ. Geology*, vol. 62, pp. 316–330, 1967. Analyses of brines and siliceous scale deposited by the brines from a well drilled for geothermal power.

Stoiber, R. E., and W. I. Rose, Fumarole incrustations at active Central American volcanoes, *Geochim. et Cosmochim. Acta*, vol. 38, pp. 495–516, 1974.

Weissberg, B. G., Gold-silver ore-grade precipitates from New Zealand thermal waters, *Econ. Geology*, vol. 64, pp. 95–108, 1969. Analyses of water and precipitates in hot springs containing unusual amounts of Au, Ag, As, Sb, Hg, and Tl.

White, D. E., J. D. Hem, and G. A. Waring, Chemical composition of subsurface waters, chap. F in Data of geochemistry, 6th ed., *U.S. Geol. Survey Prof. Paper* 440-F, 1963. Analyses of underground waters of many kinds, and an attempt to set up criteria for distinguishing waters from different sources.

White, D. E., and G. A. Waring, Volcanic emanations, chap. K in Data of geochemistry, 6th ed., *U.S. Geol. Survey Prof. Paper* 440-K, 1963. Compilation of analyses of gases and sublimates from many volcanoes, with brief critical and interpretive comments.

Zies, E. G., The Valley of Ten Thousand Smokes, *Natl. Geog. Soc. Contributed Tech. Papers, Katmai Series*, vol. 1, no. 4, 1929. Analyses of sublimates, review of an earlier paper on analyses of fumarolic gases, and speculations about gaseous transport of metals commonly found in sulfide ores.

SEVENTEEN

ORE-FORMING SOLUTIONS

The origin of ore deposits is one of the great unsolved problems in geology. More accurately, it is a large group of unsolved problems, for ore deposits can form in a variety of ways. Some methods of ore accumulation are obvious: there is no mystery about the mechanical process that leads to accumulation of gold in stream placers, or about the chemical reactions that cause iron to precipitate in bogs. But the problem of origin for the great majority of ore deposits bristles with difficulties. The difficulties are especially serious for deposits formed at temperatures higher than the normal temperatures of the earth's surface, and it is to these in particular that we turn our attention now.

Deposits laid down at medium and high temperatures take many forms. In some the ore minerals are concentrated, along with quartz and other gangue minerals, in clearly defined veins; in others the ore minerals are widely disseminated through large volumes of rock; in still others the textures show that the ore minerals and in part the gangue minerals have replaced a preexisting rock. The evidence of widespread replacement is one of the cogent arguments for assuming that the solution which carried the ore metals was liquid rather than gaseous, for the removal of large amounts of either carbonates or silicates by a gas at reasonable temperatures is hard to imagine. Additional evidence, especially from textures and from temperature measurements, corroborates this guess that the solution responsible for actual deposition of ore was liquid. Because high temperatures and liquid aqueous solutions are involved, this picture of ore deposition goes by the name of the *hydrothermal* hypothesis.

Although actual deposition of ore is generally from a liquid medium, the volatility of metal compounds, especially the chlorides, together with the theoretical likelihood that a water-rich gas would separate at intervals from a cooling magma, makes it probable that gaseous transportation of the metals plays a role in the early history of metal concentration. Very likely the final deposition of ore is often only the last step in a complex process during which the metal or metals are vaporized, precipitated, dissolved, and reprecipitated. A sequence of events like this can sometimes be observed in progress in the sublimates around fumarole orifices.

With the data and speculations on magmatic volatiles in the last two chapters as a background, we set out now to explore the available chemical facts bearing on the possibilities of transporting and concentrating the common ore metals by gases and by hot aqueous solutions.

17-1 METAL COMPOUNDS IN MAGMATIC GASES

Whether or not a gas phase appears during the crystallization of a magma depends, as we have seen, on the confining pressure and on the amount of volatile material dissolved in the magma. Very probably some magmas freeze without a gas phase ever forming, their small content of volatiles being used up entirely in the crystallization of such minerals as mica and hornblende. Other magmas, as is evident from volcanic eruptions, are accompanied by abundant volatiles and almost certainly would evolve considerable gas even at depths of several kilometers. The gas would consist largely of water vapor, at temperatures well above the critical point of pure water but possibly not above the critical point of the complex solution in equilibrium with it.

Volatility versus Solubility of Metal Compounds

A metal may be present in such a gas for two reasons: (1) because its compounds may be volatile at magmatic temperatures, and (2) because its compounds may dissolve in the highly compressed water vapor. (The *volatility* of a solid refers to the amount that will vaporize into an evacuated space; its *solubility* is the amount that is taken into the gas by interaction with the gas molecules.) The amount present in a gas due to volatility can be calculated from standard thermodynamic data, as we shall see in a moment. The amount resulting from dissolution, on the other hand, can be only roughly estimated.

The ability of a gas to dissolve substances depends on the closeness of spacing of its molecules, hence on the pressure (Sec. 15-2). For a given pressure, the closeness of spacing increases as the temperature falls, so that we might expect solubility to be relatively less important at high temperatures. At 800° and 1,000 atm pressure, for example, the density of water is only about 0.24 g/cm^3; its diminished effect on solutes is shown by the fact that the dissociation constant for KCl $(a_{K^+} \cdot a_{Cl^-}/a_{KCl})$ is only about 10^{-6}, in contrast to a value of 1 or more under ordinary conditions. It seems a reasonable assumption that the content of metals in magmatic gas at the highest temperatures is largely a question of volatility and that solubility becomes increasingly important as the temperature falls.

Since volatilities may be easily calculated, it is a matter of interest to see what concentrations of various metals might exist in a gas phase as a result of the volatilities of their possible compounds.

Composition of a Magmatic Gas Phase

To make the calculation, we look at conditions in a magmatic gas at two temperatures, 627 and 827°, under a pressure of roughly 1,000 atm. At the higher temperature a granitic magma would be almost completely molten, at the lower temperature largely crystallized (Fig. 15-2). The pressure corresponds to depths of 4 to 10 km, depending on whether the pressure is chiefly lithostatic or chiefly hydrostatic. We assume that the gas consists of water vapor at a fugacity of 1,000 atm, HCl at 10 atm, HF at 0.3 atm, and sulfur gases ($H_2S + S_2 + SO_2$) giving a total of 10 atm. (For H_2O, a fugacity of 1,000 atm is roughly equivalent to a partial pressure of 1,200 atm at 827° and 1,700 atm at 627°; for the smaller amounts of the other gases, fugacity is approximately the same as partial pressure.) These figures are chosen as a rough average of many analyses of volcanic gases; considerable variation would not change the character of the results. Estimates for the carbon and nitrogen gases need not be made, because they would have little effect on metal compounds at these temperatures.

It is a simple matter now to set up equations involving these gases and possible derivatives, to calculate free energies at 627 and 827° from the data in Appendix VIII, and from the free energies to derive the equilibrium constants in Table 17-1.

Before actual partial pressures can be calculated from the equilibrium constants, we need one additional assumption: a reasonable value to use for the partial pressure of oxygen. The chain of reasoning is too long to give in full, but

Table 17-1 Free energies and equilibrium constants for reactions among magmatic gases at 627 and 827°C†

Reaction	Free-energy change, kcal		Logarithm of equilibrium constant	
	At 627°	At 827°	At 627°	At 827°
$2H_2 + O_2 \rightleftharpoons 2H_2O(g)$	−94.7	−89.3	+23.0	+17.8
$2H_2 + S_2(g) \rightleftharpoons 2H_2S(g)$	−22.2	−17.6	+5.4	+3.5
$H_2 + Cl_2 \rightleftharpoons 2HCl(g)$	−47.8	−48.5	+11.6	+9.6
$H_2 + F_2 \rightleftharpoons 2HF(g)$	−130.8	−131.0	+31.8	+26.0
$S_2(g) + 2O_2 \rightleftharpoons 2SO_2(g)$	−141.8	−134.8	+34.4	+26.8
$S_2(g) + 3O_2 \rightleftharpoons 2SO_3(g)$	−148.6	−132.8	+36.1	+26.4

† Tabulated values are slightly more accurate than those that would be calculated directly from Appendix VIII, because variations in ΔH and ΔS with temperature have been included. The maximum difference is 3 kcal in ΔG and 0.5 in log K.

the oxygen pressure is restricted to a fairly narrow range by the kinds of metallic minerals commonly found in and adjacent to intrusive rocks. Primary hematite, for example, is an uncommon mineral in high-temperature deposits; hence evidently the oxygen pressure is usually not high enough to oxidize magnetite:

$$\tfrac{1}{2}O_2 + 2Fe_3O_4 \rightleftharpoons 3Fe_2O_3 \tag{17-1}$$

Again, metallic lead is extremely rare, so that the oxygen pressure must remain high enough to prevent the reduction of lead compounds. A number of such reactions give as extreme ranges of O_2 partial pressure the values 10^{-14} to 10^{-23} atm at 627° and 10^{-8} to 10^{-17} atm at 827°. With these values and with the assumed partial pressures for the principal gases from the last paragraph, the equilibrium constants lead to the numbers in Table 17-2. Note that free chlorine and fluorine are negligible under all the assumed conditions; that SO_2 is the principal sulfur gas under oxidizing conditions and H_2S under reducing conditions, S_2 and SO_3 remaining very minor; and that the dominance of SO_2 increases as the temperature rises. The changes with oxygen pressure shown in the table are similar to those described in the last chapter for Kilauea gases at much lower total pressures (Fig. 16-2).

It should be recalled from a previous discussion (Sec. 10-4, Fig. 10-3) that very small oxygen pressures like those in Table 17-2 and Fig. 16-2 have no significance as measurable gas pressures. For ordinary purposes they tell us, in effect, that free oxygen is simply not present in magmatic gases. But the numbers

Table 17-2 Calculated compositions of magmatic gas at 627 and 827° C, exclusive of carbon and nitrogen compounds†‡

	627°				827°			
	Oxidizing		Reducing		Oxidizing		Reducing	
	←		→		←		→	
O_2	−14	−17	−20	−23	−8	−11	−14	−17
H_2O	+3.0	+3.0	+3.0	+2.7	+3.0	+3.0	+3.0	+2.9
H_2	−1.5	0.0	+1.5	+2.7	−1.9	−0.4	+1.1	+2.5
HCl	+1.0	+1.0	+1.0	+1.0	+1.0	+1.0	+1.0	+1.0
Cl_2	−8.1	−9.6	−11.1	−12.3	−5.7	−7.2	−8.7	−10.1
HF	−0.5	−0.5	−0.5	−0.5	−0.5	−0.5	−0.5	−0.5
F_2	−31.3	−32.8	−34.3	−35.5	−25.1	−26.6	−28.1	−29.5
H_2S	−1.0	+1.0	+1.0	+1.0	−4.6	−0.1	+1.0	+1.0
S_2	−4.4	−3.4	−6.4	−8.8	−8.8	−2.9	−3.7	−6.5
SO_2	+1.0	−1.5	−6.0	−10.2	+1.0	+0.95	−2.5	−6.9
SO_3	−5.2	−9.2	−15.2	−20.9	−3.2	−4.8	−9.7	−15.6

† Assumed fugacities: H_2O, 1,000 atm; HCl, 10 atm; HF, 0.3 atm; total sulfur gases, 10 atm.

‡ Numbers are logarithms of fugacities expressed in atmospheres (approximately equal to partial pressures except for H_2O).

are meaningful as a measure of the state of oxidation of the gas, i.e., of the general extent to which equilibria like those in Table 17-1 are displaced toward the oxidizing or reducing side.

Calculation of Metal Concentrations

To find the amount of a metal that can volatilize into this gas mixture, we must know first of all what stable solid or liquid compound the metal would form under magmatic conditions. Lead, for example, might conceivably exist as a sulfide, PbS; a sulfate, $PbSO_4$; a silicate, $PbSiO_3$ or Pb_2SiO_4; a chloride, $PbCl_2$; or an oxide, PbO (higher oxides decompose at these temperatures). To test these possibilities, we calculate free energies and equilibrium constants for reactions like

$$PbS + 2O_2 \rightleftharpoons PbSO_4 \tag{17-2}$$

$$PbS + H_2O + SiO_2 \rightleftharpoons PbSiO_3 + H_2S \tag{17-3}$$

and use the fugacities from Table 17-2 to see in which direction each equilibrium is displaced. For lead it turns out (as we might have predicted from its geologic behavior) that the sulfide is the stable solid under nearly all the conditions represented in Table 17-2. Likewise for zinc, copper, silver, and molybdenum; manganese is most stable as the silicate, tin and tungsten as the oxides, again in accordance with their usual modes of occurrence in nature.

The final problem is to test the volatilities of various possible compounds that might be in equilibrium with the most stable solid or liquid. Continuing to use lead as an example, we test first the volatility of its sulfide,

$$PbS(s) \rightleftharpoons PbS(g) \tag{17-4}$$

then of metallic lead in equilibrium with the sulfide,

$$PbS(s) \rightleftharpoons Pb(g) + \tfrac{1}{2}S_2(g) \tag{17-5}$$

and then of the chloride in equilibrium with the sulfide,

$$PbS(s) + 2HCl \rightleftharpoons H_2S + PbCl_2(g) \tag{17-6}$$

Again the calculations involve finding free energies, deriving the equilibrium constants, and then substituting where necessary for $[S_2]$, $[HCl]$, and $[H_2S]$ from Table 17-2. For lead, as for most of the metals, it turns out that the chloride is responsible for almost the entire volatility. Molybdenum, on the other hand, is volatile chiefly as the oxide, and tungsten as the hydrated oxide. The results of the volatility calculations are given in Table 17-3. From the table we read, for example, that magmatic gas in contact with galena and having an oxygen partial pressure of 10^{-17} atm would contain $10^{-1.1}$ atm of $PbCl_2$ at 627° and $10^{-0.7}$ atm at 827°.

Table 17-3 Maximum vapor pressures of metal compounds in equilibrium with the most stable solids in magmatic gas at 627 and 827°C

	627°				827°			
O_2	-14	-17	-20	-23	-8	-11	-14	-17
$FeCl_2$	-3.7	-3.2	-2.9	-2.9	-3.2	-2.7	-2.2	-2.0
$MnCl_2$	-3.9	-3.9	-3.9	-3.6	-3.7	-3.7	-3.7	-3.6
$ZnCl_2$	-0.3	-2.3	-2.3	-2.3	$+3.9$	-0.1	-1.1	-1.1
$PbCl_2$	$+0.9$	-1.1	-1.1	-1.1	$+2.4$	$+0.5$	-0.4	-0.7
$CuCl\ddagger$	-3.7	-6.7	-6.7	-6.7	$+2.8$	-3.9	-5.4	-5.4
$AgCl$	-5.8	-6.8	-6.8	-7.4	-3.0	-4.2	-4.8	-5.2
$SnCl_2\S$	-3.8	-2.4	-0.9	$+0.1$	-3.9	-2.5	-1.0	$+0.3$
MoO_3	-4.6	-7.1	-8.6	-10.1	-1.9	-1.9	-4.2	-5.7
$WO_3 \cdot H_2O$	-12.0	-12.0	-12.0	-12.0	-4.7	-4.7	-4.7	-4.7

† Numbers are logarithms of partial pressures expressed in atmospheres.
‡ Cu is present as CuCl and Cu_3Cl_3. The numbers are total copper expressed as CuCl.
§ The numbers for log $O_2 = -14$ and -8 include a small contribution from $SnCl_4$.

Significance of the Calculations

What do the numbers in Table 17-3 mean? More specifically, how high must the volatility of a metal be in order for appreciable amounts of it to be transported in the gaseous form? There is no simple answer to this question, but a rough calculation will indicate the orders of magnitude involved. If we imagine a body of magma measuring 100 by 100 km in surface area and 10 km deep, containing 1% of dissolved water, then the total weight of water would be about 3×10^{18} g (Prob. 15-2). This is about 10^{17} moles, which would occupy about 10^{16} liters at 800° and 1,000 atm. For a metal whose compounds have a total vapor pressure of 10^{-7} atm at 800°, and which has an atomic weight of about 100, 10^{16} liters would contain about 1,000 tons of the metal. For this amount to form an ore deposit, all the gas from the magma would have to precipitate its metal content in one small area, which is not very likely. But perhaps 10^{-7} atm may be considered an extreme lower limit of partial pressure, beyond which the amount of metal transported would be too small to matter. Probably 10^{-6} atm would be a more reasonable lower limit for purposes of discussion.

On this basis our calculations show that all the metals listed can be present in a gas phase at 827° in amounts large enough to be important for ore accumulation. For most metals the amount decreases markedly as the temperature falls, but at 627° all except copper, silver, molybdenum, and tungsten are still above the 10^{-6} atm limit. Possibly some ore deposits are formed by precipitation out of such a high-temperature gas phase: deposits of magnetite, chalcopyrite, or scheelite immediately adjacent to an intrusive contact would be likely examples. But probably most deposits have a longer history, the metals originally carried as gases becoming part of a liquid solution and being carried in this form to regions of lower temperature.

As in all such theoretical calculations, the limitations of the figures should be kept in mind. The numbers are based on chemical data, and the data are not always reliable, as is shown by the fact that many of them in recent years have been very considerably revised. The numbers for tin seem particularly suspect, because the high computed volatilities and the lack of change of volatility with temperature are difficult to reconcile with the preferred occurrence of this metal in high-temperature deposits. The calculated partial pressures are *maximum* figures in the sense that a metal is assumed to be present in sufficient quantity to exist in a separate solid or liquid phase in equilibrium with the gas; they are *minimum* figures in that they do not include the amounts of metal compounds dissolved by the gas. The calculations are made for a specific, hypothetical gas composition and would change for different assumed proportions of water, HCl, and sulfur. Perfect-gas behavior is assumed, in that pressures are tabulated rather than fugacities; this is another way of saying that interaction between gas molecules is assumed to be negligible. Data are lacking for some compounds that might influence the volatilities, particularly the fluorides, but extrapolation from scanty data suggests that for most metals the fluorides are not likely to be as volatile as the chlorides. Despite all these reservations, the calculations do show clearly that volatility is an important factor in the high-temperature behavior of the common ore metals.

17-2 HYDROTHERMAL SOLUTIONS: THE DILEMMA

We cannot follow the history of any metal from its possible original presence in a magmatic gas to its ultimate deposition from a low-temperature or medium-temperature aqueous solution, because we lack data on the solubilities of metals and metal compounds in solutions at high temperatures and pressures. We had best jump to the other end of the temperature scale, to deposits formed at temperatures of the order of 50 to 250°, where geologic evidence and laboratory experiments have given us some insight into the behavior of the ore-depositing solutions. Even here, however, the problems are by no means solved.

The major difficulty is the extreme insolubility of the metal sulfides. Most of the familiar ore metals—lead, zinc, copper, silver, molybdenum, mercury—occur chiefly as sulfides, or as complex minerals containing arsenic or antimony in addition to sulfur. Geologic occurrences indicate that these compounds were deposited from solution in water at moderate temperatures; yet these same compounds in the laboratory are among the most insoluble with which a chemist has to deal. What sort of solutions can be imagined to carry these very insoluble substances in nature? This is the essential dilemma of the hydrothermal hypothesis of ore deposition.

To illustrate how serious the dilemma is, let us try some rough calculations for zinc sulfide, which is by no means the most insoluble of the sulfides. The solubility product of crystalline sphalerite is about 10^{-24}; measurements are not in complete agreement, but this figure is probably correct within an order of magnitude.

The solubility in pure water is not the simple square root of this number, because of hydrolysis of the sulfide ion:

$$ZnS + H_2O \rightleftharpoons Zn^{2+} + HS^- + OH^- \tag{17-7}$$

To calculate the solubility, we follow the same procedure that we used for $CaCO_3$ in Sec. 3-1: we note that $[Zn^{2+}] = 10^{-24}/[S^{2-}]$; that in pure water the concentration of Zn^{2+} is equal to the sum of the concentrations of the different forms of sulfur, $[S^{2-}] + [HS^-] + [H_2S]$; and that the distribution of sulfur is determined by the equilibria

$$H_2S \rightleftharpoons H^+ + HS^- \qquad \frac{[H^+][HS^-]}{[H_2S]} = K_1 = \text{approximately } 10^{-7} \text{ at } 25°$$

$$HS^- \rightleftharpoons H^+ + S^{2-} \qquad \frac{[H^+][S^{2-}]}{[HS^-]} = K_2 = \text{approximately } 10^{-13} \text{ at } 25°$$

Combination of these equations gives for the solubility

$$[Zn^{2+}] = \sqrt{K_{ZnS}\left(1 + \frac{[H^+]}{K_2} + \frac{[H^+]^2}{K_1 K_2}\right)}$$

$$= \sqrt{10^{-24}\left(1 + \frac{[H^+]}{10^{-13}} + \frac{[H^+]^2}{10^{-20}}\right)}$$

At pH 7 at 25°C the solubility is therefore $1.4 \times 10^{-9}M$, or about 10^{-7} g/liter of Zn^{2+}, nearly 1,000 times larger than the square root of the solubility product.

At pH 5 the calculated solubility is $10^{-7}M$, and at pH 9 it is $10^{-10}M$. Additional S^{2-} in the solution would diminish the solubility; the presence of other dissolved electrolytes would raise the solubility slightly (Secs. 1-5 and 3-5).

The effect of temperature on the solubility can be estimated from Eq. (8-3),

$$\log K = \frac{-\Delta H}{2.303RT} + C$$

on the assumption that ΔH is constant for all three reactions. This assumption is probably reasonable up to 100°, shaky at 200°, and a sheer guess at higher temperatures. The calculation cannot be made more accurate with presently available data, because heat capacities for the ions are not known. The results of the calculation with constant ΔH are expressed in Table 17-4 as grams of Zn in a liter of solution.

Table 17-4

	25°	100°	200°
pH 5	8×10^{-6}	1×10^{-5}	2×10^{-5}
pH 7	1×10^{-7}	3×10^{-7}	9×10^{-7}
pH 9	1×10^{-8}	4×10^{-8}	1×10^{-7}

Now what meaning can we attach to such figures? If a solution contains amounts of Zn as small as these figures indicate, could it deposit enough sphalerite to make an ore deposit of reasonable size? Given enough solution and enough time, any solution, no matter how dilute, could accomplish the purpose. But what quantity of solution and what quantity of time would be geologically reasonable?

This is similar to the question we tried to answer in the last section, as to the minimum concentration of metal in a magmatic gas that would be significant for the formation of ore deposits. We proceed in the same way, using rough numbers to establish a limit of reasonableness. Suppose, for example, that an ore solution carries 10^{-7} g/liter of zinc. To deposit 1 ton of metal would require a minimum of 10^{10} cubic meters of solution, approximately the volume of water carried to the sea each year by the Hudson River (average flow approximately 10,000 sec-ft). Such a solution traversing a vein at a rate of 10 ft^3/sec could deposit 1 ton of zinc in a thousand years, provided that *all* the dissolved zinc precipitates. The amount of water and the amount of time seem excessive, by comparison with scanty data on the flow of hot springs and on the geologic times required for the formation of ore bodies. Thus 10^{-7} g/liter can be taken as an absolute minimum, below which the concentration of metal is too small to be of interest. For most purposes a somewhat larger figure, say 10^{-5} g/liter, is a more reasonable minimum.

By this criterion the solubility of ZnS is barely high enough to be of interest at a temperature of 200° and a pH as low as 5. The calculated solubilities of the sulfides of some other common metals (Mn, Fe, Co, Pb) have a similar order of magnitude, but those for the sulfides of Ag, Cu, and Hg are much smaller. Thus the amounts of metal that can be carried by hot sulfide solutions seem far too small, except for a few metals under the most favorable assumed conditions, to account for the origin of ore deposits. This is the long-standing difficulty with the classical hydrothermal hypothesis.

17-3 HYDROTHERMAL SOLUTIONS: GEOLOGIC DATA

Speculative possibilities for circumventing the difficulty discussed in the preceding section are not hard to suggest. If the solutions are strongly acid, for example, or if all ore deposition takes place well above 200°C, then simple solution of the sulfides might be an adequate explanation. Before embarking on such speculations, however, let us see what limitations are imposed on the nature of ore-forming solutions by geologic data.

Composition

On general theoretical grounds we could guess that an ore solution would contain the common substances that water dissolves out of rocks, plus possible additions from a magma: chiefly silica, the cations Na^+, K^+, Ca^{2+}, Mg^{2+}, and the anions Cl^-, SO_4^{2-}, HCO_3^-. To these would be added the metals and sulfur necessary to form the ore minerals. Some confirmation of this theoretical guess about composition is provided by hot springs, which generally contain the list of common ions just mentioned and sometimes traces of ore metals as well. Hot-spring water at best, however, is a poor sample of an ore solution, because it is greatly diluted with meteoric water and because it has lost most of whatever metals it may have contained somewhere below (Sec. 16-5).

A better source of data on the composition of ore solutions is the fluid contained in tiny cavities in ore and gangue minerals (Sec. 16-6), fluid that should represent the solution with which the crystals were in contact at the time of crystallization (Roedder, 1967). Working with such tiny amounts of fluid is a difficult experimental problem, but enough analyses have been obtained to establish that the principal dissolved substances are the ions listed above. A surprising observation is that the liquid is generally a fairly concentrated salt solution. Sodium chloride is by far the most abundant salt, sometimes showing concentrations in excess of $2M$, much greater than its concentration in seawater. The fluid from cavities has the same disadvantage as hot-spring water, that it can represent at best only the spent solution from which ores have been removed, and not the original metal-containing fluid.

Analyses of wall rock adjacent to ore deposits often indicate that the ore fluids have penetrated the rock and changed its composition profoundly. Among the elements commonly added are Si, C, Mg, Na, K, Ca, and Fe. This list corroborates the analyses of hot-spring water and fluid from cavities; the ions missing from the list (Cl^- and SO_4^{2-}) would not be expected in large amounts, because their compounds are for the most part very soluble.

In summary, meager evidence indicates that the dissolved material of ore-forming solutions, aside from their content of heavy metals, consists largely of silica and the common ions of groundwater. The most abundant ions are Na^+ and Cl^-, and in some ore solutions these are highly concentrated.

Temperature

To get a fix on the temperatures of ore deposition has been the goal of much recent geochemical research. Many methods have been explored, and in favorable cases some of these permit estimates to within a few tens of degrees. The general problem is far from solved, however. A common sequence of events has been the proposal of a new method, its enthusiastic application, and then the discovery, after detailed study, that the method is affected by so many variables that its results are far less reliable than at first appeared. Detailed discussion would be out of place here, but the following paragraphs describe a few of the possible methods and illustrate the difficulties involved.

TRANSITION TEMPERATURES

Phase transitions in minerals, either from solid to liquid or from one solid crystal form to another, generally take place at sharply defined temperatures. The melting point of a mineral obviously fixes its maximum temperature of crystallization; the difficulty is that melting points are influenced by pressure, by other substances in solution, and often very greatly by the presence of volatiles. One need only recall the familiar effect of pressure on the melting point of ice, the effect of anorthite on the melting point of diopside (Sec. 13-3), and the drastic effect of a little water on the melting point of silica (Sec. 15-3), to appreciate how unreliable melting temperatures can be. A solid-solid inversion point should give a maximum temperature reading if the low-temperature form of a mineral is present, a minimum if the high-temperature form is present. This involves the assumption that the high-temperature form inverts slowly enough that it can be cooled through the transition point without changing form, or else that recognizable characteristics of the high-temperature form survive inversion; this assumption is known to be valid for a few minerals (e.g., the high-low transition in quartz) but is probably not safe for most. Another source of confusion in using inversion points is the ability of some high-temperature polymorphs to crystallize metastably at low temperatures; an often-cited example is the appearance in low-temperature environments of the high-temperature modifications of silica, tridymite and cristobalite. The effects of pressure on the temperature of phase changes can be estimated, if the enthalpy changes and specific volumes are known, but the effects of other substances and of metastability are essentially unpredictable. Melting points and inversion points are often called upon to set broad limits on ore-forming temperatures, but only a very few minerals are suitable (Table 17-5) and the information they give may be equivocal.

Table 17-5 Melting points and inversion points (under 1000°C) of common minerals in and near ore deposits†

		Melting points, °C	
Silver	Ag	960	
Arsenic	As	817 (at 36 atm)	
Antimony	Sb	630	
Bismuth	Bi	271	
Argentite	Ag_2S	788	
Cinnabar	HgS	Sublimes appreciably over 250°	
Realgar	AsS	320	
Orpiment	As_2S_3	310	
Stibnite	Sb_2S_3	546	
Pyrargyrite	Ag_3SbS_3	485	
Cerargyrite	AgCl	455	
		Inversion points, °C	
Chalcocite	Cu_2S	105	Orthorhombic to hexagonal
Acanthite	Ag_2S	175	Orthorhombic to isometric (argentite)
Realgar	AsS	267	To a black polymorph
Orpiment	As_2S_3	170	To a red polymorph
Anatase	TiO_2	800	To rutile
Quartz	SiO_2	573	Low to high quartz; increases with pressure, 28°/1,000 atm
Quartz	SiO_2	870	High quartz to tridymite

† *Source: Ingerson, Econ. Geology, 50th Anniversary vol., pp. 350–357, 1955.*

MINERAL ASSOCIATIONS

The most common basis for qualitative guesses about geologic temperatures is the presence or absence of certain minerals and mineral assemblages. Kaolinite, for example, is not stable above about 350°, serpentine above 500°, pyrite above 740°; garnet-pyroxene is commonly regarded as a high-temperature metamorphic assemblage, albite-epidote-chlorite a low-temperature assemblage. Such statements, in more formal language, mean that chemical equilibria involving minerals are shifted in one direction or the other by changes in temperature. To say this, however, indicates at once the great difficulty in using minerals and mineral associations as temperature indicators: equilibria are so sensitive to other influences besides temperature, particularly to pressure and to the presence of other substances, that their use is seldom free from uncertainty. Pressure is particularly important for reactions involving a gas phase, such as dehydration and decarbonation processes; the temperature indicated by a hydrated mineral may vary by as much as 500°, depending on whether the pressure of water vapor at the time of its formation is assumed to be high or low. Mineral associations often give valuable qualitative clues about ore-forming temperatures but are seldom by themselves suitable for making quantitative estimates.

DISAPPEARANCE OF GAS BUBBLES IN FLUID INCLUSIONS

The commonest fluid inclusions in ore minerals and gangue minerals consist of liquid plus a bubble of gas. Presumably an inclusion at the time of its formation contained only a single phase, which has separated into two on cooling. Hence the minimum temperature of crystallization of the host mineral should be determinable by heating until the phase boundaries in its inclusions just disappear. One must be sure that the inclusions examined are primary, in the sense that they were formed during crystallization of the host and not long afterward; one must also be able to make a rough estimate of pressure during crystallization, since pressure has an appreciable effect on the temperature at which two phases can exist. In a good many ore deposits these conditions are fulfilled, and the fluid-inclusion method has proved to be one of the most generally useful ways to estimate temperature.

COMPOSITION OF MIX CRYSTALS

Limited isomorphous substitution is common in ore minerals, and the extent of substitution generally increases as the temperature rises. Seemingly this phenomenon should be a good basis for temperature determination; one need only find experimentally how much of a compound can mix with another over a range of temperature and then analyze natural minerals to see where their compositions lie on the experimental curve. A moment's thought, however, suggests that the method requires fulfillment of some pretty stringent conditions. Suppose, for example, that a mineral with the formula AB is crystallizing from a solution containing a metal Z that can substitute isomorphously for A; the amount of Z that can be taken up by AB depends on the temperature, but it also depends on several other variables—the concentration of Z in the solution, the pressure, and the concentration of other elements that may substitute for A. If enough Z is present during crystallization so that another mineral ZB appears simultaneously with AB, then one may be sure that AB contains the maximum possible amount of Z for a particular temperature; pressure can often be estimated within reasonable limits; and the disturbing effects of other elements can be studied experimentally. Hence it should not be impossible to meet the necessary conditions for applying this method, but in practice it has proved extraordinarily difficult to find suitable minerals.

For several years the mineral sphalerite, ZnS, seemed to be a good candidate. Most sphalerite contains appreciable iron substituting for zinc, so that the formula is often written (Zn,Fe)S. The depth of color of the mineral increases roughly with its iron content, and in a very general way dark-colored sphalerites are known to occur in higher-temperature deposits. To make these observations quantitative seemed to require only a careful study of the Zn-Fe-S system, in particular a determination of the amount of iron in sphalerite at various temperatures when the mineral crystallized simultaneously with FeS. Then in natural deposits where textures showed that sphalerite and pyrrhotite had formed together, the iron content of the sphalerite would reveal its temperature of deposition. The method seemed particularly promising because pressure had only a small effect, because other common impurities in sphalerite had little influence on the substitution of iron, and because the analysis of sphalerite for iron could be carried out quickly and easily. It was recognized, of course, that naturally occurring pyrrhotite does not have precisely the composition FeS, but the reproducibility of experimental results seemed to show that the slight difference in composition was unimportant. A careful reexamination of the method, however, showed that the fugacity of sulfur vapor (which controls the composition of pyrrhotite) is an important variable in determining how much iron goes into sphalerite and that the apparent reproducibility of earlier results was largely an accidental consequence of the experimental techniques employed. Then it appeared that the method might still be used in deposits where pyrite was present in addition to sphalerite and pyrrhotite, because the combination pyrite-pyrrhotite would fix the sulfur fugacity. Even this hope has proved illusory, because it turns out that the iron content of sphalerite with this combination of minerals is practically constant at temperatures below 550°C (Scott and Barnes, 1971).

The sad history of the sphalerite geothermometer has cast doubt on the possibility of using mix-crystals for temperature determination, despite the theoretical soundness of the idea.

DISTRIBUTION OF TRACE ELEMENTS AND ISOTOPES

A related method that also involves isomorphous substitution depends on measurements of concentrations of a trace element in two minerals that have formed at the same time. The rare metal cadmium, for example, is nearly always present in detectable amounts in sphalerite and galena. If crystals of these two minerals form together from a hydrothermal solution that contains a small concentration of cadmium, the cadmium will distribute itself between the two in a ratio (called the *distribution coefficient* or *partition coefficient*) that depends on temperature. Analyses for cadmium in the sphalerite and galena of an ore deposit, therefore, should provide data from which temperature can be calculated. In the few deposits where the method has been tried, it gives fairly satisfactory checks with temperatures found from fluid inclusions. Wide usefulness of the method seems doubtful, however, because its success depends on a demonstration that the two minerals crystallized simultaneously and because the cadmium ratio depends to some extent on additional variables, such as pressure and amounts of other trace metals present.

Besides trace metals, other substances always present in minor amounts that distribute themselves in definite ratios among crystallizing minerals are the heavy isotopes of some of the elements, especially oxygen and sulfur (Sec. 21-4). The ratio $^{18}O/^{16}O$ in an oxide, carbonate, or silicate mineral depends on bond strengths, the heavy isotope being favored in lattice positions with the strongest bonds to other ions, and the degree of favoring depends on temperature. When two minerals crystallize from the same solution, the heavy isotope distributes itself between them in a ratio determined by relative bond strengths and temperature. In quartz and magnetite that have formed together, for example, the concentration of ^{18}O is considerably larger in the quartz, and the difference is greater at low temperatures than at high temperatures. Measurements of $^{18}O/^{16}O$ ratios for different minerals, therefore, give a means for estimating temperatures, and this method has proved particularly useful for finding temperatures of crystallization of igneous rocks. It is less satisfactory for hydrothermal ores, because in an ore deposit, isotope ratios have often been changed by partial re-equilibration with solutions that moved through the deposit after the original minerals were formed.

Somewhat more promising results have been obtained with ratios of $^{34}S/^{32}S$ in sulfide minerals that crystallized together, particularly sphalerite and galena (e.g., Rye 1974). Obvious causes for uncertainty are the difficulty of establishing simultaneous crystallization and the possibility of isotopic re-equilibrium with later solutions; whether the method will be useful despite these uncertainties can only be determined by further testing it against other methods in a variety of deposits.

In summary, several methods are used for estimating temperatures of ore deposition, the most generally reliable being the measurement of homogenization temperatures of fluid inclusions. For most vein and replacement deposits the estimated temperatures are in the range 50 to 550°C. Commonly a deposit forms not at a single temperature but over a temperature interval; for some the interval is narrow, but for a few it covers most of the 50-to-550° range.

Pressure

Pressure has figured largely in the above discussion of temperature, and it should be obvious that the two variables are closely related. Most methods of determining one will necessarily involve some sort of guess about the magnitude of the other; or, to state the same thing in different words, most determinations give actually a relation between temperature and pressure rather than either variable by itself. For pressure there are no good independent methods. The usual procedure is simply to estimate the pressure during ore deposition by reconstructing the geology and guessing at the depth of burial. A maximum pressure can be estimated by noting that most ore deposits are formed at depths of less than about 20 km; this limit is set by the facts that temperatures at 20 km are of the order of 600° and that fractures cannot long remain open because of the flow of rocks under the weight of overlying material. If ore solutions are assumed to move in a network of fissures at least partly communicating with the surface, the pressure at a depth of 20 km can be calculated roughly as that at the base of a 20-km column of water, or 2,000 atm. This rather modest figure is thus the upper limit of pressures for deposition in veins. Probably most ores are formed at much lower pressures, of the order of a few hundred atmospheres.

Acidity

Of the many acrimonious debates that have sprung up over the character of ore solutions, perhaps the bitterest has concerned the question of acidity versus alkalinity. The question is important, or seemingly important, because at ordinary temperatures some of the heavy metal sulfides are soluble in acid solutions and others in alkaline solutions. If the solubilities are enhanced at higher temperatures, then (so runs the argument) changes in pH may be a critical factor in the formation of ore deposits.

Perhaps it should be recalled once more that we are using the terms "acidity" and "alkalinity" exclusively with reference to pH. A solution with abundant Na^+ and K^+, for example, is often called "alkaline" by geologists simply because these are ions of the alkali metals; actually it may be either acid or alkaline, depending on whether H^+ or OH^- is in excess.

Geologic data on the question of pH are very scanty and inconclusive. The strongly acid character of volcanic emanations has been cited as evidence for acid ore solutions, but the volcanic environment, where pressure is suddenly released near the surface, has little resemblance to the deeper environments where most ore deposits form. On general principles one would expect ore solutions to be pretty close to neutral, say within 2 pH units of neutrality, because strongly acid solutions would be quickly neutralized by reaction with silicate and carbonate minerals, and strongly basic solutions would be neutralized by silica. Hot-spring waters seem to corroborate this statement, in that they are mostly near the neutral point, but their characteristics are so obviously modified by their immediate surroundings and by reactions with the atmosphere that this evidence can have little weight. The fact that the fluid from tiny cavities in minerals of ore deposits is approximately neutral provides more convincing evidence. Wall-rock alteration to chlorite and montmorillonite is commonly thought to signify alkaline solutions, while kaolinite and alunite suggest acid solutions, and experimental work gives some support to these conclusions; but the degree of acidity or alkalinity required is not certain and probably is not large. Still less diagnostic is sericite, which is often cited as an indicator of alkaline solutions but which can be produced experimentally in either an acid or an alkaline environment, provided that enough potassium is present. The minerals wurtzite, marcasite, and metacinnabar, if formed metastably at low to moderate temperatures, are often said to indicate acid solutions; but this conclusion has been called in question by recent experimental work, and at best these minerals can give no quantitative information about pH.

From a theoretical standpoint, a discussion of acidity and alkalinity at high temperatures is complicated by the fact that the dissociation constant of water changes as the temperature rises. As shown in Figs. 2-1 and 17-1, the constant increases to a maximum of $10^{-11.0}$ at 275°, then decreases. This means that the concentrations of both H^+ and OH^- in a neutral solution can be as high as $10^{-5.5}M$, in other words that water at first becomes *both* a better acid and a better base as the temperature rises. The pH of a solution at 275° depends, just as it does at ordinary temperatures, on how much acid is supplied to it, on the dissociation constants of the acids present, and on the extent to which adjacent minerals hydrolyze. These are generally unpredictable from available data, and it seems pointless to attempt even rough calculations. Probably reaction with the wall rock is sufficient to keep the pH not far from neutral.

At still higher temperatures the dissociation of water and of electrolytes dissolved in it becomes critically dependent on the density, hence on the pressure, to a greater extent than on the temperature. Estimates of the dissociation constants of water at high temperatures and various densities are shown on the right-hand side of Fig. 17-1. Salts like KCl and strong acids like HCl become practically undissociated when the density falls to 0.3 or 0.4 g/cm^3. Under such conditions the ordinary concepts of acid and base are difficult to apply: a mixture of HCl and H$_2$O having, say, a density of 0.4 g/cm^3 at 600° would be potentially " acid," in the sense that the HCl could react with adjacent rocks and that the HCl would dissociate when the solution cooled; but as long as the density stayed low, the pH would be only a trifle below that of pure water.

Thus there is no simple answer to the question. Are ore solutions acid or alkaline? Presumably they may be either, but the weight of evidence both from geologic observation and from theory indicates that the pH usually does not stray far from the neutral point. In the high-temperature part of the range of ore deposition, it would be possible even to have a somewhat alkaline solution containing undissociated acid gases. In the low-temperature part of the range, with which we are particularly concerned here, we probably need not consider solutions with pH's more than 2 units different from neutral.

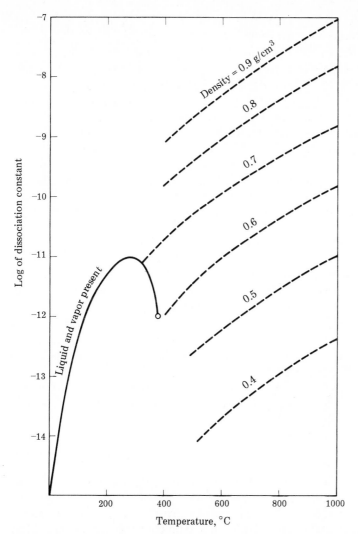

Figure 17-1 Ionic dissociation constant of water in high-temperature fluids of various densities. The solid line is the curve of Fig. 2-1, for liquid water under its own vapor pressure. The dashed lines show values of the constant for single-phase fluid water under sufficient pressure to maintain the indicated densities. (*Sources: Solid curve, J. R. Fisher and H. L. Barnes, Jour. Phys. Chemistry, vol. 76, pp. 90–99, 1972. Dashed curves, A. S. Quist, ibid., vol. 74, pp. 3396–3402, 1970.*)

Sequence of Minerals

Another hint as to the nature of ore solutions comes from observations on sequences of minerals in ore deposits. The sequence of deposition may be shown by (1) location in a vein, the earlier minerals being deposited nearer the wall rock, (2) replacement textures, the later minerals surrounding and embaying earlier ones,

(3) distance from the supposed parent intrusive, the nature of the minerals changing as the zone of mineralization is followed outward into the country rock. The order is variable from one deposit to another, and may vary within a single deposit, but shows sufficient regularity that a sequence of minerals can be set up which is followed at least roughly in most individual deposits and in most areas of ore deposition. The regularity suggests that the sequence is an expression of physico-chemical factors generally operative in ore deposition, most likely a decrease in temperature and pressure, either with time or with distance from the parent intrusive or both. This is the basis for Lindgren's (1933) classification of ore deposits into *hypothermal* (early-formed minerals, closest to the igneous source, presumably formed at the highest temperatures and greatest depths), *mesothermal* (formed at intermediate temperatures), and *epithermal* (late-formed minerals, farthest from the source, formed at lowest temperatures). Other evidence, from types of wall-rock alteration, nature of the gangue minerals, and structures in the vein, confirms this general classification according to decreasing temperature and pressure.

Minerals characteristic of the three zones are listed in Table 17-6. The table is not intended to be complete, nor to describe accurately the mineral assemblages in particular ore bodies; it gives only a generalized picture of the commoner minerals found in typical deposits of the various types. Notable are the presence of oxides and silicates in hypothermal deposits, and minerals of tungsten, molybdenum, tin, and bismuth; the abundance of copper, lead, and zinc minerals in mesothermal deposits, and of magnesium and iron carbonates among the gangue minerals; and the prevalence of minerals of silver, gold, antimony, and mercury in epithermal deposits. Within any one range there may be further regularities in order of deposition. For example, in mesothermal deposits the usual sequence is sphalerite—copper minerals—galena, the galena not only appearing later than the zinc and copper minerals at any one point in a vein but also being present at greater distances from the igneous source.

This is a very brief and generalized treatment, lumping together relationships which economic geologists distinguish as "zoning" (space relations of minerals over an area) and "paragenesis" (time order of mineral deposition as determined by layering and replacement structures). These relations seemingly should tell us something about the nature of the solutions responsible for deposition of the ores. We might guess, for example, that the ore minerals precipitate in the order of their solubility, the least soluble at high temperatures and the most soluble at low temperatures. The order observed in nature, however, shows little relationship to the order of solubilities calculated by the methods of the last section. In this respect also, then, the simple hydrothermal hypothesis fails to give a reasonable interpretation of geologic facts.

Summary From this survey of geologic data, we can guess that the solutions from which ores are formed have temperatures in the range 50 to 550° and pressures up to 2,000 atm; that they are usually saturated with silica and often contain high concentrations of NaCl; that they are neither strongly alkaline nor strongly acid;

Table 17-6 Principal minerals in ore deposits formed in different temperature zones†

────Temperature and depth decreasing ──────────→		
Hypothermal	Mesothermal	Epithermal
Ore minerals		
Wolframite, (Fe, Mn)WO$_4$	Chalcopyite, CuFeS$_2$	Argentite, Ag$_2$S
Cassiterite, SnO$_2$	Bornite, Cu$_5$FeS$_4$	Silver, Ag
(Hematite, Fe$_2$O$_3$)	Tetrahedrite, Cu$_{12}$Sb$_4$S$_{13}$	Proustite, Ag$_3$AsS$_3$
Magnetite, Fe$_3$O$_4$	Enargite, Cu$_3$AsS$_4$	Pyrargyrite, Ag$_3$SbS$_3$
Arsenopyrite, FeAsS	Pyrite, FeS$_2$	Pyrite, FeS$_2$
Chalcopyrite, CuFeS$_2$	Galena, PbS	Marcasite, FeS$_2$
Gold, Au	Sphalerite, ZnS	Gold, Au
Molybdenite, MoS$_2$	Gold, Au	Cinnabar, HgS
Pyrrhotite, Fe$_{1-x}$S	Arsenopyrite, FeAsS	Stibnite, Sb$_2$S$_3$
(Bismuth, Bi)	(Argentite, Ag$_2$S)	(Galena, PbS)
(Galena, PbS)		(Sphalerite, ZnS)
(Sphalerite, ZnS)		(Chalcopyrite, CuFeS$_2$)
Gangue minerals		
Quartz	Quartz	Quartz
Tourmaline	Carbonates	Adularia
Garnet	Barite	Chalcedony
Topaz		Opal
Micas		Fluorite
Apatite		Alunite
		Calcite

† Parentheses indicate minerals found locally or in small amounts.

and that they are sufficiently similar from one place to another to show a fairly definite order of crystallization of ore minerals. There is little in all this to support a hypothesis of transportation of the ore metals in solution as the simple sulfides, for the sulfide solubilities would not be greatly increased by the temperatures, the pressures, the electrolytes, or the permitted variations in pH; and the order of solubilities provides no explanation for the observed order of deposition.

17-4 HYDROTHERMAL SOLUTIONS: POSSIBLE ANSWERS

If the hydrothermal hypothesis is to be saved, some way must be found out of the dilemma posed by low sulfide solubilities. We review here some of the suggestions that have been offered.

Sulfur and Metals in Separate Solutions

The essential difficulty is the transportation of metal ions and sulfide ions in the same solution. The problem would vanish if the metal ions could be supplied by one solution and sulfide ions by another, the ore minerals precipitating at the point where the two solutions meet. The obvious objection to such an idea is the special nature of the assumptions: it is just not credible that every one of the thousands of known sulfide ore deposits marks the place where two solutions of appropriate composition fortuitously came together. A more plausible variant of this suggestion is to imagine that sulfur-bearing solutions invade an area first, reacting with iron minerals of the country rock to form abundant pyrite, and that later solutions bring the ore metals which displace iron from its sulfide. Chemically this mechanism is feasible, for other metals can be shown to react with pyrite to form sulfides. Geologically it is not very hopeful, since only a few ore deposits show by their textures the expected replacement of early iron sulfides.

Colloidal Dispersion

Some ore deposits show features that are plausibly ascribed to colloidal phenomena: extremely fine dispersions of gold in quartz or of cinnabar in opal, layering of fine-grained material reminiscent of Liesegang rings, smooth colloform surfaces like the rounded surfaces of laboratory gels. Extrapolating from this hint, we can suppose that ore-forming solutions in general contain metal sulfides chiefly in the form of sols rather than in true solution; perhaps at very high temperatures the sulfides are dissolved, and when the solutions cool, the sulfides do not precipitate immediately but are carried into zones of low temperature as colloidal particles. So little is known about the behavior of colloids at high temperatures that it is difficult to say how much merit this hypothesis may have. Certainly, in solutions containing only small concentrations of metal ions, precipitation might well be delayed, and a colloidal dispersion might persist into a temperature zone below that at which crystallization would theoretically take place.

On the other hand, the behavior of colloids is so erratic at any temperature, so subject to the influence of minor impurities and surface characteristics, that the regular sequence of sulfide deposition from one deposit to another is hard to explain if colloids play an important role. It is questionable also whether colloids could be stable in solutions containing concentrated electrolyte moving through small cracks and interstices, and whether the large crystals so often found in ore deposits could be formed from colloidal dispersions. More than likely colloidal phenomena play a minor part in the formation of some deposits, but their importance as a general explanation of ore transportation seems very dubious.

Disequilibrium

Similar remarks can be made about the suggestion that the sulfides remain in solution metastably. The calculated solubilities, according to this argument, depend on the assumption of equilibrium, and strict maintenance of equilibrium is most unlikely in a liquid moving through rock and changing rapidly in temperature and pressure. Hence it is sometimes conjectured that the assumption of equilibrium is completely fallacious, that the metals and sulfur can remain in supersaturated solution down to low temperatures. Definitive experimental data are again lacking, but we can reflect that extensive supersaturation is unlikely in a hot liquid continually and intimately exposed to rock surfaces, and that the vagaries in behavior of supersaturated solutions are hard to reconcile with the definite order of deposition so often observed in nature. Until clear evidence is presented to the contrary, an assumption of approximate equilibrium seems amply justified.

Complex Sulfide Ions

One of the steps in standard qualitative analysis is the dissolving of the sulfides of tin, arsenic, and antimony with ammonium polysulfide. If alkali sulfide or polysulfide is used instead of the ammonium compound, mercury sulfide also is dissolved. The process may be symbolized, using mercury as an example,

$$HgS + S^{2-} \rightarrow HgS_2^{2-} \tag{17-8}$$

This rather spectacular reaction, in which the most insoluble of all sulfides is brought into solution by a not very drastic reagent, has led to another suggestion for modifying the hydrothermal hypothesis: perhaps ore-bearing solutions contain an excess of S^{2-}, and the metals remain in solution as complex sulfide ions like HgS_2^{2-}. A solution that contains appreciable S^{2-} is necessarily alkaline (Fig. 2-3), and a solution capable of dissolving HgS is therefore an alkaline sulfide solution. The conjecture that such solutions play an important role in ore transportation is called the "alkaline-sulfide hypothesis."

As a general explanation for the origin of sulfide ores this hypothesis has little to recommend it. Only the four metals just mentioned have sulfide complexes sufficiently stable to be important under geologic conditions, and even for these the degree of alkalinity required is so high that its common occurrence may be questioned. Dickson (1964) notes that cinnabar has been recently deposited on the apron of a hot spring in eastern California whose water has a pH of 9, and here the complex sulfide ion may indeed play a role in the transportation of mercury. On the other hand, cinnabar and stibnite can be caused to precipitate from the water at Steamboat Springs in Nevada (White, 1955), where the pH is close to neutral and the concentration of sulfide ion is negligible. Thus the alkaline-sulfide hypothesis offers an explanation for the solubility of a very few metals under special circumstances, but its complete inability to account for the solubility of other common metals makes it inadmissible as a general answer to the dilemma of hydrothermal solutions.

Other Complexes

The existence of stable sulfide complexes of a few metals, however, has pointed the way to a promising solution for the dilemma of hydrothermal ore formation. Many other kinds of complexes have been studied in recent years, and some of these are probably stable enough at high temperatures to keep appreciable amounts of the ore metals dissolved even in the presence of sulfide ion. Such investigations are experimentally difficult, and have been carried out almost entirely by geologists and geochemists; research on inorganic complexes at temperatures of 100 to 600°C has attracted little interest among chemists.

One kind of complex that probably plays a role in the transportation of some metals is a complex involving the hydrosulfide ion, HS^-. Existence of such complexes is established by experiments showing that metals can be kept in solution at concentrations well above those allowed by the solubility products if the solution is near neutral and contains abundant H_2S. Under such conditions the principal sulfur ion is HS^-, and the relation of metal concentration to pH and total sulfur makes it possible to pick a complex of this ion as the most likely form of the metal in solution. Zinc, for example, shows an enhanced solubility in H_2S solutions between 25 and 200° that can be accounted for by formation of the ion $ZnS(HS)^-$; the solubility reaches values up to $0.03M$ under geologically reasonable conditions. In similar fashion, solubilities high enough to be of geologic interest have been demonstrated for Hg [as $Hg(HS)_2$ and $HgS(HS)^-$], for Ag (as AgHS), and for Pb [as $PbS(HS)^-$]. The major objection to such complexes is the high concentration of H_2S and HS^- required to keep them stable, a concentration much larger than that usually found in hot-spring waters and in fluid inclusions.

The high concentrations of sodium chloride often observed in analyses of fluid inclusions lead to the conjecture that chloride complexes might be important in preventing premature precipitation of sulfides. Many chloride complexes have long been known; for example, silver chloride dissolves readily in a solution of sodium chloride:

$$AgCl + Cl^- \rightleftharpoons AgCl_2^- \tag{17-9}$$

If this complex is sufficiently stable, it seems possible that Ag_2S might be kept from precipitating by the reaction

$$Ag_2S + 4Cl^- \rightleftharpoons 2AgCl_2^- + S^{2-} \tag{17-10}$$

This possibility can be explored for silver and some other metals on the basis of equilibrium data for the complexes. (See, for example, the stability constants for a few chloride complexes in Table VII-4, Appendix VII.) Such calculations show that chloride complexes for most metals at ordinary temperatures do not have the necessary stability to overcome the insolubility of the sulfides, in other words that reactions like Eq. (17-10) are strongly displaced toward the left for any conceivable concentrations of S^{2-} and Cl^-. Although calculations at ordinary temperatures are not encouraging, there is good evidence that chloride complexes become more stable with respect to sulfides at higher temperatures. Unfortunately, reliable experimental data are scarce on the stabilities of halide complexes at temperatures of interest in ore formation.

Despite the meagerness of quantitative data. Helgeson (1969) has shown that combinations of chloride, sulfide, and hydrosulfide complexes are probably adequate to account for transportation of the common ore metals in hydrothermal solutions. His work involves the amassing of all available experimental data, some reasonable extrapolations, and ingenious methods for estimating entropies, heat capacities, and activity coefficients so that equilibrium constants at high temperature can be calculated. The calculated solubilities for many metals at different temperatures show a satisfying correspondence with sequences of crystallization observed in nature. Additional experimental work is needed to back up Helgeson's brilliant reconstruction of the compositions of ore-forming solutions, but he has demonstrated beyond much doubt that complexes play a central role in the movement of ore metals.

Summary Of the various suggestions for modifying the hydrothermal hypothesis to fit geologic observations, the increase in metal solubility resulting from the formation of complexes is by far the most promising. The more extreme modifications of the hydrothermal idea, depending on multiple solutions or colloidal behavior or supersaturation, encounter serious objections that make them untenable as general explanations of ore transportation. Such phenomena may be important, of course, in special situations. But the general solution to the dilemma of transporting metals together with sulfur, even though all details are not yet worked out, lies very clearly in the high-temperature stability of complex ions and molecules.

17-5 GENERAL REFLECTIONS

This long discussion of magmatic fluids, which has occupied the better part of three chapters, has provided us with the broad outlines of a possible explanation for the origin of at least some high- and moderate-temperature ore deposits. On the basis of volatilities we can understand how ore metals might separate themselves from a magma, and how they can be transported at high temperatures in a gas phase. Some metallic ores may have been laid down directly from a gas phase in the form we find them today, but most ore deposits have had a longer history. Perhaps the gas cools slightly and condenses, retaining its metal content in the liquid (or in the high-density supercritical fluid); perhaps the metals are deposited and dissolved again repeatedly. About these processes in the high-temperature part of the ore-forming range we know little, because they involve liquids and supercritical fluids at high temperatures and pressures, a set of conditions for which experimental data are scarce. By a series of steps essentially hidden from us, the ore-forming fluid becomes a liquid at more moderate temperatures, in which the ore metals are dissolved and from which they somehow precipitate. Into the nature and behavior of this liquid we have gained considerable insight from the experimental work and theoretical studies of recent decades.

The simple fact that ore deposits the world over show a certain regularity in the sequence of their minerals, a sequence that has enabled geologists to classify most deposits in a temperature-depth pattern, is testimony that the deposition of ore takes place according to physicochemical rules that have wide application.

This means, in turn, that we can eliminate processes of ore formation which depend on unique or fortuitous circumstances, and that we can largely discount the effects of colloidal processes and metastability. The most promising explanation for the observed regularities in ore-mineral associations and sequences has come from a study of stabilities of complex ions and molecules, which offer a means for keeping the metals in solution at appreciable concentrations in the presence of sulfur, and for permitting their crystallization in an orderly sequence in response to changes in temperature, pressure, pH, or oxidation state.

The explanation seems satisfying as a general logical framework, but many details of the actual mechanics of ore formation at specific deposits remain elusive. Untouched also are some of the deeper questions about the origin of ores. We have assumed that the metals came originally from an igneous magma, concentrated in the residual fluids by fractional crystallization. Is this the only possible source? If, as suggested in Sec. 15-5, much of the water in hydrothermal solutions is meteoric, could the metals have had their source in sedimentary rocks adjacent to an intrusive, dissolved by ground water that moves through the sediments and ultimately becomes part of the intrusive? Or is the intrusive really necessary at all—can some hydrothermal solutions originate as ground water heated by deep circulation, without ever coming near an actual magma? If, as seems probable, at least some deposits do consist of metals derived from a cooling magma, how do we account for the great differences in kinds of metal concentrations associated with very similar granitic bodies, and the complete absence of metal concentrations near most granites? These are questions that have long vexed students of ore deposits, and that still lack satisfactory answers. The questions are broader than the field of geochemistry, and chemical arguments can do no more than aid in their solution.

From a still broader perspective, we can pose some queries relating to plate tectonics. Can deposits of the metals characteristically found with mafic and ultramafic rocks be formed during the rise of mafic magmas at spreading centers? Are metals concentrated by the processes of metamorphism and magma generation along crustal plates moving downward in subduction zones? Can the apparent rough distribution of ore deposits of different metals in belts parallel to orogenic zones be related to distance from a nearby ocean trench, or to the depth of a moving plate beneath? These questions also are only in part geochemical.

. The great usefulness of chemistry in the study of ore deposits is less in attacking the deeper questions than in limiting possibilities. We know, for example, as a result of this discussion of magmatic fluids, that the volatility of metals and metal compounds may play a role at the high-temperature end of the ore-forming range but not at lower temperatures; that colloids and supersaturation and alkaline sulfide solutions can have restricted roles at best; that the pH of ore solutions is limited to a narrow range near the neutral point; and that the oxidation state is limited by the kinds of minerals in the rocks through which the solutions pass. Conclusions of this sort can serve at least to eliminate the more extreme hypotheses about ore-forming solutions and to guide the experimental work with which the final answers can be approached.

PROBLEMS

1 In an elaborate study of the solubility of HgS in excess sulfide ion, Dickson (1964) has shown that the solubility decreases when the temperature rises from 25 to 100°C. If the mercury is dissolved in an ore solution as HgS_2^{2-}, say at 100°, this means that cinnabar cannot precipitate by simple cooling of the solution. Suggest other possible causes for precipitation of HgS from such a solution.

2 Using the data of Appendix VIII, calculate the solubility of chalcocite in a solution of NaCl, due to the formation of the chloride complex ion $CuCl_2^-$. Assume a temperature of 25°C and $[OH^-] = 10^{-7}M$, and make the calculation for various concentrations of NaCl. Is it possible for significant amounts of copper to be dissolved as $CuCl_2^-$ in a solution that contains sulfide ion? Would the solubility of copper in this form be dependent on pH? Explain.

3 Lead is known to form a stable hydroxy complex ion, $HPbO_2^-$ or $Pb(OH)_3^-$. How alkaline would a solution have to be in order for appreciable PbS to dissolve to give this ion? Would this be a geologically reasonable method of transporting lead in ore solutions?

4 Prove that molybdenite is more stable than molybdates at 627° and an oxygen partial pressure of 10^{-17} atm. (Use data for MoO_3 rather than a molybdate, and use the assumptions on which Table 17-2 is based.)

5 It has been postulated that metals exist in ore solutions at temperatures above 200° because sulfur is largely oxidized to SO_2, and that metal sulfides are precipitated below this temperature because the following equilibrium shifts to the right as temperature falls:

$$SO_2 + 3H_2 \rightleftharpoons 2H_2O + H_2S$$

Make the necessary calculations to evaluate this suggestion. Use an arbitrary figure for the partial pressure of hydrogen, say 10^{-3} atm.

6 Limits can be set for the ratios of various ions present in ore solutions on the basis of the occurrence and nonoccurrence of various compounds. For example, the fact that magnesite occurs as a gangue mineral and sellaite (MgF_2) does not means that the ratio $[CO_3^{2-}]/[F^-]^2$ must always be high enough to prevent sellaite from forming:

$$MgCO_3 + 2F^- \rightleftharpoons MgF_2 + CO_3^{2-}$$

Calculate this limiting ratio at 25°. In similar fashion calculate the limiting ratio of $[SO_4^{2-}]/[S^{2-}]$ from the fact that galena rather than anglesite is the common primary lead mineral, and the limiting ratio of $[SO_4^{2-}]/[CO_3^{2-}]$ from the facts that barite $(BaSO_4)$ is a commoner vein mineral than witherite $(BaCO_3)$ and that calcite is a commoner vein mineral than anhydrite.

7 From Fig. 17-1, what is the pH of neutral supercritical water at a temperature of 500°C and a density of 0.5 g/cm³? Assuming that the fluid behaves as a perfect gas, calculate the pressure needed to maintain this density. Do you think the assumption is justified?

REFERENCES AND SUGGESTIONS FOR FURTHER READING

Barnes, H. L., and G. K. Czamanske, Solubilities and transport of ore minerals, chap. 8 in "Geochemistry of Hydrothermal Ore Deposits," H. L. Barnes, ed., pp. 334–381, Holt, Rinehart and Winston, New York, 1967. Calculations of equilibria in sulfide-sulfate systems to 250°C, and summary of work on sulfide and hydrosulfide complexes.

Barton, P. B., P. M. Bethke, and P. Toulmin, Equilibrium in ore deposits, *Mineralog. Soc. America Spec. Paper* 1, pp. 171–185, 1963. A critical discussion of evidence for equilibrium and lack of equilibrium during ore deposition.

Dickson, F. W., Solubility of cinnabar in Na₂S solutions at 50–250°C and 1–1800 bars, *Econ. Geology*, vol. 59, pp. 625–635, 1964.

Helgeson, H. C., Thermodynamics of hydrothermal systems at elevated temperatures and pressures, *Amer. Jour. Sci.*, vol. 267, pp. 729–804, 1969.

Lindgren, W., "Mineral Deposits," 4th ed., McGraw-Hill Book Company, New York, 1933. Chapter 16 describes Lindgren's classification of ore deposits.

Nash, J. T., Fluid-inclusion petrology—data from porphyry copper deposits, *U.S. Geol. Survey Prof. Paper* 907-D, pp. 1–16, 1976. An excellent review of the use of fluid inclusions for estimating temperatures, pressures, and compositions of ore-forming fluids, especially as applied to one kind of copper deposit.

Ohmoto, H., Systematics of sulfur and carbon isotopes in hydrothermal ore deposits, *Econ. Geology*, vol. 67, pp. 551–578, 1972. Isotopic compositions of sulfur and carbon are shown to be strongly controlled by oxygen fugacity and pH as well as temperature. Isotope data can give information about physicochemical parameters, origin of fluids, and mechanisms of ore deposition.

Roedder, E., Fluid inclusions as samples of ore fluids, chap. 12 in "Geochemistry of Hydrothermal Ore Deposits," H. L. Barnes, ed., pp. 515–574, Holt, Rinehart and Winston, New York, 1967. An excellent summary of the kinds of information obtainable from detailed study of the tiny inclusions of fluid found in some ore and gangue minerals.

Rye, R. O., A comparison of sphalerite-galena sulfur isotope temperatures with filling temperatures of fluid inclusions, *Econ. Geology*, vol. 69, pp. 26–32, 1974. Data for six ore deposits show good correspondence between temperatures determined from sulfur-isotope ratios and fluid inclusions in the range 370 to 200°C. Deviations below 200° indicate isotope disequilibrium at low temperatures.

Scott, S. D., and H. L. Barnes, Sphalerite geothermometry and geobarometry, *Econ. Geology*, vol. 66, 653–669, 1971. When sphalerite is present with pyrite and pyrrhotite, the iron content of the sphalerite is nearly constant below 550°C, hence useless for temperature measurement. The iron content decreases as pressure rises, however, so may serve as a measure of pressure during ore formation.

Seward, T. M. Thio complexes of gold and the transport of gold in hydrothermal ore solutions, *Geochim. et Cosmochim. Acta*, vol. 37, pp. 379–399, 1973. Experimental data on sulfide, bisulfide, and chloride complexes of gold at 160–300°C.

Sheppard, S. M. F., R. L. Nielsen, and H. P. Taylor, Hydrogen and oxygen isotope ratios in minerals from porphyry copper deposits, *Econ. Geology*, vol. 66, pp. 515–542, 1971. An example of the application of isotope ratios to the study of an important kind of ore deposit.

Sillitoe, R. H., The tops and bottoms of porphyry copper deposits. *Econ. Geology*, vol. 68, pp. 799–815, 1973. A model for the development of an important kind of copper deposit, relating such deposits to the combined activity of meteoric and magmatic water, in a granitic body intruded at shallow levels and surmounted by a volcano.

Taylor, H. P., Oxygen isotope studies of hydrothermal mineral deposits, chap. 4 in "Geochemistry of Hydrothermal Ore Deposits," H. L. Barnes, ed., pp. 109–142, Holt, Rinehart and Winston, New York, 1967. A description of techniques and a critical review of the kinds of information obtainable from studies of oxygen isotope ratios.

White, D. E., Thermal springs and epithermal ore deposits, *Econ. Geology*, 50th Anniversary vol., pp. 99–154, 1955. Evidence about conditions of ore deposition from a detailed study of five hot springs.

White, D. E., and C. S. Roberson, Sulfur Bank, California: a major hot-spring quicksilver deposit, "Petrologic Studies (Buddington Volume)," pp. 397–428, Geological Society of America, New York, 1962. Report of a long and detailed investigation of a major mercury deposit associated with a present-day hot spring.

EIGHTEEN

OXIDATION OF ORE DEPOSITS

After an ore deposit has formed, erosion may eventually bring it close to the earth's surface. How do its minerals respond to the agents of chemical weathering? To answer this question requires no more than an extension of earlier discussions in Chaps. 4 and 8, but the weathering reactions of ore minerals are generally more complicated than those of rock-forming minerals, because most of the metals involved show more than one oxidation state in normal weathering environments. The reactions, in contrast to those we have been discussing in the past several chapters, take place under ordinary temperatures and pressures and are therefore accessible to detailed observation in the field and to easy experimental study in the laboratory. For this reason, and also because weathering processes often have great economic importance in concentrating the valuable metals of low-grade deposits, the weathering of ores has been the subject of long and intensive investigation. From the great wealth of accumulated data we select for discussion only a few examples to illustrate general principles.

From what we know of the chemistry of sulfur, we should expect the sulfide ore minerals to fall easy prey to chemical weathering. The sulfur, we might predict, would be oxidized to sulfate ion, SO_4^{2-}; presumably the oxidation goes through the intermediate stages of free sulfur and SO_2, but these substances are seldom detectable in oxidized zones in nature. The metals may be converted into insoluble compounds stable under surface conditions (oxides, carbonates, sulfates, silicates) or may be taken into solution. The dissolved metal ions may be removed completely in streams and ground water or may be in part carried down into the unoxidized portion of a sulfide deposit and there precipitated by reaction with the sulfide minerals. The details of such processes, which may be followed both in the field and in the laboratory, will be the subject of this chapter. Behind the details we shall see once again how basic chemical principles can bring order and a measure of predictability into seemingly complex natural phenomena. Sulfides will occupy the center of attention, simply because the heavy metals so commonly occur in or with sulfide minerals, but we shall also look briefly at some other kinds of ores in which sulfides are nearly or completely absent.

18-1 OXIDATION OF SULFIDES

It is easy enough to write a symbolic equation for the oxidation of a simple sulfide like galena:

$$PbS + 2O_2 \rightarrow PbSO_4 \qquad (18\text{-}1)$$

But how does the reaction actually take place? Almost certainly it is not a simple matter of collisions between oxygen molecules and lead sulfide, as the equation would suggest, because shiny specimens of galena can be kept in contact with ordinary air indefinitely without showing the slightest sign of change. The agency of water in promoting the change is suggested by the observation that moist galena surfaces gradually tarnish in the laboratory. Precisely how water acts is not known, but one possibility is a series of reactions like those listed in Sec. 4-7:

$$H_2O + CO_2 \rightleftharpoons H_2CO_3$$

$$PbS + 2H_2CO_3 \rightleftharpoons Pb^{2+} + 2HCO_3^- + H_2S$$

$$H_2S + 2O_2 \rightarrow SO_4^{2-} + 2H^+$$

$$Pb^{2+} + SO_4^{2-} \rightleftharpoons PbSO_4$$

$$H^+ + HCO_3^- \rightleftharpoons H_2O + CO_2$$

If these reactions are added together, the result is Eq. (18-1). According to this mechanism water serves as a catalyst, its catalytic activity consisting of the formation of carbonic acid, which in turn dissolves a tiny amount of PbS and permits O_2 to react with dissolved H_2S rather than with the solid sulfide. Reactions of O_2 at ordinary temperatures are slow, but it is known to react more rapidly with molecules and ions in solution than with solids, so these steps are at least plausible.

But this is far from the only possible mechanism. Sato (1960) has suggested, for example, that water is oxidized in trace amounts to hydrogen peroxide and that the peroxide is the active agent in oxidizing the sulfide:

$$2H_2O + O_2 \rightarrow 2H_2O_2$$

$$4H_2O_2 + PbS \rightarrow PbSO_4 + 4H_2O$$

The first reaction is presumably very slow and accounts for the slowness of reactions involving O_2. Some justification for this mechanism is provided by Sato's observation that measured Eh values in mine waters are well below the theoretical maximum for the O_2-H_2O couple (Sec. 9-4):

$$2H_2O \rightleftharpoons O_2 + 4H^+ + 4e^- \qquad Eh = 1.22 - 0.059pH$$

but are close to values for the O_2-H_2O_2 couple:

$$H_2O_2 \rightleftharpoons O_2 + 2H^+ + 2e^- \qquad Eh = 0.68 - 0.059pH$$

Sato has also presented evidence that the oxidation of sulfur in the sulfide does not go directly to sulfate but proceeds by a series of partial oxidations.

The overall oxidation reaction is easy to write for sulfates, but less easy for carbonates. Let us try, for example, to write as accurate an equation as possible for the conversion of sphalerite to smithsonite in the oxidized zone of an ore deposit. We can eliminate at once equations like

$$ZnS + \tfrac{3}{2}O_2 + CO_2 \rightarrow ZnCO_3 + SO_2$$

and

$$ZnS + 2O_2 + CO_2 \rightarrow ZnCO_3 + SO_3$$

simply on the grounds that neither SO_2 nor SO_3 is found in such an environment. The equation

$$ZnS + 2O_2 + CO_2 + H_2O \rightarrow ZnCO_3 + SO_4^{2-} + 2H^+$$

would be more reasonable, inasmuch as ground water near oxidizing sulfide deposits generally is somewhat acid and contains sulfate ion. This formulation is somewhat objectionable on purely chemical grounds, in that it shows an insoluble carbonate and a strong acid being produced at the same time. A better possibility is

$$ZnS + 2O_2 + 2HCO_3^- \rightarrow ZnCO_3 + H_2CO_3 + SO_4^{2-}$$

This equation is realistic in the sense that all substances shown are known to exist in the environment of oxidizing sulfides, and that those on the right-hand side are chemically compatible as long as H_2CO_3 is not concentrated. Perhaps even better would be to write the reaction in steps, the successive steps taking place in somewhat different environments; for example, the reaction

$$ZnS + 2O_2 \rightarrow Zn^{2+} + SO_4^{2-}$$

might be followed by precipitation of the carbonate whenever the Zn^{2+} moves into a more alkaline environment. Or in places where smithsonite replaces calcite, the $CaCO_3$ may be shown as reducing the acidity and so favoring precipitation:

$$ZnS + O_2 + H_2O + 2CaCO_3 \rightarrow ZnCO_3 + 2Ca^{2+} + 2HCO_3^- + SO_4^{2-}$$

There is thus no single "correct" way to write the equation for the alteration of sphalerite to smithsonite. The equation must be tailored so as to describe as realistically as possible the chemical situation at individual smithsonite occurrences.

The oxidation of any insoluble sulfide, as noted in Sec. 4-7, leads to the formation of acid solutions. The increased acidity may be caused by simple hydrolysis of the metal ion or by precipitation of an insoluble hydroxide. The degree of acidity depends on the nature of the metal ion, particularly on the extent of its hydrolysis or on the insolubility of its hydroxide. Most acid of all are solutions resulting from oxidation of sulfides containing iron, because the ferric ion produced is hydrolyzed to an extremely insoluble oxide or hydroxide:

$$2FeS_2 + \tfrac{15}{2}O_2 + 4H_2O \rightarrow Fe_2O_3 + 4SO_4^{2-} + 8H^+ \tag{4-12}$$

Here it is reasonable to show the free acid in the equation, because ferric oxide can exist in contact with fairly high acid concentrations.

Thus the weathering of sulfides may be described by a variety of equations, all of them only approximate representations of a complex natural process. The net results are (1) to get the metal ion into solution or into the form of an insoluble compound stable under surface conditions, (2) to convert the sulfur to sulfate ion, and (3) to produce relatively acid solutions.

18-2 SOLUBILITIES OF OXIDIZED METAL COMPOUNDS

The compounds that may appear in the gossan, or oxidized outcrop, of a sulfide vein include a variety of chemical types, of which the following are examples:

Hydrated oxides (limonite, psilomelane)
Carbonates (smithsonite, cerussite)
Sulfates (anglesite, brochantite)
Chlorides (cerargyrite, atacamite)
Phosphates (pyromorphite)
Silicates (chrysocolla, hemimorphite)
Native metals (mercury, silver)

The minerals that form in any particular case will depend, obviously, on what anions are present and on the solubilities of the possible metal compounds.

In older geologic literature one finds frequently statements like "copper travels as the sulfate," or "zinc is mobile as the chloride." Phrases of this sort are half-truths. Since $CuSO_4$ and $ZnCl_2$ are soluble salts, the metals can indeed be present in solution in the oxidized zone accompanied by the ions SO_4^{2-} and Cl^-. But they can also be present with other ions as well, or with combinations of ions. The thing that determines the behavior of the metals is not the anions with which they form soluble salts, but the anions which can precipitate them. The older statements are faulty in that they put the emphasis in the wrong place: our chief concern is not the hypothetical soluble compounds, but the possible insoluble compounds, which the oxidized metals can form. These insoluble compounds set limits on the amounts of the various metals that can be transported away from an oxidizing ore deposit.

Representative data on the solubilities of common metal compounds are assembled in Table 18-1. The numbers indicate, for each metal, the concentrations that could be present in solution with stated amounts of various anions. In the first five columns, concentrations of the anions and the redox potential are chosen so as to represent possible natural situations in the oxidized zone. The S^{2-} concentration in the last column is typical of conditions at the base of the oxidized zone, a subject to which we shall return in a later section. The numbers are obtained from solubility products, oxidation potentials, and stability constants for complex ions, according to methods we have used in previous chapters. For columns 3 and 4 (SO_4^{2-} and Cl^-), the assumption of low pH should be particularly noted: if the pH is higher than 6, formation of basic sulfates and chlorides would cut down the concentrations of some of the metals (for example, the basic copper sulfate, brochantite, and the basic copper chloride, atacamite). The table is necessarily incomplete, because reliable solubility data are lacking for most silicates, phosphates, arsenates, and other compounds which may be important for particular metals.

Table 18-1 Maximum concentrations of metals in equilibrium with common anions at 25°C

The numbers in each column show the maximum activity of each metal in equilibrium with the anion at the head of the column. Activities in moles per liter; for these metals an activity, or concentration, of $10^{-5}M$ is approximately equivalent to 1 ppm.

Assumed conditions:

Column 1: Activity of $OH^- = 10^{-8}M$ (pH 6) and Eh = 0.7 volt.
Column 2: Total dissolved carbonate = 0.001M, pH 6, Eh = 0.7 volt. Under these conditions $a_{CO_3^{2-}} = 10^{-7.7}M$.
Column 3: Activity of SO_4^{2-} = 0.01M, Eh = 0.7 volt, pH low enough to prevent hydrolysis.
Column 4: Activity of Cl^- = 0.001M, Eh = 0.7 volt, pH low enough to prevent hydrolysis.
Column 5: Eh = 0.3 volt, pH low enough to prevent hydrolysis. These figures show maximum concentrations of metal ions in equilibrium with native metals.
Column 6: Activity of $S^{2-} = 10^{-20}M$. This is roughly the concentration in equilibrium with sphalerite in a solution with pH = 3.

	(1) OH^-	(2) CO_3^{2-}	(3) SO_4^{2-}	(4) Cl^-	(5) Eh	(6) S^{2-}
Cu	$10^{-4.3}$	$10^{-5.0}$	>1	>1	$10^{-1.3}$	$10^{-14.3}$
Ag	>1	>1	>1	$10^{-6.7}$	$10^{-8.5}$	$10^{-15.0}$
Hg	$10^{-6.8}$	$10^{-6.8}$	$10^{-3.9}$	$10^{-9.5}$	$10^{-16.3}$	$10^{-33.3}$
Pb	>1	$10^{-5.4}$	$10^{-5.8}$	>1	>1	$10^{-7.5}$
Zn	0.2	$10^{-3.3}$	>1	>1	>1	$10^{-4.7}$
Cd	>1	$10^{-6.0}$	>1	>1	>1	$10^{-7.0}$
Sn	$10^{-7.8}$	$10^{-7.8}$	>1	>1	>1	$10^{-5.9}$
Ni	0.2	0.2	>1	>1	>1	$10^{-6.6}$
Co	>1	$10^{-3.3}$	>1	>1	>1	$10^{-5.6}$
Mn	$10^{-6.0}$	$10^{-6.0}$	>1	>1	>1	>1
Fe	10^{-13}	10^{-13}	>1	>1	>1	>1

Despite the incompleteness and the uncertainty of some of the data, the table shows a gratifying correspondence with the observed behavior of metals in the oxidized zone of sulfide deposits. Iron and manganese, under the assumed Eh and pH conditions, go into solution scarcely at all because of the insolubility of oxides and hydroxides of their higher oxidation states; these compounds remain behind at the outcrop to form the familiar brown and black colors of gossans. Under more reducing conditions iron and manganese could be carried away in large quantities, since in acid solution they are not precipitated by the common anions nor reduced to the native metals. Tin also commonly remains in the oxidized zone as an oxide of higher oxidation state (SnO_2, cassiterite), either residual from the ore or formed by oxidation of other tin minerals. Silver, according to the table, should be readily precipitated as the chloride or reduced to the metal, a prediction that agrees with the common occurrence of cerargyrite (AgCl) and native silver in oxidized silver deposits. Mercury is readily reduced to the metal even under fairly oxidizing conditions, as is shown by the presence of mercury globules in oxidized mercury ores. The common anions also form insoluble compounds with mercury, corresponding to such oxidized minerals as calomel (HgCl), montroydite (HgO), and eglestonite (Hg_4OCl_2). Lead and cadmium have fairly insoluble carbonates (cerussite and otavite), and lead also has an insoluble sulfate (anglesite). Zinc carbonate (smithsonite) is somewhat less insoluble. Copper can be partly retained in the oxidized zone as a basic carbonate (malachite or azurite); the solubility figure ($10^{-4.9}M$) in the table is based on somewhat questionable data for malachite. Under slightly less acid conditions copper can also be deposited as an oxide, as basic sulfates and chlorides, or as the native metal. In contrast to the other metals so far mentioned, zinc, copper, and cadmium should be "mobile" metals in the oxidized zone, since under the assumed conditions the common anions permit considerable amounts to remain in solution. Nickel and cobalt, according to the table, should also be mobile elements, but here the figures are misleading, because these elements form insoluble arsenates, and nickel an insoluble silicate, which the table does not show.

18-3 Eh-pH DIAGRAMS FOR WEATHERING OF ORE DEPOSITS

This sort of discussion can be refined by making use of Eh-pH plots like those described in Chap. 9. Detailed diagrams for all the common ore metals have been prepared by Garrels and his colleagues (Garrels and Christ, 1965, chap. 7). As a single example, we consider here an Eh-pH plot for lead (Fig. 18-1).

Like the diagrams of Chap. 9, this one shows (by dashed lines) the field of Eh and pH values commonly found in nature. In the environment of a weathering sulfide deposit the acid limit (pH = 4) is not very realistic, since much lower pH's are frequently encountered. Of the four lead minerals shown on the diagram, galena is the most stable (or the least soluble) at low values of Eh, regardless of

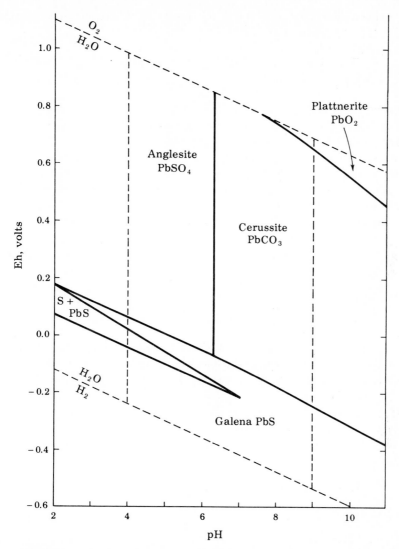

Figure 18-1 Eh-pH diagram for lead minerals at 25°C and 1 atm. $\sum CO_2 = 10^{-3}$ M and $\sum S = 10^{-2}$ M. (*After Garrels and Christ, 1965.*)

acidity; when conditions are oxidizing, anglesite is the most stable in an acid environment and cerussite in most alkaline environments; plattnerite becomes stable only if conditions are highly oxidizing as well as alkaline. These relations, of course, are just what would be expected from the chemistry of lead, and the diagram serves only to express the relations more quantitatively.

Boundaries between fields on the diagram are located by finding Eh and pH values where the Pb^{2+} concentrations from two compounds are identical. For example, the anglesite-cerussite boundary is drawn through points where the two equilibria

$$PbSO_4 \rightleftharpoons Pb^{2+} + SO_4^{2-} \qquad K = 10^{-7.8}$$

$$PbCO_3 \rightleftharpoons Pb^{2+} + CO_3^{2-} \qquad K = 10^{-13.1}$$

are simultaneously satisfied, in other words where the ratio $a_{SO_4^{2-}}/a_{CO_3^{2-}}$ is equal to $10^{5.3}$. Under the conditions specified on the diagram (total dissolved sulfur $= 0.01M$, total dissolved carbonate $= 0.001M$), the activity of SO_4^{2-} at moderate and high Eh values can be set equal to the total sulfur, and $a_{CO_3^{2-}}$ on the boundary must therefore be $10^{-2}/10^{5.3}$, or $10^{-7.3}M$. From Fig. 2-2, this activity for CO_3^{2-} requires a pH of 6.3. Hence the boundary is a vertical line at pH 6.3.

For boundaries with the PbS field the calculation is more troublesome. The PbS-PbCO$_3$ line is obtained by equating Pb^{2+} activities from the two equilibria

$$PbS \rightleftharpoons Pb^{2+} + S^{2-} \qquad K = 10^{-27.5}$$

$$PbCO_3 \rightleftharpoons Pb^{2+} + CO_3^{2-} \qquad K = 10^{-13.1}$$

from which the ratio $a_{S^{2-}}/a_{CO_3^{2-}}$ must be $10^{-14.4}$. Both sulfide ion and carbonate ion are dependent on pH, and sulfide ion is also determined in part by Eh, so that a formal analytical expression relating Eh to pH would be rather complicated. The simplest procedure is to work out the relation for specific points. Take first the point at the upper end of the line, where the fields of PbS, PbCO$_3$, and PbSO$_4$ come together. From the last paragraph we know that here pH $=$ 6.3 and $a_{CO_3^{2-}} = 10^{-7.3}M$. The above ratio tells us that $a_{S^{2-}}$ must be $10^{-14.4} \times 10^{-7.3}$, or $10^{-21.7}M$. If $a_{S^{2-}}$ is this low, neither it nor a_{HS^-} can be quantitatively important compared with $a_{SO_4^{2-}}$, so again we set the sulfate activity equal to the total dissolved sulfur, or $10^{-2}M$. From the half-reaction relating S^{2-} to SO_4^{2-},

$$S^{2-} + 4H_2O \rightleftharpoons SO_4^{2-} + 8H^+ + 8e^- \qquad E^\circ = +0.16 \text{ volt}$$

we write for Eh, using Eq. (9-9),

$$Eh = E^0 + \frac{0.059}{n} \log Q = 0.16 + 0.0074 \log \frac{a_{H^+}^8 \cdot a_{SO_4^{2-}}}{a_{S^{2-}}}$$

$$= 0.16 + 0.0074 \log \frac{10^{-50.4} \times 10^{-2}}{10^{-21.7}}$$

$$= -0.07 \text{ volt}$$

For another point on the line, take pH $= 11$, high enough so that practically all the dissolved carbonate is CO_3^{2-}. If $a_{CO_3^{2-}}$ is set equal to $10^{-3}M$, then $a_{S^{2-}} = 10^{-17.4}M$, still low enough so that $a_{SO_4^{2-}}$ remains effectively equal to the total dissolved sulfur, or $10^{-2}M$. Repeating the calculation from the half-reaction then gives Eh $= -0.38$ volt. Other Eh-pH pairs are located similarly.

The diagram does not give directly the activities of Pb^{2+} in equilibrium with the solid phases, but these are easily calculated and may be added as contour lines in the different fields. Over the $PbSO_4$ field, $a_{Pb^{2+}}$ is constant at $10^{-5.8}M$. because the sulfate-ion concentration is not appreciably affected by either Eh or pH in this range. In the $PbCO_3$ field the Pb^{2+} activity drops off as the pH rises, since $a_{CO_3^{2-}}$ steadily increases; at pH 10.3, for example, where $a_{CO_3^{2-}} = a_{HCO_3^-} = \frac{1}{2} \times 10^{-3}M$, $a_{Pb^{2+}}$ would be $10^{-9.8}M$. Over the PbS field the Pb^{2+} activity would fall as pH rises and as Eh decreases, since both of these changes would cause $a_{S^{2-}}$ to increase.

Different assumptions as to total dissolved sulfur and total carbonate would change the boundaries somewhat, but the alterations are surprisingly minor as long as the concentrations remain within the range commonly found in natural solutions.

Such diagrams represent an improvement over the figures of Table 18-1 in that they give solubility and stability relations for the whole range of Eh and pH values rather than for a few points. A skeptic might legitimately inquire, however, whether the refinement represented by Fig. 18-1 is actually much more informative or more useful for discussions of ore deposits than are simple qualitative predictions from the solubility figures. Boundaries of fields on the diagram refer to pure compounds, in contact with solutions of fixed composition at a specified temperature and pressure. None of these conditions, obviously, would be duplicated in nature. Nor would it be easy to check the diagram by measurements in the field, because reliable determinations of Eh are formidably difficult in a situation where many of the reactions are slow and where gas pressures would have to be carefully controlled. Of course it is philosophically satisfying to know that field observations conform with an idealized reconstruction, but the question is always pertinent as to how far it is worthwhile carrying refinements of the reconstruction. Galena alters on weathering to a carbonate, a sulfate, and rarely a dioxide; this has been known for a long time, and agrees qualitatively with laboratory measurements of solubility and stability. Does it add to our useful knowledge of natural phenomena, or to our ability to make testable predictions, to have precise values of solubility under idealized conditions? Such questions are common in geochemistry, and opinions are often sharply divided. Whatever be the judgment about a simple diagram like Fig. 18-1, more complex diagrams of the same sort are assuredly useful in revealing permissible and expectable mineral associations. We shall consider one such diagram in a later section (Fig. 18-2), and many other examples can be found in the work of Garrels and his colleagues (Garrels and Christ, 1965).

18-4 SUPERGENE SULFIDE ENRICHMENT

When a sulfide deposit is followed downward beneath the oxidized outcrop, the richest ore is often found just at the top of the unoxidized zone. This level in many places corresponds with the regional water table, or with a position of the water table at some time in the recent geologic past. The rich ore commonly consists of sulfide minerals, which from replacement relations are clearly of a later generation than the primary sulfides found at deeper levels in the deposit. It is a reasonable inference that these secondary sulfides have been formed by downward-moving solutions containing metal ions derived from relatively soluble compounds in the oxidized zone, these metal ions having reacted with and replaced sulfides in the unoxidized rock. This phenomenon, long recognized by students of ore deposits and often of great economic importance, goes by the name of supergene sulfide enrichment.

Of all the metals, copper shows supergene enrichment most commonly and most conspicuously. The simple sulfides covellite (CuS) and especially chalcocite (Cu_2S) are concentrated in thick blankets and lenses near the top of many copper deposits, often making up the richest part of the ore. The more complex sulfides bornite (Cu_5FeS_4) and chalcopyrite ($CuFeS_2$) are reported in smaller amounts, but their supergene origin is less clearly established. The formation of supergene copper sulfides is complex in detail, but the general nature of the process is easy to understand. Near the surface, copper minerals are oxidized, and much of the metal goes into solution, since the possible oxidized compounds are not very insoluble as long as the solutions are slightly acid (Table 18-1). The dissolved Cu^{2+}, carried downward by ground water, comes in contact with primary sulfide minerals below the zone of oxidation. Because the copper sulfides are less soluble than most other common sulfides (last column of Table 18-1), reactions like the following are possible:

$$Cu^{2+} + ZnS \rightarrow CuS + Zn^{2+}$$

$$8Cu^{2+} + 5ZnS + 4H_2O \rightarrow 4Cu_2S + 5Zn^{2+} + SO_4^{2-} + 8H^+$$

$$14Cu^{2+} + 5FeS_2 + 12H_2O \rightarrow 7Cu_2S + 5Fe^{2+} + 3SO_4^{2-} + 24H^+$$

The iron, zinc, and other metals displaced by the copper are carried away in solution, perhaps to be deposited later as limonite, smithsonite, and so on, if the solutions reach an oxidizing and less acid environment. Whether chalcocite or covellite is preferentially deposited would depend on the Eh and pH at any particular point.

The displacement may be indirect as well as direct. Acid solutions working down from the oxidized zone may dissolve the more soluble primary sulfides with formation of H_2S or HS^-, and these may in turn precipitate the copper sulfides:

$$ZnS + 2H^+ \rightarrow Zn^{2+} + H_2S$$

$$Cu^{2+} + H_2S \rightarrow CuS + 2H^+$$

$$8Cu^{2+} + 5HS^- + 4H_2O \rightarrow 4Cu_2S + SO_4^{2-} + 13H^+$$

The chalcocite and covellite formed in such reactions would not necessarily replace the original sulfides but might form separate crystals.

Now what characteristics of a metal would make it more suitable than another for supergene enrichment? We might guess that two properties have special significance: very low solubility of its sulfide, and relatively high solubility of compounds it might form with the common anions of the oxidized zone. These qualifications copper satisfies better than any other metal in Table 18-1, so that the particular importance of supergene enrichment in copper deposits is hardly surprising.

Inspection of the table suggests that silver also has the necessary combination of properties to show supergene enrichment. It is true that silver may be caught in the oxidized zone as the chloride, or may be reduced to the native metal if the Eh is fairly low, but solutions in the oxidized zone should often have low enough chloride ion and high enough Eh to permit considerable downward migration of Ag^+ to levels where the very insoluble sulfide can precipitate. This conclusion fits the well-established fact that silver deposits are second only to copper deposits in the prominence of secondary ore minerals just below the zone of oxidation. The simple sulfide argentite (Ag_2S), and the more complex sulfosalts proustite (Ag_3AsS_3) and pyrargyrite (Ag_3SbS_3), are common secondary minerals. Native silver is also formed during supergene enrichment, more commonly than native copper because Ag^+ is more easily reduced than is Cu^{2+} (see Eh values in Table 18-1).

No unquestioned examples have been reported of notable supergene sulfide enrichment of lead, zinc, nickel, cobalt, or mercury. For lead and mercury a likely explanation is the low solubility of their compounds in the oxidized zone. In addition to the carbonate and sulfate noted in Table 18-1, lead forms a very insoluble vanadate, arsenate, and molybdate. Mercury should also remain in the oxidized zone as the native metal, since its ion is easily reduced even at relatively high Eh. For zinc, nickel, and cobalt, supergene enrichment is unlikely because of the fairly high solubilities of their sulfides. In other words, these metals are often largely removed by surface water and ground water without being trapped either by the common anions of the oxidized zone or by displacement of other metals from their sulfides.

To say that a metal is not known to undergo marked supergene enrichment does not mean that its sulfide may never occur as a secondary mineral. Galena and sphalerite crystals are sometimes found in old mine workings that have been flooded for many years, and even mercury has been reported in secondary cinnabar. Unusual conditions of acidity, oxidation, or water movement which might lead to such abnormal occurrences can readily be imagined.

The reactions involved in secondary sulfide enrichment, like those in the oxidized zone, are for the most part easily understandable on the basis of fundamental chemical properties. As might be expected, the reactions have been studied extensively in the laboratory as well as in the field. Laboratory results, theoretical predictions, and field observations are here in remarkable agreement. As a generalization from all three, we can summarize the discussion with the statement that copper and silver have sufficiently insoluble sulfides and sufficiently soluble oxidized compounds to permit them to form conspicuous deposits of secondary sulfides, while other metals form secondary sulfides only in small amounts and under unusual circumstances.

18-5 OXIDATION OF GOLD

Gold, although it forms no stable sulfide in nature, commonly is found associated with sulfide minerals. As the sulfide minerals undergo weathering, the gold is subject to the action of solutions containing oxygen plus sulfuric acid from the oxidation of sulfur. The metal is so inert that even such corrosive solutions have little effect on it, and native gold persists indefinitely under surface conditions—as placer deposits eloquently testify. Nevertheless, slight supergene enrichment of gold and occasional reports of gold in films and minute crystals on pebbles suggest that traces of the metal are oxidized and dissolved during weathering. Both experimental and theoretical inquiries into the solubility of gold support this inference from observation.

The two simple ions of gold, Au^+ and Au^{3+}, have such high oxidation potentials that they cannot exist in appreciable amounts in natural solutions (Appendix IX). The chloride complex $AuCl_4^-$, however, is sufficiently stable that its potential is less than that of the oxygen electrode; the oxidation of gold, then, requires the simultaneous presence of a powerful oxidizing agent and a solution of chloride ion. In basic solutions other complex ions like AuO_2^- and AuS^- may also play a role, but such solutions will only rarely be encountered in weathering processes.

The electrode potentials in Appendix IX suggest only two substances which occur widely in nature and which are stronger oxidizing agents than $AuCl_4^-$: MnO_2 and free oxygen. Two others, Fe^{3+} and Cu^{2+}, have potentials close enough to that of $AuCl_4^-$ so that in high concentration they might bring about some solution. These various possibilities can be studied quantitatively by the methods of Chap. 9; as a single example we shall look into the reaction with MnO_2.

Combining the two electrode reactions gives an equation for the oxidation:

$$2Au + 12H^+ + 3MnO_2 + 8Cl^- \rightleftharpoons 3Mn^{2+} + 2AuCl_4^- + 6H_2O$$

From the electrode potentials:

$$E° = 1.00 - 1.23 = -0.23 \text{ volt}$$

$$\Delta G° = nf\,E° = 6 \times 23.1 \times (-0.23) = -31.8 \text{ kcal}$$

$$\log K = \frac{-\Delta G°}{1.364} = 23.3 \quad \text{and} \quad K = 10^{23.3}$$

Evidently the reaction is displaced far to the right; in other words, gold should be readily soluble in a mixture of MnO_2 and $1M$ HCl. The large coefficients of H^+ and Cl^-, however, tell us that the solubility is very sensitive to changes in the concentrations of these ions. To calculate the effect of concentration, we set up the equilibrium constant:

$$K = \frac{[Mn^{2+}]^3[AuCl_4^-]^2}{[H^+]^{12}[Cl^-]^8} = 10^{23.3}$$

We assume that no Mn^{2+} is present except what is produced in the reaction, in other words that $\frac{3}{2}$ moles of Mn^{2+} will be present for every mole of $AuCl_4^-$. With this substitution,

$$[AuCl_4^-]^5 = \tfrac{8}{27} \times 10^{23.3} \times [H^+]^{12}[Cl^-]^8$$

From this equation the concentrations in Table 18-2 are calculated for various choices of $[H^+]$ and $[Cl^-]$. The figures are only approximate, since we are concerned here only with orders of magnitude.

Table 18-2 shows that gold is appreciably soluble (i.e., 10^{-5} g/liter or more) in the presence of MnO_2 in HCl as dilute as $0.001M$, or in still smaller concentrations of either H^+ or Cl^- provided the other is present in large amount. Concentrations of these ions of $0.001M$ and higher have been reported from the oxidized parts of ore deposits, but are sufficiently uncommon so that appreciable transportation of gold should be a rare and local phenomenon—which agrees well with geologic observation.

In older literature the fact that HCl dissolves gold in the presence of MnO_2 but not in its absence was often accounted for by supposing that MnO_2 liberates "nascent" chlorine from the acid, the chlorine at the moment of its formation allegedly having special solvent properties. This sort of language, a heritage from the mystical formulas of alchemy, has no place in a modern description. The falseness of such an explanation here is evident from the figures in Table 18-2, which show that gold is dissolved by concentrations of HCl far too small to produce appreciable chlorine—nascent or otherwise—by reaction with MnO_2. In modern terminology we say that gold dissolves primarily because of the great stability of the complex ion $AuCl_4^-$; the function of the MnO_2 is to oxidize the gold, and the function of the Cl^- is to tie up the oxidized gold in this ion. The solution must be acid because the combining of hydrogen ion with the oxygen of MnO_2 to form water makes MnO_2 a far better oxidizing agent than it could be in neutral or basic solution.

From Appendix IX it appears that free oxygen should be just as effective as MnO_2 in dissolving gold. Study of the reaction shows, however, that the diluted oxygen of the atmosphere would have an appreciable effect only at concentrations of H^+ and/or Cl^- higher than are generally encountered in nature. Even where

Table 18-2 Solubility of gold in the presence of MnO_2 at 25°C

$[H^+]$, moles/liter	$[Cl^-]$, moles/liter	Concentration of $AuCl_4^-$	
		moles/liter	g/liter of Au
10^{-1}	10^{-1}	4	800
10^{-2}	10^{-2}	$10^{-3.4}$	$10^{-1.1}$
10^{-3}	10^{-3}	$10^{-7.4}$	$10^{-5.1}$
10^{-4}	10^{-1}	$10^{-6.6}$	$10^{-4.3}$
10^{-2}	10^{-4}	$10^{-6.6}$	$10^{-4.3}$

such concentrations exist, the reaction would be very slow, as are most processes involving molecular oxygen. The other two possible agents for oxidizing and dissolving gold, Fe^{3+} and Cu^{2+}, must be present together with Cl^- in abnormally high concentrations, especially so since their electrode potentials are lower than that of the Au-$AuCl_4^-$ couple. Again we find that dissolution of gold is possible, but that it requires such a special set of circumstances that it should be noted only rarely in nature.

All the reactions just described for dissolving gold have been demonstrated in the laboratory except the reaction with free oxygen. Agreement is excellent between the laboratory results, thermodynamic predictions, and geologic observations. Laboratory work confirms also theoretical predictions that other solvents suggested in the literature—notably $Al_2(SO_4)_3$ and $Fe_2(SO_4)_3$ without Cl^-—can have no effect on gold.

In contrast to acid solutions, theoretical data on alkaline solutions are meager. Several experimenters have reported that gold dissolves appreciably in NaHS solutions at room temperature, a difficult reaction to understand because NaHS seems to contain no oxidizing agent of sufficient power to attack gold. Conceivably hydrogen might be liberated,

$$Au + HS^- \rightarrow AuS^- + \tfrac{1}{2}H_2$$

but this would be possible only if AuS^- (or some similar complex sulfide or hydrosulfide ion) is exceptionally stable. The existence of such ions has been demonstrated experimentally in solutions simulating hydrothermal conditions, but not in oxidizing solutions at low temperatures.

18-6 OXIDATION OF URANIUM ORES

A discussion of oxidation processes could be extended to ores of many other metals. A particularly interesting one is uranium, whose complex and varicolored oxidation products have become well known because of the search for uranium deposits to provide fuel for nuclear reactors.

The chief primary compound of uranium in vein deposits is the dioxide, UO_2, which occurs in the well-crystallized variety uraninite and the microcrystalline form pitchblende. Incipient oxidation and loss of uranium by radioactive decay may increase the oxygen-uranium ratio, so that uraninite and pitchblende seldom show precisely the composition UO_2, often approaching a composition symbolized by U_3O_8. In the zone of weathering, pitchblende and uraninite are converted to one or more of the bright-colored oxidized uranium minerals, such as carnotite $[K_2(UO_2)_2(VO_4)_2 \cdot 3H_2O]$, tyuyamunite $[Ca(UO_2)_2(VO_4)_2 \cdot nH_2O]$, autunite $[Ca(UO_2)_2(PO_4)_2 \cdot nH_2O]$, and rutherfordine (UO_2CO_3). These minerals are slightly soluble, so that their uranium can be carried by surface water or ground water into reducing environments (a bed of lignite or black shale, for example) and precipitated as pitchblende or coffinite $(USiO_4 \cdot nH_2O)$.

Thus uranium, like iron or copper, is an element showing changes from one oxidation state to another in geologic environments. Details of its chemistry provide an explanation for much of its behavior. Uranium has many oxidation states ($+2$, $+3$, $+4$, $+5$, $+6$), but only the $+4$ and $+6$ states are of geologic interest. In its two lowest oxidation states uranium is such a powerful reducing agent that it can liberate hydrogen from water, and the $+5$ state in the presence of water is unstable with respect to $+4$ and $+6$:

$$2UO_2^+ + 4H^+ \rightleftharpoons UO_2^{2+} + U^{4+} + 2H_2O \qquad \Delta G^\circ = -13.0 \text{ kcal}$$

The transition from $+4$ to $+6$ has a redox potential within the normal range for geologic environments:

$$U^{4+} + 2H_2O \rightleftharpoons UO_2^{2+} + 4H^+ + 2e^- \qquad E^\circ = +0.33 \text{ volt}$$

so we would expect to find compounds of these two oxidation states in nature.

Uranous ion, U^{4+}, reacts with bases to form an extremely insoluble hydroxide, $U(OH)_4$. The low dissociation constant, about 10^{-50} at 25°, means that in neutral solution the U^{4+} concentration is negligibly small and even at a pH of 4 would be only $10^{-10}M$. The hydroxide is somewhat unstable with respect to dehydration,

$$U(OH)_4 \rightleftharpoons UO_2 + 2H_2O \qquad \Delta G^\circ = -8.4 \text{ kcal}$$

so that the maximum concentrations of U^{4+} in equilibrium with UO_2 are even lower than for the hydroxide. Small wonder the pitchblende and uraninite are so stable in reducing environments!

The uranyl ion, UO_2^{2+}, on the other hand, forms a considerably more soluble hydroxide:

$$UO_2(OH)_2 \rightleftharpoons UO_2OH^+ + OH^- \qquad K = 10^{-14.2}$$
$$UO_2OH^+ \rightleftharpoons UO_2^{2+} + OH^- \qquad K = 10^{-8.2}$$

In a solution at pH 7 the concentration of UO_2OH^+ would be $10^{-7.2}M$, and of UO_2^{2+} $10^{-8.4}M$. At a pH of 4 the concentration of UO_2^{2+} would rise to $10^{-2.4}M$, which gives a total of nearly 1 g/liter of uranium. In this case the hydroxide is more stable than the oxide:

$$UO_2(OH)_2 \rightleftharpoons UO_3 + H_2O \qquad \Delta G^\circ = +10.0 \text{ kcal}$$

At least two other hydrates or hydroxides of UO_3 are known, all having roughly similar stabilities. In nature, hydrates of UO_3 are sometimes found as minerals, but more commonly the uranyl ion unites with other anions: carbonate, phosphate, and vanadate more often than others. The number of known complex minerals of uranium in the sexivalent state is enormous.

Uranyl hydroxide is slightly soluble in alkaline solutions as well as in acid:

$$UO_2(OH)_2 + OH^- \rightleftharpoons UO_2(OH)_3^- \qquad K = 10^{-3.6}$$
$$UO_2(OH)_2 + 2OH^- \rightleftharpoons UO_2(OH)_4^{2-} \qquad K = 10^{-3.8}$$

(These are the diuranate and uranate ions, respectively, often written without H_2O as HUO_4^- and UO_4^{2-}.) The low values of the constants mean that solubilities become appreciable only in strongly alkaline solutions. Minerals containing uranium in these anions are not known.

In solutions containing carbonate ion, the solubility of compounds of sexivalent uranium is greatly increased by the formation of carbonate complexes:

$$UO_2(CO_3)_2^{2-} \rightleftharpoons UO_2^{2+} + 2CO_3^{2-} \qquad K = 10^{-14.6}$$

$$UO_2(CO_3)_3^{4-} \rightleftharpoons UO_2(CO_3)_2^{2-} + CO_3^{2-} \qquad K = 10^{-3.8}$$

The first equation means, for example, that a solution at pH 7 with $0.01M$ total carbonate would contain about 10,000 times as much $UO_2(CO_3)_2^{2-}$ as UO_2^{2+} From the calculation above for a carbonate-free solution, we can estimate that the total dissolved uranium would now consist of about $10^{-7}M$ UO_2OH^+, $10^{-8}M$ UO_2^{2+}, and $10^{-4}M$ $UO_2(CO_3)_2^{2-}$, or 1,000 times as much as we estimated for the carbonate-free solution. The contribution of the carbonate complex would increase rapidly as the amount of CO_3^{2-} is increased, either by dissolving more carbonate or by raising the pH. The second carbonate complex becomes significant only in strongly alkaline solutions.

The general geochemistry of uranium in near-surface environments, therefore, can be readily described. Primary minerals are oxidized to uranyl ion, which is somewhat mobile in weakly acid solutions, and also in neutral and alkaline solutions if CO_3^{2-} is present. Surface waters in contact with uranium minerals should contain a few parts per million of uranium ordinarily, and up to a few thousand parts per million in exceptional situations. From such solutions uranium may be precipitated in the sexivalent state by a variety of anions, forming the familiar oxidized minerals; or it may be reduced by any one of a number of reducing agents, notably by organic matter, commonly forming UO_2 or one of its hydrates. Processes of these general kinds are probably taking place, or have taken place recently, in the famous uranium deposits of the Colorado Plateau and central Wyoming, where oxidized minerals are disseminated in sandstones near the surface, and at lower levels black unoxidized ore is found in sediments containing much organic material. Especially conspicuous is the concentration of uranium minerals around buried logs, fragments of bone, and isolated lenses of dark shale.

Note, however, that an understanding of the chemistry of the uranium occurrences does not in itself give an explanation of the origin of these deposits. As far as chemistry is concerned, it is equally plausible to regard the surface ore as derived by oxidation of original deposits in the organic-rich sediments, or to think of the unoxidized ore as derived by downward leaching and reduction from original deposits of oxidized material in the sandstone. The *mechanism* of oxidation, dissolution, precipitation, and reduction is explained by the chemistry of uranium, but the *sequence* of events must be worked out from geologic relations.

To give a complete account of the oxidation of uranium ores in deposits like those of the Colorado Plateau requires consideration of the geochemistry of vanadium, because this element is intimately associated with uranium in the Plateau deposits. In its chemical behavior vanadium somewhat resembles uranium, but details are more complicated, because vanadium has three oxidation states rather than two which are stable in the normal range of geologic environments, and because the element forms a bewildering variety of complex ions. Rather than discuss the geochemistry of vanadium in detail, we reproduce an Eh-pH diagram worked out by Garrels and his colleagues (Fig. 18-2), showing the relations of some uranium and vanadium compounds in contact with water at 25° and 1 atm total pressure. For the specified concentrations of vanadium (total dissolved = $10^{-3}M$), carbonate (total = $0.1M$), and potassium (total = $10^{-3}M$), the diagram shows that unoxidized ores should contain uraninite and montroseite, minerals with both metals in their lowest naturally occurring oxidation states. This agrees

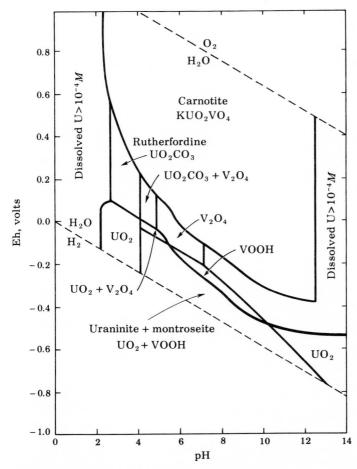

Figure 18-2 Eh-pH diagram for uranium and vanadium compounds at 25°C and 1 atm total pressure. Total dissolved $V = 10^{-3}M$, carbonate = $10^{-1}M$. *(Source: Garrels and Christ, 1965, page 393.)*

with experience, for these minerals are common in the lower parts of many Colorado Plateau deposits, especially in the vicinity of organic matter. Increasing oxidation changes the uranium to uranyl compounds and vanadium to compounds in which it shows an oxidation number of 4. Maximum oxidation gives carnotite, in which both elements have their highest oxidation states; this mineral occupies most of the upper part of the diagram, corresponding with its widespread occurrence in oxidized ores. At both sides of the diagram uranium shows a total solubility exceeding $10^{-4}M$, on the acid side taking the form chiefly of uranyl ion and on the alkaline side appearing chiefly in the carbonate complexes. The diagram shows general relationships of the oxidation states clearly, but it includes only a handful of the compounds found in natural occurrences.

These various examples illustrate how simple chemical ideas can be applied in explaining the reactions that occur when ore deposits are exposed to oxidation. This is one place in geochemistry where predictions from thermodynamic data accord beautifully with observed mineral associations. It should be emphasized once more, however, that our chemical explanations are largely *ex post facto:* given a mineral assemblage, we can construct plausible hypotheses for its derivation from other assemblages. But we are still a long way from being able to select the only possible hypothesis of origin, and still farther from being able to predict just what mineral assemblage must follow from a given set of geologic conditions. The oxidized and unoxidized ores of the Colorado Plateau, for example, are apparently related by simple chemical reactions, but these reactions do not tell us whether the dominant process has been in the direction of oxidation or of reduction. And they certainly cannot tell us with any degree of certainty whether in similar environments elsewhere we may expect to find other deposits of uranium and vanadium minerals.

PROBLEMS

1 In the oxidized zone of lead deposits, it is common to find a crystal of galena surrounded by a layer of anglesite, and the anglesite surrounded by an outer layer of cerussite. Suggest an explanation.

2 The metal cadmium often occurs as a minor constituent of sphalerite in primary sulfide ores. In partially oxidized ores the sulfide of cadmium (greenockite, CdS) is sometimes found as tiny yellow crystals on the surface of sphalerite. Suggest an explanation.

3 Using free energies from Appendix VIII, show how the line separating the PbO_2 and $PbCO_3$ fields in Fig. 18-1 is located.

4 Construct a diagram similar to Fig. 18-1 showing the stability fields of the zinc minerals sphalerite, smithsonite, and zincite. Include "contour" lines showing concentrations of Zn^{2+} in the part of the diagram representing acid, oxidizing conditions. Use any reasonable figures for total dissolved carbonate and total dissolved sulfur.

5 Explain why supergene sulfide enrichment is more important for copper than for iron, and more important for silver than for nickel.

6 Calculate the concentration of all dissolved species containing uranium in a solution in contact with UO_2 and having a pH of 6, an Eh of 0.0 volt, and total dissolved carbonate equal to $10^{-3}M$.

7 Using the oxidation potentials in Appendix IX, investigate the possibility of oxidizing and dissolving the metal palladium in the oxidized zone of ore deposits.

8 Several ferrous sulfate and ferric sulfate minerals are known [e.g. melanterite, $FeSO_4 \cdot 7H_2O$, and coquimbite, $Fe_2(SO_4)_3 \cdot 9H_2O$]. Under what geologic conditions would you expect these minerals to form?

9 One of the minerals found in the oxidized zone of cobalt deposits is the hydroxide, stainierite. A controversy has arisen as to whether this mineral is cobaltic hydroxide, $Co(OH)_3$, or cobaltous hydroxide, $Co(OH)_2$. Using free energies or electrode potentials, see if $Co(OH)_3$ would be stable under geologically reasonable conditions.

10 A common method of geochemical prospecting is to analyze soil for traces of metals released to soil solutions during the oxidation of concealed sulfide deposits below. Such analyses in areas known to have deposits of chalcopyrite, sphalerite, and galena often show anomalously large concentrations of copper and zinc in the soil, but seldom unusual concentrations of lead. Suggest an explanation.

11 At some ore deposits oxidized minerals have been the principal source of the metals recovered, the weathering process apparently having concentrated the metals from low-grade primary ores. Suggest what geologic, topographic, and climatic conditions might lead to such concentrations in the oxidized zone. Would the same conditions necessarily favor supergene sulfide enrichment?

REFERENCES AND SUGGESTIONS FOR FURTHER READING

Anderson, C. A., Oxidation of copper sulfides and secondary sulfide enrichment, *Econ. Geology*, 50th Anniversary vol., pp. 324–340, 1955. An excellent brief review of the voluminous literature.

Garrels, R. M., and C. Christ, "Solutions, Minerals, and Equilibria," Harper & Row, Publishers, Incorporated, New York, 1965. Chapter 7 has Eh-pH diagrams and partial-pressure diagrams showing relations among the oxidized products of ores of most of the heavy metals.

Hansuld, J. A., Eh and pH in geochemical prospecting, *Geol. Survey of Canada Paper* 66-54, pp. 172–187, 1966. Geochemical prospecting is a method of exploration for ore deposits that depends on analyses for small amounts of metals that have weathered from ore minerals and are dispersed in soil, water, and vegetation. This paper emphasizes the importance of Eh and pH in drawing conclusions from such analyses.

Huff, L. C., A geochemical study of alluvium-covered copper deposits of Pima County, Arizona, *U.S. Geol. Survey Bull.* 1312-C, 1970. Contrasts in the behavior of copper and molybdenum in soil, gravel, vegetation, and ground water near an oxidizing copper deposit.

Jarrell, O. W., Oxidation at Chuquicamata, *Econ. Geology*, vol. 39, pp. 251–286, 1944. A description of the complex and spectacular oxidation zone of a famous copper deposit in an arid climate.

Langmuir, D., Uranium solution-mineral equilibria at low temperatures with applications to sedimentary ore deposits, *Geochim. et Cosmochim. Acta*, vol. 42, pp. 547–569, 1978. Thermodynamic data applied to the movement and deposition of uranium in surface water and ground water.

Levinson, A. A., "Introduction to Exploration Geochemistry," Applied Publishing, Ltd., Calgary, 1974. Chapter 3 is an excellent treatment of the weathering of ores and the dispersal of various metals into adjacent soils and waters.

Park, C. F., and R. A. MacDiarmid, "Ore Deposits," 3d ed., W. H. Freeman and Company, San Francisco, 1975. Chapters 18 and 19 discuss the weathering of ore deposits and sulfide enrichment, with good descriptions of numerous examples.

Sato, M., Oxidation of sulfide ore bodies. II. Oxidation mechanisms of sulfide minerals at 25°C, *Econ. Geology*, vol. 55, pp. 1202–1231, 1960. An experimental study of the detailed mechanism of oxidation of some common sulfide minerals, showing that the reaction proceeds in steps leading to the formation of metal ion in solution plus free sulfur, followed by oxidation of the sulfur to sulfate ion.

NINETEEN

METAMORPHISM

Near intrusive granite contacts, rocks commonly show pronounced changes in color and texture. Shale and sandstone become hard, brittle rocks, and limestone is converted into coarsely crystalline marble. New minerals may appear in the altered rocks, evidently resulting from the high temperatures produced by the intrusive. Elsewhere, over broad areas not visibly associated with igneous activity, rocks may show a different kind of alteration: shales are transformed into slates and mica schists, lavas into chlorite and amphibole schists, limestones again into marbles. All such changes, which may be reasonably ascribed to the action of heat and pressure beneath the earth's surface, go by the name of metamorphism.

To frame a more precise definition than this is difficult. On the one hand, we need a distinction between metamorphic processes and sedimentary processes; yet sedimentary processes too may involve conspicuous changes in the original material: cementation, recrystallization, dehydration, ion exchange, and several others. We can assert that these changes take place at low temperatures and shallow depths, in contrast to the high temperatures and deep burial required for metamorphism, but to specify the limiting temperatures and depths means drawing an arbitrary boundary where no boundary exists in nature. At the opposite extreme we need a criterion to differentiate metamorphic processes from igneous processes. This also is troublesome, because rocks heated sufficiently must melt, and the melting generally takes place over a considerable range of temperature. At what point do we separate metamorphic from igneous phenomena as a rock gradually becomes molten? Recognizing that the definition must be somewhat arbitrary, we set out in this chapter to study the chemistry of metamorphic processes—processes which grade into sedimentary reactions at one end of the temperature scale and into igneous reactions at the other.

19-1 CONDITIONS OF METAMORPHISM

Metamorphic reactions differ from igneous processes in that they are largely reactions in the solid state. No actual melting is involved in metamorphism except at the very highest temperatures. That this must be true is shown by the common preservation of earlier structures in metamorphic rocks: slates and quartzites often show clearly the original stratification of shales and sandstones, marbles sometimes have recognizable fossils, and greenish altered lavas may betray their origin· by well-preserved phenocrysts, vesicles, and flow structures. More thoroughly recrystallized metamorphic rocks like gneisses and amphibolites show the essential preservation of the solid state by an absence of intrusive relations with their surroundings: no dikes or stringers branching from the main mass, no decrease in grain size at contacts, no xenoliths or intrusive breccias. Interstitial fluids may indeed play a role in metamorphism, but the bulk of a rock remains solid while the transformation is taking place.

What is known about actual conditions of temperature and pressure during metamorphism? The upper limit of temperature, defined as the point at which melting becomes extensive, can be specified fairly precisely. Experimental work shows that the range of melting temperatures depends on the composition of the rock material, on overall pressure, and on the nature and concentration of accompanying fluids. For most ordinary kinds of rock, under reasonable pressures of water vapor, the temperatures of incipient melting fall in the interval 650 to 800°C. The lower temperature limit is harder to set, partly because of the simple difficulty of definition and partly because observational evidence is conflicting. Sedimentary rocks from deep drill holes, where bottom temperatures are over 150°, often show no sign of metamorphic change; yet in hot-spring areas water near 100° can evidently cause extensive alteration. The conversion of kaolinite to muscovite, a familiar metamorphic change in the early stages of alteration of shales, according to experiment can take place at a variety of temperatures from 350° down at least as far as 200°, depending on the acidity and potassium content of interstitial solutions (Sec. 7-4, Fig. 7-6). Epidote, another mineral characteristic of early stages of metamorphism, is known from experiment and observations at hot springs to form at temperatures in the range 130 to 240°, depending on pressure and composition of solutions. Perhaps temperatures of the order of 100 to 150° would be reasonable for the low-temperature end of the metamorphic scale. Alteration below such temperatures would come under the heading of diagenesis, or weathering, or possibly hydrothermal activity.

Pressure during metamorphism is due chiefly to weight of overlying material. Most metamorphic rocks have not been more than 30 km below the surface (at greater depths granitic material would begin to melt), a depth at which the lithostatic pressure is about 10,000 atm. Locally and temporarily pressures resulting from orogenic movement may exceed this figure, and some rocks have certainly come from depths greater than 30 km, but 10,000 atm or 10 kilobars serves well as a rough upper limit for most metamorphic changes. The lower limit of pressure goes back to the question of definition: ordinarily we think of metamorphism as taking place below the surface, at pressures of at least a few hundred atmospheres, but if we include the sort of alteration found near fumaroles, the lower limit would be the ordinary pressure of the atmosphere.

Extreme figures can thus be set down with some confidence. More difficult is the problem of determining the temperature and pressure under which any particular sample of metamorphic rock has formed. In the field we apply various rule-of-thumb criteria derived from geologic observation: coarse grains suggest higher temperatures than fine grains, biotite a higher temperature than muscovite, hornblende a higher temperature than actinolite, and so on. Such mineralogical observations can be refined so as to classify metamorphic rocks roughly according to their relative temperatures and pressures of formation (Sec. 19-4). Increasingly, experimental work is providing more quantitative data on stabilities of mineral assemblages, on distribution of elements and isotopes between coexisting minerals, on composition and degree of filling of fluid inclusions, from which absolute values of temperature and pressure can be guessed. Estimated temperatures and pressures depend strongly on the amount and nature of fluids which have long since vanished, so that figures on the conditions of metamorphism for a particular rock can seldom be given with an accuracy greater than $\pm 100°$ or $\pm 1,000$ atm.

With respect to pressure, an added complication is the necessity for distinguishing between uniform and directed pressure. Uniform pressure is the kind resulting from weight of overlying material, the only kind that can be sustained in a fluid medium. In water this uniform or nondirected pressure is called *hydrostatic*, and the same term is often used to refer to nondirected pressure in other media, even to pressures in rocks. A more specific term like *lithostatic* is preferable, to avoid ambiguity. In contrast to lithostatic pressure, *directed* pressure in rocks is pressure stronger in one direction than in another. Evidence for the action of directed pressure is plentiful: folded strata, thrust faults, granulation of rocks, and mineral orientation in schists and gneisses are common results. In geologic literature directed pressure is often loosely referred to as *stress*. Technically stress should describe the internal condition of a solid, the forces acting across any surface within it, but as a matter of convenience the meaning may be extended without serious ambiguity to the external directed pressure which sets up the internal forces. Regarding the numerical magnitude of directed forces operative in metamorphism we have practically no data, except that such pressures over a long period could not exceed lithostatic pressures by more than a small fraction because of the limited strength of rocks.

A final important condition that determines the rate and often the nature of metamorphic reactions is the presence of interstitial fluids. Practically all rocks undergoing metamorphism contain such fluids, either as actual water in pores and surface films or as water and carbon dioxide in minerals which can be released as the rock is heated. Fluid is often added to the original supply during metamorphism, either distilled out of adjacent rocks or emanating from a cooling body of magma. Whatever their source, the fluids can be expected to bear considerable resemblance to the emanations from volcanoes that we discussed in Chap. 16. A long argument has gone on in geologic circles as to whether metamorphism can take place at all in the absence of fluids. Certainly reactions between dry solids are possible, and certainly some metamorphic rocks retain very little evidence in their composition that fluids were ever present. On the other hand solid-solid reactions are generally slow, especially reactions between silicates, and quite possibly the reactions necessary for metamorphism would not take place appreciably even in geologic time without the help of fluids. The argument hinges on reaction rates, about which quantitative information is meager. In a practical geologic sense the argument does not seem very profitable, since at least a little fluid is almost universally present in rocks. Most metamorphic reactions are speeded up by fluids, and for some reactions the nature and amount of fluid determines what kind of metamorphic minerals will form.

Temperature, lithostatic pressure, directed pressure, and nature and amount of interstitial fluids: these are the factors we must know, at least in a qualitative sense, to be able to predict how a given rock will respond to metamorphic conditions.

19-2 CLASSIFICATION OF METAMORPHIC ROCKS

The difficult problem of classifying metamorphic rocks belongs in the province of petrography rather than geochemistry. Here we need no more than definitions of a few terms that are useful in discussing the chemistry of metamorphic processes.

A convenient twofold division of metamorphic rocks separates them, largely on the basis of field observation, into (1) *thermally metamorphosed rocks*, those found locally near the contacts of intrusive bodies, and (2) *regionally metamorphosed rocks*, those occurring in larger areas and having no apparent relation to intrusive rocks. As a rule the former are *unfoliated* (lacking in directional structures due to mineral orientation), apparently because pressures were largely lithostatic, whereas the latter are typically *foliated*, with minerals like micas and amphiboles conspicuously oriented in planes and lines, presumably as a result of directed pressure during metamorphism. The distinction between foliated and unfoliated rocks is not quite synonymous with that between regionally and thermally metamorphosed rocks, since areas of regional metamorphism commonly include unfoliated varieties, especially rocks like quartzite and marble whose composition does not permit the formation of platy or needlelike minerals. Contact metamorphism and regional metamorphism cannot always be clearly distinguished, because intrusive magma may invade rocks already regionally metamorphosed and superpose thermal effects on the original metamorphism. Despite these ambiguities the distinction between thermally and regionally altered rocks is a useful one, because it not only is readily applied in the field but also has considerable genetic significance.

Some metamorphic rocks show evidence of addition of much foreign material during metamorphism. Tactites with garnet and diopside, greisens with abundant tourmaline, replacement ores with sulfide minerals, all come to mind as examples.

These could be regarded as a third major group of metamorphic rocks, but generally are considered as varieties of the two major classes.

Subdivision of the two major groups is based either on chemical composition (for example, quartzite, marble, andalusite-cordierite hornfels) or on the concept of change in mineral composition with increasing temperature and pressure (for example, slate, schist, gneiss). Compositional changes in response to temperature and pressure are the basis for the facies classification, which is the subject of Sec. 19-4.

19-3 EQUILIBRIUM RELATIONS

General

Metamorphism is a progressive adjustment of rocks to changing temperatures and pressures. Ideally, from a rock of given initial composition, metamorphic reactions should give a sequence of mineral assemblages, each representing the stable equilibrium assemblage for a certain range of temperature and pressure. Slow reactions may prevent attainment of some equilibria, and permeation of the rock by fluids may change its bulk composition and so alter the equilibrium assemblages, but despite these complications it is useful to discuss metamorphic processes in terms of idealized equilibria.

The Mineralogic Phase Rule

How complex can a metamorphic rock be? The phase rule should help us fix the maximum number of minerals that can exist at equilibrium for a given composition. Theoretically the number of phases would be a maximum when the number of degrees of freedom is zero, and would be two more than the number of components:

$$p = c + 2 - f \qquad \text{whence } p_{max} = c + 2 - 0$$

This maximum could be attained only at a fixed point, a point where both temperature and pressure have unique values. To find such a special assemblage preserved in a metamorphic rock would be possible but extremely unlikely; in general, an actual rock must represent an assemblage stable over a considerable range of both temperature and pressure. In other words, the assemblage must have a minimum of two degrees of freedom, so that the maximum number of solid phases is equal to the number of components. This generalization was first suggested by Goldschmidt, and it often bears his name. It holds for igneous rocks as well as metamorphic rocks, hence is commonly called the *mineralogic phase rule*.

As a simple example, consider the three-component system CaO-MgO-SiO_2. At various temperatures and compositions metamorphism might give assemblages like enstatite-wollastonite-quartz, enstatite-diopside-quartz, diopside-forsterite-periclase, but the number of coexisting minerals would never be greater than three. In an ordinary shale or arkose or lava the number of principal components is larger, generally seven or eight, and the possible number of coexisting minerals would be correspondingly greater.

Actually the mineralogic phase rule gives scant help in studying natural systems. One reason is the difficulty of deciding on the number of components because of isomorphous substitution. The very extensive substitution of iron for magnesium, for example, would in many rocks reduce the number of independent components by one. Volatiles pose another stumbling block: a volatile substance would count as a component if it is present in limited amount and if it is confined within the system (a "closed system"), but not if it is free to move in and out of the system (an "open system"). Suppose, for example, that CO_2 is added to the three-component system considered in the last paragraph. If only a small amount is present and if the system is closed, so that the pressure of CO_2 is determined by equilibrium in reactions like

$$CaCO_3 + SiO_2 \rightleftharpoons CaSiO_3 + CO_2 \tag{19-1}$$

then calcite, magnesite, or dolomite could appear as a fourth solid phase together with the silicate minerals (for example, enstatite-wollastonite-quartz-calcite). If, on the other hand, the system is open to fluids and the pressure of CO_2 has a fixed value determined by external factors, the reaction of Eq. (19-1) would be displaced in one direction or the other at a given temperature; any pair of the three substances $CaCO_3$, SiO_2, $CaSiO_3$ could appear among the solid phases, but not all three. To use more technical language, in a system open to fluids, CO_2 is a *mobile component*, and the phase rule may be modified to read

For the general case: $\qquad\qquad p = c - m + 2 - f$

For the mineralogic phase rule: $\quad p_{max} = c - m$

where c in each expression is the total number of components and m is the number of mobile components. Now in trying to work out the history of a metamorphic rock, we seldom know how many mobile components were present or how nearly conditions approached the ideal closed or open system; hence the number of phases to be predicted from the mineralogic phase rule is always to some extent indeterminate.

Aside from these theoretical considerations, the application of Goldschmidt's rule commonly serves little purpose because of the simple fact that most metamorphic rocks have only a small number of minerals anyway, generally smaller than the maximum permitted by the rule. In fact one of the striking characteristics of most metamorphic rocks is the relative simplicity of their mineral assemblages. Such very abundant rocks as quartz-mica schists and amphibolites, for example, have only two or three minerals of any importance despite the presence of seven or eight components.

ACF Diagrams

Some shorthand method is clearly desirable for representing possible metamorphic assemblages obtainable from different proportions of starting materials. The problem is one we have encountered before: how to represent variations in seven or eight components on a two-dimensional diagram. A convenient solution, here as elsewhere, is to reduce the number of significant variables by combining the components in groups and selecting the more important ones. Various ways of doing this have been tried, of which the most generally useful is the ACF diagram devised by Eskola (1939).

Constructing an ACF diagram from a chemical analysis involves the following steps:

1. Recalculate weight percentages of oxides to molecular percentages.
2. Disregard constituents of accessory minerals and Al_2O_3 in alkali feldspars (in other words, subtract from the Al_2O_3 percentage an amount equivalent to that combined with Na_2O and K_2O in $Na_2O \cdot Al_2O_3 \cdot 6SiO_2$ and $K_2O \cdot Al_2O_3 \cdot 6SiO_2$).
3. Add the remaining Al_2O_3 to Fe_2O_3 to get a quantity called A.
4. Let CaO be a quantity C.
5. Let F be the sum $MgO + FeO + MnO$.
6. Recalculate $A + C + F$ to 100%, and plot on a triangular diagram.

Additional rules are needed for more exact computation to take account of the accessory minerals, and for rocks of unusual composition. The calculation assumes that silica and alkali feldspar are present in excess. This assumption is justified by the observation that these substances are present in a majority of metamorphic rocks, hence are not significant as phases differentiating one mineral assemblage from another. For rocks which do not have excess silica and alkali feldspar, a different choice of independent variables can be made.

A classic example of the use of ACF diagrams is provided by the lime-silicate hornfelses of the Oslo region in southern Norway. Here quartz and K-feldspar are in excess, so that possible equilibrium assemblages may be adequately represented by variations in A, C, and F (Fig. 19-1). The ten varieties of hornfels that can be distinguished are shown by number on the figure and in the following list (quartz and K-feldspar are also present in each variety):

1. Andalusite-cordierite
2. Andalusite-cordierite-plagioclase
3. Cordierite-plagioclase
4. Cordierite-plagioclase-hypersthene
5. Plagioclase-hypersthene
6. Plagioclase-hypersthene-diopside
7. Plagioclase-diopside
8. Plagioclase-diopside-grossularite
9. Diopside-grossularite
10. Diopside-grossularite-wollastonite (plus idocrase)

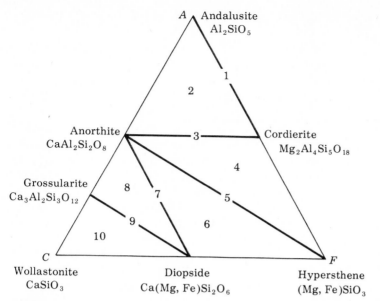

Figure 19-1 ACF diagram for hornfelses of the Oslo region. (Pyroxene hornfels facies.) *(After Turner, 1968, page 227.)*

From these compositions we might guess that the original rocks which underwent metamorphism were shales, ranging in composition from lime-free (high A) to lime-rich (high C).

Note carefully what a diagram like Fig. 19-1 signifies. It shows the possible equilibrium assemblages for different initial compositions in a certain range of temperature and pressure—no more than this. It cannot be interpreted as showing possible paths of crystallization, like the triangular diagrams of Chaps. 12 and 13. Nor can it serve as a basis for predictions except in a negative sense; one could venture a guess on the basis of Fig. 19-1, for example, that the combinations cordierite-diopside and hypersthene-grossularite would not occur in the Oslo hornfelses. (If such combinations were found, they would be interpreted as equilibrium assemblages in a different pressure-temperature range, for which a new ACF diagram would have to be constructed.) In a word, these diagrams are simply a means of summarizing concisely a mass of analytical data.

It should be noted that ACF diagrams, or similar triangular diagrams with other variables, are useful only for groups of metamorphic rocks whose compositional relations can be adequately expressed by variations in three components. A great many metamorphic rocks meet this requirement, but by no means all.

Recognition of Equilibrium Assemblages

We have been assuming that the minerals of metamorphic rocks form equilibrium assemblages, but the assumption may well be questioned on the basis that metamorphic reactions are slow. How is the assumption justified? And how can we distinguish observationally between equilibrium and nonequilibrium assemblages?

Such questions can be approached both by experimental duplication of metamorphic mineral combinations and by study of petrographic data. One petrographic indication that equilibrium is attained in most metamorphic rocks is the fact that the number of minerals does not exceed the number permitted by the phase rule. Inasmuch as the number is often well below the maximum, however, this is hardly a rigorous test. Another observation suggesting the attainment of equilibrium is the usual lack of composition zoning in metamorphic plagioclase and pyroxene. In igneous rocks these minerals commonly betray failure of the melt to maintain equilibrium on cooling by showing a different composition between the centers and margins of crystals, but in metamorphic rocks their generally uniform composition must mean that the crystals adjusted themselves completely to the temperature and pressure of their environment. Perhaps the most convincing evidence for equilibrium is the presence of the same mineral assemblage in many different rocks of similar overall composition, regardless of the ages of the rocks or the localities from which they come. Surely, if nonequilibrium were common, one would expect rocks of similar composition to show a considerable variety of minerals depending on how closely equilibrium was approached.

Rather surprisingly, then, we find that equilibrium assemblages are pretty much the rule among metamorphic rocks. Exceptions are by no means lacking, especially exceptions in which relics of original minerals can still be recognized, but such cases are rare in comparison with rocks in which the metamorphic transformation has gone to completion.

Preservation of High-Temperature Equilibrium Assemblages

If equilibrium is so much the general rule, a troublesome question presents itself. After a rock has been metamorphosed at high temperature and pressure, it must undergo a gradually decreasing temperature and pressure in order to appear finally in surface outcrops. Why doesn't its composition readjust itself so as to be in equilibrium with the lower temperature-pressure conditions? How are the high-temperature mineral assemblages preserved? Why, to put it baldly, do we ever find metamorphic rocks at all?

Metamorphic reactions are known to be sluggish and are known to increase in rate as the temperature rises. If a rock is heated slowly, say to 500°C, and then cooled rapidly, an assemblage characteristic of 500° might form which would not have time to readjust itself to low-temperature conditions. But this seems like special pleading; nothing in current knowledge of metamorphism would suggest that temperatures always or even commonly rise more slowly than they fall. Why do we not find at least incipient adjustments to lower temperatures? Partial conversion of high-temperature minerals to low-temperature minerals is indeed sometimes noted ("retrograde metamorphism"), but it is rare rather than commonplace. Differing rates of temperature rise and fall thus seem a poor explanation for preservation of high-temperature assemblages.

For regional metamorphism a better explanation is suggested by the fact that this kind of metamorphism is generally an accompaniment to orogenic movement. Perhaps reactions can occur only during the movement itself, in response to intimate crushing and granulation of the rocks; perhaps reaction ceases when movement ceases, preserving the mineral assemblage formed during the orogeny. For thermally metamorphosed rocks a possible explanation lies in the observation that metamorphic zones at intrusive contacts are widest where quartz veins and abundant new minerals show addition of fluids during the alteration. Perhaps metamorphism *always* requires the presence of at least minor amounts of fluid, and the metamorphic reactions cease abruptly when the supply of fluid is cut off. One or both of these explanations—the speeding up of metamorphic reactions either by granulation due to orogenic movement or by the presence of fluids—probably accounts for the formation and subsequent preservation of metamorphic rocks.

Effect of Lithostatic Pressure

Since metamorphic reactions involve mostly changes from one assemblage of solid silicate minerals to another and since silicates do not differ greatly among themselves in density, we might expect that the influence of lithostatic pressure on most metamorphic equilibria would be small. This follows from Le Chatelier's rule, or from the algebraic expression of the rule given by Eq. (8-21):

$$\frac{d\,\Delta G}{dP} = \Delta V$$

Thus the rate at which the free-energy change for a reaction varies with pressure is equal to the volume change in the reaction. This means that the variation of free energy, and hence of the position of equilibrium, must be small for reactions where the overall change of density is slight. Reactions involving volatiles in a closed system would be exceptions to this rule. The reaction shown by Eq. (19-1), for example, would obviously be very sensitive to pressure if the CO_2 cannot easily escape. Even reactions with volatiles, however, are not much affected by changes in rock pressure if the volatiles are free to move.

Pressure does have a pronounced effect in some solid-solid reactions that produce minerals of exceptionally high or exceptionally low density. The conversion of the light form of carbon (graphite, density 2.1 g/cm^3) to the dense form (diamond, 3.5) is a familiar example; the rocks in which diamonds are generally found (kimberlites) give evidence of an origin below the base of the crust, in the upper mantle, where pressures would be very great. Another example is the formation of the dense varieties of silica, coesite and stishovite, by the momentary enormous pressures produced when a meteor strikes the earth. A more important example from a petrographic standpoint is the metamorphic rock called eclogite.

We have met this rock in an earlier discussion (Sec. 14-5), where it was described as a possible result of crystallization of a basaltic liquid under very high pressure, or alternatively as the product of high-pressure alteration of solid basalt. The metamorphic conversion of basalt to eclogite is represented roughly by Eq. (14-1), which shows a change from a plagioclase-pyroxene-olivine mixture, with an average density about 3.0 g/cm^3, to a rock containing garnet, soda pyroxene, and quartz, with a density of about 3.4. Experimentally the change requires pressures in the range 12,000 to 16,000 atm; in agreement with these figures, field observation shows that most eclogites are associated with rocks that have been deeply buried in the crust, or have come from the upper part of the mantle.

At the other extreme of pressure, rocks containing the relatively low-density minerals sanidine, andalusite, and cordierite may be suggested as indicating a low-pressure metamorphic environment. Sanidine in particular is limited (in metamorphic rocks) to inclusions caught up in lava flows and shallow intrusives, where metamorphism takes place at high temperatures and practically at atmospheric pressure.

Except where one of these rather unusual minerals is present, unfoliated metamorphic rocks give little hint as to the magnitude of the pressure under which they formed. Lithostatic pressure, in general, is a less important determiner of metamorphic equilibria than is temperature.

Directed Pressure

Would directed pressure be capable of influencing metamorphic equilibria to a greater extent than lithostatic pressure? This is a difficult question, from whatever point of view it is approached—theoretical, experimental, or observational. We had best begin by reviewing current ideas about the more obvious effects of directed pressure on metamorphic textures.

Just how directed pressure operates to produce the characteristic textures of rocks like slates and schists has long been argued and is still not entirely clear. Purely mechanical effects of directed pressure are obvious enough: granulation, fracturing and bending of grains, streaking out of fine material, rotation of large crystals. But the difficult thing to understand is the development of *foliation*, the orientation of newly formed crystals, particularly the arrangement of platy crystals like mica in planes and of elongated crystals like amphibole either in planes or in linear patterns. Foliation has not yet been produced in the laboratory, so that our only evidence comes from geologic observation; and unfortunately the observations permit a variety of interpretations.

The foliation of some rocks, particularly the type of foliation called slaty cleavage, is often approximately parallel to the axial planes of folds. This has led to the suggestion that platy minerals grow in a metamorphic rock with their long dimensions perpendicular to the direction of maximum stress. It is hard to see, however, how by this rather passive mechanism the original random orientation of grains can be transformed into the ordered orientation of the metamorphic product—an orientation of grains not only according to dimensions but also according to crystal structure.

Foliation is by no means always parallel to axial planes of folds. More commonly it appears to follow old structures of the original rock (such as bedding planes or flow lines) or else planes of movement (minor thrust planes, planes of pronounced granulation and streaking) formed by the action of the deforming force. This suggests another possible explanation: perhaps foliation develops as a result of intimate movement in a rock undergoing deformation, the new minerals as they form being pushed or rolled into orientations that will offer the least resistance to movement, and the orientations being guided by any planes of weakness, either original or developed by movement, which the rock may possess. Strong support for this hypothesis has come from demonstrations that the equi-dimensional grains (quartz, feldspar, calcite) as well as the micas and amphiboles of a foliated rock have preferred crystallographic orientations, not megascopically visible but readily measured with a universal stage. These orientations are most easily explained by differences in strength, in ability to twin or to develop glide planes, along particular crystallographic directions—in other words, differences that would be accentuated by pervasive movement through the rock. This is the generally accepted explanation for most foliation at the present time, but argument persists over details.

From a chemical viewpoint the details of development of foliation have only minor interest. Our chief concern is the conclusion implied in the above reasoning, that a foliated rock must have been intimately granulated, so that minor movements have occurred all through its fabric. Granulation and movement are precisely the factors that would speed up the solid-solid reactions required for metamorphism or that would expose a maximum of solid surface to the action of interstitial fluids. Hence we could say immediately that one chemical effect of directed pressure is a catalytic action, a speeding up of reactions that might otherwise be very slow.

Now the crucial chemical question is this: Does directed pressure have *more* than a catalytic effect in determining the mineral composition of a metamorphic rock? Does nonuniform pressure permit the formation of minerals and mineral assemblages which could not form in its absence? In more symbolic language, can we consider directed pressure as an additional thermodynamic variable determining the state of a system? Instead of representing fields of stability by a simple temperature-pressure diagram, should we use a three-dimensional diagram with nonuniform pressures shown on a third axis, and expect to find some stability fields which do not touch the temperature-pressure plane at all?

To approach the problem theoretically, we note that any stress can be considered as made up of two parts, a nondirected pressure coupled with a pure shear stress. The nondirected part would behave like any other lithostatic pressure and would influence equilibrium reactions according to their volume changes. Pure shear, best visualized as the tendency of one layer to slide over another in response to a force couple, can have little effect on volume as long as the rock is elastic, because elongation in one dimension would be compensated in large part by compression in other dimensions. From a straight thermodynamic standpoint, then, the influence of directed pressure cannot be very different from that of uniform pressure; part of it is equivalent to the action of a certain uniform pressure, and the remaining part has only negligible effect on possible free-energy changes. Conceivably, however, the effect of stress may lie outside the realm of orthodox thermodynamics. The incipient movement of one layer past another, for example, might bring atom groups into contact in certain orientations that would be more favorable for one reaction than for another. Or stress may be nonhomogeneous, different from one part of a grain to another part, which could conceivably permit the growth of two different minerals where only one would be stable under lithostatic pressure. Or minerals may be sheared beyond their elastic limit, producing permanent deformation and greater change in volume than below the elastic limit. In summary, on the basis of theory a marked effect of directed pressure on equilibrium relations is unlikely, but the possibility of slight effects cannot be altogether excluded.

A better test than theoretical argument would be an examination of metamorphic rocks themselves, to see if some minerals are present in foliated rocks which are absent in rocks that have not been under stress. For a long time it was thought that such minerals existed. In the early years of this century, the great British authority on metamorphism, Alfred Harker, tried to distinguish between *stress* minerals, those found only in rocks subjected to directed pressure, and *antistress* minerals, those which are never observed in such rocks. To the first category Harker assigned minerals like kyanite, chloritoid, staurolite, almandite; to the second such minerals as andalusite, cordierite, nepheline, and leucite. In recent years, however, the list of stress minerals has steadily shrunk. One by one they have been observed in cavities, veins, or thermally metamorphosed rocks where nonuniform pressure could not have been effective. These minerals may be most common in foliated rocks, but are not restricted to them. The antistress minerals are indeed not found in rocks subjected to nonuniform pressure, but they are likewise probably absent from thermally metamorphosed rocks that have been subjected to high lithostatic pressure; they are best interpreted simply as minerals of low density that normally would not occur in environments where any kind of pressure is high. It seems doubtful, then, that there is any valid evidence from geologic observation for the existence of minerals whose stability is determined largely by the presence or absence of directed pressure.

Both theory and observation suggest that nonuniform pressure is important during metamorphism in producing foliated textures and in catalyzing chemical reactions, but not important as a determiner of stable equilibrium assemblages of metamorphic minerals.

19-4 METAMORPHIC FACIES

So far we have been discussing equilibrium relations in metamorphic processes from a very general standpoint. Implicit in the discussion was the idea that the mineral makeup of a metamorphic rock changes progressively as temperature and pressure rise, but we have not examined such changes in detail. We turn our attention now to specific examples of metamorphic reactions, and in particular to the sequences of rocks that have been ascribed to increasing temperature and pressure.

A striking feature of the thermally metamorphosed rocks near an intrusive contact is the change in their appearance and their mineral composition as the contact is approached. Shales, for example, first become harder, more brittle, less fissile; closer to the contact they may develop small rounded spots as the first sign of growth of new minerals; then flakes of sericite and biotite become prominent, and still closer to the contact andalusite and feldspar may appear. Through all these changes the overall chemical composition may have remained practically constant, the different minerals simply representing alternative combinations of the same chemical components. It is difficult to avoid the inference that we see here the effect of higher and higher temperatures as we go from unaltered shale to the andalusite rocks near the contact.

A zoning of rock types is often equally prominent in regional metamorphism. If the original rock is andesite or andesite tuff, for example, the first sign of change may be a greenish color due to development of chlorite, and chlorite schists may form a band that can be followed for miles across country. Going across the band, one may find adjacent to it a zone of actinolite schist, and beyond that an area of amphibolite. Again, the overall composition of the rock need not have changed at all through this sequence. The progression from unaltered tuff to amphibolite will also generally be in the direction of increasing deformation, of increasing grain size, and of increasing frequency of quartz veins and small intrusive bodies. The field relations seem most plausibly explained by assuming that the different mineral assemblages have developed in response to temperatures and pressures increasing toward the amphibolite.

Metamorphic sequences in nature are seldom as simple and regular as those just described. Outcrops all too often are inadequate to show the entire sequence; locally the sequence may be reversed or one of its members may be lacking; initial rock composition may be far from uniform; and complications arise where fluids have made notable alterations in composition. Nevertheless, sequences of this sort are sufficiently well defined and sufficiently similar from one area to another so that they must be regarded as a normal phenomenon of metamorphism. They are universally recognized as due to progressive increase in one of the factors of metamorphism, but just which factor or which group of factors is most important may be not entirely certain. Often a sequence is merely ascribed to a change in "intensity" of metamorphism—this being a conveniently vague term that has connotations of temperature, pressure, depth of burial, action of fluids, and degree of deformation, without specifying any one of them.

From the observation that overall composition remains approximately constant through a sequence, one might guess that the role of fluids in adding or subtracting material is usually minor. From the general considerations of temperature and pressure examined in the last section, one might further guess that of these two variables temperature would be the more important. It would seem, therefore, that in a sequence of progressively metamorphosed rocks we are dealing primarily with readjustments of mineral equilibrium to rising temperatures. This is the assumption underlying the most widely accepted classification of metamorphic zones: Eskola's (1939) classification according to metamorphic facies.

A *metamorphic facies* includes all rocks with equilibrium mineral assemblages in a certain range of temperature and pressure, temperature being the more critical variable. Successive layers of shale and tuff, for example, may be altered to quartz-sericite schist and chlorite-albite-epidote schist, respectively; these two mineral combinations would belong to one facies, because they represent adjustments of two different rocks to the same conditions of temperature and pressure. Interbedded limestone might be converted to fine-grained marble, and sandstone to quartzite; these would also belong to the same facies. A biotite-garnet schist, on the other hand, would belong to a different facies representing a higher temperature. Facies are named from characteristic minerals or rocks that are most sensitive to changes in temperature.

Metamorphic grade is a term used, often rather loosely, to describe the relation of one facies to another. Low-grade metamorphic rocks are rocks belonging to facies formed at low temperatures and pressures, high-grade rocks those formed at higher temperatures and pressures. In approaching an igneous contact, we often say that metamorphism increases in intensity and that the rocks represent assemblages of increasing metamorphic grade.

One suggested grouping of metamorphic facies is given in Table 19-1. The primary breakdown is into four temperature ranges, but the ranges overlap extensively, because limiting temperatures depend somewhat on total pressure and water-vapor pressure. The facies are subdivided further according to pressure, the lower pressures being those characteristic of contact metamorphic zones and the higher pressures those typical of regional metamorphism. Note that differences between typical high-pressure and low-pressure assemblages are not very great; at high pressures one finds relatively dense minerals like garnet and sillimanite, at low pressures lighter minerals like cordierite and andalusite. The principal changes from the first temperature range to the next are the formation of Ca-plagioclase from albite and epidote, the change of actinolite to hornblende, the disappearance of chlorite and the formation of garnet; from the second range to the third, the disappearance of mica in favor of orthoclase, cordierite, andalusite, and sillimanite, and the conversion of hornblende to pyroxene; from the third range to the fourth, partial melting to glass, the appearance of high-temperature forms like tridymite and mullite, and the incorporation of soda in pyroxene to form omphacite. As might be expected, hydrous minerals become progressively less prominent in the higher temperature facies.

An additional facies is often added at the beginning of such a table, the *zeolite facies*, to include mineral assemblages formed at temperatures only slightly above those near the earth's surface. Shales in this temperature range might have illite formed from kaolinite or montmorillonite, plus quartz and chlorite; mafic rocks would have chlorite and one or more zeolite minerals (laumontite, heulandite, analcite, rarely others). All rocks of the zeolite facies represent a low-pressure

Table 19-1 Common metamorphic facies, with examples of typical mineral assemblages†

Limits of temperature and pressure ranges are not known; the figures given are rough approximations only. Sh means metamorphic assemblages derived from shale, and Ma means metamorphic assemblages derived from mafic igneous rocks

Temperature, °C	*Low pressure* [Lithostatic pressure generally less than 3,000 atm (depth < 10 km). Water pressure variable. Conditions typical of contact metamorphic zones]	*Moderate to high pressure* [Lithostatic pressure and water pressure approximately equal, generally 3,000–12,000 atm (depth 10–40 km). Conditions typical of regional metamorphism]
250–550	*Albite-epidote hornfels facies* Sh: quartz-albite-muscovite-biotite	*Greenschist facies* Sh: quartz-albite-chlorite-muscovite and quartz-albite-muscovite-biotite
	Ma: albite-epidote-actinolite-chlorite	Ma: albite-epidote-actinolite-chlorite
400–650	*Hornblende hornfels facies* Sh: quartz-plagioclase-microcline-biotite-muscovite	*Amphibolite facies* Sh: quartz-plagioclase-muscovite-biotite-almandine
	Ma: plagioclase-hornblende	Ma: plagioclase-hornblende
Over 500	*Pyroxene hornfels facies* Sh: quartz-plagioclase-orthoclase-cordierite-andalusite	*Granulite facies* Sh: quartz-plagioclase-orthoclase-garnet-sillimanite
	Ma: plagioclase-diopside-hypersthene	Ma: plagioclase-garnet-hypersthene-quartz
Over 600	*Sanidinite facies* Sh: tridymite-cordierite-mullite-glass	*Eclogite facies* Sh: Not found
	Ma: plagioclase-diopside-hypersthene	Ma: omphacite-garnet

† Taken largely from Fyfe, Turner, and Verhoogen (1958), chap. 7, but in places grossly simplified.

environment, so there would be no distinction between contact and regional metamorphism.

Relations between facies can be represented conveniently by means of *ACF* diagrams, or similar diagrams with other variables. Two typical plots for rocks of the greenschist and amphibolite facies are shown in Figs. 19-2 and 19-3. Point X in Fig. 19-2 might represent the composition of a shale, with quartz and clay minerals as the chief original constituents, and with a low content of calcium, iron, and magnesium. In the greenschist facies such a rock would consist chiefly of muscovite and quartz, with a little albite, chlorite, and epidote or clinozoisite. In the amphibolite facies the mineral composition would have changed (point X' of

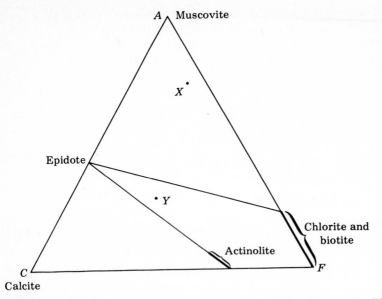

Figure 19-2 Some mineral assemblages of the greenschist facies. Quartz and albite are possible additional phases.

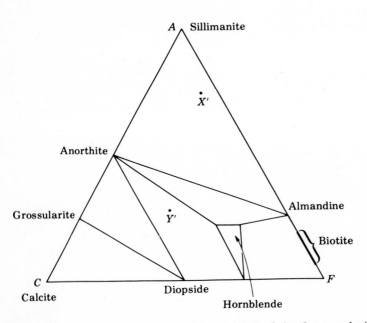

Figure 19-3 Some mineral assemblages of the amphibolite facies. Quartz and microcline are possible additional phases.

Fig. 19-3) to sillimanite, quartz, and microcline, with a little plagioclase and almandine. A metamorphosed tuff might be represented by points Y and Y': chiefly epidote, actinolite, chlorite, and albite in the greenschist facies, and plagioclase, hornblende, and diopside in the amphibolite facies. Formally the changes from one facies to the other can be represented by equations like

$$KAl_3Si_3O_{10}(OH)_2 \rightarrow Al_2SiO_5 + KAlSi_3O_8 + H_2O$$

$$6Ca_2Al_3(SiO_4)_3OH + Mg_5Al_2Si_3O_{18}(OH)_8 \rightarrow Ca_2Mg_5Si_8O_{22}(OH)_2$$
$$+ 10CaAl_2Si_2O_8$$

but such formalism gives little information not already conveyed by the diagrams.

Not all geologists, by any means, would separate facies in precisely the manner shown by Table 19-1 and Figs. 19-2 and 19-3. The general principle of facies classification has won wide acceptance, but there remains much divergence of opinion about the specific minerals or mineral combinations best suited to establish boundaries between facies. The rocks commonly used to define boundaries are shales and mafic lavas or tuffs, because these are both widespread and sensitive to changes in temperature and pressure; but it is not always certain that boundaries based on a metamorphic sequence derived from shale will coincide with boundaries based on metamorphosed basalt. Such difficulties are inevitable in a classification based entirely on observation of complex natural materials. The difficulties have proved so troublesome, especially in assigning rocks to finer subdivisions (subfacies), that one prominent student of metamorphism (Winkler, 1974) has suggested that the facies classification be abandoned altogether.

19-5 EXAMPLES OF EXPERIMENTAL STUDIES ON METAMORPHIC REACTIONS

The facies classification might be put on a sounder basis if it were possible to establish experimentally the phase relations in metamorphic systems, in other words to determine the conditions of temperature, pressure, and composition under which various mineral assemblages are stable. Given this information, we could define facies unequivocally in terms of pressure and temperature limits, and designate specific equations as showing the transition from one facies to another. Such experimental study of metamorphic systems has lagged behind similar work on igneous rocks, for the understandable reason that common metamorphic reactions involve solids at relatively low temperatures, hence are slow and sensitive to small changes in the amount and character of associated fluids. Nevertheless progress in recent years has been rapid, and the chemistry of many metamorphic phase changes is known in satisfactory detail. The following paragraphs give a few examples of well-studied systems.

Calcite-Wollastonite

A very simple metamorphic reaction that has been examined in great detail is the formation of wollastonite from calcite and silica [Eq. (19-1)]. The occurrence of this reaction in nature is shown by the bands of wollastonite rock often found at contacts between granite and limestone. The silica in the wollastonite comes either from quartz grains in the original limestone or from silica-rich solutions permeating the contact rock from the molten granite. The formation of wollastonite is endothermic, hence should be favored by high temperatures. Of the four substances that appear in the reaction, three are solids and CO_2 is a gas, so that the formation of wollastonite should be inhibited by high partial pressures of CO_2. The effect of these variables is shown in Fig. 19-4. Note that the three solids (heavy line) can exist together at equilibrium over a considerable temperature range, depending on the CO_2 pressure. With pressures near atmospheric, wollastonite can form at temperatures even below 400°C, but at a CO_2 pressure of 6,000 atm (corresponding to a depth of about 20 km) the combination calcite-quartz is stable up to well over 800°. The presence of wollastonite at an igneous contact, in other words, tells us only that the maximum temperature must have been somewhere above about 400°—unless, of course, we have information from another source about the pressure of CO_2, which would fix the temperature more precisely.

The heavy line in Fig. 19-4 shows equilibrium for the reaction in a closed system, where the pressure of CO_2 is the same as the pressure on the entire system. Suppose now that the system is more open, so that CO_2 is free to escape, i.e., so that the gas is a mobile component, in the language of Sec. 19-3. The pressure on the three solids is still determined by the weight of overlying rocks, but the pressure on the gas is much less; in the extreme case, if fissures are open to the surface, it may be no greater than the pressure of the atmosphere. In this case the effect of pressure on the reaction would be determined by the relative densities of the solid reactants and product; since wollastonite is denser than either calcite or quartz, high pressure should enable it to form at lower temperatures. Equilibrium for a completely open system is shown approximately by the light dashed line of Fig. 19-4. In nature the system would probably be neither completely open nor completely closed, so that equilibrium would be reached at points between the two lines. This, of course, introduces further uncertainty into any attempt to use the presence or absence of wollastonite as an indicator of temperature at an intrusive contact.

Kaolinite-Mica-Feldspar

In the progressive metamorphism of shale, the behavior of the clay minerals is particularly important. The simplest of the clays, kaolinite, is known to break down on heating into pyrophyllite, alumina, and water:

$$2Al_2Si_2O_5(OH)_4 \rightleftharpoons Al_2Si_4O_{10}(OH)_2 + Al_2O_3 + 3H_2O \qquad (19\text{-}2)$$

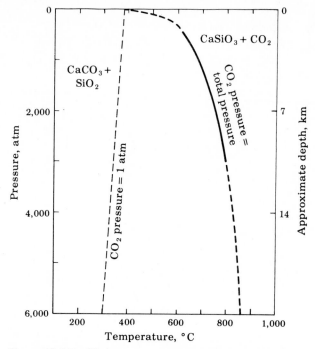

Figure 19-4 Equilibrium curves for the reaction $CaCO_3 + SiO_2 \rightleftharpoons CaSiO_3 + CO_2$ at $P_{CO_2} = P_{total}$ (heavy line) and $P_{CO_2} = 1$ atm (light line). Heavy solid line experimentally determined *(Harker and Tuttle, 1955)*. Dashed lines extrapolated or theoretical. At P-T values between the two lines, calcite and quartz are stable if $P_{CO_2} = P_{total}$, and wollastonite is stable if $P_{CO_2} = 1$ atm. *(After Barth, 1962.)*

The alumina may form corundum, or alternatively one of the hydrates (boehmite or diaspore), depending on temperature; another complication is the appearance in experiments of an intermediate product which is not known in nature. These complicating factors make it impossible to draw a simple phase diagram for the system, but the appearance of a single volatile substance on the right-hand side of the equation indicates that the equilibrium temperature for the decomposition of kaolinite would rise with increasing pressure of H_2O, just as the temperature of the calcite-wollastonite reaction rises with increasing CO_2 pressure. The measured upper stability limit for pure kaolinite is 405° at 680 atm and 415° at 1,700 atm.

Under natural conditions the reaction would ordinarily not be this simple. If silica is present with kaolinite, no free alumina is produced:

$$Al_2Si_2O_5(OH)_4 + 2SiO_2 \rightleftharpoons Al_2Si_4O_{10}(OH)_2 + H_2O \qquad (19\text{-}3)$$

Or if the clay contains adsorbed potassium ions, as most clays do, muscovite would form instead of pyrophyllite:

$$3Al_2Si_2O_5(OH)_4 + 2K^+ \rightleftharpoons K_2Al_4(AlSi_3O_{10})_2(OH)_4 + 2H^+ + 3H_2O$$

Kaolinite   Muscovite

$$(19\text{-}4)$$

Equilibrium in this reaction depends not only on temperature, total pressure, and partial pressure of H_2O, but on the concentrations of K^+ and H^+ as well. The effects of these variables have been studied by Hemley and his colleagues in some experiments (summarized in Hemley and Jones, 1964) that we discussed previously in connection with stabilities of the clay minerals (Sec. 7-4).

The results of Hemley's experiments are summarized in Fig. 7-6. In addition to data for Eq. (19-4), this figure shows similar experimental results for the further change of muscovite to K-feldspar:

$$K_2Al_4(AlSi_3O_{10})_2(OH)_4 + 4K^+ + 12SiO_2 \rightleftharpoons 6KAlSi_3O_8 + 4H^+ \quad (19\text{-}5)$$

Still another variable appears in this reaction, the activity of silica, which was kept constant in the experimental work by including quartz in the reaction mixtures. Hemley has extended his work to systems including compounds of calcium and sodium instead of potassium and has obtained strikingly similar results. The diagram for the sodium system is shown in Fig. 19-5; the corresponding phase boundaries for the potassium system, transferred from Fig. 7-6, are shown as light lines in this diagram. Note that the stability field of the Na-mica paragonite is much smaller than that of K-mica, as would be expected from the fact that muscovite is a far commoner mineral. The boundary between kaolinite and pyrophyllite for both systems, shown by dashed lines, is placed at about 350°.

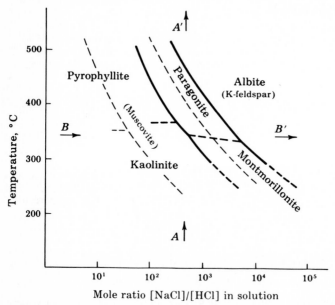

Figure 19-5 Equilibrium curves for the system Na_2O-Al_2O_3-SiO_2-H_2O. (Light lines and mineral names in parenthesis refer to the corresponding K_2O system, superposed from Fig. 7-6.) Total pressure, chiefly H_2O vapor, 15,000 psi (approx. 1,000 atm). Quartz present in the reaction mixture. Arrows AA' show possible metamorphic changes kaolinite-mica-feldspar brought about by increasing temperature at constant [NaCl]/[HCl] and [KCl]/[HCl] ratios; arrows BB' show the same progression caused by change in composition alone. (*Source: Hemley and Jones, 1964, page 549.*)

Equilibrium was not attained for this reaction, and the lines are only approximate. As mentioned above, other experiments with simpler systems suggest that the equilibrium boundary may be as high as 400°.

In these experiments Hemley and his colleagues have, in effect, isolated some of the essential reactions in the facies changes from unaltered shale to greenschist facies (kaolinite → muscovite), and from greenschist to amphibolite facies (muscovite → K-feldspar). Particularly striking is the wide temperature range over which these facies changes can occur: both of the alkali feldspars are stable down to at least 200° if the ratios $[K^+]/[H^+]$ and $[Na^+]/[H^+]$ are high, and both micas are stable to well over 500° if the ratios are small. Thus facies changes represented by the progression kaolinite → mica → feldspar could be brought about either by a change in temperature alone (arrows A and A' on Fig. 19-5) or by a change in composition alone (arrows B and B'), or by many combinations of these factors. Changes in other variables also—total pressure, H_2O pressure, SiO_2 activity— would cause the field boundaries to shift and hence could bring about changes of facies.

Temperature may well be the most important determiner of facies in nature, as it is commonly thought to be, but certainly other variables can have a profound influence locally.

Experimental Metamorphism of Shale

Hemley's work is nicely supplemented by experiments on natural materials described by Winkler (1957, 1958). Instead of pure minerals, Winkler started with natural clays, fairly simple clays, to be sure, but much closer than Hemley's materials to the usual makeup of shales. The clays were various mixtures of quartz, illite, and kaolinite, with traces of several other minerals; to keep complications within bounds, Winkler was careful to select material low in calcium. Experimental temperatures were in the range 400 to 750°, and water-vapor pressure was maintained at 2,000 atm. K^+ was present as adsorbed ions and as an essential constituent of the illite, but no attempt was made to control or determine the $[K^+]/[H^+]$ ratio. When a clay containing both illite and kaolinite was heated, the former recrystallized to muscovite and set free some of its silica; the kaolinite reacted with this silica and with original quartz to form pyrophyllite, the reaction completing itself below 420°. No K-feldspar appeared until the temperature reached 665°. Starting with a sample whose clay was almost entirely illite, Winkler obtained evidence for other reactions: as the illite changed to muscovite, iron and magnesium were set free in addition to silica, and these formed chlorite; at about 550° the chlorite and some of the muscovite reacted to form biotite and cordierite. The temperature of appearance of K-feldspar was a trifle lower than in the first clay sample, about 620°, and the temperature of final disappearance of muscovite about 665°. On the basis of these data, for the particular conditions of his experiment Winkler sets the upper boundary of the greenschist facies (marked by disappearance of chlorite) at 550°, and the upper boundary of the amphibolite facies (disappearance of muscovite) at 665°.

In a further series of experiments Winkler came still closer to natural conditions by adding a few percent of NaCl to the clay samples, on the grounds that pore spaces of deeply buried shales are commonly filled with saline solution. This made possible the formation of albite from kaolinite and quartz at temperatures lower than 400°; the albite also took up whatever small amount of calcium the clay contained to form plagioclase. K-feldspar appeared at a slightly lower temperature than in the absence of NaCl, about 600°, and included a good deal of albite in solid solution. The boundary between the greenschist and amphibolite facies is unaffected by the presence of NaCl, but the amphibolite–pyroxene hornfels boundary is moved down to about 600°.

Winkler's experiments lack the close control of variables that Hemley's had, but by working with more complex systems he comes closer to duplicating natural metamorphic assemblages. Like Hemley, he finds support for the generalization that the principal reactions separating one facies from another are strongly dependent on temperature but are affected by several other variables as well.

Experimental Study of Mafic Rocks

Similar laboratory studies on natural materials of mafic composition have been carried out by several investigators. Particularly informative is work by Liou, Kuniyoshi, and Ito (1974) on the transition between the greenschist and amphibolite facies for rocks with the original composition of basalt. The materials used were samples of greenschist (albite-epidote-chlorite-actinolite) and amphibolite (plagioclase-hornblende) from an area of progressive metamorphism on Vancouver Island. Separate samples and mixtures of the two rocks were sealed with excess water in capsules that could be held at pressures of 2 kilobars and 5 kilobars and at various temperatures between 450 and 800°C for periods up to three months. Changes of minerals from those typical of one facies to those of the other could be observed in both directions: in the direction of increasing temperature, albite and epidote reacted to form intermediate plagioclase (An_{40-52} in the amphibolite), and hornblende was formed by reactions among actinolite, chlorite, and epidote. At 2 kilobars, chlorite decreased sharply at 475°, but did not disappear completely until 550°, in general agreement with Winkler's experiments.

The change from one facies to the other, according to this study, is complex. A temperature of 475° can be taken as a rough upper limit for the greenschist facies at 2 kilobars, and 550° as a lower limit for the amphibolite facies; between the two is a transitional assemblage plagioclase-actinolite-chlorite. At higher pressures the transition zone probably has the composition albite-epidote-hornblende, characteristic of an "epidote amphibolite facies" which is often distinguished in natural occurrences between greenschist and amphibolite. Liou et al. point out the many factors on which the width and the temperature boundaries of the transition zone depend: the overall pressure, the fraction of the pressure due to water vapor, and the fugacity of oxygen (controlled in their experiments by a quartz-fayalite-magnetite buffer). They suggest that an increase in oxygen fugacity could lead to narrowing and disappearance of the transition zone, hence a sharp break between the greenschist and amphibolite facies, as is sometimes observed in nature.

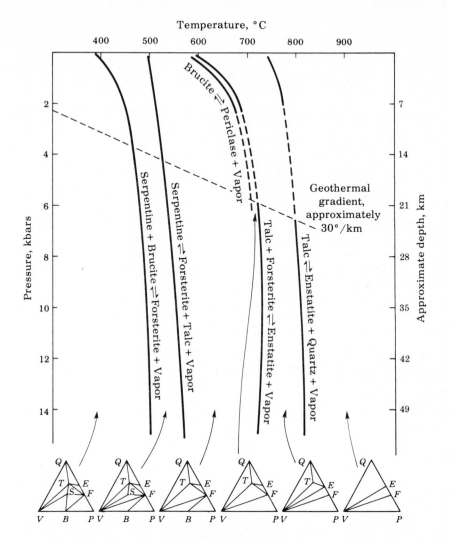

Figure 19-6 Equilibrium curves for reactions in the system $MgO\text{-}SiO_2\text{-}H_2O$. Each univariant curve shows equilibrium for the reaction specified. Triangles show stable mineral assemblages in the divariant areas between curves. The triangles are similar to ACF diagrams, except that water vapor is substituted for CaO at the lower left. The light dashed line shows temperatures within the earth, on the assumption of an average gradient of 30°C/km. Abbreviations: Q = quartz, P = periclase (MgO), B = brucite [$Mg(OH)_2$], V = vapor (chiefly H_2O), E = enstatite (or pyroxene), F = forsterite (or olivine), S = serpentine, T = talc. [*Sources: Solid lines at low pressures (below about 5 kilobars) from Bowen and Tuttle, 1949. Solid lines at high pressures from Kitahara et al., 1966. Curve for brucite dehydration from Weber and Roy, 1965. Dashed lines show interpolations and extrapolations of experimental data.*]

Serpentine

One further example of the application of experimental work to metamorphic equilibria is provided by the MgO-SiO_2-H_2O system, which includes the principal constituents of ultramafic igneous rocks and the metamorphic rock serpentine. On the pressure-temperature plot in Fig. 19-6, the lines show equilibrium conditions for specific reactions, and the triangles drawn for each field show the phases that can be present at equilibrium over a range of temperature and pressure. The reactions are all essentially dehydration processes, with a single gas (water vapor) as one of the products. In this respect they are similar to the calcite-wollastonite reaction [Eq. (19-1)], and the curves accordingly resemble the solid-line curve of Fig. 19-4. All show equilibrium temperatures increasing with pressure, the rate of increase falling off at high pressures.

The curves can be taken as a model for the conversion of an ultramafic rock (pyroxene plus olivine) to serpentine on cooling and reaction with water, or for the opposite metamorphic change in serpentine when it is heated at an igneous contact. Particularly significant is the demonstration that enstatite and forsterite are stable in the presence of water vapor over a wide temperature range. Even at temperatures well over 1000°C the two solid minerals remain unaffected by high water-vapor pressures, in sharp contrast to the behavior of the quartz-feldspar mixtures of granitic rocks. This means that an ultramafic magma cannot exist at temperatures normally encountered in the earth's crust, no matter what the amount of water. The familiar conversion of ultramafic rocks to serpentine can occur only at temperatures below about 500°; serpentine cannot exist, even at high pressures, if the temperature is greater than this, and a "serpentine magma" is impossible. Some of the consequences of these experimental conclusions we have explored in an earlier discussion (Sec. 14-7).

19-6 HIGH-PRESSURE FACIES

It is commonly said, as we have noted, that temperature is a more important variable than pressure in determining what metamorphic assemblage will form from a given rock composition. This is a very loose generalization that needs refinement.

The most often observed effects of pressure are embodied in the two columns of Table 19-1: mineral assemblages formed by thermal metamorphism at shallow intrusive contacts represent conditions of lower pressure than the foliated rocks that have undergone regional metamorphism at deeper levels. As the table shows, common assemblages from the two environments are not very different, except for low-density minerals like cordierite and andalusite in the former and higher density minerals like garnet, kyanite, and sillimanite in the latter. The general similarity, of course, is the basis for the above generalization about the importance of temperature.

In the usual sequence of regionally metamorphosed rocks (zeolite, greenschist, amphibolite, granulite facies) we think of pressure and temperature as

increasing together, presumably in rough correspondence with the normal geothermal gradient of about 30° per kilometer of depth. In many places, however, the
geothermal gradient strays far from the average, and facies different from the
"normal" ones can develop. Where the gradient during metamorphism was
markedly greater than 30°/km, transitional facies may appear between those typical of contact zones and those of normal regional metamorphism. More interesting are the facies produced where the gradient was exceptionally low, where
conditions of high pressure were attained at relatively low temperatures. Under
these conditions a very different set of metamorphic minerals can develop: preh-
nite and pumpellyite at low temperatures, glaucophane, lawsonite, and jadeite at
higher temperatures. Correspondingly, new facies can be distinguished, of which
the most striking is the *blueschist facies*, named from the color of its characteristic
mineral glaucophane.

Pressure-temperature relationships among metamorphic facies can be conveniently displayed in a diagram like Fig. 19-7, a so-called "petrogenetic grid."
The various low-pressure facies (different varieties of hornfels) appear across the
top of the diagram, and the "normal" series of regional metamorphism occupies a
zone running diagonally down from the upper left corner. The high-pressure,

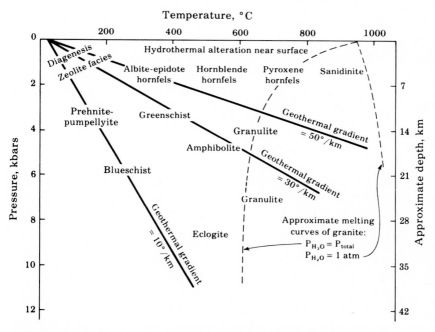

Figure 19-7 Approximate pressure-temperature fields of the principal metamorphic facies. The three
solid lines show possible values of the geothermal gradient: the mean of measured values is about
30°/km, the maximum about 50°/km, the minimum about 10°/km. The dashed lines show temperatures
of incipient melting of granite, under water-vapor pressure equal to total pressure and water-vapor
pressure equal to 1 atm. Between the two lines is the region where differential melting of high-grade
metamorphic rocks may occur.

low-temperature facies form a separate sequence near the left margin. Several other sequences representing intermediate pressure-temperature combinations have been distinguished in recent studies of metamorphic terrains, so that a complete petrogenetic grid becomes a complicated diagram with very numerous facies and subfacies.

Rocks of the high-pressure blueschist facies are not common. They are restricted in time to the Phanerozoic (with a couple of questionable exceptions in the late Precambrian), and mostly to the Mesozoic and Cenozoic eras; they are restricted in occurrence to orogenic zones, and often show strong folding and fragmentation. These facts, plus the observation that such rocks commonly occupy a belt parallel to an adjacent belt of more normal metamorphic rocks, suggest that their origin is related to plate tectonics, particularly to subduction of crustal plates at convergent plate boundaries.

In imagination we can think of a plate moving downward along an inclined Benioff zone, carrying cool rock and sediments down to depths below the bottom of the normal crust. The moving slab heats up gradually, of course, but if its descent is fast enough the cool material becomes subject to enormous pressure before its temperature has risen markedly. Because the top of the plate is being sheared and crushed as it moves, conditions are ideal for high-pressure, low-temperature metamorphism. In the crust above the moving plate, where the geothermal gradient is normal (or greater than normal because of bodies of magma moving upward), the usual facies of medium-pressure, medium-temperature conditions can form. Then many millions of years later, when vertical faulting and erosion have exposed these rocks at the surface, a belt of deformed high-pressure metamorphic rocks will be found paralleling a medium-pressure belt. A good example is the pair of metamorphic belts in California, with high-pressure assemblages in the Coast Range and more normal metamorphic types in the Sierra Nevada foothills, probably produced by motion of an oceanic plate eastward under the continental margin in the late Mesozoic and early Cenozoic.

These inferences from plate tectonics can be carried a step further. The apparent restriction of most blueschist-facies rocks to the Mesozoic and Cenozoic suggests that the character of lithospheric plates and the nature of their movement may have changed during geologic time. The typical high-pressure, low-temperature assemblages can form only if the moving plate is thick enough, and its motion is fast enough, so that its temperature does not rise rapidly. Perhaps, then, thick lithospheric plates and rapid plate movement have become common only in post-Paleozoic time (Ernst, 1973). Such a conclusion is highly speculative, but it is consistent with other inferences about the earth's past.

19-7 ULTRAMETAMORPHISM

As temperature rises in the normal course of metamorphism, many sedimentary rocks are converted into mineral assemblages that resemble more and more closely the assemblages in igneous rocks. With igneous rock as the starting material, metamorphism produces aggregates of hydrous, "low-temperature" minerals at first, then, at higher grade, assemblages that resemble increasingly the original mineral composition. The distinction between igneous rocks and the products of high-grade metamorphism is based more on texture than on mineral composition, although significant differences in composition may exist, particularly if the initial rock was at all out of the ordinary chemically. In texture a metamorphic rock consists typically of rounded, intergrowing grains, no mineral showing much tendency to develop euhedral crystal outlines; igneous textures, on the other hand, are characterized by good or fairly good crystal outlines for some minerals, notably mafic minerals and plagioclase, and complete lack of euhedral shape for others, particularly quartz and K-feldspar. There are obvious and numerous exceptions to these generalizations, but not enough to destroy their validity for typical specimens of the two kinds of rock. The generalizations, furthermore, seem to have genetic significance: the differences in crystal shape found in igneous rocks are what one might expect during the freezing of a melt, where some constituents crystallize before others, and the lack of crystal form in metamorphic textures could well result from reactions in a solid where all constituents crystallize simultaneously and interfere with each other's growth.

The typical "metamorphic" texture, however, may not be the end result of metamorphism. The highest temperatures of metamorphism overlap the lowest temperatures of igneous activity; sufficient heating would certainly produce an interstitial melt with a quartz-plus-alkali-feldspar composition; only a small percentage of the rock would need to liquefy to make it lose its coherence and become a mush of crystals; and this material could ultimately freeze to give a typical "igneous" texture. Thus the distinction between metamorphic and igneous rocks becomes hazy. Can we assume that many rocks, apparently igneous, have actually formed by extreme metamorphism?

This question we have encountered before, in discussing the formation of granitic rocks (Secs. 14-4 and 14-6). From a purely chemical standpoint, the generation of granite as an end stage of metamorphism is certainly possible. It is equally possible, as far as chemistry is concerned, to produce granite by fractional crystallization of a more mafic melt rising out of the mantle. Which of these processes is more likely to account for the majority of granite bodies in the earth's crust? To this question there is still no definitive answer.

Rocks showing an apparent transition between metamorphic and igneous varieties are common in areas of medium- and high-grade metamorphism. The descriptive literature about the processes and products in such areas is cluttered with long words of Greek derivation, as is usual in parts of geology where basic ideas are hard to define and decisions between rival hypotheses are difficult. *Ultrametamorphism* refers to any alteration that takes place at temperatures higher than those in the metamorphic range, hence alteration that involves the production of igneous-appearing textures and partial or complete melting. *Anatexis* and *palingenesis* emphasize the production of new magma by the partial or complete melting of preexisting rocks; in recent literature anatexis is the commoner term. *Granitization* may be used in a narrow sense to describe the conversion of a solid rock to a rock of granitic composition and texture solely by the action of heat and tenuous fluids, the material remaining essentially solid throughout; or it may refer more loosely to any process, whether accompanied by melting or not, in which preexisting material is changed to granite. *Migmatitization* signifies the formation of a migmatite, or " mixed rock," consisting partly of igneous material and partly of high-grade metamorphic remnants; the process may involve either anatexis in place, or the injection of fluid material along layers or cracks in an adjacent metamorphic rock. *Mobilization* is a marvelously indefinite word that refers to any process by which either ions or the products of partial fusion are put in a condition that enables them to move from one place to another. These are the commonest terms in general use, but the list by no means exhausts the number of long words invented to describe vaguely defined aspects of extreme metamorphism.

The relation of anatexis to high-grade metamorphism is shown on the right-hand side of Fig. 19-7, where melting curves for granite are drawn through the high-grade metamorphic assemblages. The left-hand curve is the temperature of initial melting of granite in contact with water vapor at a pressure equal to the total pressure (taken from Fig. 15-2); under these conditions the melting temperature falls as pressure rises. The right-hand curve shows the temperature of incipient melting for granite in a system open to water vapor, in which the change of melting temperature with pressure depends simply on the difference in density between solid and liquid granite; since the liquid has the lower density, the melting point rises slightly as pressure increases. In nature, melting of granite must follow a path between these extremes, and the pressure-temperature conditions of anatexis would be represented by the area between the curves. Note that the areas of the granulite and pyroxene hornfels facies straddle the melting-point curve for granite under hydrous conditions: rocks of these facies can be strictly metamorphic if conditions are relatively dry, or they can be constituents of a migmatite if much water is present and the rock has partly melted. The high-temperature boundary of the amphibolite facies also grazes the hydrous melting curve, so that migmatites may be formed with a considerable variety of metamorphic rocks.

Thus the possible products of ultrametamorphism can be easily imagined from a theoretical standpoint, but recognizing such products in the field and sorting out the processes that have operated in a particular area can be extraordinarily difficult.

19-8 METAMORPHIC DIFFERENTIATION

As a rule, metamorphic processes tend toward homogenization. Elements partially separated by weathering and erosion are recombined into complex silicates. Fluids move ions from one part of a rock to another. Crushing and granulation of rocks destroy original structures and mix original constituents. All such processes tend to produce rocks more homogeneous, more uniform both in structure and in composition, than the original material. One of the outstanding characteristics of metamorphic rocks, in fact, is their monotonous sameness over large areas.

Are there any processes in metamorphism that lead the other way, toward differentiation of the original material rather than homogenization? A few can be suggested. An obvious one is differential anatexis, the melting of a silica-and-alkali-rich fraction when a rock becomes sufficiently hot. The migration of such material into regions of least pressure is often suggested to explain pockets of pegmatite in metamorphic rocks along fissures and at the crests of folds. Another possible mechanism of differentiation is the small-scale diffusion of ions to crystallizing nuclei. If garnet crystals have begun to form in one layer of a rock, for example, it seems reasonable that they would draw material for further growth in part from movable ions in adjacent layers; in this way a layered rock would result with much sharper chemical differences between adjacent layers than were present originally. Still another possibility is mechanical deformation: if a rock is undergoing shear, mica flakes may be somewhat concentrated into layers where sliding is active, and garnet crystals may aggregate into layers where shear can be taken up by rotation of grains.

Except for anatexis, the possible mechanisms of differentiation can hardly act on a scale of more than a few centimeters. It seems a safe generalization, then, that except at extreme temperatures metamorphism is much more effective in smoothing out original heterogeneities in a rock than in causing its materials to differentiate.

PROBLEMS

1 What kinds of rock would you expect to form from the progressive metamorphism of (a) a rhyolite consisting chiefly of quartz, potassium feldspar, silica-rich glass, and a little biotite, (b) a shale consisting chiefly of montmorillonite and quartz?

2 Equilibrium pressure-temperature relations for the reaction

$$CaCO_3 + SiO_2 \rightleftharpoons CaSiO_3 + CO_2$$

are shown in Fig. 19-4. Consider mixtures in which CO_2 pressures are equal to total pressures.

(a) What combinations of minerals (calcite-quartz, calcite-wollastonite, quartz-wollastonite) are stable at 100°C and 1,000 atm? At 700°C and 2,000 atm? At 900°C and 3,000 atm?

(b) What is the sign of the free-energy change, for the reaction as written above, at 600°C and 3,000 atm?

(c) Show that the shape of the curve in Fig. 19-4 is qualitatively what you might expect from the law of mass action.

(d) If a dike intrudes limestone and if geologic evidence permits you to guess that the intrusion took place at a depth of about 4 km, what conclusions, if any, could you draw from (1) the presence of wollastonite, (2) the absence of wollastonite, at the contact?

3 Explain why the solid-line curve in Fig. 19-4 has the same general shape as the curves in Fig. 19-6. Note that curves for the incipient melting of albite and granite (Fig. 15-2) have a similar shape but the opposite slope, although both involve reactions with essentially a single substance in the gas phase. Account for the difference in slopes.

4 In general, how would the chemical composition of a typical shale differ from that of a diorite or quartz diorite? How, then, might you distinguish a gneiss formed by the metamorphism of shale from a gneiss formed by flow in a crystallizing quartz-diorite magma? Point out possible criteria of a chemical, mineralogical, and textural nature.

5 What substances must be added to (*a*) limestone to convert it into grossularite-diopside hornfels; (*b*) pyroxenite to convert it into serpentine; (*c*) shale to convert it into tourmaline-lepidolite rock; (*d*) gabbro to convert it into scapolite-epidote rock?

6 Proponents of large-scale granitization point out that big xenoliths and roof pendants in a batholith commonly consist chiefly of marble, quartzite, or siliceous hornfels rather than of hornfels derived from more ordinary shales, lavas, or tuffs. The argument is that shale and volcanic rocks are more easily converted into granite during ultrametamorphism than are rocks like limestone or quartz sandstone. Is this argument reasonable from a chemical standpoint?

7 Amphibolites consisting chiefly of hornblende and plagioclase (andesine), often with minor epidote and sphene, are a very common type of metamorphic rock. From what kinds of sedimentary or igneous rocks might such a metamorphic rock originate? What criteria could you use to decide in a particular case which origin is most probable?

8 The green color so common in the low-grade metamorphic rocks derived from lavas and tuffs is generally due to one of the four minerals chlorite, epidote, actinolite, or antigorite. From what original minerals are these derived? Express the formation of these minerals by symbolic equations, using oxides to represent constituents added from or removed by interstitial fluids.

9 Would the following be possible equilibrium mineral assemblages in the metamorphism of a shaly limestone consisting initially of kaolinite and calcite?

(*a*) Calcite-grossularite
(*b*) Wollastonite-anorthite-quartz-andalusite
(*c*) Anorthite-grossularite-calcite
(*d*) Andalusite-grossularite-quartz
(*e*) Corundum-wollastonite-grossularite

Represent the possible mineral combinations by means of a triangular diagram with CaO (or calcite) as one corner, Al_2O_3 as a second, and SiO_2 as the third.

REFERENCES AND SUGGESTIONS FOR FURTHER READING

Bowen, N. L., and O. F. Tuttle, The system $MgO-SiO_2-H_2O$, *Geol. Soc. America Bull.*, vol. 60, pp. 439–460, 1949. A classical paper in experimental metamorphism.

Ernst, W. G., Blueschist metamorphism and pressure-temperature regimes in active subduction zones, *Tectonophysics*, vol. 17, pp. 255–272, 1973. Experimental evidence regarding the physical and chemical conditions of blueschist formation, and speculation about the relation of blueschist to subduction at convergent plate junctions.

Eskola, P., "Die Entstehung der Gesteine" (Barth, Correns, Eskola), Springer-Verlag OHG, Berlin, 1939; reprinted 1960. A summary of Eskola's many years of work on metamorphic rocks. Eskola's concept of facies is described on pages 336 to 368, and the construction of *ACF* diagrams is explained on page 347.

Fyfe, W. S., F. J. Turner, and J. Verhoogen, Metamorphic reactions and metamorphic facies, *Geol. Soc. America Memoir* 73, 1958. A clear exposition of the thermodynamics of metamorphic processes.

Harker, R. I., and O. F. Tuttle, Studies in the system $CaO-MgO-CO_2$, *Am. Jour. Sci.*, vol. 253, pp. 209–244, 1955.

Hemley, J. J., and W. R. Jones, Chemical aspects of hydrothermal alteration with emphasis on hydrogen metasomatism, *Econ. Geology*, vol. 59, pp. 538–569, 1964. Experimental work on clay-mica-feldspar transitions.

Kitahara, S., S. Takenouchi, and G. C. Kennedy, Phase relations in the system MgO-SiO_2-H_2O at high temperatures and pressures, *Am. Jour. Sci.*, vol. 264, pp. 223–233, 1966.

Liou, J. G., S. Kuniyoshi, and K. Ito, Experimental studies of the phase relations between greenschist and amphibolite, *Am. Jour. Sci.*, vol. 274, pp. 613–632, 1974.

Miyashiro, A., Metamorphism and related magmatism in plate tectonics, *Am. Jour. Sci.*, vol. 272, pp. 629–656, 1972. The relation of plate tectonics to paired metamorphic belts and to associated igneous rocks.

Morgan, B. A., Mineralogy and origin of skarns in the Mount Morrison pendant, Sierra Nevada, California, *Am. Jour. Sci.*, vol. 275, pp. 119–142, 1975. High-temperature contact-metamorphic rocks at a granodiorite-limestone contact; effects of oxidation state and of retrograde metamorphic reactions.

Turner, F. J., "Metamorphic Petrology," McGraw-Hill Book Company, New York, 1968. A standard textbook, particularly good in relating theoretical concepts to field observations.

Vance, J. A., and M. A. Duncan, Formation of peridotites by deserpentinization in the Cascade Mountains, Washington, *Geol. Soc. America Bull.*, vol. 88, pp. 1497–1508, 1977. A field and laboratory study of an unusual kind of metamorphism in the amphibolite facies.

Weber, J. N., and R. Roy, Complex stable-metastable solid reactions illustrated with the $Mg(OH)_2$-MgO reaction, *Am. Jour. Sci.*, vol. 263, pp. 668–677, 1965.

Winkler, H. G. F., Experimentelle Gesteinsmetamorphose, *Geochim. et Cosmochim. Acta*, vol. 13, pp. 42–69, 1957 and vol. 15, pp. 91–112, 1958.

Winkler, H. G. F., "Petrogenesis of Metamorphic Rocks," 3d ed., Springer-Verlag, New York, 1974. A standard textbook. Chapter 18 is an excellent treatment of anatexis.

TWENTY

DISTRIBUTION OF THE ELEMENTS

Up to this point we have been engaged in a review of geologic processes as seen from a chemical standpoint. We turn now to a different kind of problem, the distribution of elements in geologic environments and the reasons underlying this distribution. Elementary facts about the distribution of elements are already familiar: igneous rocks have a characteristic group of major elements, sulfide ores another group, carbonate sediments another group, and salt deposits still another. Our purpose now is to refine such generalizations, and in particular to see how the rarer elements are distributed in various environments. According to one definition, this is the central problem of geochemistry.

To begin such a study, we need first of all a scheme of classification, a way of grouping the elements so that their properties may be related to their geologic behavior.

20-1 THE PERIODIC CLASSIFICATION

The most fundamental grouping of the elements is the periodic classification, first proposed by Mendeleev in 1869. Based originally on chemical properties as observed in ordinary laboratory reactions, especially on the property of valence, Mendeleev's classification has proved also to be an expression of regularities in the arrangement of electrons in atomic structures. Its importance in chemistry has increased through the years, and it is scarcely less important in geochemistry. Some geochemical relationships, to be sure, are not expressed in the periodic law, but in general it is the basic classification to which all others are referred.

The intricate relationships among the electronic structures of atoms, which the periodic law expresses, can be displayed in a variety of ways. Table 20-1 shows an arrangement that is convenient for geochemical purposes. The horizontal sequences of elements are the *rows*, or *periods*, of the table, and the vertical sequences are the *columns*, or *groups*. The number under each symbol is the atomic number, which is equal to the total number of electrons in the neutral atom (and also to the total number of positive charges in the atomic nucleus). For the *main-group* elements (those at either side of the table), the Roman numerals at the top of the columns indicate the number of electrons in the outer shells of the atoms, or the *valence* electrons, the electrons which are shared or transferred during the formation of chemical bonds (Sec. 5-3) and which therefore are most important in determining the chemical behavior of an element. These same numbers are also used as designations of the groups of elements; thus the elements N, P, As, Sb, Bi belong to Group V and have atoms containing 5 valence electrons.

The principal relationships of the main-group elements are clear immediately. The inert gases form a group at the extreme right. Exclusive of these, active metals stand at the left and active nonmetals at the right. Within each row properties show a gradual change with increasing atomic number, metallic activity decreasing and nonmetallic activity increasing. Down each column is a similar but less pronounced change toward elements of increasing metallic character. This means that the most active metals of all are in the lower left corner and the most active nonmetals in the upper right.

Relationships of the *transition metals* in the middle of the table are more complicated. These elements have atoms with 1, 2, or 3 valence electrons, and with 8 to 18 electrons in the shell below the valence shell. Since the number of valence electrons is small, the elements are all metals. Many of them show several oxidation states in their chemical compounds, because some of the electrons in the second shell ("*d*" electrons) have nearly the same energy as the outer electrons and in some compounds can act as additional valence electrons. Chemical properties of elements in the vertical columns of the transition group show fairly close relationships. Some of these elements have faint relationships with main-group elements, as indicated by the Roman numerals designating A and B "subgroups." In general, however, these elements show less regularity in their properties than elements in the main groups, and their properties are less readily predictable from their position in the table.

The *lanthanide*, or *rare-earth*, elements and the *actinide* elements, which form part of the transition group but for convenience are shown at the bottom of the table, present still further complications. All the rare earth elements and most of the actinide elements have atoms with 3 valence electrons and 8 electrons in the shell below the valence shell. Each atom differs from the one preceding it by 1 electron in the third shell ("*f*" electrons), two shells below the valence shell. Since the outer electron structures are so similar, the elements in each of the two groups show very similar chemical properties, so similar that the elements are difficult to separate in the laboratory and generally occur intimately associated in nature. This generalization is less true for the actinides than for the rare earths, because

Table 20-1 Periodic classification of the elements

Group / Period	IA	IIA	IIIB	IVB	VB	VIB	VIIB	VIII	VIII	VIII	IB	IIB	IIIA	IVA	VA	VIA	VIIA	0
1	H 1																	He 2
2	Li 3	Be 4											B 5	C 6	N 7	O 8	F 9	Ne 10
3	Na 11	Mg 12											Al 13	Si 14	P 15	S 16	Cl 17	Ar 18
4	K 19	Ca 20	Sc 21	Ti 22	V 23	Cr 24	Mn 25	Fe 26	Co 27	Ni 28	Cu 29	Zn 30	Ga 31	Ge 32	As 33	Se 34	Br 35	Kr 36
5	Rb 37	Sr 38	Y 39	Zr 40	Nb 41	Mo 42	(Tc) 43	Ru 44	Rh 45	Pd 46	Ag 47	Cd 48	In 49	Sn 50	Sb 51	Te 52	I 53	Xe 54
6	Cs 55	Ba 56	† 57–71	Hf 72	Ta 73	W 74	Re 75	Os 76	Ir 77	Pt 78	Au 79	Hg 80	Tl 81	Pb 82	Bi 83	Po 84	(At) 85	Rn 86
7	(Fr) -87	Ra 88	‡ 89															

Transition Metals

† Rare-earth metals	La 57	Ce 58	Pr 59	Nd 60	(Pm) 61	Sm 62	Eu 63	Gd 64	Tb 65	Dy 66	Ho 67	Er 68	Tm 69	Yb 70	Lu 71
‡ Actinide metals	Ac 89	Th 90	Pa 91	U 92	(Np) 93	(Pu) 94	(Am) 95	(Cm) 96	(Bk) 97	(Cf) 98	(E) 99	(Fm) 100	(Md) 101	(No) 102	(Lw) 103

Elements whose symbols are enclosed in parentheses do not occur in nature, but have been prepared artificially by nuclear reactions.

Handwritten annotations:
- ⊗ = siderophile
- ▣ = chalcophile
- △ = lithophile
- ⊗(cloud) = Atmophile
- active metals
- active nonmetals
- metallic activity increasing
- metallic activity decreasing

some of the former have stable higher oxidation states (additional electrons coming from the shell under the valence shell) which make separation easier.

On the basis of the periodic classification a great many successful predictions can be made about the chemical behavior of the rarer elements. Predictions about geologic behavior are also possible but are generally less satisfactory. One might expect, for example, that the rare elements selenium and tellurium would occur with sulfur, that gallium would be found in aluminum minerals, that molybdenum would occur with chromium and tantalum with vanadium. The first two of these predictions are accurate, the last two not. The periodic grouping is useful in geochemistry, but it cannot serve by itself to account for the distribution of the elements.

20-2 GOLDSCHMIDT'S GEOCHEMICAL CLASSIFICATION

V. M. Goldschmidt, the pioneer investigator of the rules of distribution of the elements, suggested another sort of classification, in part based directly on observed distributions. This classification is an attempted answer to the theoretical question: If the earth at some time in the past was largely molten and if the molten material on cooling separated itself into a metal phase, a sulfide phase, and a silicate phase, how would the elements distribute themselves among the three phases? An answer can be sought from theoretical arguments and also from three kinds of observation: (1) from the composition of meteorites, on the assumption that meteorites have an average composition similar to that of the primitive earth and underwent a similar differentiation; (2) from analyses of metal, slag (silicate), and matte (sulfide) phases in metallurgical operations; (3) from the composition of silicate rocks, sulfide ores, and the rare occurrences of native iron found in the earth's crust.

From a theoretical standpoint, we might consider first the expected distribution of elements between metallic iron and silicates, in a system with iron in excess. Metals more chemically active than iron would presumably combine with silica, and the remaining silica would react so far as possible with iron. Metals less active than iron would have no chance to form silicates but would remain as free metals with the uncombined iron. In other words, the fate of any given metal should depend entirely on the free energy of formation of its silicate. For most metals the free energies of silicates are not known, but to a good approximation we can use instead the free energies of oxides, since in the formation of a silicate the energy of the reaction

$$Me + \tfrac{1}{2}O_2 \rightarrow MeO$$

is always much larger than the energy of the reaction

$$MeO + SiO_2 \rightarrow MeSiO_3$$

Free energies of formation of representative oxides are listed in Table 20-2. From the list we could predict that the elements above iron would go preferentially into

Table 20-2 Free energies of oxides†

	Free energy of formation, kcal	
	At 25°	At 827°
CaO	− 144.4	− 124.7
ThO$_2$	− 139.7	− 121.8
MgO	− 136.1	− 115.2
Al$_2$O$_3$	− 126.0	− 105.7
ZrO$_2$	− 123.9	− 105.8
UO$_2$	− 123.3	− 107.0
TiO$_2$	− 106.2	− 89.0
SiO$_2$	− 102.3	− 85.2
VO	− 91.4	− 74.9
MnO	− 86.7	− 72.8
Cr$_2$O$_3$	− 84.4	− 67.8
K$_2$O	− 77.0	− 48.6
ZnO	− 76.1	− 56.1
WO$_2$	− 62.3	− 45.8
SnO$_2$	− 62.1	− 42.1
FeO	− 60.1	− 48.0
MoO$_2$	− 59.2	− 45.5
CoO	− 51.4	− 37.6
NiO	− 50.6	− 33.6
PbO	− 45.1	− 25.9
Cu$_2$O	− 35.0	− 21.8
PdO	− 16.8	− 1.1
HgO	− 14.0	
Ag$_2$O	− 2.5	

† Free energies of formation per oxygen atom, in kilogram calories. The oxides are arranged in order of decreasing free energy at 25°C.

Source: Robie, R. A., and D. R. Waldbaum, Thermodynamic properties of minerals and related substances, *U.S. Geol. Survey Bull.* 1259, 1968.

the silicate (or oxide) phase, and those below iron into the metallic phase. The grouping is not entirely unambiguous, in that two elements (tin and tungsten) appear to stand above iron at low temperatures and below iron at high temperatures, but in general the elements sort themselves out pretty much as would be expected from their chemical activities.

Similar lists could be set down showing the expected distribution of metals between a metal phase and a sulfide phase with iron in excess, and between a sulfide phase and a silicate phase with silica in excess. So many additional assumptions are required, however, that the numerical values are not very helpful.

On the observational side, the compositions of the three phases (metal, sulfide, silicate) in meteorites and in smelter products are well known. Some general averages for meteorites, and an example of analyses of products of smelting copper ore, are shown in Table 20-3. The resemblance is far from perfect, but in general the behavior of the minor elements is similar.

Table 20-3 Concentrations of some elements in the metallic, sulfide, and silicate phases of meteorites and in the metallurgical products of Mansfeld copper ores†

	Meteorites			Metallurgical products			
	Metal phase	Sulfide phase (troilite)	Silicate phase	Metal (pig iron)	Sulfide (copper matte)	Silicate (slag)	Flue dust
Si	0.015	0	21.60	0.02	0.05	22.09	4.03
Al			1.83	0.05	0	9.11	
Fe	88.60		13.25	73.58	22.92	3.0	5.3
Mg			16.63	0	0.05	7.46	
Ca			2.07	0.003	~ 0.001	13.50	
Na			0.82	~ 0.1	0.1	0.64	
K			0.21		0.49	3.28	
P	1,800	3,000	700	18,400	0	300	~ 50
Cr	300	1,200	3,900	0	0	40	0
Ni	84,900	1,000	3,300	17,200	2,800	500	20
Co	5,700	100	400	24,400	2,500	40	0
V	6		50	800	~ 100	200	70
Ti	100	0	1,800	20	20	300	0
Zr	8	0	95				
Mn	300	460	2,050	0	6,400	2,000	~ 10
Cu	200	500	2	64,400	462,000	2,340	~ 30,000
Pb	56	20	2	20	2,200	200	~ 100,000
Zn	115	1,530	76	8	16,800	3,700	~ 400,000
Ag	5	19	0	150	2,520	0	300
Au	2	0.5	0	8	0	0	0
Pt	16	3	0	8	0	0	0
Sn	100	15	5	80	0	0	0
W	8	Trace	18	0	0	30	0
Mo	17	11	3	66,400	0	20	~ 5

† Major elements in weight percent (of the elements, not the oxides), minor elements in parts per million. Data on major elements in meteorites represent analyses of stony meteorites, from H. Brown and C. Patterson, The composition of meteoritic matter, *Jour. Geology*, vol. 55, pp. 405 and 508, 1947. Data on minor elements in meteorites are taken from a compilation by K. Rankama and T. G. Sahama ("Geochemistry," University of Chicago Press, 1950), p. 87, and represent averages for meteorites in general; the original data are largely from the work of Goldschmidt, but figures from several other sources have been added. Data on metallurgical products of the Mansfeld copper ores, also from a table in Rankama and Sahama (page 85), are based on analyses by A. Cissarz and H. Moritz, Untersuchungen über die Metallverteilung in Mansfelder Hochofenprodukten und ihre geochemische Bedeutung, *Metallwirtschaft*, vol. 12, p. 131, 1933. The symbol ~ indicates approximate values.

Occurrences of elements in silicate rocks, sulfide ores, and native iron agree fairly well with the distributions observed in meteorite and smelter analyses. The lack of perfect agreement is not surprising, as Goldschmidt pointed out, since the conditions of formation of sulfide ore deposits are quite different from the conditions under which sulfides would separate from an artificial melt.

On the basis of these considerations, Goldschmidt suggested that the elements could be usefully grouped into those which preferentially occur with native iron and which probably are concentrated in the earth's iron core (*siderophile elements*); those concentrated in sulfides and therefore characteristic of sulfide ore deposits (*chalcophile elements*); and those that generally occur in or with silicates (*lithophile elements*). For completeness, elements that are prominent in air and other natural gases can be put in a fourth group, the *atmophile elements*. This classification, shown in Table 20-4, is clearly consistent with the data of Tables 20-2 and 20-3. Lithophile elements are for the most part those toward the top of Table 20-2, and those concentrated in the silicate phases of Table 20-3; siderophile elements are those near the bottom of Table 20-2, and those prominent in the metal phases of Table 20-3. Chalcophile elements are not distinguished in Table 20-2, but are scattered through the lower half; in Table 20-3 they are the elements prominent in the sulfide phases.

As might be expected, Goldschmidt's classification is closely related to the periodic law. Comparison of Tables 20-1 and 20-4 shows that in general siderophile elements are concentrated in the center of the table, lithopile elements to the left of center, chalcophile elements to the right, and atmophile elements on the extreme right. The classification can also be correlated with electrode potentials: siderophile elements are dominantly noble metals with low electrode potentials; lithophile elements are those with high potentials; and chalcophile elements have an intermediate position.

Table 20-4 Goldschmidt's geochemical classification of the elements

Siderophile	Chalcophile	Lithophile	Atmophile
Fe Co Ni	Cu Ag (Au)†	Li Na K Rb Cs	H N (C) (O)
Ru Rh Pd	Zn Cd Hg	Be Mg Ca Sr Ba	(F) (Cl) (Br) (I)
Re Os Ir Pt Au	Ga In Tl	B Al Sc Y REE‡	Inert gases
Mo Ge Sn C P	(Ge) (Sn) Pb	(C) Si Ti Zr Hf Th	
(Pb) (As) (W)	As Sb Bi	(P) V Nb Ta	
	S Se Te	O Cr W U	
	(Fe) (Mo) (Re)	(Fe) Mn	
		F Cl Br I	
		(H) (Tl) (Ga) (Ge) (N)	

† Parentheses around a symbol indicate that the element belongs primarily in another group, but has some characteristics that relate it to this group. For example, gold is dominantly siderophile, but (Au) appears in the chalcophile group because gold is often found in sulfide veins.

‡ REE = rare-earth elements.

Such a grouping of elements can at best express only tendencies, not quantitative relationships. The different groups overlap, as is shown by the occurrence of many elements in more than one category. Iron, for example, is not only the principal element of the earth's core but is also common in sulfide deposits and in igneous rocks. Such overlaps are inevitable in a classification that is based partly on distributions in very-high-temperature processes and partly on distributions under ordinary surface conditions. Various refinements have been suggested, but there seems little point in trying to make the classification more quantitative or more detailed. It is useful simply as a rough qualitative expression of the geologic behavior of the elements.

20-3 DISTRIBUTION OF ELEMENTS IN IGNEOUS ROCKS

Point of View

The origin of many igneous rocks, as noted in previous discussions, is a subject of lively debate. Some granites have formed by the cooling of a melt that has moved as a fluid body, others by anatexis of sedimentary and volcanic rocks without much movement. For most granites, or at least for large parts of most granites, the evidence is not decisive one way or the other. Granitic magma is known to form in small amounts by the differentiation of basaltic magma, and such a process may be responsible for some larger granite masses also. Alkalic rocks may form by normal differentiation, by assimilation of country rocks, or by the action of volatiles. Ultramafic rocks may be a product of crystal settling during differentiation, or samples of magma from beneath the crust, or a result of metamorphism of partly assimilated country rock. Clearly, a discussion of the distribution of elements in igneous rocks could involve us in endless arguments over these hypotheses of origin unless we adopt some definite point of view at the beginning and then modify it as specific cases require.

The easiest procedure is to start with the "classical" viewpoint that most igneous rocks are formed by the differentiation of basaltic magma, mafic minerals settling out first to form ultramafic rocks and the remaining melt changing in composition through the series gabbro–diorite–granodiorite–granite–pegmatite. We can speak then of "early-formed minerals" of high melting point, such as olivine and calcic plagioclase, "late-formed minerals" like quartz and biotite, and "residual fluids" of pegmatitic composition. We can work out semiempirical rules of behavior for various elements during such a differentiation process and compare our predictions with analyses. When this procedure is adopted, it happens that agreement between predictions and analyses is surprisingly good, and surprisingly uniform from one set of igneous rocks to another.

This agreement, however, cannot be considered evidence in favor of the assumed differentiation process. Suppose that a sequence of igneous rocks is formed in the reverse order, by progressive partial melting of a shale or graywacke, proceeding from the composition of alkali granite through granodiorite, tonalite, and gabbro, as more and more mafic material is incorporated into the melt. The same distribution of elements would be expected at comparable stages in this series as in the sequence formed by crystallization differentiation, provided that the composition of the starting material was reasonably close to the crustal average and provided that equilibrium was maintained during the melting. If marked departures from equilibrium occur, during either the fractional crystallization of a mafic magma or the progressive melting of average silicate material, the distribution of minor elements might be considerably different in corresponding parts of the two series. But ordinarily a close enough approach to equilibrium can be assumed so that the rare elements would be expected to distribute themselves according to uniform rules.

Thus the point of view that most igneous rocks have formed by crystallization differentiation, a point of view that will underlie much of the subsequent argument, is presented *not as a hypothesis to be proved but as a framework to unify the discussion*. When we speak of "early-formed minerals," we shall mean minerals that *would* form early during differentiation, but that in a particular case might instead have formed late during metamorphism. This method of presentation is the usual one in geochemical literature, and it has the virtue of brevity: we shall not have to repeat on every page the alternative hypotheses of igneous-rock formation. But it must be kept clearly in mind that *nothing is implied about the correctness of the differentiation hypothesis*, except of course in particular cases where evidence of other kinds is available.

Rules of Distribution

The sequence of minerals that separate during differentiation of a silicate melt is already familiar (Chap. 14). Most commonly olivine and calcic plagioclase appear first; as the temperature falls, part or all of the olivine reacts with the melt to form pyroxene, then amphibole, then biotite; the composition of the plagioclase meanwhile becomes increasingly sodic; near the end of the crystallization, quartz and potash feldspar appear along with sodic plagioclase. When oxygen pressures are low, this scheme is altered, in that abundant iron remains longer in the melt; this means that iron silicates continue to crystallize until a late stage and that the accumulation of silica in the residual fluid is less pronounced. Other variations are possible, but in general during differentiation the major elements separate out according to a fairly uniform pattern: most of the magnesium and calcium leave the melt in early stages, and most of the alkali metals later; iron may be largely concentrated in early minerals or may appear at all stages; aluminum drops out in feldspars all during the differentiation process, and also in micas toward the end. Our problem now is to see how the behavior of the less common elements is related to these changes.

For an element to crystallize in a mineral of its own—a mineral in which it is a major constituent—requires that the element be present in the melt in appreciable amounts. If only a few ions of the element are present, they can be taken up by the crystal structures of the major silicates, either as isomorphous replacements of an abundant element or as random inclusions in the holes of a crystal lattice. *How much* of a given element can be accommodated in the silicate structures depends on the characteristics of its ions. The rare alkali metal rubidium, for example, is so similar to potassium that several tenths of a percent can be accommodated as replacements of Rb^+ for K^+ in micas and feldspars; but the rare metal zirconium cannot fit easily into common silicate structures, and even very small amounts of it go into separate crystals of the accessory mineral zircon. Some elements, like beryllium, boron, copper, and uranium, are capable neither of forming their own high-temperature minerals nor of substituting appreciably for common ions in silicate structures, and so are concentrated in the residual solutions that give rise to pegmatites and sulfide veins. What specific characteristics of an element determine which way it will behave?

The question takes us back to the discussion of isomorphism in Sec. 5-7. We recall the rules of isomorphous replacement, rephrased a little to apply to the present discussion:

1. A minor element may substitute extensively for a major element if the ionic radii do not differ by more than about 15%.
2. Ions whose charges differ by one unit may substitute readily for one another, provided their radii are similar. Substitution is generally slight when the charge difference is more than one unit.
3. Of two ions that can occupy the same position in a crystal structure, the one that forms the stronger bonds with its neighbors is the one with the smaller radius or the higher charge (or both).
4. Substitution of one ion for another may be very limited, even when the size criterion is fulfilled, if the bonds formed differ markedly in covalent character.

Leaving aside the question of bond character for the moment, we find abundant illustrations of the first three rules. We could predict, according to rules 1 and 2 and the table of ionic radii in Appendix IV, that Ba would commonly substitute for K, Cr for Fe^{3+}, Y for Ca; and these predictions fully accord with analytical data. From the third rule we could generalize that, in an isomorphous pair of compounds, the one containing ions of smaller radius and higher charge would have the higher melting point (because of the stronger bonding) and therefore would appear earlier in a crystallization sequence. Hence a minor element like Li, which substitutes extensively for Mg because of the similarity in their ionic radii, should be concentrated in late-forming Mg minerals rather than early ones because its single charge forms weaker bonds than the double charge on Mg^{2+}; this is borne out by the near absence of Li in olivine and its common presence as a substitute for Mg in the micas of pegmatites. Similarly Rb, having an ion similar in charge to K^+ but somewhat larger, should be enriched in late K minerals rather than early ones, a prediction that agrees with the observed concentration of Rb in the feldspars and micas of pegmatites.

The same rules, it should be noted, are illustrated beautifully by some of the major elements. In the isomorphous series of the olivines, forsterite has a higher melting point than fayalite and is enriched in early crystals (Prob. 13-3), corresponding with the fact that Mg^{2+} is a smaller ion than Fe^{2+}. Early crystals of plagioclase are calcium-rich, late crystals sodium-rich (Sec. 13-3), in accordance with the higher charge of the Ca^{2+} ion. The rules are also similar to generalizations we have formulated regarding ion-exchange processes (Sec. 6-5), as they should be, because in both cases we are dealing with the relative strengths of bonds formed between ions and crystal structures.

The rules of substitution according to ionic radius work admirably as long as the discussion is limited to elements in the first three groups of the periodic table, but with the remaining elements agreement between prediction and observation is often much less satisfactory. Cuprous ion, for example, is similar in size and charge to Na^+, but copper shows no enrichment in sodium minerals. Cadmium ion closely resembles Ca^{2+} in size and charge but is not concentrated in calcium minerals. The difficulty goes back to rule 4, the difference in bond character (or polarization): the ions Cu^+ and Cd^{2+} form bonds of markedly less ionic character with the anions of a crystal structure than do Na^+ and Ca^{2+} (last column of Appendix IV). This failure of the rules of substitution is reminiscent of the failure of predictions from simple geometry to account for crystal structures when the structures involve strongly covalent bonds (Sec. 5-3).

To get a feeling for the effectiveness of the distribution rules in predicting the behavior of minor elements in igneous processes, we look now at some analytical data for typical rock sequences.

Examples

Table 20-5 shows average concentrations of major and minor elements in four rock series. The first series consists of worldwide averages for the major igneous-rock types, and the other three give analyses for rocks in specific localities. The worldwide averages, of course, do not represent a sequence necessarily related by a differentiation process, but they are pertinent to the present discussion in the light of our preliminary guess that trace elements would distribute themselves in much the same manner whether rocks form by differentiation or by extreme metamorphism. For each of the other three series there is good geological and chemical evidence that the rocks are actually members of a differentiation sequence. The Irish lavas show a fairly typical trend from olivine basalt to rhyolite, with a marked increase in silica and alkalies and a decrease in iron; the Skaergaard rocks show a notable increase in iron and fairly constant silica until the very end of the differentiation process; the Hawaiian lavas are rich in soda and potash and show only a modest increase in silica. Thus the four sets of analyses enable us to compare the distribution of trace elements in three different kinds of differentiation sequences with the average distribution for igneous rocks in general.

Cursory examination of the data shows that the distribution has a certain amount of regularity, but also much apparently random variation. As so often happens in geochemistry, the numbers seem to follow a vague general rule that is subject to many exceptions. A little reflection will show that this is what we might expect. We are dealing, of course, with a very complicated process. The distribution is affected not only by the various trends of differentiation, but also by details of the way the trace elements enter into individual minerals. Furthermore, the steps in differentiation do not necessarily correspond from one set of analyses to another, nor do the extremes represent necessarily equivalent stages in differentiation; we have, in fact, no good way even of defining what we mean by "equivalent stages of differentiation." Recognizing these reasons for lack of complete regularity in the data, we can nevertheless draw useful conclusions from the analyses about general patterns of trace-element behavior.

Trends shown in the table can be visualized with the diagrams in Fig. 20-1. Lines on the figure represent analyses for various elements in the four categories of the world averages—ultramafic rocks, mafic rocks, intermediate rocks, and felsic rocks. These four rock types are shown, from left to right, by four points on each line. The heavy lines represent major elements, and the light lines in each column represent possibly related minor elements. Scales are not uniform, so that absolute values cannot be compared from diagram to diagram. The lines simply show trends, the relative increases or decreases of different elements in going from ultramafic to felsic rocks.

The first column of Fig. 20-1 gives trends for three large cations that might be expected to substitute for K^+ because of their size. The similarity in the lines is evident, suggesting that Rb, Ba, and Pb do indeed enter potassium minerals in small amounts and hence, like potassium, are concentrated in rocks formed late in the crystallization sequence. The similarity in trend is most marked for Rb^+; because of their double charge and smaller size, Ba^{2+} and Pb^{2+} tend to enter early K minerals, hence are not as strongly enriched at the felsic end of the series as is Rb. For these elements the rules of substitution work beautifully.

The second column shows the trend for calcium and the related minor element strontium. Here agreement with the rules is less clear. Sr^{2+}, with a radius between those of Ca^{2+} and K^+, can substitute for both, so that its trend is a compromise between the trends for the two major elements. Except for its low abundance in ultramafic rocks, strontium remains fairly constant through the differentiation sequence, because calcium decreases as potassium increases. Copper, with an ionic radius close to that of calcium, has a somewhat similar trend line; but the similarity is probably fortuitous, because a chalcophile element like copper, which forms dominantly covalent bonds with oxygen, cannot readily substitute for an element whose bonds are strongly ionic.

Also in the second column are curves for aluminum and gallium, illustrating the great similarity of these two elements and the consequent extensive substitution of Ga^{3+} for Al^{3+} in aluminosilicate minerals.

In the third column of Fig. 20-1 are trend lines for several minor elements with ionic radii similar to those of iron and magnesium. The two major elements show a markedly decreasing abundance as differentiation proceeds, and most of the minor-element trend lines follow this pattern except that some have low concentrations at the ultramafic end of the series. Trends for Cr and Ni mimic that

Table 20-5 Average content of major and minor elements in four rock series†

	Ionic radii, Å	World averages				Tertiary lavas of NE Ireland				Skaergaard intrusive rocks				Hawaiian lavas				
		Ultra-mafic rocks	Mafic rocks	Inter-mediate rocks	Felsic rocks	Oli-vine basalt	Tho-leite basalt	Quartz tra-chyte	Rhyo-lite	Gabbro-picrite	Hyp-olivine gabbro	Ferro-gabbro	Grano-phyre	Picrite-basalt	Basalt	Andes-ine andesite	Oligo-clase andesite	Tra-chyte
Si^{4+}	0.34	19.0	24.0	26.0	32.3	21.1	24.3	30.0	35.2	19.3	21.6	20.3	31.2	22.3	24.0	23.7	24.4	29.0
Al^{3+}	0.61	0.5	8.8	8.9	7.7	7.8	7.5	7.5	6.5	4.6	8.9	7.7	6.7	4.9	6.8	8.7	8.8	9.9
Fe^{3+}	0.73	9.9	8.6	5.9	2.7	1.9	1.8	3.2	0.6	1.9	1.1	2.3	2.6	1.0	1.2	4.1	2.2	3.0
Fe^{2+}	0.86					6.7	6.5	2.2	0.3	8.2	8.1	17.1	3.9	8.1	7.3	4.8	5.7	0.1
Mg^{2+}	0.80	25.9	4.5	2.2	0.6	6.8	3.5	0.5	0.1	16.3	5.8	1.5	0.3	11.4	4.6	2.6	1.9	0.2
Ca^{2+}	1.08	0.7	6.7	4.7	1.6	7.1	7.1	1.5	0.6	4.7	8.1	6.5	2.0	5.5	7.5	4.6	4.5	0.6
Na^+	1.10	0.6	1.9	3.0	2.8	1.5	1.9	3.1	2.0	0.5	1.8	2.1	3.0	1.2	1.5	3.5	4.1	5.1
K^+	1.46	0.03	0.8	2.3	3.3	0.2	0.9	3.5	3.8	0.1	0.2	0.3	2.6	0.3	0.3	1.8	1.8	4.1
P^{5+}	0.25	170	1,400	1,600	700	970	1,260	535	90	90	260	5,000	1,500	1,000	1,140	740	4,060	1,050
Ga^{3+}	0.70	2	18	20	20	33	35	40	43	8	19	20	33	18	23	20	23	25
Cr^{3+}	0.70	2,000	200	50	25	1,600	150	50	<1	1,500	230	<1	11	1,750	470	<1	<1	20
Li^+	0.82	<1	15	20	40	8	18	65	80	2	2	3	20	1	2	15	20	30
Ni^{2+}	0.77	2,000	160	55	8	900	50	65	13	1,000	120	<2	9	950	85	7	13	15

Ion	Radius (Å)																	
Co^{2+}	0.83	200	45	10	5	140	110	8	2	90	48	20	5	75	35	14	6	2
V^{3+}	0.72	40	200	100	40	630	750	55	2	120	220	4	9	250	280	70	19	< 5
Ti^{3+}‡	0.75	300	9,000	8,000	2,300	7,000	6,700	3,200	700	9,000	5,000	15,000	5,000	12,000	20,000	16,000	15,000	2,000
Zr^{4+}	0.80	30	100	100	200	180	700	3,000	2,000	30	33	20	1,200	75	100	350	1,250	1,500
Mn^{2+}	0.91	1,500	2,000	1,200	600	1,380	1,360	1,200	130	1,200	700	3,300	800	860	1,010	1,710	1,010	1,170
Sc^{3+}	0.83	5	24	3	3	19	25	25	< 10	< 10	20	10	< 10	< 10	10	< 10	< 10	< 10
Cu^{+}‡	0.96	20	100	35	20	400	180	50	66	100	67	400	200	150	170	< 10	< 10	< 10
Sr^{2+}	1.21	10	440	800	300	680	1,250	650	170	100	600	400	450	300	800	2,500	3,500	100
Pb^{2+}	1.26	< 1	8	15	20	< 20	25	43	43	No data						No data		
Ba^{2+}	1.44	1	300	650	830	310	1,350	2,000	2,300	10	18	60	1,100	110	120	600	1,000	800
Rb^{+}	1.57	2	45	100	200	3	90	1,000	930	< 20	< 20	< 20	110	< 20	< 20	35	55	300

† Major elements in weight percent (of the elements, not the oxides), minor elements in parts per million.

‡ The elements Ti and Cu may also be present in part as other ions, Ti^{4+} (radius 0.69 Å) and Cu^{2+} (radius 0.81 Å).

Sources of data:

Ionic radii: Appendix IV.

World averages: A. P. Vinogradov, Sredniye soderzhaniya khimicheskikh elementov v glavnykh tipakh izverzhennykh gornykh porod zemnoi kory, Geokhimiya, vol. 1962, pp. 560–561, 1962.

Irish lavas: E. M. Patterson, A petrochemical study of the Tertiary lavas of northeast Ireland, Geochim. et Cosmochim. Acta, vol. 2, p. 291, 1952.

Skaergaard rocks: L. R. Wager and R. L. Mitchell, Distribution of trace elements during strong fractionation of a basic magma, Geochim. et Cosmochim. Acta, vol. 1, p. 199 and Table F, 1951.

Hawaiian lavas: L. R. Wager and R. L. Mitchell, Trace elements in a suite of Hawaiian lavas, Geochim. et Cosmochim. Acta, vol. 3, p. 218, 1953.

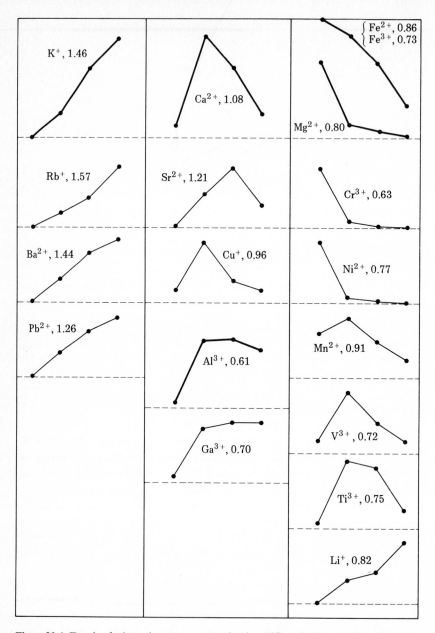

Figure 20-1 Trends of minor-element concentration in a differentiation sequence. Data from columns headed "world averages" in Table 20-5. The four points on each line represent concentrations in (from left to right) ultramafic rocks, mafic rocks, intermediate rocks, and felsic rocks. Scales are not uniform; lines show trends only. Heavy lines are trends for major elements, light lines for minor elements. Numbers beside symbols are ionic radii in angstroms.

of Mg, as would be expected from the extensive substitution of these metals in early-formed Mg minerals (pyroxenes and olivines). Mn, V, and Ti substitute more readily for Fe, and their trend lines are correspondingly similar except at the left-hand ends. Li^+, with its single positive charge, substitutes readily for Mg^{2+} only in the latest formed magnesium minerals (especially micas), hence increases in abundance during the course of differentiation.

Analyses for the three differentiation series in Table 20-5 (northern Ireland, Skaergaard, and Hawaii) show trends similar to those for the world averages. There are a few conspicuous deviations, as might be expected because the rock series have different overall compositions and different degrees of differentiation, but nevertheless the trends displayed in Fig. 20-1 apparently represent a general pattern widely applicable to igneous rocks.

Regularities of Distribution in Igneous Rocks

Important features of this general pattern, as revealed by many analyses of rock series like those in Table 20-5, are the following:

1. Cations with large radii and low electric charges tend to substitute for potassium, hence are concentrated in felsic rather than mafic rocks. These elements (Rb, Cs, Ba, Pb, Tl) are often called the "large-ion lithophile" group (LIL); their abundance in a rock series is a good indication of the extent to which differentiation has sorted out constituents of the original igneous material. The rare-earth elements are often included here, although their concentration in late-forming rocks is for a different reason.
2. Several cations with smaller radii and mostly with higher charges (U, Th, B, Be, Mo, W, Nb, Ta, Sn, Zr) also are concentrated at the felsic end of the series, not because of extensive substitution but rather because their size and charge make it difficult for them to substitute for any major ions in common silicate minerals. Because they do not fit into the usual positions available in silicate structures, this group is often referred to as "incompatible" elements. Differentiation segregates them in the late residual solutions, and if present in appreciable amounts they may form minerals of their own (uraninite, beryl, columbite, zircon, etc.).
3. Many elements with intermediate radii, especially metals of the transition groups, substitute readily for iron and magnesium, hence are abundant in the earlier members of the differentiation sequence. Some (Cr, Ni, Co) are strongly enriched with Mg in ultramafic rocks; others (Mn, V, Ti) have maximum abundance in gabbros and basalts.
4. Among the chalcophile elements, a few substitute to some extent for major cations in silicate structures (Pb and Tl for K, Zn for Mg and Fe, Bi for Ca), but for the most part these metals are left to accumulate in the residual solutions that may eventually form sulfide ores. In a magma that contains abundant sulfur, part of the chalcophile elements may separate early in differentiation as

an immiscible sulfide liquid. The traces of chalcophile metals commonly reported in ordinary igneous rocks may be present in large part as tiny sulfide grains rather than as substitutes for major elements; this would explain, for example, the apparently erratic values for copper in the analyses of Table 20-5.

5. Some minor elements are so similar in size and chemical properties to major elements that normal differentiation cannot separate them effectively from their major relatives. Gallium is a good example: it is always present in aluminum minerals, and very seldom becomes sufficiently segregated to appear in a mineral of its own. Other closely related pairs are Rb and K, Hf and Zr, Cd and Zn. The rare members of these pairs are not particularly scarce metals, but they are little known because minerals in which they appear separated from their relatives are scarce or nonexistent.

Thus the principal features of minor-element distribution in igneous rocks can be explained by the simple rules of substitution based on ionic charge and radius, supplemented by empirical statements about the behavior of elements whose bonds are largely covalent. The simple rules, however, have many exceptions, and details of the distribution would require much more sophisticated analysis. We have spoken of distribution, for example, as if it were influenced only by mineral structures in the crystallizing solids; a complete description would have to consider also structures in the silicate liquid with which the solids are in contact. Again, the transition elements show many apparent anomalies in the relative amounts that substitute for Mg^{2+} and the two ions of iron; the anomalies can be in large part clarified by considering effects on the strengths of directed bonds exerted by electric fields in different positions in crystal structures ("crystal-field effects"). Detailed study shows that relationships between major and minor elements in crystal structures are highly complex, so it seems remarkable that the simple rules work as well as they do.

Summary The distribution of minor elements in igneous rocks can be explained fairly satisfactorily by assuming the slow crystallization of an orderly sequence of minerals from a melt, usually leading to differentiation. Chalcophile elements may separate early, either in an immiscible sulfide liquid (like the matte of smelting operations) or as sulfide crystals. Some elements, like chromium and zirconium, may form separate minerals even when only small amounts are present. Most of the trace elements are taken up by the crystallizing silicates, substituting for the major elements largely on the basis of ionic size. Many details of the distribution, both in individual minerals and in rocks formed at different stages of differentiation, can be correlated with such ionic properties as size, charge, and tendency to form covalent bonds. Trace metals markedly dissimilar to major elements are largely left out of the principal minerals and concentrate in the residual solutions that ultimately form pegmatites and sulfide veins.

An equally effective explanation for the distribution could be given by assuming that igneous rocks form by progressive melting of average rock material during extreme metamorphism. Differentiation and progressive melting under equilibrium conditions would lead to about the same sorting out of minor elements; emphasis on differentiation in the preceding discussion is a matter of convenience and does not imply a choice of one assumption over the other.

20-4 DISTRIBUTION IN SEDIMENTARY ROCKS

General

The processes of weathering and sedimentation, it is often noted, act like a huge and inefficient sort of chemical analysis, breaking down the assemblages of elements in igneous rocks and regrouping them into assemblages that are generally simpler. Locally the "analysis" may be very effective, isolating silica in the form of pure quartz sandstone or chert, alumina in bauxite, iron in residual laterite or in sedimentary oxides, carbonates, and silicates, calcium in limestone or gypsum, sodium and potassium in salt deposits. More commonly the breakdown is incomplete, giving only a preponderance of silica in sandstone, of alumina and silica in clays, and of calcium and magnesium in carbonates. In the various operations of this huge analytical procedure, what happens to the minor elements? Can they be separated completely from the main elements, or with what main elements are they usually concentrated?

For detrital sediments the answer is easy. Some trace elements can indeed be separated almost completely from others, simply on the basis of the resistance of their minerals to solution and abrasion. Placer deposits of gold, platinum, monazite, and zircon come to mind. Minerals that are less resistant to attack often show at least partial concentration; magnetite-rich streaks in sandstone are a good example.

Much more complicated are the chemical processes that may lead to preferential precipitation of minor elements out of solution. Differences in solubility of compounds, adsorption processes, and the activity of organisms all must play a role. In general these processes are not very effective in separating minor elements from major ones. With the exception of phosphates, borates, nitrates, some manganese deposits, and accumulations of copper, vanadium, and uranium with organic matter, the concentrating of rare elements by purely sedimentary processes is not notable.

Some rough averages showing the distribution of minor elements in the principal kinds of sedimentary rocks are listed in Table 20-6.

Explanation of the Distribution

One generalization about the distribution stands out immediately: most of the rarer elements are more abundant in shales than in sandstones and limestones. The outstanding exceptions are strontium and manganese, which are markedly enriched in carbonate sediments, and zirconium and the rare earths, which are

Table 20-6 Average concentrations of minor elements in shales, sandstones, and carbonate rocks, in parts per million

	Shales	Sand-stones	Carbo-nates		Shales	Sand-stones	Carbo-nates
†Li	60	15	5	Ge	1.5	0.8	0.2
‡B	100	35	20	‡As	10	1	1
F	600	270	330	‡Se	0.6	0.05	0.08
‡P	750	170	400	Br	5	1	6.2
Cl	180	10	150	†Rb	140	60	3
Sc	15	1	1	†Sr	400	20	610
Ti	4,600	1,500	400	Y	35	40	30
V	130	20	20	Zr	180	220	19
Cr	100	35	11	‡Mo	2	0.2	0.4
Mn	850	X0	1,100	I	2	1.7	1.2
Co	20	0.3	0.1	†Ba	600	X0	10
Ni	80	2	20	Ce	70	92	11.5
Cu	50	X	4	Pb	20	7	9
Zn	90	16	20	Th	12	1.7	1.7
Ga	25	12	4	U	3.5	0.45	2.2

substitutes for Ca

substitutes for Aluminum

Notes: 1. X means between 1 and 10; X0 means between 10 and 100. 2. Elements marked with a dagger (†) have low ionic potentials (<2.5); those marked with a double dagger (‡) have high ionic potentials (>9.5). Unmarked metallic elements have ionic potentials between 2.5 and 9.5.

Source: For shales, Appendix III; for sandstones and carbonates, K. K. Turekian and K. H. Wedepohl, Distribution of the elements in some major units of the earth's crust, *Geol. Soc. America Bull.*, vol. 72, pp. 175–192, 1961.

concentrated in sandstone. The enrichment in sandstone is easily explained by the mechanical concentration of the resistant minerals zircon and monazite. The enrichment of strontium in limestones is accounted for by assuming that Sr^{2+} substitutes readily for the very similar ion Ca^{2+}, just as it does in the minerals of igneous rocks. The smaller but appreciable concentration of manganese, yttrium, and the rare-earth metals in carbonate sediments is probably also attributable to similarities in ionic size.

Now why should most of the minor elements show such a marked preference for the fine-grained sediments? In part the answer may again lie in ionic substitution, this time of elements like manganese and zirconium for the magnesium of montmorillonite, or of barium for potassium in illite, or of gallium for aluminum in any of the clay minerals. In part the answer rests on reactions of some of the minor elements with organic material, which is more abundant in fine-grained sediments than in other kinds. But probably the chief process that leads to enrichment of rare elements in clays and shales is adsorption.

From an earlier discussion (Sec. 6-5) we recall the rules of adsorption: that small ions are more strongly adsorbed than big ones, multivalent ions more than univalent ions, and polarizing ions more than nonpolarizing ions. We recall also the many exceptions to these rules, particularly the apparent reversal in the order of adsorption of the alkali-metal ions, probably because hydration changes the relative sizes. The order of adsorption is somewhat dependent on the nature of the adsorbent and on the conditions under which adsorption takes place, but in general we might exect a fine-grained sediment to have a sufficient variety of small particles, and to have experienced enough changes of environment during deposition and diagenesis, so that rare elements of all kinds can be firmly held.

Ionic Potential

An alternative explanation often suggested for the general slight enrichment of minor elements in fine-grained sediments is that some of these elements may be precipitated with clay as hydroxides. The tendency of an ion to precipitate as a hydroxide can then be related, at least theoretically, to a quantity called *ionic potential*, which is defined as the quotient of the positive charge on a simple ion divided by its radius. The name is an unfortunate one because it is so easily confused with a very different concept, ionization potential.

The idea of ionic potential is simple. Any positive ion in water is somewhat attracted to the negative ends of the polar water molecules in its vicinity. We can think of this attraction as setting up a competition between the positive ion and hydrogen ion for the oxygen of water:

$$Me-O-H$$

If the attraction between Me and O is weak compared with the attraction between H and O, the Me ion remains free in solution, merely surrounding itself with a loosely held layer of water molecules; Na^+ is a good example. If the $Me-O$ bond is strong compared with the $H-O$ bond, the Me appropriates one or more O ions, forming an anion and leaving H^+ free; a good example is the hypothetical ion S^{6+}, which would react with water to form SO_4^{2-}:

$$S^{6+} + 4H_2O \rightarrow SO_4^{2-} + 8H^+$$

If the two bonds are of comparable strength, the structure forms an insoluble hydroxide, for example, $Zn(OH)_2$. Now the strength of the $Me-O$ bond should be greatest for ions of high positive charge and small ionic radius; hence the ionic potential should be a measure of the tendency of an ion to remain free, to form an anion with oxygen, or to precipitate as a hydroxide. The general validity of this correlation is shown by the list of elements in Table 20-6.

As usually happens when ionic properties are expressed as functions of simple geometric quantities, the agreement is excellent for elements on both sides of the periodic table but less satisfactory for those in the middle. Elements in the middle, because of their tendency to form covalent bonds, distort the large O^{2-} anion so that the relative strength of the $Me-O$ and $O-H$ bonds is no longer a simple matter of charge and radius. Various attempts have been made to correct the ionic potential for the covalent character of $Me-O$ bonds, but none are very successful. Ionic potential is significant only as a qualitative device for describing the behavior of hydroxides.

In connection with the behavior of rare elements in sediments, the importance of ionic potential is dubious. The elements that form insoluble hydroxides (which include the vast majority) are indeed somewhat enriched in fine-grained sediments, but there is not the slightest evidence that they are present *as* hydroxides. Small amounts of such hydroxides would be difficult to detect, but there seems no point in assuming their existence when adsorption is a more straightforward explanation. Moreover, as Table 20-6 shows, there are no conspicuous differences in the extent of enrichment of elements with low, intermediate, and high ionic potentials.

Possibly ionic potential has more bearing on the behavior of minor elements in the hydroxide sediments, such as bauxite and sedimentary iron ores. Similarities in ionic potential have been suggested, for example, as an explanation of minor concentrations of beryllium and titanium in bauxite. Even here, however, as long as no separate rare-metal hydroxides are detectable, an explanation in terms of adsorption or of replacement of aluminum and iron in their hydroxides seems preferable.

Precipitation, Oxidation, and Reduction

The separation of rare elements by direct precipitation of insoluble compounds is not common. One can suggest phosphorite (Sec. 3-11), manganese carbonate (Sec. 3-9), and evaporite deposits containing minerals of boron, nitrogen, and iodine. Somewhat more common is the precipitation of rare elements following oxidation or reduction.

Manganese is the most important minor element that is often precipitated in sedimentary environments as a result of oxidation (Sec. 10-2). Arsenic and antimony could also be mentioned here, but the insoluble arsenates and antimonates that form by the oxidation of ore deposits make a negligible contribution to sedimentary rocks.

Reduction, usually brought about by organic matter, is responsible for the formation of some uranium and vanadium deposits in sediments, since compounds of these elements in their lower oxidation states are in general much more insoluble than those of the higher oxidation states. It is hardly proper to call this a "sedimentary" process, however, since most such uranium-vanadium deposits give clear evidence that the metals were introduced by solutions long after the sedimentary rocks were formed. Native copper has been observed in modern swamps, under conditions suggesting strongly that it is forming by reduction where dilute copper-bearing solutions encounter organic matter. A similar mechanism is often proposed to account for minor amounts of native copper sometimes found in black shales and associated with organic matter in sandstones, but for such occurrences it is usually difficult to prove that the copper-bearing solutions were contemporaneous with the sedimentation. Where reduction takes place in the presence of sulfide ion, many metals can theoretically precipitate as sulfides, but the only metal known to be precipitating at the present time in this manner, in ordinary sedimentary environments, is iron. Deposits of other metal sulfides in sedimentary rocks (most commonly chalcopyrite, sometimes galena and sphalerite) are often explained by postulating reduction and precipitation during sedimentation, but the evidence is seldom sufficient to eliminate the possibility that the metals were introduced long after the sediment was formed. How important reduction and precipitation are for the concentration of minor elements during sedimentation remains a much-disputed point.

Reactions with Organic Matter

The organic matter of sedimentary rocks is known to be slightly enriched in many of the rarer elements, particularly V, Mo, Ni, Co, As, Cu, Br, I, and locally many others. These elements may form specific minerals, but more commonly are present in the organic matter itself. How the concentration has occurred is seldom entirely clear. In part the metals have doubtless been reduced, in the manner just described, from solutions carrying more soluble compounds of higher oxidation states; theoretically, at least, this could happen either during sedimentation or much later. Some of the elements are used by organisms, and remain when the organisms die and become part of a sediment. Vanadium and nickel, whatever their ultimate source may be, are known to form metal porphyrins, in which they have replaced the magnesium and iron of original porphyrins in living substance. Probably other elements similarly form definite compounds with the organic matter (Sec. 11-1). Simple adsorption on the surface of organic particles may account for some of the metal content. Much remains unknown about the precise mechanisms of concentration, but certainly organic matter has played a role in some of the most important examples of enrichment of trace elements in sedimentary rocks: vanadium and molybdenum in asphalt, uranium and copper in black shales, and germanium in coal are familiar illustrations.

In summary, the major features of the distribution of minor elements in sedimentary rocks can be related, just as in igneous rocks, to ionic size, charge, and bond character. But the influence of these ionic properties is different: substitution of minor-metal ions for major ions in crystal structures is of lesser importance, and adsorption of rare-metal ions on the surfaces of particles in fine-grained sediments plays a major role. Most of the minor elements are more abundant in fine-grained detrital sediments than in sandstones or carbonate rocks, probably largely because of adsorption. Other processes helping to determine the distribution of elements in sediments are precipitation following oxidation or reduction, and various reactions with organic matter. In general, notable concentrations of minor elements formed by sedimentary processes alone are not common.

less important

20-5 DISTRIBUTION IN METAMORPHIC ROCKS

Information about minor-element distribution in metamorphic rocks is relatively meager. What data we have indicate that the distribution will hold few surprises. Metamorphism of fine-grained rocks to hornfelses or phyllites produces no detectable change in rare-metal content, unless the rocks have been permeated by solutions during the metamorphic process. At higher grades of metamorphism the minor elements redistribute themselves locally among the growing crystals of new minerals, but again the overall concentrations do not change markedly unless movement of solutions has played an important role. Since some sedimentary rocks have unique assemblages of trace elements, a study of these elements in metamorphic rocks provides a possible way to guess at the nature of the premetamorphic material, but the trace elements seldom give a better basis for guessing than the more obvious major elements. When metamorphism reaches the ultimate stage of partial melting, the minor elements go into the melt, and then recrystallize from the melt according to the pattern we have outlined for igneous rocks.

The distribution of elements in various special kinds of geologic materials—soils, evaporites, unusual igneous rocks, sulfide minerals of ore deposits—is a fascinating and productive study but involves too much detail for the present discussion. In general, the theoretical side of such a study means only refinement of the ideas we have outlined in this chapter, particularly the dependence of distribution on such ionic properties as radius, charge, and bond character. Details of distribution in many geologic environments remain elusive, but the general pattern follows theoretical expectations remarkably well.

PROBLEMS

1 Explain what is meant by the statement that copper is a more chalcophile element than either zirconium or platinum.

2 Vanadium is a more abundant element than boron, yet boron minerals are more common than vanadium minerals in both igneous and sedimentary rocks. Why?

3 For each of the following ratios, indicate whether you think it would usually be higher in mafic igneous rocks or in felsic igneous rocks, and give reasons for your answers: Rb/Sr, Sr/Ba, B/Mn, Li/Mg, Pb/Rb, Cr/Al, Y/Ca.

4 Of the following ratios, which would you expect to be greater in evaporites than in shales? Why? Ca/Ba, Si/Na, K/Rb, Mg/Mn.

5 "In general, an element is more chalcophile in its lower valences than in its higher valences." Is this statement true? Discuss the proposition, using examples to back up your opinion.

6 Lithium and cesium are both somewhat concentrated in the late micas of pegmatites, although their ionic radii are very different. Explain.

7 Account for the lack of extensive isomorphous replacement of Na by Li, of Fe^{3+} by Li, of Mg^{2+} by Nb, of Mn^{2+} by Pt^{2+}.

8 Why are gallium minerals so exceedingly rare, whereas minerals of the much less abundant elements Sn, U, and W are well known?

9 Account for the fact that the six most abundant elements in seawater are Na, K, Mg, Ca, S, and Cl.

10 Why are uranium and thorium much more abundant in granites than in ultramafic rocks?

11 Compute free energies of formation of oxides and sulfides for several metals, and show how these free energies are correlated with the chalcophile or lithophile character of the metals.

REFERENCES AND SUGGESTIONS FOR FURTHER READING

Burns, R. G., and W. S. Fyfe, Trace element distribution rules and their significance, *Chemical Geology*, vol. 2, pp. 89–104, 1967. Review and critique of the distribution rules, with emphasis on the importance of knowing bonding forces not only in crystalline solids but in the liquid surrounding the solids during crystallization.

Goldschmidt, V. M., "Geochemistry," Oxford University Press, Fair Lawn, N.J., 1954. The first six chapters contain general principles of element distribution, and later chapters give details about the distribution and geochemical behavior of each element.

Mason, B., "Principles of Geochemistry," 3d ed., John Wiley & Sons, Inc., New York, 1966. An excellent critical discussion of the behavior of trace elements in geological processes, and compilations of analytical data on many terrestrial and extraterrestrial materials.

McCarthy, T. S., and R. A. Hasty, Trace element distribution patterns and their relationship to crystallization of granitic melts, *Geochim. et Cosmochim. Acta*, vol. 40, pp. 1351–1358, 1976. Theoretical study of the effects of equilibrium and nonequilibrium crystallization on trace-element distribution.

Nockolds, S. R., The behavior of some elements during fractional crystallization of magma, *Geochim. et Cosmochim. Acta*, vol. 30, pp. 267–278, 1966. Goldschmidt's rules of distribution can be made more precise by calculating bond energies for metal-oxygen bonds.

Wedepohl, K. H., ed., "Handbook of Geochemistry," Springer-Verlag, Berlin, 1969–1974. A four-volume compilation of data on the distribution of the elements in geologic materials, including critical discussion of many of the data.

TWENTY-ONE

ISOTOPE GEOCHEMISTRY

Not only the distribution of the elements themselves but the distribution of their varieties or *isotopes* depends on changing geologic environments and can be used to draw inferences about geologic history. Some isotopes, furthermore, provide our most reliable method of measuring the ages in years of ancient rocks and minerals. We have mentioned isotopes briefly in previous chapters—the heavy isotopes of oxygen and hydrogen as clues to the source of hot-spring waters and ore-forming solutions, the isotopes of sulfur and carbon as indicators of the origin of salt-dome sulfur deposits, and the isotopes of strontium as a possible guide to the history of granitic magma. These are examples of a kind of study that has become one of the most active fields of geochemical research. It is time that we give isotopes and their uses a closer look.

21-1 KINDS OF ISOTOPES

All atoms of any one element have the same number of electrons outside the nucleus and the same number of protons within the nucleus (this is the *atomic number*), but their nuclei may contain differing numbers of neutrons. Thus the atoms may differ in mass (*atomic mass* or *atomic weight* = number of protons plus number of neutrons), and hence differ slightly in properties. An element, in other words, is not necessarily a simple substance in the classical sense, but may consist of a mixture of substances with slightly different chemical and physical behavior. The separate varieties of an element are its *isotopes*. Some elements in nature consist of a single isotope (e.g., F, Na, Co); most have at least two, and a few have eight or more (e.g., Sn, Xe, Te). Artificially, by bombarding nuclei with fast-moving particles, additional isotopes of all elements except hydrogen can be prepared.

The isotopes of an element have different chemical and physical properties, but only very slightly different. Chemical properties depend largely on nuclear charge and arrangement of electrons, especially the outermost electrons; since atoms of isotopes differ only in nuclear mass and have the same electron distribution, their chemical properties are almost identical. Properties are so very similar that ordinary chemical reactions, either in the laboratory or in nature, lead to practically no separation of isotopes. Only a few of the lighter elements (e.g., H, C, O, S) have isotopes with a sufficient percentage difference in atomic mass to cause detectable separation in nature. Laboratory separation of isotopes of the heavier elements can be accomplished, but only with extreme difficulty.

Some isotopes are *radioactive*, meaning that their atomic nuclei give out radiation and thereby change to nuclei of other elements. Various kinds of radiation are observed, the commonest being alpha particles (helium nuclei, consisting of 2 protons and 2 neutrons) and beta particles (electrons); either of these may be accompanied by gamma radiation (high-energy x-rays). Nuclei produced by radioactive decay are spoken of as *daughter products*, and are said to constitute *radiogenic* isotopes. Daughter products may be much more radioactive than their parents (e.g., ^{226}Ra, ^{222}Rn, ^{227}Ac), and the quantity of such isotopes in nature is maintained by a balance between their rates of formation and decay. Some radioactive isotopes are continually being formed by reactions of stable nuclei with high-energy particles in the atmosphere (^{14}C, tritium). Most artificially produced isotopes are radioactive, and a considerable quantity has been added to the surface of the earth in recent decades by nuclear reactors and nuclear bombs. The great majority of naturally occurring isotopes are not radioactive, and are spoken of as *stable* isotopes.

The accumulation of radiogenic isotopes by the decay of radioactive parents has provided a powerful means of establishing absolute ages of earth materials. Study of the radiogenic isotopes of strontium and lead has yielded also information on the differentiation of planetary substance, both in the early part of the earth's history and in later episodes. Distribution of stable isotopes has proved to be a useful tool for measuring temperatures and drawing inferences about sources of rocks and fluids. These three aspects of isotope geology are described in the following sections, but the treatment here of a complicated subject is necessarily brief. Fuller discussions can be found in the references at the end of the chapter.

21-2 RADIOACTIVE AND RADIOGENIC ISOTOPES

General

Because radioactive decay goes on at a constant rate and is unaffected by the temperatures, pressures, or chemical combinations found in geologic environments, it can be made to serve as a clock for the measurement of geologic time.

Once a radioactive isotope is imprisoned in the structure of a growing crystal, its atoms decay to atoms of its daughter element or elements at a fixed rate. The ratio of daughter to parent thus steadily increases, and measurement of the ratio gives a number from which the time elapsed since the crystal was formed can be calculated. One must assume, of course, that the mineral has not been altered since its formation; if any amount of either parent or daughter has been added to or subtracted from the mineral, the calculated age would be false. Commonly, freedom from alteration can be inferred from geologic evidence or from the consistency of ages calculated from more than one parent-daughter pair. When alteration is evident or suspected, say as a result of metamorphism during the rock's history, the isotope ratio may not give a valid age but nevertheless often provides useful information about the time and nature of the metamorphic event.

The following paragraphs describe first the formulas used in calculating ages for the ideal case of a single parent-daughter pair and then the four methods most commonly used for age measurements. These methods depend on the systems Rb-Sr, U-Th-Pb, K-Ar, and ^{14}C.

Equations of Decay

In any large number of atoms of a radioactive isotope, the decay follows a statistical rule: during any fixed time interval, a definite proportion of the atoms changes to the daughter product. Thus the rate of decay, or the number of atoms that decay, is simply proportional to the total number of parent atoms present:

$$-\frac{dP}{dt} = \frac{dD}{dt} = \lambda P \tag{21-1}$$

where P is the number of parent atoms remaining at any time t, dD/dt is the rate of formation of daughter atoms, and λ is a proportionality constant called the *decay constant*. Integration gives

$$-\ln P = \lambda t + C \tag{21-2}$$

C is the integration constant, which may be expressed in terms of the original number of parent atoms when $t = 0$:

$$C = -\ln P_0$$

Substitution of this value gives

$$\ln P - \ln P_0 = -\lambda t$$

Thus
$$P = P_0 e^{-\lambda t} \quad \text{or} \quad P_0 = P e^{+\lambda t} \tag{21-3}$$

The number of daughter atoms produced by the time t is the difference between P_0 and P:

$$D^* = P_0 - P = P e^{\lambda t} - P = P(e^{\lambda t} - 1)$$

Now at $t = 0$ some daughter atoms may have been present, and the total number existing at time t is therefore this original number plus those produced by radioactive decay. If the original number is D_0, the total will be

$$D = D_0 + D^* = D_0 + P(e^{\lambda t} - 1) \qquad (21\text{-}4)$$

Rearrangement gives an explicit formula for t:

$$t = \frac{1}{\lambda} \ln\left(\frac{D - D_0}{P} + 1\right) \qquad (21\text{-}5)$$

This is the time during which an amount of the daughter product represented by D has accumulated, leaving undecayed an amount of the parent represented by P. Values for D and P can be found by analyzing the rock or mineral in which the radioactive isotope occurs. If, then, we can also find values for λ and D_0, the equation will give us the age of the rock or mineral in years.

The decay constant λ is found by laboratory measurement of decay rate. For isotopes whose radioactivity is feeble, and whose decay is therefore very slow, precise determination of λ may be difficult. As methods of measurement improve, values of λ for the geologically important isotopes are refined, and calculated ages that depend on λ must be revised. Some of the discrepancies among ages published at different times can be ascribed to changes in the accepted values for decay constants.

More easily visualized than λ is a related quantity called the *half-life*, which is the time required for half of any given amount of an isotope to decay. If P_0 has decreased to $\frac{1}{2}P_0$ in a time $t_{1/2}$, we can substitute in Eq. (21-3):

$$\tfrac{1}{2}P_0 = P_0 e^{-\lambda t_{1/2}} \qquad \text{or} \qquad t_{1/2} = \ln\frac{P_0}{\frac{1}{2}P_0} \cdot \frac{1}{\lambda} = \frac{\ln 2}{\lambda} = \frac{0.693}{\lambda}$$

As an example, the measured value of λ for ^{238}U is 1.537×10^{-10} year^{-1}, and the corresponding half-life is 4.51×10^9 years. The radioactivity of this isotope is feeble, and its half-life is enormously long—roughly equal to the age of the earth.

The quantity D_0 in Eq. (21-5) (the amount of daughter product that may have been present initially when the parent isotope started decaying) can be estimated in various ways. If the parent isotope is known, on geochemical grounds, to be present in a mineral that ordinarily contains little or none of the daughter product, D_0 can be set equal to zero or given an arbitrary low value; or if a nonradiogenic isotope of the daughter product is present, its initial ratio to the radiogenic isotope can be estimated if the age of the rock or mineral is roughly known. Another procedure depends on the fact that Eq. (21-4) has the form $y = a + bx$: if D is plotted against P (y against x) for a number of rocks or minerals formed at the same time, the values should lie on a straight line whose intercept on the D axis is D_0 and whose slope is $e^{\lambda t} - 1$.

The fact that a plot of D versus P for samples of rock material formed at the same time is a straight line with slope $e^{\lambda t} - 1$ provides a graphical method for finding t from analyses for the radioactive isotope and its daughter. The line is called an *isochron* (because it represents samples of the same age), and the steepness of its slope is a measure of the length of time during which decay has been going on.

Use of the above formulas for calculating ages depends on two major assumptions: (1) The value of λ has not changed over geologic time. This seems a reasonable assumption since λ is a property of atomic nuclei, unaffected by extremes of temperature and pressure or by chemical combination in any ordinary geologic environment. (2) The rock or mineral has been a closed system since its formation; in other words neither parent nor daughter has been added to or lost from the system. This assumption is often questionable, and many anomalies of apparent ages can be traced to its failure. Ages based on the assumption when its validity is uncertain are commonly called *model ages*, meaning that the ages refer only to a particular model of geologic history which may or may not be consistent with other geologic evidence.

Rubidium-Strontium Dates

The heavier of the two Rb isotopes, ^{87}Rb, decays by emission of an electron to stable ^{87}Sr. For this pair of isotopes Eq. (21-4) would be

$$^{87}Sr = {}^{87}Sr_0 + {}^{87}Rb(e^{\lambda t} - 1) \tag{21-6}$$

Because ratios of isotopes are more easily measured than numbers of atoms, it is customary to rewrite an equation of this sort in terms of ratios by dividing through by the amount of a stable and nonradiogenic isotope. For this system the isotope selected is ^{86}Sr, so that Eq. (21-6) becomes

$$\frac{^{87}Sr}{^{86}Sr} = \left(\frac{^{87}Sr}{^{86}Sr}\right)_0 + \frac{^{87}Rb}{^{86}Sr}(e^{\lambda t} - 1) \tag{21-7}$$

($^{86}Sr_0$ is the same as ^{86}Sr, because this isotope is nonradiogenic and does not change with time). Equation (21-5) for this system is then

$$t = \frac{1}{\lambda} \ln \left[\frac{\dfrac{^{87}Sr}{^{86}Sr} - \left(\dfrac{^{87}Sr}{^{86}Sr}\right)_0}{\dfrac{^{87}Rb}{^{86}Sr}} + 1 \right] \tag{21-8}$$

The value of λ commonly used is 1.39×10^{-11} year^{-1}, corresponding to a half-life of 5.0×10^{10} years, but because the decay energy is low a precise value is difficult to obtain. For a mineral rich in ^{87}Sr (meaning a mineral of considerable antiquity that contained abundant Rb initially), the value of t is not very sensitive to the initial strontium isotope ratio $(^{87}Sr/^{86}Sr)_0$; a figure often chosen is 0.704, an average of ratios found in recent volcanic rocks that have presumably come directly from the earth's mantle. On the assumptions that this selection of an initial ratio is reasonable and that the mineral has remained a closed system with respect to Sr and Rb since its formation, Eq. (21-8) gives the mineral's age. It is particularly useful for Rb-rich minerals (lepidolite, muscovite, biotite, K-feldspar) of Paleozoic age or older. Rubidium forms no minerals of its own, but substitutes extensively for potassium in these minerals because of the close similarity of Rb^+ and K^+ (Sec. 20-3).

Unfortunately the assumption of a closed system, i.e., that a particular mineral has neither gained nor lost Rb or Sr, is often not valid because some migration of one or both elements has been caused by later heating of the rock. It is often safer to assume that the entire rock has remained a closed system, and thus to use rock analyses rather than mineral analyses for determining ages. In this calculation, however, we can no longer choose an arbitrary low value for $(^{87}Sr/^{86}Sr)_0$, because in the whole rock the concentration of Rb is small and the value of the present ratio $^{87}Sr/^{86}Sr$ may not be very different from the initial ratio. We resort, then, to the plotting of an isochron by obtaining isotope analyses from several rock specimens. If, for example, we are examining a suite of igneous rocks which on geologic grounds we know to be of the same age, different specimens will contain differing amounts of ^{87}Sr and ^{87}Rb, simply because the amounts of Rb originally present in the various minerals were different. Hence we can plot $^{87}Sr/^{86}Sr$ against $^{87}Rb/^{86}Sr$, and we should obtain a straight line because Eq. (21-7) has the form $y = ax + b$ when t is constant. The line can be extrapolated back to the $^{87}Sr/^{86}Sr$ axis to give the initial value of this ratio (Fig. 21-1) Furthermore the slope of the line is equal to $e^{\lambda t} - 1$, and from the slope the value of t (the age of the igneous suite) can be calculated.

Uranium-Thorium-Lead Dates

Natural uranium consists chiefly of the isotope ^{238}U, with a small amount (about 0.72%) of ^{235}U. Present in addition is a tiny quantity of ^{234}U, formed as one of the steps in the decay of ^{238}U. Both ^{238}U and ^{235}U decay by series of short-lived intermediate radioactive isotopes to form stable isotopes of lead, ^{206}Pb from ^{238}U and ^{207}Pb from ^{235}U. Once U is incorporated in a mineral and decay is well under way, each intermediate substance decays as fast as it forms, so that for dating purposes we may disregard the intermediates and write the overall decay reactions as

$$^{238}U \rightarrow {}^{206}Pb + 8He$$

$$^{235}U \rightarrow {}^{207}Pb + 7He$$

The helium represents the alpha particles emitted in some of the decay steps. The first reaction, as noted previously, is very slow (half-life of $^{238}U = 4.51 \times 10^9$ years); the second reaction is faster (half-life of $^{235}U = 0.71 \times 10^9$ years). Most of the earth's original ^{235}U has vanished after 4.55 billion years of planetary history, but about half of the original ^{238}U remains.

Uranium minerals generally contain at least a little ^{232}Th, which also decays to form an isotope of Pb:

$$^{232}Th \rightarrow {}^{208}Pb + 6He$$

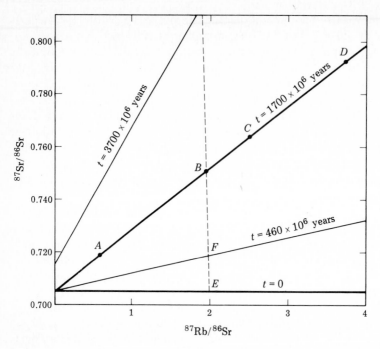

Figure 21-1 Rb-Sr isochrons. The heavy line is an isochron constructed from four hypothetical isotope analyses of a suite of igneous rocks represented by points A, B, C, D. Projection of the line to the $^{87}Sr/^{86}Sr$ axis gives an initial value for this ratio of 0.705, and the slope of the line ($= e^{\lambda t} - 1$) gives an age of 1700 m.y. The lower light line is an isochron for a younger suite of rocks with the same initial Sr ratio; the upper light line is an isochron for rocks formed 3700 m.y. ago with an initial ratio of 0.715. The horizontal line is an isochron at $t = 0$ for rocks with an initial Sr ratio of 0.705; in other words, it shows possible values of $^{87}Rb/^{86}Sr$ for such rocks at the time of their formation. For one such rock having an initial $^{87}Rb/^{86}Sr$ ratio of 2.0, the dashed line shows the changes in the two ratios over time, to point F after 460 m.y. and to point B after 1700 m.y. Note that the slope of this line must be -1, since ^{87}Sr is formed as fast as ^{87}Rb disappears. (*After Faure, 1977.*)

Thus complete analysis of a U mineral can in principle provide data for three independent age determinations: ^{238}U-^{206}Pb, ^{235}U-^{207}Pb, and ^{232}Th-^{208}Pb. The helium produced in all three reactions would seemingly be the basis for still another age estimate, but the helium method has proved unreliable because the gas escapes from minerals too readily.

The amounts of Pb, U, and Th are customarily expressed as ratios of the isotopes to the nonradiogenic isotope of lead, ^{204}Pb. Thus Eq. (21-4) for the decay of ^{238}U to ^{206}Pb is written

$$\frac{^{206}Pb}{^{204}Pb} = \left(\frac{^{206}Pb}{^{204}Pb}\right)_0 + \frac{^{238}U}{^{204}Pb}\left(e^{\lambda t} - 1\right) \qquad (21\text{-}9)$$

and the corresponding expression for Eq. (21-5) is

$$t = \frac{1}{\lambda} \ln \left[\frac{\frac{^{206}Pb}{^{204}Pb} - \left(\frac{^{206}Pb}{^{204}Pb} \right)_0}{\frac{^{238}U}{^{204}Pb}} + 1 \right] \tag{21-10}$$

Similar equations can be set up for the other two pairs of isotopes, and from each of the three equations corresponding to Eq. (21-5) an age can be calculated—provided that a reasonable value for the isotopic composition of any lead initially incorporated in the uranium mineral can be estimated.

To minimize possible inaccuracy due to initial lead, it is desirable to use a mineral which contains substantial amounts of U and Th but which is known to exclude most Pb during its crystallization. The mineral that best fits this description is zircon, and many U-Th-Pb dates are obtained using zircon separated from igneous rocks. Other minerals that give satisfactory ages include uraninite, sphene, apatite, and monazite.

The three independent ages calculated for a sample of zircon (or other mineral) should agree, provided that no U, Th, or Pb has been added to or subtracted from the mineral during its history. The ages often do not agree, suggesting that the system has not remained entirely closed. In this case a possible recourse is to calculate an age simply from the ratio of ^{207}Pb to ^{206}Pb. These two isotopes are produced at different rates from their parents, so that the ratio $^{207}Pb/^{206}Pb$ increases with time at a rate that can be calculated from the equations for the two U isotopes. This ratio is not sensitive to loss of Pb, because the two isotopes behave the same chemically and would be lost at the same rate. So if the discrepancy in calculated ages is due to loss of Pb, an age calculated from $^{207}Pb/^{206}Pb$ should be closer to the true age than any of the others.

An alternative and preferable method of obtaining the age of a mineral from discordant results is based on calculated values for the two ratios

$$\frac{^{206}Pb^*}{^{238}U} \quad \text{and} \quad \frac{^{207}Pb^*}{^{235}U}$$

where $^{206}Pb^*$ is the amount of radiogenic ^{206}Pb produced from ^{238}U (in other words, $^{206}Pb - ^{206}Pb_0$), and $^{207}Pb^*$ is the amount of radiogenic ^{207}Pb produced from ^{235}U. These two quantities increase with time at different rates, and the increase may be represented by a curve drawn by plotting one against the other (Fig. 21-2). For an ideal system in which no Pb is lost, calculated ages would be concordant and a point representing present values of $^{206}Pb^*/^{238}U$ and $^{207}Pb^*/^{235}U$ would lie on the curve—called a *concordia* curve (Wetherill, 1956). If Pb has been lost at some time in the mineral's history, so that the calculated ages are discordant, a point for the two ratios would fall below the concordia. Lead loss would presumably not be the same for all samples of the mineral, so that several points can be plotted showing ratios for the different samples. It is not immediately obvious that such points will lie on a straight line (called *discordia*), but

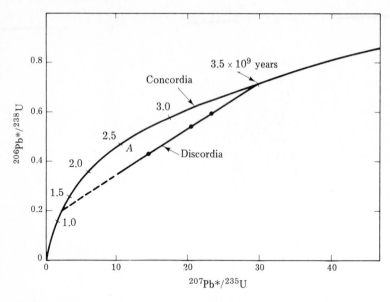

Figure 21-2 Concordia diagram for U-Pb. The concordia curve is the locus of points representing $^{206}Pb*/^{238}U$ and $^{207}Pb*/^{235}U$ for concordant U-Pb systems (systems that have remained closed to U and its daughters since their formation). The ages of such systems are indicated by the numbers along the curve (in billions of years); thus the two ratios for any concordant system 2.5×10^9 years old would be represented by point A. The discordia line is drawn through points plotted from (hypothetical) isotope analyses for samples of a mineral that have lost varying amounts of Pb. Extension of the line upward to its intersection with concordia gives the age of the mineral, in this case 3.5×10^9 years. The lower intersection is harder to interpret; if Pb loss occurred at only one brief period in the mineral's history, the point gives the time of that event, but the history is seldom so simple. *(After Wetherill, 1956.)*

this can be demonstrated mathematically. If the line is extended until it intersects the concordia at two points, the upper intersection gives the age of the mineral. A similar treatment, using the same concordia, permits calculation of useful ages from samples in which varying amounts of uranium have been lost during weathering.

Still another method of handling U-Th-Pb data is to construct isochron diagrams for whole-rock samples, like the isochrons for the Rb-Sr system (Fig. 21-1). Such diagrams for the U-Pb pair are generally not useful because U is too easily lost by oxidation and leaching from near-surface rock samples, but isochrons for Th-Pb and ^{207}Pb-^{206}Pb have proved satisfactory for age calculations.

Potassium-Argon Dates

The isotope ^{40}K, which makes up a minute fraction (about 0.01%) of naturally occurring potassium, undergoes several kinds of radioactive decay. The two principal kinds are decay to ^{40}Ca by electron emission (about 89% of the ^{40}K nuclei) and decay to ^{40}Ar by electron capture (about 11%). The decay to ^{40}Ca has little importance in dating because ^{40}Ca is the common isotope of calcium, and enough ordinary Ca is present in most K minerals so that the tiny amount added by radioactive decay would be undetectable. The decay to ^{40}Ar, on the other hand, is the basis of one of the most useful isotopic methods of age determination. The great advantage of the method is its applicability to very common and abundant minerals, since potassium is one of the major elements in the earth's crust; its major drawback is the fact that the radiogenic product is a gas, which may in part escape from the mineral in which it forms.

The total decay constant for ^{40}K may be written as the sum of the constants for the decay paths to Ca and Ar, and Eq. (21-4) for the decay to Ar alone is then

$$^{40}\text{Ar} = {}^{40}\text{Ar}_0 + \frac{\lambda_a}{\lambda} {}^{40}\text{K}(e^{\lambda t} - 1)$$

where λ is the total decay constant and λ_a the constant for decay to ^{40}Ar. Commonly used values are 5.305×10^{-10} year^{-1} for λ and 0.585×10^{-10} year^{-1} for λ_a, giving 0.110 for λ_a/λ. Since most K minerals have no original Ar, the value of ^{40}Ar$_0$ is generally zero, and the equation simplifies to

$$^{40}\text{Ar} = 0.110{}^{40}\text{K}(e^{\lambda t} - 1) \tag{21-11}$$

The equation giving the age explicitly [from Eq. (21-5)] is then

$$t = \frac{1}{\lambda} \ln \left[\frac{{}^{40}\text{Ar}}{{}^{40}\text{K}} \left(\frac{1}{0.110} \right) + 1 \right] \tag{21-12}$$

The age of a mineral calculated from this equation is the true age only if (1) no radiogenic Ar has escaped from the mineral since its formation, (2) no excess ^{40}Ar was present in the mineral when it was formed, and none was introduced later, and (3) no K was added or removed. The first of these provisos is generally the most critical.

Escape of Ar is possible from any mineral, but experience has shown that minerals differ greatly in their ability to retain the gas. Among common minerals the one that generally gives the most reliable ages is hornblende—which at first sight seems strange, because the K content of hornblende is very low. Biotite and muscovite retain Ar fairly well, but ages determined for micas are often somewhat lower than for hornblende in the same rock. Sanidine from volcanic rocks is highly retentive, but K-feldspar from plutonic rocks is not. Escape of Ar from any mineral is hastened by high temperature; the temperature at which rapid escape begins varies from one mineral to another, but for most minerals is of the order of a few hundred degrees. If an episode of metamorphism has temporarily maintained temperatures high enough for all Ar to escape from a rock, the ^{40}K geochronometer will be "reset," and age determinations will record the time of metamorphism rather than time of formation.

In addition to dating minerals, the K-Ar method can be used for dating some volcanic rocks, particularly basalts. The fine-grained aggregate of plagioclase and pyroxene in basalt contains only a little K, but the Ar produced by its decay is effectively retained by the two minerals. K-Ar ages for basalts have proved particularly useful in dating times of reversal of the earth's magnetic field. Lavas more felsic than basalt, especially if they contain devitrified glass, give less reliable ages because of Ar loss.

Carbon-14 Dates

Atoms of ^{14}N in the atmosphere are converted to ^{14}C by reaction with neutrons from cosmic-ray collisions. Most of the ^{14}C is quickly oxidized to CO_2, which disperses through the atmosphere. The ^{14}C is radioactive, decaying with a half-life of 5,730 years (to ^{14}N, by emission of an electron), and its amount in air is maintained at a constant value by the balance between rates of formation and decay. Carbon dioxide is consumed by plants in photosynthesis, and animals acquire ^{14}C both by eating plant material and by absorbing CO_2 from air and water. Thus organisms maintain a steady-state concentration of ^{14}C as long as life processes continue. When an organism dies the interaction with the atmosphere ceases, and the ^{14}C it contained at the time of death steadily decreases. The concentration of this isotope in the carbon of dead organic matter is therefore a measure of the time since death occurred. More broadly, the carbon in any compound formed in contact with air and then kept from further reaction with CO_2 can be dated by the proportion of ^{14}C in its isotopic makeup.

This method of age measurement differs from other isotope-based methods in that the activity of a particular isotope is measured rather than the ratio of concentrations of a parent and daughter. If the activity of C in organic matter (or another carbon compound) in equilibrium with atmospheric CO_2 is represented by A_0, and the activity of material that has not interacted with air for t years is represented by A, the equation for decay is

$$A = A_0 e^{-\lambda t} \tag{21-13}$$

The A and A_0 are measured as numbers of disintegrations per minute per gram of carbon. Rearrangement of the equation and insertion of a numerical value for λ gives a simple expression for t in years:

$$t = 19,035 \log_{10} \frac{A_0}{A} \tag{21-14}$$

Because of the short half-life of ^{14}C, its activity in dead organic matter remains measurable for only about 50,000 years. For dates within this period, dates that are especially important for archeological and anthropological studies, the method has proved extremely valuable.

Dates determined by ^{14}C are subject to minor corrections of several sorts. The dates can be valid only if the intensity of cosmic rays striking the upper atmosphere has remained constant (so that A_0 may be considered constant); since the cosmic-ray flux comes in large part from the sun, whose activity is known to fluctuate, and since it is influenced by the earth's magnetic field, which is by no means steady, the assumption of a constant rate of production of ^{14}C cannot be strictly correct. By comparing ^{14}C dates with dates determined in other ways it is possible to estimate the variations in cosmic-ray intensity over the last few thousand years, hence to correct ^{14}C dates for this period. Radiocarbon dates would also be affected by marked changes in the amount of organic activity on the earth's surface, or by the introduction of unusual amounts of inert CO_2 into the atmosphere. Such changes from natural causes would presumably be small, but the marked increase of "dead" CO_2 in air over the past century due to the burning of fossil fuel has produced a detectable decrease in the amount of ^{14}C incorporated by organisms during the twentieth century. On the other hand, since about 1945 considerable amounts of ^{14}C have been added to the atmosphere by the explosion of nuclear devices. Thus radiocarbon dates are in principle simple to obtain, but making them precise requires a good deal of adjustment.

21-3 STRONTIUM AND LEAD

Besides their use in dating rocks and minerals, isotope ratios for strontium and lead can give other kinds of geologic information. Both elements consist of radiogenic and nonradiogenic isotopes, and amounts of the radiogenic isotopes in the world as a whole have increased over geologic time. The ratio of radiogenic to nonradiogenic isotopes in any given sample of strontium or lead, however, will not have a fixed value, because it depends on the history of the sample—how much of the radioactive precursor was present in the sample originally, and how much of the radioactive element or Sr or Pb has been added to or removed from the sample at later times. Differences from the average ratio are very small, as is attested by the fact that atomic weights for the two elements are given in standard tables to two decimal places. But the differences are readily detectable, and isotope ratios for these elements have proved to be useful tools in studying the history of rocks and minerals that contain them.

Strontium

The isotope geology of Sr concerns the increases over time in the amount of ^{87}Sr produced by the radioactive decay of ^{87}Rb. The nonradiogenic isotope commonly used for comparison is ^{86}Sr, so that changes in composition are expressed as changes in the ratio $^{87}Sr/^{86}Sr$.

To study Sr compositions, we need to know the original value of $^{87}Sr/^{86}Sr$ in the material that made up the primitive earth. This figure must be obtained indirectly, because the earth's original material is nowhere preserved. The best approximation we have to such material is meteorites—although even meteorites may have had a complex history, so that their validity as samples of the original solar nebula is not entirely without question. Some meteorites, however, have an extremely low content of Rb, so that their Sr cannot have changed greatly since they crystallized from a melt. Analyses of many such meteorites agree on a figure of 0.699 as the most probable $^{87}Sr/^{86}Sr$ ratio for the earth's original Sr.

On the general assumption that the bulk of the earth's crust differentiated from the mantle fairly early in geologic time, we would expect that the behavior of Rb and Sr during the differentiation would be markedly different. The ion Rb^+ is large and singly charged, hence would be concentrated, like its close relative K^+, in the silica-rich differentiates that formed the crust. The smaller and doubly charged Sr^{2+} resembles Ca^{2+} in its geochemical behavior, and accordingly would distribute itself more evenly between crust and mantle. The greater concentration of Rb in the crust means that production of ^{87}Sr has been faster in crustal material than in the mantle, and samples of rock that has been part of the crust for most of its history should in general have higher $^{87}Sr/^{86}Sr$ ratios than samples that come from long residence in the mantle. This difference in isotope ratios, then, is a means for distinguishing igneous rocks that have formed by partial melting of crustal rocks from those that have their origin in differentiation or partial melting of mantle material.

The present $^{87}Sr/^{86}Sr$ ratio for mantle rock can be estimated by analyses of recent basalts and gabbros from oceanic environments, where direct origin from the mantle can be assumed and contamination by continental material would be absent or nearly so. Most such analyses lie in the range 0.702 to 0.706, with an average about 0.704; the fact that the values cover a considerable range is an indication that mantle material may not be entirely homogeneous. To produce a ratio of 0.704 from the original 0.699 during the 4.55×10^9 years of earth history would require an average Rb concentration in mantle rock of about 0.025%. Whether this amount has remained approximately constant through most of geologic time, or has decreased as additional crust was formed, is uncertain. For many purposes a simple linear interpolation between 0.699 and 0.704 gives a reasonable estimate of the Sr isotope ratio in the mantle at any time in the geologic past.

To draw inferences about the parentage of other igneous rocks, their isotope ratios are compared with these interpolated values for mantle material. The ratios needed are the *initial* $^{87}Sr/^{86}Sr$ values, the values at the time the rocks crystallized, rather than present-day values; these initial values can be obtained by constructing isochrons, as explained earlier (Fig. 21-1). If, for a suite of igneous rocks, the initial value is close to the corresponding mantle value (between 0.699 and 0.704), it is a reasonable inference that the magma was derived by differentiation or partial melting of mantle material (or of crustal material not long separated from the mantle). If it is much higher than mantle values, a conclusion seems safe that the magma originated by melting of crustal material, or else was contaminated by assimilating crustal rocks after it was formed in the mantle.

Studies of $^{87}Sr/^{86}Sr$ ratios are obviously pertinent to the old question about the origin of granite (Sec. 14-6). Measured values for various granites range from numbers identical with mantle values to numbers higher than 0.730. The high values clearly suggest derivation of the granite magma by partial melting of crustal material, and the low values indicate origin from the mantle or from material with a short history in the crust. A great many values lie between the extremes, and for these a variety of interpretations is possible. Granitic rocks of the Sierra Nevada, for example, show initial ratios ranging from 0.703 to 0.709. The low values could represent material directly from the mantle, and the higher numbers suggest varying amounts of contamination or varying proportions of crustal material in the original mixture that melted to form the granites (Kistler and Peterman, 1973). Since initial ratios for different crustal rocks cover a wide range, the amount of crustal material involved in forming the magma cannot be estimated without additional information.

One complication in the interpretation of Sr ratios is the mixing of crustal and mantle material that is a necessary consequence of subduction in the theory of plate tectonics. The material carried down into the mantle by a subducting plate consists of (1) a layer of the uppermost mantle, (2) a layer of basalt derived from the mantle at a spreading center, and (3) a layer of ocean-floor sediment. The basalt has Sr ratios similar to those of the mantle, but the sediment generally has considerable amounts of material with higher ratios derived from the continents. A major unanswered question is just how much of the sediment is carried down with the moving plate: the sediment may be in large part scraped off as the plate begins its descent, in which case there would be little mixing of crust and mantle material, or it may be moved down in considerable quantity and hence may contribute to the molten material formed by partial melting at the upper surface of the plate (Sec. 14-6). Recent measurements of initial ratios for andesites of island arcs, presumably representing melts formed along or above subducting plates, give generally low values, so low that the andesite magma was apparently generated largely by anatexis of basalt, with a maximum contribution of less than 10% of sedimentary material.

Besides helping to unravel the history of igneous rocks, Sr isotope ratios are useful as tracers for identifying sources of other geologic materials. A good example is the metal-bearing hot brine in geothermal wells near the Salton Sea in southern California. Originally the surprisingly high metal content (more than 80 ppm of Pb and 500 ppm of Zn) was thought to be derived from rhyolitic magma at depth, related to small rhyolite domes in the vicinity. Isotopic analyses for dissolved Sr and Pb in the brine, however, showed a much closer resemblance to ratios in the sediments through which the brine circulated than to those in the rhyolite. Very probably the brine derives its heat from the underlying magma, but at least a large part of the metal content comes from sediments rather than igneous material. (Doe et al., 1966).

Lead

The three heavier isotopes of Pb have increased through geologic time with respect to the nonradiogenic isotope ^{204}Pb, so that the ratios $^{206}Pb/^{204}Pb$, $^{207}Pb/^{204}Pb$, and $^{208}Pb/^{204}Pb$ for a lead sample should provide information about its history. The Pb isotopes and their radioactive parents are less useful than Sr and Rb in identifying possible sources of igneous rocks, because they are less different in geochemical behavior: Pb, U, and Th are all somewhat con-centrated in felsic magmas during igneous differentiation, so that no cleancut difference between U/Pb and Th/Pb in mantle material as opposed to crustal rocks would be expected. But in many other ways the Pb isotopes have been very informative.

As for Sr isotopes, the initial isotopic composition of Pb incorporated into the primitive earth can be estimated from meteorite analyses. The troilite (FeS) of iron meteorites contains Pb but practically no U or Th, so that its Pb presumably has the original isotopic composition of earth and meteoritic material unmodified by later addition of radiogenic isotopes. Analyses of such Pb do indeed show lower 206/204, 207/204, and 208/204 ratios than any terrestrial leads, and are commonly accepted as representing the primeval isotope distribution.

Stony meteorites, on the other hand, contain U and Th as well as Pb, and radioactive decay since the meteorites were formed has increased the amounts of the heavier Pb isotopes. Since the rates of decay are well known, data from the two kinds of meteorites permit calculation of the time when the meteorites, and presumably also the earth, were formed. This calculation leads to the generally accepted figure of $4.55 \pm 0.05 \times 10^9$ years for the earth's age. The calculation involves plotting an isochron (analogous to Sr isochrons, Fig. 21-1) for the $^{206}Pb/^{204}Pb$ and $^{207}Pb/^{204}Pb$ ratios of different meteorites, and finding the age from the slope of the line (Patterson 1956). Support for the assumption of a common age for earth and meteorites comes from a demonstration that similar ratios for an average sample of modern Pb (obtained from recent marine sedi-ments, which should represent a well mixed sample of earth materials) locate a point which lies on the isochron (Fig. 21-3).

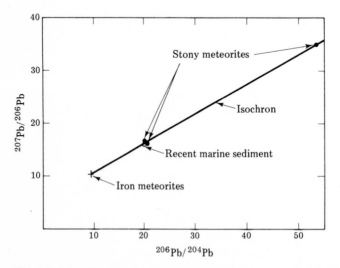

Figure 21-3 Lead isochron for meteorites and recent marine sediment. The slope of the line indicates an age of $4.55 \pm 0.07 \times 10^9$ years for the meteorites. *(After Patterson, 1956.)*

The ratios $^{207}Pb/^{204}Pb$ and $^{206}Pb/^{204}Pb$ should be usable for determining ages not only of meteorites but also of ordinary lead in earth materials. The Pb in a deposit of galena, for example, has presumably been out of contact with U or Th since the deposit was formed; before that time we may imagine that the Pb was disseminated in rock material, either in the mantle or in the lower crust, along with U and Th, and that the amounts of the three radiogenic isotopes increased steadily from radioactive decay. When the Pb was concentrated and segregated to form the ore deposit, its isotopic evolution ceased. The relation of the Th-derived ^{208}Pb to the other two isotopes would depend on the relative amounts of U and Th in the source material; but the relation between ^{207}Pb and ^{206}Pb, since both come from isotopes of the same element, would depend *only* on the length of time that decay had been going on (hence on the age of the deposit), and *not* on the amount of U present. In other words, the decay of ^{235}U and ^{238}U in the course of earth history has produced a steadily increasing ratio of radiogenic ^{207}Pb to ^{206}Pb, which at any one time should be the same for any U-bearing material. The increasing ratio, referred to ^{204}Pb, can be plotted as a "growth curve" showing the development of radiogenic Pb in the earth. We have here a possible method of dating an ore deposit simply from the isotopic composition of its Pb, without any reference to radioactive parents. We need only measure $^{207}Pb/^{204}Pb$ and $^{206}Pb/^{204}Pb$ for an ore sample, subtract from them the ratios in primeval Pb (which would make up part of the ore) to obtain the amounts due solely to radioactive decay, and then compare these ratios with the calculated growth curve to find the date when the ore was formed.

This procedure can give an accurate age only under rigidly prescribed conditions. In the source material from which the Pb came, U must have been quietly decaying since the earth was formed, without any disturbance that would add to or decrease the amount of either U or Pb; then in a short time the Pb must have been separated cleanly to form the ore; and afterward the Pb must have remained isolated from any contact with U or additional Pb. Obviously such special conditions are seldom realized in nature. Because this method of dating depends on an idealized model, the numbers obtained are *model ages*, which may or may not have geologic significance.

Out of the many ore deposits for which Pb-isotope ages have been obtained, only a few give model ages that are probably close to real ages. Indications that conditions of the model were approximately fulfilled are constancy of isotope ratios in different samples from a deposit, agreement of ages calculated for different samples, and general agreement of the model age with ages calculated in other ways for the deposit and its surroundings. Pb from such deposits is often referred to as "ordinary lead," in contrast to "anomalous lead" in deposits for which the model ages are in apparent conflict with geologic evidence.

Anomalous leads may show model ages either greater or less than the true age. If the age is apparently too high, meaning a $^{207}Pb/^{206}Pb$ ratio that is too low, an obvious possibility is that the Pb has been remobilized from an earlier accumulation, so that the calculated age represents the time of this earlier event. Low ages are harder to explain, but suggest changes in amounts of U and Pb in the source material at some time, or more probably at several times, before the Pb was segregated into the ore. If isotope ratios are obtained for many samples of anomalous lead from a given deposit, they can often be handled mathematically to obtain a probable age and to draw inferences about the history of the ore, just as discordant ages for samples of a U-bearing mineral can often be treated to reveal events in the mineral's past.

Study of Pb-isotope ratios is not limited to the lead of ore deposits, but can be extended to the small amounts of lead present in many ordinary rocks. Such studies have been especially provocative for the Pb in recent volcanic rocks from oceanic islands: the unexpected variation in isotope ratios both from one island to another and among the rocks of a single island is good evidence for considerable heterogeneity of upper mantle materials and for a multistage history of much of the lead.

21-4 STABLE ISOTOPES OF LOW ATOMIC NUMBER

General

Elements with isotopes suitable for dating, except for carbon, show practically no isotope separation as a result of ordinary geologic processes. For example, we may confidently assume that U or Sr removed from a rock by weathering or hydrothermal alteration has the same proportion of isotopes as in the rock from which it came. Again, the reliability of the $^{207}Pb/^{206}Pb$ method of dating altered rocks depends on the fact that the ratio of these isotopes does not change in response to any geologic accidents that may befall the original rock material. For a few elements of low atomic number, however, whose stable isotopes have a large *proportional* difference in atomic mass, detectable changes in isotope ratios may result from very ordinary geologic processes. By accurate measurement of isotope ratios, conclusions can be drawn about the conditions and reactions that produced the changes.

The difference in properties between two isotopes that may lead to slight separation is largely a result of the different frequencies of vibration of heavy and light atoms in a molecule or crystal structure. Atoms of a light isotope vibrate with higher frequencies, hence in general are less strongly bonded to other atoms, than atoms of a heavy isotope. The differences in bond strength are appreciable only for isotopes whose atoms have a large relative difference in mass: thus detectable separation might be expected for hydrogen and its compounds, since the stable hydrogen isotopes differ in atomic mass by a factor of 2 (^1H and ^2H), but would not be expected for tungsten, whose heaviest and lightest naturally occurring isotopes have a relative mass difference of only 1.03 (^{180}W and ^{186}W). This means that separation of stable isotopes is important in geologic environments only for elements of low atomic number. The most important of these elements are:

Hydrogen: Two stable isotopes, atomic numbers 1 and 2, symbolized either ^1H and ^2H or H and D (the D standing for deuterium). A third isotope, ^3H or T (tritium), also exists in nature, but is radioactive with a short half-life (12.26 years).

Carbon: Two stable isotopes, ^{12}C and ^{13}C, plus the radioactive isotope ^{14}C.

Oxygen: Three stable isotopes, with atomic numbers 16, 17, and 18. The proportion of ^{17}O is very small, and ordinarily only variations in ^{16}O and ^{18}O are considered.

Sulfur: Four stable isotopes, with atomic numbers 32, 33, 34, and 36. Commonly only variations in ^{32}S and ^{34}S are studied, because amounts of the others are small.

Differences in vibrational frequencies of particles of different masses become smaller at higher temperatures, and the separation of isotopes is correspondingly less pronounced. The ratio of isotopes of one of the light elements in a mineral, or the distribution of isotopes between two minerals that have formed at the same time, is therefore a possible measure of geologic temperatures. Some isotope distributions have been found to be characteristic of particular geologic environments or processes, so that measurements of isotope ratios can give clues about sources of minerals and fluids. For these reasons the study of stable isotopes has become a valuable adjunct to many geologic investigations.

For simple molecules, especially molecules of gases, the equilibrium distribution of isotopes between different substances can be calculated from experimentally determined vibrational frequencies by the methods of statistical mechanics. Approximate calculations of this sort have been carried out for liquids and solids also, but cannot be made quantitative because of the complex effects of adjacent particles on vibration frequencies. The calculations are sometimes useful, but conclusions about isotope separation are based largely on empirical data rather than theoretical calculations. Such experimental data have been obtained in large quantity, so that accurate predictions can be made about the degree of separation to be expected in many geologic situations.

Three mechanisms of isotope separation can be distinguished:

1. Mechanisms depending on physical properties, for example evaporation or diffusion. Evaporation of water, for example, leads to concentration of the light isotopes ^1H and ^{16}O in the vapor phase and of the heavy isotopes in the liquid, because H_2O molecules containing the light isotopes move more rapidly and thus have a higher vapor pressure.

2. Exchange reactions resulting in isotopic equilibrium between two or more substances. For example, if CO_2 containing only ^{16}O is mixed with water containing only ^{18}O, exchange will take place according to the reaction

$$\tfrac{1}{2}C^{16}O_2 + H_2{}^{18}O \rightleftharpoons \tfrac{1}{2}C^{18}O_2 + H_2{}^{16}O$$

until equilibrium has been reached among the four species. At equilibrium the ratio $^{18}O/^{16}O$ will be nearly the same in CO_2 and H_2O, but not quite, because bond strengths in the two compounds are different.

3. Separation depending on reaction rates. This is similar to mechanism 1, but refers to rates of chemical rather than physical processes. It is particularly important for reactions catalyzed by bacterial activity. In the bacterial reduction of sulfate ion, for example, the production of sulfide (S^{2-}, HS^-, and H_2S) is faster for the light isotope of sulfur than for the heavy isotope, so that ^{32}S becomes concentrated in the sulfide species and ^{34}S is enriched in the residual SO_4^{2-}.

Nomenclature

Regardless of mechanism, the extent of isotope separation can be expressed by a ratio called the *fractionation factor*:

$$\alpha = \frac{R_A}{R_B} \tag{21-15}$$

where R_A is the ratio of concentrations of the heavy to the light isotope in phase A ($^{18}O/^{16}O$ in liquid water, for example), and R_B is the same ratio in phase B ($^{18}O/^{16}O$ in water vapor). If equilibrium is established between water vapor and liquid at 25°C, the value of α is about 1.0092. Similar numbers, very slightly greater or very slightly less than 1, are obtained for other examples of isotope separation.

Because differences among such numbers are small and hard to visualize, another kind of symbolism is commonly used to describe isotope separations. The isotope ratio in a given sample is compared with the ratio in a standard by the expression

$$\delta_{\text{heavy}} = \frac{\text{heavy/light in sample} - \text{heavy/light in standard}}{\text{heavy/light in standard}} \times 1000\%_0 \tag{21-16}$$

For water, as an example, the usual standard for both O and H is a sample of sea water, Standard Mean Ocean Water or SMOW; hence ^{18}O for any other water sample is

$$\delta^{18}O = \frac{(^{18}O/^{16}O)_{\text{sp}} - (^{18}O/^{16}O)_{\text{SMOW}}}{(^{18}O/^{16}O)_{\text{SMOW}}} \times 1000\%_0$$

The symbol $\%_0$ stands for per mil, or parts per thousand. A positive value of $\delta^{18}O$ indicates enrichment of a sample in ^{18}O relative to SMOW, and a negative value indicates depletion. Most samples of fresh water have negative values of $\delta^{18}O$ (because the light isotope is concentrated in vapor escaping from the sea), ranging down to about -60 per mil; oxygen in the air has a high positive value, $+23.5\%_0$, and CO_2 in air a still higher value, $+41\%_0$. The relation between δ and α is given by the expression

$$\alpha = \frac{R_A}{R_B} = \frac{\delta^{18}O_A + 1000}{\delta^{18}O_B + 1000} \tag{21-17}$$

where A and B refer to different samples of an oxygen-containing compound. For example, in the condensation of water vapor to liquid a typical pair of values might be $\delta^{18}O = -5\%_0$ for the liquid and $-14\%_0$ for the vapor; the corresponding fractionation factor would be

$$\alpha = \frac{-5 + 1000}{-14 + 1000} = 1.0092$$

A few examples of geologic problems in which isotope studies have proved useful are described in the following paragraphs.

Oxygen

When calcium carbonate precipitates from seawater under equilibrium conditions, exchange of oxygen isotopes between $CaCO_3$ and H_2O results in an equilibrium that can be expressed

$$1/3\ CaC^{16}O_3 + H_2{}^{18}O \rightleftharpoons 1/3\ CaC^{18}O_3 + H_2{}^{16}O$$

The equilibrium constant for this reaction is

$$K = \frac{([CaC^{18}O_3]/[CaC^{16}O_3])^{1/3}}{[H_2{}^{18}O]/[H_2{}^{16}O]}$$

which is equal to the fractionation factor for isotope distribution between $CaCO_3$ and H_2O:

$$\alpha = K = \frac{{}^R CaCO_3}{{}^R H_2O}$$

The value of α at 25°C from experiment is 1.0286, and the corresponding $\delta^{18}O$ value referred to a seawater standard is $+28.6\%_0$ [Eq. (21-17)]. The fractionation factor depends on temperature; since the ${}^{18}O/{}^{16}O$ ratio for seawater is approximately constant, this means that the isotope composition of the oxygen in $CaCO_3$ is also a simple function of temperature. Measurement of the ratio in calcareous sediments has given a detailed record of fluctuations in ocean temperature, and hence presumably of atmospheric temperature, over the last few hundred thousand years. The measurements depend on assumptions that (1) the ${}^{18}O/{}^{16}O$ ratio of seawater has been constant (which is not quite true, but probable variations can be at least roughly estimated), (2) $CaCO_3$ secreted as shell material by various organisms has oxygen of the same isotopic composition, and (3) the $CaCO_3$ has not changed in isotope composition since it was deposited. The third assumption would not be valid for $CaCO_3$ that has recrystallized since deposition, so that unaltered shells provide the best material for temperature measurements. Temperatures obtained for $CaCO_3$ older than Pleistocene become increasingly uncertain because of questions about the validity of these assumptions.

The isotopic compositions of other oxygen-containing minerals likewise are temperature dependent, and the amount of change with temperature varies from one mineral to another. If two minerals are formed at the same time, with the same access to a source of oxygen, the distribution of light and heavy oxygen between them should thus give a measure of the temperature of origin. Temperatures indicated by distributions between representative mineral pairs are shown in Fig. 21-4. This method of geothermometry has the great advantage over most other methods that the results are not influenced by pressure, because pressure has no appreciable effect on isotopic equilibria. Success of the method depends on the assumptions that isotope equilibrium was maintained during formation of the minerals, and was not altered by isotope exchange later. The assumptions do not always hold, but enough $^{18}O/^{16}O$ temperatures have been checked with temperatures obtained in other ways to indicate general reliability of the results, especially for mineral pairs in igneous and metamorphic rocks.

Besides their use in geothermometry, oxygen isotope ratios in combination with hydrogen ratios have proved spectacularly useful as indicators of the source of hydrothermal fluids (Secs. 15-5 and 16-5). The long-standing argument among economic geologists as to a magmatic source versus a meteoric source for the

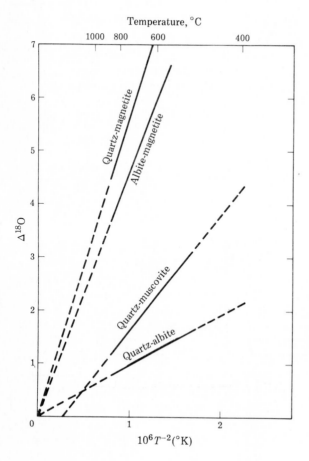

Figure 21-4 Variations with temperature of ^{18}O distribution between mineral pairs. The vertical axis shows values of $\Delta^{18}O$, which are the differences in δ values for the various mineral pairs. Thus for the quartz-magnetite pair, a difference in $\delta^{18}O$ values of 5‰ indicates a temperature of about 800°C for isotopic equilibrium. (*Plotted from data of Bottinga and Javoy, 1975.*)

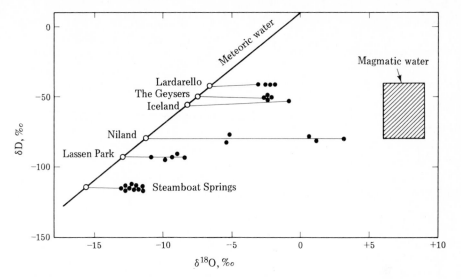

Figure 21-5 Values of δD and $\delta^{18}O$ for meteoric water, magmatic water, and several hot-spring waters. The diagonal line shows the relation between deuterium and oxygen 18 for many samples of water from rivers, lakes, rain, and snow; points for individual samples are not shown. The open circles on this line are average values for the meteoric water in the different hot-spring areas. The solid dots represent analyses of hot-spring waters, and the light lines connecting them show the trend of isotope variation due to exchange of ^{18}O with adjacent rocks. The box for magmatic water encloses the area where analyses of biotite and hornblende from igneous rocks cluster. (*After Craig et al., 1956; magmatic water from Taylor, 1974.*)

water in ore-forming solutions is in large degree settled: both kinds of water play a role, but in very many deposits the role of meteoric water is dominant.

The evidence for this conclusion is based on differences in the ratio

$$\frac{\delta D}{\delta^{18}O}$$

between water contained in the minerals of igneous rocks and water in various solutions obtained near the earth's surface. In minerals of unaltered igneous rocks such as biotite and hornblende the ratio has a fairly restricted range, as shown in Fig. 21-5. Meteoric water, on the other hand, shows a wide range of values representing the extent of isotope separation during successive episodes of evaporation and condensation of water originally evaporated from the sea. Because hydrogen and oxygen have been affected simultaneously, the points for pairs of δ values for the most part distribute themselves along a straight line; in general the highest negative δ-values for both D and ^{18}O are obtained for samples from high latitudes, where the prevailing low temperatures lead to large fractionation factors between liquid and vapor. Besides the line for meteoric water samples and the box for magmatic water, Fig. 21-5 shows values obtained for waters from hot-spring areas, in several of which deposition of sulfides is known to occur. Such measurements in a given area fall close to a horizontal line, indicating that the hot water contains an excess of ^{18}O over the meteoric water of the same region, but roughly the same deuterium content. This suggests that the hot water has reacted to varying degrees with the rocks through which it has traveled, exchanging some of its ^{16}O for ^{18}O from the minerals of the rock; a similar exchange of hydrogen isotopes is insignificant because most minerals contain only a little of this element. If the spring waters contained any considerable amount of magmatic water in addition to meteoric water, the mixing should be represented by values along lines connecting the meteoric-water points with the composition of magmatic water—which the diagram does not show. Thus the horizontal lines are good evidence for the derivation of most hot-spring waters chiefly from rainfall in the vicinity of the springs.

Further support for the meteoric origin of solutions responsible for many ore deposits has come from isotope analyses of minerals in alteration zones around ore accumulations (Taylor, 1974). The analyses commonly show lower $^{18}O/^{16}O$ ratios than those for unaltered rocks, and the difference often increases with the extent of alteration. This is not proof that the added water is all or mostly meteoric, because other sources of water are possible—connate water imprisoned in adjacent sediments, for example, or water released during low-grade metamorphism. But at least alteration entirely by magmatic water is ruled out. The extent to which meteoric water has taken part in alteration and ore deposition varies widely from one deposit to another. The significant result of isotope study is not to exclude magmatic water as an often important agent of ore formation, but simply to show that meteoric water (and perhaps other kinds of water) is far more important than was commonly thought by students of ore deposits a few decades ago.

Sulfur

Sulfur isotope concentrations are expressed by $\delta^{34}S$ values, analogous to the expressions for $\delta^{18}O$. The commonly used standard is the sulfur in the troilite of meteorites, on the assumption that the composition here should be close to that of the original sulfur in the material that formed the earth. Recorded values of $\delta^{34}S$ range from about $-30\%_{oo}$ for some samples of coal and H_2S to values more positive than $+60\%_{oo}$ for the sulfur in some sulfate minerals.

The sulfur of sulfide minerals shows slight differences in isotope ratios, the enrichment of ^{34}S in minerals precipitated from the same solution depending on relative bond strengths. The degree of enrichment depends also on temperature, and the change with temperature is different for different minerals. Thus the isotopic composition of sulfur in coexisting sulfides can serve as a measure of formation temperatures, just as oxygen isotopes provide temperature estimates for minerals that contain oxygen. Checks of temperatures obtained with sulfur isotopes against those from filling of fluid inclusions have not been wholly satisfactory, but this method of geothermometry has great promise for future development.

The principal reaction leading to separation of sulfur isotopes in nature is the reduction of sulfate ion by anaerobic bacteria. The reaction produces H_2S or one of its ions, and in these substances ^{32}S is markedly concentrated. Because sulfur in near-surface environments undergoes frequent change between the oxidized and reduced forms, it can acquire a wide variety of $^{34}S/^{32}S$ ratios. Equilibrium would not be expected in environments at ordinary temperatures, because SO_4^{2-} is not reduced at an appreciable rate in the absence of bacteria. At higher temperatures reactions are faster and equilibrium is possible between sulfate ions and sulfide, either in solution or in the form of sulfide minerals. Whether isotope separation is a consequence of equilibrium at high temperatures or differential reaction rates at low temperatures, the heavy isotope is enriched in the oxidized forms and the light isotope in the reduced forms.

Because the sulfur of magmatic ores associated with mafic and ultramafic rocks shows a narrow range of isotope ratios close to the ratio in meteoritic sulfur, while sulfur that has undergone oxidation and reduction in near-surface processes shows a wide spread of values, it has long been assumed that a determination of $^{34}S/^{32}S$ in the sulfide minerals of an ore deposit could give information about the source of the sulfur. If values of the ratio were found to be widely divergent, so the argument ran, the sulfur must have taken part in near-surface reactions, hence was not derived directly from a magma. If, on the other hand, the ratios clustered in a narrow range near that of meteoritic troilite, there was a fair presumption of magmatic origin. The presumption was by no means certain, because sulfate and sulfide from the surface might have been mixed during ore formation to give isotope ratios approaching the primitive ratio. Divergent ratios seemed a sure indication of source, uniform ratios a somewhat ambiguous one.

Serious doubts about the validity of this reasoning have been raised by recent work of Ohmoto (1972), who has demonstrated that equilibrium isotope separation at the temperatures of sulfide ore formation depends on pH, oxygen fugacity, and ionic strength of the ore-forming solutions as well as on temperature and relative bond strengths. Thus at some values of acidity and oxidation potential sulfur derived from a magmatic source can acquire a considerable range of ^{34}S values, which would not reflect a history of near-surface oxidation and reduction. Although Ohmoto casts doubt on simple reasoning about the source of sulfur, his work suggests that measurements of sulfur isotope ratios may give other kinds of information about ore deposits, particularly estimates of the acidity and oxidation potential of ore-forming solutions.

Carbon

Relative concentrations of the two stable isotopes of carbon, ^{12}C and ^{13}C, are expressed by $\delta^{13}C$ values like the δ values for oxygen, hydrogen, and sulfur. The usual standard is the isotope ratio found in belemnites of the Cretaceous Peedee formation of South Carolina, abbreviated PDB. Relative to this standard most

$\delta^{13}C$ values for terrestrial materials are negative, ranging down to $-70\%_0$ for samples of methane, but high positive values have been reported for carbonates in some meteorites.

Many reactions can lead to slight separation of the carbon isotopes in nature, the most effective ones being, as for sulfur, oxidation-reduction processes in which the heavier isotope is concentrated in the more oxidized forms. Particularly important is the reaction of photosynthesis in the leaves of green plants, in which the carbon of CO_2 is reduced and incorporated into organic compounds. Because the carbon in organic matter is enriched to varying degrees in ^{12}C, an abundance of light and isotopically variable carbon in bituminous material of doubtful origin in ancient rocks can serve as an indication of its source in living organisms rather than in "juvenile" carbon gases from an igneous source. Similarly, the extremely low $\delta^{13}C$ in the $CaCO_3$ that forms the caprock of salt domes ($-36\%_0$, as opposed to values in the range 0 to -5 for most carbonate rocks) has been cited as evidence for the origin of this carbonate by oxidation of methane from organic material in the adjacent sediments (Sec. 10-4).

At higher temperatures, equilibrium fractionation of isotopes may occur between CH_4 and CO_2 (or carbonate ions formed by dissolution of CO_2). Ohmoto has noted that this isotope separation, like the fractionation of sulfur isotopes between sulfate and sulfide species, is sensitive to changes in pH, oxygen fugacity, and ionic strength as well as to changes in temperature. Evidently carbon isotopes as well as sulfur isotopes may prove useful as a tool for estimating the acidity and oxidation potential of ore-forming solutions.

Thus isotope ratios can be made to yield a variety of geologic information, from absolute ages to temperatures to sources of rocks and fluids. It is small wonder that the study of isotopes has attracted so much recent attention from geochemists.

REFERENCES AND SUGGESTIONS FOR FURTHER READING

Bottinga, Y., and M. Javoy, Oxygen isotope partitioning among the minerals in igneous and metamorphic rocks, *Reviews of Geophysics and Space Physics*, vol. 13, pp. 401–418, 1975.

Craig, H., G. Boato, and D. E. White, Isotopic geochemistry of thermal waters, *Proc. 2d Conf. Nuclear Processes in Geologic Settings, National Research Council, Nuclear Science Series Report 19*, pp. 29–38, 1956.

Dalrymple, G. B., and M. A. Lanphere, "Potassium-Argon Dating," W. H. Freeman and Company, San Francisco, 1969. An authoritative review of theory, methods, and results of K-Ar dating.

Doe, B. R., C. E. Hedge, and D. E. White, Preliminary investigation of the source of lead and strontium in deep geothermal brines underlying the Salton Sea geothermal area, *Econ. Geology*, vol. 61, pp. 462–483, 1966.

Faure, G., "Principles of Isotope Geology," John Wiley & Sons, New York, 1977. An excellent treatment of all aspects of the study of isotopes as applied to geologic problems, with a very complete bibliography. This is the standard intermediate-level reference work on isotope geology.

Kistler, R. W., and Z. E. Peterman, Variations in Sr, Rb, K, Na, and initial $^{87}Sr/^{86}Sr$ in Mesozoic granitic rocks and intruded wall rocks in central California, *Geol. Soc. America Bull.*, vol. 84, pp. 3489–3512, 1973. An interesting attempt to relate strontium isotope ratios in granitic rocks of the Sierra Nevada to the kinds of rocks which they intrude and from which they may have been formed by partial melting.

Lipman, P. W., B. R. Doe, C. E. Hedge, and T. A. Steven, Petrologic evolution of the San Juan volcanic field, southwestern Colorado: Pb and Sr isotope evidence, *Geol. Soc. America Bull.*, vol. 89, pp. 59–82, 1978. A good example of recent applications of isotope study to petrologic problems in mineralized areas.

Moorbath, S., and H. Welke, Lead isotope studies in igneous rocks from the Isle of Skye, *Earth and Planetary Science Letters*, vol. 5, pp. 217–230, 1969. Isotopes indicate that the igneous rocks of Skye contain mixtures of Precambrian and Cenozoic lead, and confirm evidence from strontium isotopes of the crustal origin of the granitic magma.

Ohmoto, H., Systematics of sulfur and carbon isotopes in hydrothermal ore deposits, *Econ. Geology*, vol. 67, pp. 551–578, 1972. A theoretical study of the effects on isotope distribution of the temperature, pH, oxygen fugacity, and ionic strength of hydrothermal solutions.

Patterson, C. C., Age of meteorites and the earth, *Geochim. et Cosmochim. Acta*, vol. 10, pp. 230–237, 1956. A classical paper in which the age of the earth and the isotope composition of primeval lead are derived from meteorite analyses.

Ralph, E. K., and H. N. Michael, Twenty-five years of radiocarbon dating, *American Scientist*, vol. 62, pp. 553–560, 1974. Corrections of radiocarbon dates on the basis of tree-ring counts, archeological data, and fluctuations in the earth's magnetic field.

Silver, L. T., and S. Deutsch, Uranium-lead isotope variations in zircons—a case study, *Jour. Geology*, vol. 71, pp. 721–758, 1963. Detailed experimental study of the isotope ratios of uranium, thorium, and lead in the zircons from a single body of granitic rock.

Taylor, H. P., Jr., The application of oxygen and hydrogen isotope studies to problems of hydrothermal alteration and ore deposition, *Econ. Geology*, vol. 69, pp. 843–883, 1974. One of Taylor's many papers on the use of isotopes in determining the source and history of hydrothermal fluids. This number of *Economic Geology* (No. 6 of vol. 69) has many other papers on the application of isotopes to the problems of metallic ores.

Wetherill, G. W., Discordant uranium-lead ages, *Trans. Amer. Geophys. Union*, vol. 37, pp. 320–326, 1956. A classical paper explaining the use of concordia diagrams in interpreting discordant uranium-lead ages.

TWENTY-TWO

HISTORICAL GEOCHEMISTRY

We have kept our eyes pretty well fixed, up to now, on chemical changes that occur in or on the earth's crust. By ordinary human standards this is the most important part of the universe, the part in which familiar geologic processes are taking place and have taken place for a long time in the past. But from other points of view the crust looks less significant. On a cosmic scale it is no more than the thin skin of a small planet, one of several planets circling a star which, in its turn, is only one of many billions of similar objects in the wide universe. It is time that we raise our eyes briefly from detailed study of crustal rocks and consider chemistry on this broader scale. We look now at the chemical relations between the crust and the great mass of the planet beneath it, and more sketchily at how this one planet fits into the external system of planets and stars. And as geologists we cannot limit ourselves to the cosmochemistry of the present time, but must ask how the chemistry of the earth and its neighbors may have changed down the ages.

Inquiry in these new directions, so different from the paths we have followed hitherto, requires new kinds of data. Many of the pertinent facts are drawn from the fields of astronomy and geophysics rather than from classical geology and chemistry. In a study of the universe as a whole the boundaries between the familiar sciences break down, and we find ourselves grasping for bits of evidence from any discipline that seems relevant. The data, as might be expected, are far from adequate to answer all the questions that would be of interest. Even more than in the obscure parts of crustal geochemistry, we must here depend on imaginative speculation to fill enormous gaps between well-established facts.

In so broad and speculative a field it becomes difficult to follow the pattern of our previous discussions, in which, for the most part, ideas have been traced back to well-known observations and principles from elementary science. If we cast loose from such moorings, cosmochemistry becomes a fabric of everchanging hypotheses that rest on assumptions and mathematical arguments which must be taken as items of faith by all but specialists. There is no harm, of course, in weaving such a fabric of hypotheses untestable by the ordinary reader; cosmochemistry, even in the most elementary terms, is a fine exercise for the imagination. Many popular accounts of the subject, based on imaginative appeal rather than strict examination of hypotheses and assumptions, are available in the current literature. To avoid duplicating such expositions, to maintain the purposes of a textbook, and to keep the length of the discussion within bounds, we shall pass lightly over the more speculative aspects of the subject and pay particular heed to the nature of the basic data.

22-1 THE COMPOSITION OF THE EARTH

Are the rocks of the earth's crust an average sample of the material that makes up the universe as a whole? This looks like a simple question, since it requires only a comparison of two analyses.

To obtain the analyses, however, is far from easy. Even for the simpler of the two, the composition of the crust, we face the problem of how to weight different kinds of rocks in order to strike a reasonable average. The great bulk of the crust is clearly igneous rock, or metamorphic rock that was once igneous, and the most common igneous rocks by far are granite and basalt. Hence a combination of analyses of these two ordinary rocks would seem a reasonable way to get an average. But what particular combination is appropriate? For the part of the crust that makes up the continents, on the basis of geologic observation plus seismic evidence about lower levels in the crust, a one-to-one ratio of granite to basalt should not be far wrong. An average composition calculated in this manner is given in Appendix III and, in different units, as the first column of Table 22-1. If the oceanic crust were included, basalt would be more heavily weighted; this would tend to raise the figures for iron and magnesium and to lower those for the alkali metals.

To make a guess about the composition of the universe, we depend primarily on spectrographic study of stellar atmospheres. The relatively cool gases surrounding a star absorb light from the incandescent interior, and the absorption produces dark lines in the star's spectrum. The frequencies and intensities of the lines are a great storehouse of information about the composition and state of matter in the absorbing gases. The gases in turn are in rough equilibrium with the turbulent matter of the star's interior, and enough is known about the equilibrium relations so that an analysis of the atmosphere can be translated into figures for the composition of the star as a whole. Averages for many stars of different kinds, taken together with estimates of interstellar material, give results shown in the second column of Table 22-1. No great accuracy can be claimed

Table 22-1 Relative abundances of elements in the earth and the universe (atoms per 10,000 atoms of Si)†

	Continental crust	Universe	Meteorites	Whole earth	Moon
Rock-forming elements					
Si	10,000	10,000	10,000	10,000	10,000
Al	3,000	950	740	740	740
Fe	960	6,000	9,300	11,500	3,000
Mg	940	9,100	9,700	9,700	12,400
Ca	1,020	490	520	520	430
Na	1,040	440	460	460	20
K	540	30	40	40	(3)
Mn	18	70	70	70	(20)
Ti	104	20	20	20	(20)
Ni	13	270	450	750	
P	35	100	60	60	
Cr	19	80	90	90	(20)
Volatile elements					
H	1,400	4.0×10^8		84	
O	29,000	215,000	34,300	34,000	37,000
N	1	66,000		0.2	
C	18	35,000		70	
S	9	3,750	990	1,100	(130)
F	34	16		3	
Cl	4	90		30	
Inert gases					
He		3.1×10^7		3.5×10^{-7}	
Ne		86,000		12×10^{-7}	
Ar		1,500		$5,900 \times 10^{-7}$	
Kr		0.51		0.6×10^{-7}	
Xe		0.04		0.05×10^{-7}	

† *Sources:* Continental crust: Recalculated from data in Appendix III.

Universe: H. E. Suess and H. C. Urey, Abundances of the elements, *Rev. Modern Physics*, vol. 28, p. 53, 1956.

Meteorites: Recalculated from B. Mason, "Principles of Geochemistry," 3d ed., John Wiley & Sons, New York, 1966, p. 20.

Whole earth: Recalculated from Mason, *ibid.*, p. 52.

Moon: Recalculated from J. V. Smith, Development of the earth-moon system, in "The Early History of the Earth," B. F. Windley, ed., John Wiley & Sons, New York, 1976, pp. 3–19. Parentheses indicate values that Smith regards as less reliable than the others.

for such figures, of course, and improvements in astronomical techniques lead to frequent revisions in the estimates for various elements.

Imperfect as the analyses are, a comparison of the two columns brings out at once some important information about the earth's chemistry. The abundant elements of ordinary igneous rocks are also abundant elements in the universe as a whole; to this extent the earth's crust is a fair sample of cosmic material. Examined in detail, however, the figures for the lithophile elements in the upper part of Table 22-1 show many discrepancies. The differences would be lessened by including oceanic basalts in the crustal analysis, but by no means eliminated. The discrepancies are enormously larger for the elements in the lower part of the table, those that commonly form volatile compounds under terrestrial conditions; for some of these the crust is depleted with respect to the universe by several orders of magnitude. As a sample of cosmic material, therefore, the earth's crust leaves a good deal to be desired.

Perhaps our initial question was naïve. The composition of the crust may be vastly different from that of the earth as a whole, and quite possibly the overall composition of the planet would give a more meaningful comparison with cosmic analyses. How do we go about getting a set of figures for the entire earth?

Elementary geophysical measurements of the earth's mass, volume, and moment of inertia tell us that the density of the earth's materials increases downward from the surface, reaching a figure of the order of 15 g/cm^3 near the center. The measured speeds of earthquake waves give details about the distribution of materials in the interior, details that are embodied in the familiar model of a crust averaging 35 km thick under the continents and 6 km under the oceans, a mantle extending about halfway to the earth's center, a liquid core occupying about two-thirds of the remaining distance, and a solid inner core. The nature of the material in each of these regions is specified by seismic-wave measurements only with regard to its density and elastic constants; the limitations imposed by these properties could be satisfied by substances with a wide variety of compositions. To narrow our guesses about the actual chemistry of the interior, we need data from other sources.

A possible hint comes from a study of meteorites. These objects, falling to the earth from orbits around the sun, are plausibly interpreted as fragments of a vanished planet, or as residual chunks of the material from which the earth was originally constructed. On either interpretation the average composition of meteorites should resemble the average composition of the entire earth. To obtain a reasonable average for meteoritic material involves another problem in weighting, for meteorites show a great variety of compositions. At one extreme are the "irons," which are dominantly metallic iron alloyed with a few percent of nickel; at the other extreme are "stones," which consist chiefly of silicates and resemble ultramafic rocks in composition. Many meteorites include both silicates and native metal, and some have a sulfide phase (chiefly troilite, FeS) as well. Because irons are more durable than stones, and more apt to attract attention because they look different from ordinary rocks, museum collections of meteorites commonly have a disproportionate number of irons. From counts of observed falls, however, it seems clear that stones far outweigh irons in actual numbers striking the earth. Opinions vary widely as to the best ratio to use in calculating an average analysis. The numbers in the third column of Table 22-1 are an often quoted guess.

On the assumption that the earth's average composition is similar to that of meteorites, a plausible hypothesis about the chemistry of the interior can be framed by distributing the metal and silicates of meteorites so as to satisfy the requirements of density and elastic properties deduced from the behavior of seismic waves. This gives the generally accepted model of a core consisting largely of molten iron surrounded by a mantle made up chiefly of magnesium and iron silicates. Refinements of the model in almost infinite variety have been suggested, in an effort to secure a better fit between details of the seismic data and meteoritic compositions. An average composition for the entire earth, calculated from one possible variant of the model, is given in the fourth column of Table 22-1.

One other object whose composition should be related to that of the earth is the moon. So far our sampling of the moon is meager, and estimates of the moon's average composition based on these samples are widely divergent. The last column of Table 22-1 shows one recent estimate, obtained from the mineralogy of lunar samples and reasonable guesses about the mineral composition of the interior based on geophysical data. Most of the major elements in the earth are also important in the moon's substance, but there are notable differences in composition: the moon has less iron and more magnesium than the earth, and remarkably low concentrations of the alkali metals. If the earth and moon were both formed by accretion from planetesimals with the average composition of meteorites, the different gravitational pulls of the two bodies must have led to considerable sorting of the meteoritic material as the accretion went on.

Strong support for the iron-core, magnesium-silicate-mantle hypothesis for the earth's internal structure has come from experimental work with the extreme high pressures obtainable from shock waves. Sustained pressures in the laboratory at temperatures of a few thousand degrees are still limited to a couple of hundred kilobars, corresponding to depths of only a few hundred kilometers; but momentary pressures achieved in shock-wave experiments are far higher, making possible reasonable extrapolations to conditions in the earth's core and lower mantle. From such experiments it seems clear that the magnesium silicates of the mantle, which are probably ordinary pyroxenes and olivines immediately below the crust, change downward into denser forms in response to the enormous pressure at lower levels. Silicon assumes a coordination number of 6 rather than the usual 4, permitting a transformation of olivine into a denser crystal form with the structure of spinel; at still lower depths the compound may break up into MgO and SiO_2, each oxide in a densely packed structure. Experiments with metals show that iron is the only reasonably abundant substance with high-pressure properties corresponding to conditions in the outer core. This experimental work has put the familiar hypothesis of a "meteorite model" for the earth's interior on a firm basis, much firmer than the mere assumption of a common origin for planets and meteors.

The numbers in the fourth column of Table 22-1 show that the composition of the earth as a whole is closer to that of the universe (second column) than is the composition of the continental crust alone, but there are still wide discrepancies for many elements. Particularly striking is a comparison of the estimated abundances of the inert gases: concentrations of these elements in the earth are less than concentrations in the stars by factors ranging from 10^7 to 10^{14}. If the earth's composition ever resembled that of larger bodies in the universe, some process of differentiation during its history has changed somewhat the proportions of the lithophile elements and has caused tremendous losses of the more volatile elements.

It should be remembered that the numbers in Table 22-1 are guesses at the compositions of enormous masses of material for which, just in the nature of things, no accurate analysis is possible. Many similar analyses may be found in the literature, often showing large differences in the estimates for particular elements. Constructing an analysis of the earth, or of the universe, or even of accessible parts of the crust, is an operation that requires the fitting together of data from a great variety of sources and the weighing of many alternative hypotheses. It becomes a game, like trying to put together the pieces of an enormously complex puzzle, and agreement among the players can hardly be expected. No special virtue, and certainly no great accuracy, may be claimed for the analyses reprinted in Table 22-1. They are simply representative samples of current thinking, which we shall find convenient as a basis for speculating about the earth's past.

22-2 ORIGIN OF THE EARTH

Geochemists, in common with other geologists (and, for that matter, with most of humanity), are fond of speculating about how our planet came to exist. The factual basis for such speculations is extremely meager, since processes of progressive change that are observable today and that can be confidently extrapolated into the distant past are not very numerous. A few long-term changes might be suggested—the transport of dissolved material to the sea, the addition of gas to the atmosphere by volcanoes, perhaps the differentiation of continental rocks from the mantle—but we cannot be sure that these are not parts of long cycles rather than one-way changes. Chemically as well as mechanically the earth and its sister planets seem to form a remarkably stable system. About our only recourse is to try out various hypotheses of origin, and see which one is most successful in predicting a planetary system like the observed system. We cannot here look into all the hypotheses that have been proposed and discarded, but shall limit discussion largely to the chemical aspects of the currently most favored line of speculation.

We might ask first, How can we be sure the earth actually *has* an origin? Isn't it possible (despite philosophical difficulties) that the planets have *always* existed in very much their present form? One indication that the earth cannot be infinitely old comes from astrophysical data about the process of energy production in the sun and stars: radiation from the sun originates in nuclear reactions by which hydrogen is converted into heavier elements, and the current ratio of remaining hydrogen to other elements indicates that the sun cannot have existed as a star for more than 5 or 6 billion years. The presence in rocks of the earth of appreciable amounts of radioactive elements with long half-lives is another indication of finite age, for no known process is capable of producing these elements on or in the earth today. So the question about origin is not meaningless: the materials that compose the earth may have existed in some form through an infinite past, but the earth as a planet revolving around a luminous star must have originated at a definite point in time.

For reasons that are not quite clear philosophically, all hypotheses about the earth's beginning postulate a time when its substance was part of a homogeneous sample of average cosmic matter. Fashionable hypotheses of a generation ago put the homogeneous sample in the sun or a similar star; mechanical difficulties with generating a planetary system from a fully formed star have led recent speculations back to an earlier hypothesis, in which the earth's material was once part of a huge cloud of gas and small particles spread thinly over a volume larger than the present orbit of Pluto. The cloud at first was cold, and the nucleus of solid matter that grew into the primitive earth was formed as an aggregate of cold particles and larger fragments, often called *planetesimals.*

One justification for postulating cold accretion of planetesimals, rather than a fiery beginning for the earth as a large chunk of incandescent matter torn from the sun, goes back to the analyses in Table 22-1. As we noted earlier, a striking difference between average compositions of cosmic and terrestrial matter is the far greater amount of volatile elements in the former, a difference that is especially pronounced for the inert gases. The earth's gravitational pull is sufficient at present to hold all gases except hydrogen and helium; therefore the deficiency in volatiles cannot be explained on the basis of loss from the atmosphere in recent geologic time. If, however, the earth's substance were once spread over a large volume in the form of planetesimals, the gravitational attraction of the separate chunks would be too weak for the chunks to retain most volatile materials. The loss would be greatest for light elements like neon, helium, and hydrogen, in accordance with the figures of Table 22-1. A low temperature is indicated, furthermore, by the fact that the inert gases have been lost in much greater amounts than more active volatile materials like water, carbon dioxide, and ammonia. At temperatures of several hundred degrees the active volatile compounds would be just as easily lost as the inert gases, since their molecules would be free to move and their molecular weights are similar. At low temperatures, however, the active compounds would be partly retained as mineral hydrates and carbonates and by adsorption on solid surfaces. Detailed analysis of rates of escape and relative retention indicates that the original earth cannot have had a temperature of more than a few hundred degrees centigrade.

When did the accretion of the earth take place? An upper limit for the time is set by the calculated age of the sun, 5 or 6 billion years, since the sun formed also out of the original cloud, presumably by aggregation of material in the central part. A similar rough maximum age is suggested by the abundance of the lead isotope ^{207}Pb. This isotope is being produced today at a known rate by the radioactive decay of ^{235}U; if *all* the existing ^{207}Pb had been so produced, a time of 5.5×10^9 years would have been required. Since some ^{207}Pb was probably present in the lead of the original earth, this figure is an extreme maximum. A minimum figure is the age of the oldest rocks, roughly 3.8 billion years. A number probably closer to the time of actual accretion of planetesimals was obtained by Patterson (1956) by an analysis of lead isotopes from meteorites (Sec. 21-3). Patterson obtained for the age of the meteorites 4.55×10^9 years, a number that is generally considered to be also the age of the earth. Corroboration is provided by age measurements on samples of rock from the moon, the oldest of which show ages in the range 4.5 to 4.6 billion years.

Acceptance of these ages requires the assumption, of course, that the earth, moon, and meteorites were all formed at the same time. A skeptic could argue that the age determined for meteorites is strictly the time when lead was fixed in the sulfide phase of certain iron meteorites; the time of this fixation might be either earlier or later than the accretion of planetesimals to form the earth, depending on one's ideas about the origin of meteorites. The moon could conceivably be a body that had existed long before the earth appeared, or it might have formed later. Certainly it seems most reasonable to assume that all these objects originated roughly simultaneously, but there is still room for skepticism.

22-3 EARLY HISTORY

If the earth originated by the falling together of cold planetesimals, the low temperature could not have persisted very long. At least two sources of energy were available that must have produced heat in large quantities: the kinetic energy of moving chunks and particles as they came together and collided under the influence of gravity, and the energy of radiation from the radioactive elements present in the original materials. As long as the earth was small and most of the planetesimals were far apart, heat from both sources could be radiated away into space. But once the planet had grown to appreciable size, much of the heat would be retained, for rocks are notoriously poor conductors, and the temperature of the interior would steadily climb. How fast and how far the temperature rose remain debatable, for details of the process depend on the unknown rate of infall of planetesimals and the unknown original distribution of radioactive elements.

In the earth of the present day the release of gravitational energy can no longer be a significant source of heat, but radioactive decay should still be warming the interior. The heating is offset by loss of heat from the earth's surface, and it is interesting to speculate on the effectiveness of the balance: Is the temperature of the interior at present increasing or decreasing or remaining about constant?

From available data the question cannot be definitively answered. The rate of heat production by the four principal radioactive isotopes is well known:

$$^{238}U: \quad 0.72 \text{ cal/g year}$$

$$^{235}U: \quad 4.7 \text{ cal/g year}$$

$$^{232}Th: \quad 0.21 \text{ cal/g year}$$

$$^{40}K: \quad 0.21 \text{ cal/g year}$$

But the distribution of the isotopes in various earth materials is known only approximately, and the overall heat production depends critically on the precise concentrations. The amount of heat loss from the surface is known with fair accuracy, and also the conductivity of common rocks; but it is by no means certain that conduction is the only, or even the chief, mechanism of heat transfer from the interior. So neither side of the heat-balance equation can be expressed with the necessary accuracy to determine whether the earth is becoming hotter or cooler. One can rule out certain extreme assumptions: for example, if the entire earth had anywhere nearly as much radioactivity as do rocks of the crust and if conduction were the only means of heat transfer, then the earth would be so hot that the interior would be largely molten, which contradicts the evidence from seismic waves. By piecing together evidence of many kinds a plausible picture of the earth's present thermal state can be constructed (for example, Birch, 1965), but much room remains for argument over details.

If information about the present heat balance is unsatisfactory, precise data about the earth's thermal state in the past are practically nonexistent. One gross fact gives us a hint about conditions at some early period: the earth has an iron core, and it seems impossible for the core and mantle to have separated unless a considerable part of the interior has at one time or another been hot enough to melt iron. This does not mean that the entire earth was necessarily molten, but only that various levels of the interior, perhaps at different times, became hot enough for part of the material to liquefy. In a gross mixture of metal, sulfide, and silicates, at pressures like those in the interior, the lowest melting material would be a mixture of iron plus a little iron sulfide. This liquid can be imagined to aggregate in large masses and sink toward the earth's center. The time required for heating of an originally cold earth to the point where an iron-sulfur liquid would separate and form a core has been variously estimated; some estimates are as brief as a hundred million years. An alternative explanation for the origin of the core is a difference in the kinds of planetesimals that accreted to the growing earth at different times, the early ones being chiefly heavy metallic fragments, and the later ones containing larger proportions of silicate and oxide materials; thus the earth's chemical layering would be an original feature rather than one produced by later partial melting. Quite possibly both processes played a role.

The other major large-scale differentiation process that must, at least in part, go far back in time is the separation of continental material out of the mantle. The overall chemistry of the crust, with granitic rocks at higher levels, more mafic rock toward the base, and material of peridotitic composition in the mantle beneath, is strongly reminiscent of the chemical layering observed in large differentiated masses of unquestioned igneous origin, and also of differentiation processes observed in simplified laboratory systems. An easy assumption about the origin of the crust, therefore, is an imaginative reconstruction on a planetary scale of a process of crystal settling out of molten material. The heavier magnesium and iron silicates would drop into, or remain in, the mantle, and a scum of material rich in aluminum, silicon, and the alkali metals would rise to the surface. Such a process might take place piecemeal over the earth, in response to slightly higher temperatures at one place than another, so that the entire outer part of the planet need never have been molten at any one time. Whether the early crust was uniform over the earth or contained thicker parts which might be embryo continents we do not know, for none of the original crust has been preserved.

The period between the planet's accretion and the formation of the oldest existing rocks (4.5 to 3.8 billion years ago) must have been a tumultuous one. The craters that pock the moon's surface record intense bombardment by meteorites during this period, and presumably the earth, with its larger gravitational field, would have been pelted with bigger and more numerous projectiles. Unlike the moon, the earth's much-eroded surface retains no vestige of this early bombardment. Conceivably some of the apparent chemical heterogeneity of the earth's crustal rocks (and possibly the upper mantle also) is a heritage from the variety of meteoritic masses that were incorporated into the surface material; the odd concentrations of tin ores in some parts of the earth, and of iron and copper ores in other parts, would find a convenient explanation, albeit unprovable, in such a hypothesis. Even the present lopsided distribution of continents and ocean basins over the earth's surface, according to some speculations, might possibly go back to irregularities in the addition of material during this late stage in the planet's growth.

Even larger perturbations of the earth's early surface may have been caused by the moon. The origin of the moon is still an enigma, but very likely at one time it was either very close to the earth or possibly even a part of the earth. Ejection of the moon from the earth, or tidal deformation if the moon were nearby, would have disrupted large parts of whatever crust might have formed earlier.

By the end of the earth's first 700 million years, when meteorite infall slackened, crustal material presumably covered the entire planet. Most of it was basalt, as is shown by the dominance of mafic metamorphic rocks among the oldest dated varieties. Yet the existence of granitic rocks and ocean basins even at this early time is demonstrated by the presence of quartzite and a little marble in some areas where dates go back 3.5 to 3.8 billion years. How large the granitic continents were, and how they had originated, we can only guess. It is commonly thought that the granitic material was the end product of differentiation of mafic magma, that the early continents were small and scattered, that they were moved about by slow currents in the mantle beneath, and that they coalesced and broke apart repeatedly. The details we do not know, but certainly from very early times the earth's surface material was sharply separated, just as it is today, into continental masses of roughly granitic composition and oceanic areas underlain largely by basalt.

The obvious surficial differences between continental and oceanic materials imply differences in the materials or the behavior of the mantle underlying them. Such differences in the present world are strikingly suggested by simple measurements of heat flow from continental and oceanic areas. Since granite contains much higher concentrations than basalt of the radioactive heat-producing elements, one might expect that average heat flow from continental areas would greatly exceed heat flow from material beneath the oceans. Surprisingly, figures for the two kinds of area turn out to be approximately the same, averaging 1 or 2×10^{-6} cal/cm^2 sec. If the heat comes largely from radioactivity, this means the total quantity of radioactive elements at fairly shallow levels under the oceans must be the same as that under the continents. Now analyses show that the principal heat-producing elements (U, Th, K) are so concentrated in the siliceous rocks of the continental crust that almost the entire observed heat flow can be explained by production of heat in the crust alone, with only a small contribution from the underlying mantle. The basaltic crust of the ocean basins, on the other hand, is much thinner and contains far less radioactive material (compare, for example, the amounts of U, Th, and K in the analyses of granite and basalt in Appendix III). To account for the measured heat flow through the oceanic crust requires either a substantial contribution from radioactive elements in the mantle beneath, or else a different source of heat. Since the ultramafic rocks of the mantle have only traces of radioactivity, a large thickness of mantle material (estimated at about 400 km) would be needed to account for the observed heat flow. In other words the uppermost several hundred kilometers of the mantle under continents would have to be strongly differentiated, with radioactivity concentrated almost entirely in the overlying crust, while under the oceans the mantle would remain largely undifferentiated. Such a profound difference seems hardly credible.

The theory of plate tectonics suggests another source of heat: beneath the oceans are moving currents of hot mafic and ultramafic rock, originating at mid-ocean rises and carrying the oceanic crust on their surfaces. Perhaps the heat escaping from this subcrustal flow is comparable to the heat generated by the excess radioactivity of the continents, so that no great difference would be required in the composition of upper mantle material between continental and oceanic segments of the earth. Can we extrapolate backward, and suppose that continents and ocean basins have had this same relationship from earliest times? In other words, were the original continents also just rootless islands of silica-and-alkali-rich material, floating on and pushed by currents in the mantle underlying and surrounding them? This seems a reasonable guess, although there is no assurance that plate tectonics operated then on as large a scale as it does today. More likely, as indicated by the distribution of greenstone belts and granitic gneisses in Archean rocks of many areas, the original felsic masses were much smaller than present continents, and "spreading centers" where mafic material was emerging from the mantle were more numerous and more closely spaced.

22-4 LATER GEOCHEMICAL HISTORY OF
THE CRUST AND MANTLE

By a time well before 3×10^9 years ago, then, the major geologic processes that affect the solid earth today were already well established. Mafic lava came to the surface in places from deep in the upper mantle, presumably formed by partial melting of garnet peridotite; the solidified lava became oceanic crust, and was carried as part of moving lithospheric plates away from spreading centers; at other places the lava, together with sediments that had accumulated on its surface, moved downward with the plates into the mantle, eventually becoming hot enough to generate felsic and andesitic magmas by partial melting. The new magmas made their way upward to feed volcanoes and to form batholiths that added themselves to the edges of continents—or at least to the edges of granitic islands that were to become continents. Very likely the scale was different from that of modern times, as is indicated by the great abundance of greenstone (slightly metamorphosed mafic rock) and the relatively small masses of granitic rock in the early Archean. But the essential mechanism was the same: huge convection currents in the upper mantle, powered by the earth's internal heat, bringing new molten material to the surface under the oceans and returning cooled lava and sediments from the surface back into the mantle.

From a chemical standpoint the overall effect was differentiation of the mantle peridotite, first by partial melting to form basalt, then a more radical differentiation in the downward moving plates at continental margins to produce felsic lavas. Presumably the latter step of differentiation was aided by rock weathering at the surface and sorting of the weathered debris that was deposited on top of the lava, providing relatively felsic material to be carried down into the zone of magma generation. Formation of granite batholiths steadily added to the bulk of felsic continental crust. It was a slow and inefficient process, but in the course of time much of the silica, alumina, and alkali metals originally present in the mantle was concentrated in the continents, leaving somewhere below an increasing accumulation of refractory olivine and pyroxene crystals.

Of the many questions that could be asked about this process of global differentiation, the major one in a historical context is simply whether the differentiation has continued unabated through geologic time. Are the continents growing today, or was most of their growth completed far back in the Precambrian? More accurately, we should ask whether the total mass of continental crust is increasing, for the individual continents have moved about so much on the earth's surface, joining together and pulling apart in new patterns, that it is misleading to ask about the size of any one. In the geologic record we find considerable evidence for continuing continental accretion. As examples, the Precambrian granites of North America decrease in age from the oldest ones in the Canadian Shield to younger ones near the continental margins; and the continent was enlarged in the late Mesozoic by addition of the Sierra Nevada granite and related rocks in a subduction zone at its western edge. Such evidence for particular regions, however, does not prove that the overall mass of the continents has changed greatly, at least in later geologic time.

The question is difficult to approach on theoretical grounds, because too little is known about just what happens to subducting plates. The material of a lithospheric plate consists of a large thickness of peridotite in the uppermost mantle, surmounted by a layer of basalt and layers of sediment derived in large part from erosion of nearby lands. By partial melting, some of the substance of the basalt and sediments is converted into felsic magma. For the sediments, from a chemical standpoint, this is largely a process of reconstituting the igneous rocks from which they originally came; for the basalt it may involve production of new felsic material that had not existed before, hence would represent a net addition to the continental mass. On the other hand, a part of the basalt and sediments is probably carried with the moving plate down into the mantle, there to be melted and mixed with the peridotite. Because the sediments would have a relatively felsic composition, this mixing represents a net loss of continental material to the mantle. Now which is more important, the production of new felsic material by partial melting and differentiation of the basalt, or the consumption of felsic material by addition of sediments to the mantle? Almost certainly the former predominated in the very early part of earth history, but an answer is much less clear for the later part.

A different way to ask the question is to inquire whether plate tectonics during Phanerozoic time has been a cyclic process, merely forming new crust in one place and destroying it in another, or alternatively a one-way process, steadily forming new continental crust at the expense of mantle peridotite. If the process is cyclic, it can presumably continue indefinitely, or at least until the earth's heat is exhausted to the point where convection can no longer be maintained. If the process goes chiefly in the single direction of differentiation, it must terminate when the mantle is shorn of its felsic constituents and becomes a cumulus of olivine and pyroxene from which basalt can no longer be generated.

Geologic evidence about such questions is conflicting. If the global differentiation of peridotite goes on without a reverse process to counteract it, one might expect to see a long-term decrease in the amount of granite produced per million years, or perhaps a long-term change in the composition of granitic rocks. Neither kind of change is apparent in the geologic record. Possibly granitic rocks were produced in larger quantity during parts of the Precambrian than at any time since, but this is hard to establish convincingly. From the later Precambrian to the present, production of granite has been episodic, but there is no discernible general trend toward decreasing amounts. Regarding a change in character of granitic rocks, it is hard to predict even what sort of change should be expected. Should later granites be more mafic, because felsic constituents have been steadily extracted from the mantle? Or should granites become on the average more highly differentiated, meaning richer in potassium and closer to the low-temperature minimum of the quartz-alkali-feldspar system (Sec. 14-4), because felsic sedimentary material is recycled again and again through subduction zones? Whatever the expectation, long-term trends of any kind are hard to demonstrate. In a study of such possible trends, Engel et al. (1974) note that average K_2O/Na_2O ratios in granites and other rocks show a general increase during the Precambrian and into the Paleozoic, as if differentiation were becoming more extreme, but the trend is sharply reversed in the Mesozoic. The lack of definitive long-term trends in either amount or composition of granitic rocks supports the idea that global differentiation resulting from plate movements is cyclic rather than unidirectional.

On the other hand, the evidence cited previously for growth of continents by addition of granites at their margins would weight the argument in the other direction. Available information is simply not adequate to settle the matter. We can be sure that in the distant past continental material was generated by differentiation out of the mantle, and that small continents for a time grew into larger ones. We can be equally sure that the same sort of complicated differentiation process is at work today. But from well back in the Precambrian down to the present we find no persuasive evidence for long-term change in the amount or makeup of the products of differentiation, so it may be that the differentiation is effectively reversed during subduction and the average amount of felsic material in mantle peridotite has reached a steady state.

The overall chemistry of the solid earth shows little apparent long-term change, but this is not true of the surface veneer where weathering and erosion have converted original igneous material into sediments. Sedimentation may be regarded as a special kind of differentiation, in which silica is partly segregated into sandstone and chert, calcium and magnesium into carbonate rocks, potassium and aluminum into shale, sodium into evaporites, and so on. Some aspects of sedimentation are cyclic, but superposed on the cycles are very real long-term changes in the chemical nature of the sedimentary rocks that have been dominant at different periods in geologic history. Carbonate rocks, for example, are uncommon through most of the Precambrian, but become abundant toward its close and have continued to form in large amounts down to the present; banded iron formation (a rock with conspicuous alternating layers of chert and various iron minerals, most commonly magnetite or hematite) is largely restricted to a period between 2.5 and 1.8×10^9 years ago; placer deposits of easily oxidized minerals like uraninite (Blind River in Ontario and the Witwatersrand in South Africa) are restricted to the early Precambrian, and strongly oxidized sediments (" red beds ") become abundant only in the later Precambrian; and evaporites are not found until just before the Paleozoic. These changes in rock chemistry are intimately related to long-term changes in the atmosphere and oceans, which we shall consider in the next two sections.

22-5 HISTORY OF THE ATMOSPHERE

The scarcity of the inert gases on earth compared with their cosmic abundance, as noted in Sec. 22-2, is most plausibly explained by escape of gases from the growing earth at a time when its materials were still spread through such a large volume of space that gravitational attraction was much smaller than it is today. The fact that other gases with similar molecular weights, particularly H_2O, HCl, NH_3, CH_4, CO_2, H_2S, SO_2, were not lost in nearly the same proportion as the inert gases is good evidence that the original temperature of the earth was low, so that these substances could be retained in the planetesimals either as frozen particles or in the form of compounds. If this reasoning is valid, the cold primitive earth had very little atmosphere. As gravitational contraction and radioactive decay caused heating in the newly formed planet, gases would be released from their compounds and accumulate at the surface. Gravity would now be sufficient to retain all except the two lightest gases, hydrogen and helium. Thus it is a necessary consequence of the cold-accretion hypothesis, together with the observed distribution of the inert gases, that our present atmosphere is largely of secondary origin, having formed by "degassing" of the solid body of the planet. Presumably the water of the oceans would also have come to the surface as one of the released gases.

This conclusion would be strengthened if we could find independent evidence for the degassing. If such evidence can be found, further questions present themselves: did the degassing take place all at once, in some earlier and warmer stage of the earth's history, or was degassing a gradual process that is continuing at the present time? Did gas mixtures with the present average composition of the atmosphere rise directly out of the interior, or have the primitive gases undergone changes at the earth's surface?

Good evidence regarding the degassing process comes from some simple calculations about the amounts of carbon in various geologic materials (Rubey, 1951). The carbon of sedimentary rocks was nearly all derived from CO_2 that once existed in the atmosphere: the carbon of organic materials was fixed in organic compounds by photosynthesis (Sec. 11-2), and the carbon of precipitated carbonates represents atmospheric CO_2 added to seawater either directly by solution or indirectly by the respiration and decay of organisms. If we estimate the total amount of carbon buried in sedimentary rocks, therefore, we should get a figure indicating how much CO_2 has existed in the air at one time or another. Rubey's calculations indicate that the amount of buried carbon exceeds that in the present atmosphere, oceans, and organisms by a factor of about 600 times (Table 22-2). Even if some of the analyses and estimates of volumes on which the calculations rest are greatly in error, the figure would still be startlingly large. Beyond any reasonable doubt, the amount of carbon now in the air is only a tiny fraction of the amount that has existed at some time in the geologic past.

This result can be interpreted in several ways. One extreme possibility is that the atmosphere at some early period was very dense, consisting chiefly of CO_2 at a partial pressure of about 12 atm, and that the activity of plants plus the deposition of carbonate sediments has gradually reduced the amount to its present low value, 0.0003 atm. This is an unlikely hypothesis, for it would mean that we are living at the very end of the history of life on our planet. Some CO_2 is returned to the air by respiration, rock weathering, and organic decay, but the amount is too small to make up for the carbon that is being steadily removed as precipitated carbonates and organic matter buried with sediments. A rough calculation of the carbon balance indicates that CO_2 in air will fall to a level too low to support plant life

Table 22-2 Inventory of total carbon in the atmosphere, hydrosphere, organisms, and sedimentary rocks, expressed in units of 10^{20} g of CO_2 †

Atmosphere	0.023	
Ocean and fresh water	1.30	
Living organisms and undecayed organic matter	0.145	
Total for atmosphere, hydrosphere, and organisms		1.47
Carbonate rocks	670	
Organic carbon in sedimentary rocks	250	
Coal, oil, etc.	0.27	
Total for sedimentary rocks		920

† *Source:* Rubey, 1951, p. 1124.

within a few centuries, unless some other source of the gas is available. Since the geologic record gives indisputable evidence for the continuous existence of multicellular organisms for at least 600 million years, and of unicellular life for at least 2 billion years, the CO_2 content of air cannot have dropped far below its present figure for a long time. And it is scarcely believable that the present 0.0003 atm has been reached only now after 2 billion years of steady depletion.

An obvious additional source of carbon dioxide is volcanic activity. This gas is one of the most abundant emanations from practically all volcanoes, and from hot-spring areas that persist when volcanic activity has ceased (Sec. 16-2). Can we assume that volcanic CO_2 is a manifestation of the degassing of the earth—that it is coming out of the deep interior, and being added to the atmosphere for the first time? Almost certainly some CO_2 is recycled, in the sense that it is derived by decomposition of sediments heated by contact with molten lava; but its abundance in volcanic areas where sedimentary strata are lacking seems evidence that a part at least is indeed rising from the interior. It makes no difference that some of the carbon expelled by volcanoes has the form of CH_4 and CO rather than CO_2, because these gases would be quickly oxidized by atmospheric oxygen and so would contribute to the supply of CO_2. Since geologic evidence shows that volcanoes have been active all during the earth's history, a mechanism for continuous degassing and for maintenance of adequate CO_2 in the atmosphere seems assured.

In trying to interpret Rubey's results, can we then jump to the opposite extreme and suppose that a balance between life processes and volcanic activity has maintained an approximately constant partial pressure of CO_2 throughout geologic history? This is a question on which evidence is still indecisive. The continued existence of life establishes a fairly definite minimum of CO_2 pressures, but it is much less informative about a possible maximum. Some present-day land organisms would be adversely affected by a sudden increase in CO_2, and some marine forms would suffer from the resulting decrease in the pH of seawater. But the capacity of living things to adjust to gradual changes is enormous, and if the decrease in CO_2 was slow enough, there seems no reason to deny that its content in Paleozoic and Precambrian atmospheres could have been considerably greater than at present.

Nor is it easy to pick out inorganic reactions that might be sufficiently sensitive to CO_2 variations to leave a record in the geologic column. Extreme amounts, say a partial pressure greater than 1 atm, can probably be ruled out on the grounds that surface waters would become highly acid, chemical weathering would be greatly speeded, silicates would be largely broken down to free silica, and cations from weathering would accumulate in an enormously concentrated ocean. The sedimentary rocks of the Precambrian and Paleozoic are too normal to support such a conjecture. On the other hand, the huge amounts of silica deposited with iron minerals in the banded iron formation that is so characteristic of one period in the mid-Precambrian has been adduced as evidence for at least a moderately increased CO_2 content of the Precambrian atmosphere (Sec. 10-1). Limits are difficult to set, and present knowledge would permit wide fluctuations of CO_2 partial pressures in the geologic past. The fact of continuous degassing seems established, but its rate in different eras is unknown.

One other gas for which degassing seems clearly demonstrable is argon. In ordinary air this gas consists chiefly of the isotope ^{40}Ar, whereas the abundant isotope in stellar atmospheres is ^{36}Ar. Since ^{40}Ar is a product of the radioactive decay of ^{40}K, this peculiar fact of distribution suggests that most terrestrial argon has accumulated during geologic time from the decay of potassium in rocks. Such an assumption accounts also for the anomalously great total abundance of argon on the earth as compared with the other inert gases (Table 22-1). Now if ^{40}Ar comes chiefly from ^{40}K, it is a simple matter to calculate how much ^{40}K must have decayed to give the present amount of argon in air. The result shows that rocks near the surface, in fact all the rocks in the crust, contain only a small fraction of the necessary ^{40}K. Hence argon must have escaped to the surface from the decay of potassium at much deeper levels.

For atmospheric gases other than argon and carbon dioxide, no equally direct proof of large-scale emergence from the earth's interior has been suggested. It seems eminently reasonable to suppose that if two common gases have such an origin, others do also, but this is hardly a definite proof. In the absence of evidence to the contrary, and because the hypothesis is consistent with current ideas about the earth's origin, it is common to assume that most volatile materials now at the surface, including water, have come from the interior. Very likely the degassing was most rapid early in the earth's history, but the atmosphere and oceans are probably still receiving additions from present-day volcanic activity.

The composition of volcanic gases, together with analyses of gases obtained from heating rocks and meteorites, plus spectrographic data on the makeup of stellar atmospheres, gives us a basis for conjecture about the kinds of volatile substances that have escaped from the interior. However uncertain such guesses may be, they tell us at least that the average product of degassing bears little resemblance to the composition of the present atmosphere. Water vapor, hydrogen, carbon oxides, sulfur gases, hydrogen halides, nitrogen or ammonia, possibly hydrocarbons—some such mixture is the primary material that has come out of the earth and that presumably constituted the primitive atmosphere. How, from such a mixture, could the vastly different atmosphere of our present earth evolve?

The elimination of the sulfur and halogen gases is no problem, for they would dissolve in surface waters, react with rocks, and eventually appear for the most part as constituents of seawater. The really difficult question is the evolution of large amounts of free oxygen—a highly active gas that is found in comparable amounts nowhere else in the solar system. The answer that comes to mind immediately is the activity of plants, for oxygen is set free in large amounts by the process of photosynthesis in green leaves. This is a good answer, and almost certainly the correct one, for the maintenance of the present supply of oxygen in air. But it does not suffice for the historical question of how the primitive atmosphere of a lifeless earth was transformed into a mixture that permits higher forms of life to flourish.

The geochemical evolution of the early atmosphere is a subject of great current interest, not only to geologists but to astronomers and biologists as well. If free hydrogen was a prominent gas at the beginning, the earliest atmosphere would have consisted of strongly reduced compounds like methane, ammonia, and hydrogen sulfide; if most of the initial hydrogen escaped during the accretion of planetesimals, the first atmosphere would have had more free nitrogen and oxides of carbon and sulfur. In either case water would have been a prominent constituent. At high levels in the atmosphere H_2O molecules would be decomposed by the absorption of ultraviolet radiation from the sun, producing a little free hydrogen and oxygen; the hydrogen molecules would in part escape from the earth completely, but the oxygen would be retained. Thus a tiny concentration of oxygen would have been produced, and over long periods the gas mixture would have slowly become more oxidizing, but free O_2 is used up so rapidly in a variety of processes that its steady-state concentration could not have exceeded a very small figure. Somehow in the primitive ocean under this still largely reducing atmosphere the first living forms appeared. Experiments simulating supposed primitive-earth conditions have suggested several mechanisms by which the complex organic molecules necessary for life might have formed, but the story of the origin of life is still far from complete. The first organisms, presumably, were simple anaerobic forms, and only much later did complex plants evolve that could use the energy of sunlight to effect a reduction of atmospheric carbon dioxide to organic compounds and thereby set oxygen free.

That some such general sequence of events occurred can hardly be doubted, but the individual steps and the time when major changes in the atmosphere took place are still subjects of lively controversy. Seemingly information about atmospheric composition should be obtainable by geologic study of ancient rocks, but the evidence is far from clear. It has long been a speculative possibility that the great increase in abundance and complexity of organisms between the late Precambrian and early Paleozoic might somehow be linked with atmospheric change, possibly from a CO_2-rich atmosphere to an oxygen atmosphere. Such a change might also help to explain the peculiarities of Precambrian sedimentary rocks. A high concentration of CO_2 in the atmosphere would make the ocean more acid than at present, and a low concentration of O_2 would permit extensive transportation of iron as Fe^{2+} in surface waters. These are conditions that would favor the deposition of silica in large amounts as chert, together with iron either in a reduced form as siderite or pyrite, or where oxygen was not extremely scarce, as magnetite or hematite; such sediments would eventually become the banded iron formation that is so characteristic of the middle Precambrian and so important as the world's major source of iron ore. A low pH in seawater would prevent the precipitation of calcium and magnesium carbonates except locally, hence would explain the scarcity of carbonate rocks through much of the Precambrian. A reducing atmosphere would allow the stable existence of uraninite in detrital grains, hence the formation of uranium placers like the famous deposits of the Witwatersrand. Extensive "red beds" in the later Precambrian suggest an increase in atmospheric oxygen, but the increase need not have been large to make hematite stable. In an effort to quantify the probable changes in oxygen concentration, Berkner and Marshall (1964) identified the Precambrian-Paleozoic transition with a rise in oxygen to one-hundredth of its present value, and the appearance of land organisms in the late Silurian with a further rise to one-tenth of the present value. This hypothesis rests on a detailed analysis of the interaction between various frequencies in the sun's radiation and probable constituents of the early atmosphere.

Details of atmospheric evolution remain obscure, but much present thinking inclines to an ultimate origin by gradual degassing of the earth's interior, slow increase in free oxygen accompanying the evolution of life, and persistence of high CO_2/O_2 ratios until at least the beginning of the Paleozoic era.

22-6 HISTORY OF SEAWATER

The ocean is a complex solution with six major ions (Na^+, Cl^-, SO_4^{2-}, Mg^{2+}, K^+, Ca^{2+}) and a host of minor ones (Sec. 12-3 and Appendix III). Relative amounts of the major ions are remarkably uniform in different parts of the sea, and the pH remains within narrow limits (generally 8.1 to 8.4 in surface waters and slightly lower at depth). The constancy of composition suggests that strong controls are at work to keep the amounts of different solutes at steady levels. To determine what these controls are and how effectively they have operated over geologic time is an enterprise that has attracted the interest of many geochemists.

Control of concentrations of the major ions, as we have noted previously (Sec. 12-3), can be plausibly ascribed to a balance between rate of supply from rivers and various precipitation and adsorption processes in the sea. For some of the ions, perhaps for all, control represents a steady state between rates of competing reactions rather than attainment of equilibrium. Nevertheless, equilibrium reasoning combined with simple facts about the composition of marine evaporites can be used to show that major-ion concentrations cannot have deviated far from their present values during Phanerozoic time (Holland, 1972). The concentration of Ca^{2+}, for example, is limited by the facts that gypsum would be a common rock accompanying limestone if its concentration had ever been much higher, and that sodium carbonate minerals would be abundant in marine evaporites if it had been much lower. The usual presence of dolomite as a primary mineral at the base of marine evaporite sequences requires that Mg^{2+} cannot have been much below its present value, and the absence or near absence of sepiolite and brucite in evaporites sets an upper limit. By an ingenious refinement of such generalizations, Holland shows that the concentrations of the major ions cannot have been much more than twice, or much less than half of, their present values since the late Precambrian. Controls of concentration have been so effective that seawater has not had a very exciting chemical history in recent geologic time.

In earlier periods more radical changes were possible. The Precambrian ocean, as we saw in the last section, was probably in contact with a different kind of atmosphere, and older Precambrian rocks are devoid of evaporites, so that some of the assumptions on which Holland based his reasoning for the Phanerozoic ocean are no longer valid. What guesses can we make about changes in seawater composition during this earlier time?

Almost certainly the ocean formed originally by condensation of water vapor escaping from the earth's interior during the early major degassing of the planet. Additional water from the interior escaped later, and is probably still escaping, but a large part of the ocean's volume was probably present soon after the solid crust became stable. The early ocean, and the streams that flowed into it, must have contained acid gases (CO_2, HCl, SO_2) in abundance, and attack of such gases and such solutions on exposed rock would have been rapid. Much material would have been dissolved, and in short order the ocean must have acquired solutes in concentrations comparable to those of the present sea.

The composition of the dissolved substances, however, would have been different. The ocean was almost surely acid; some of the acid would be neutralized by reaction with silicate minerals, but in the presence of a CO_2-rich atmosphere, and with a supply of acid gases continually renewed by numerous volcanoes, the pH of seawater must have been on the acid side of neutral. The dissolved material, therefore, would have had higher concentrations of Ca^{2+} and Mg^{2+} than the present ocean, since at low pH these ions can exist together with large amounts of HCO_3^-. Sulfate ion would have been scarce or absent, because sulfur would be largely in reduced form; this means that Ca^{2+} concentrations would not even be limited by the precipitation of gypsum. With no free oxygen in either air or water, iron could have been present in abundance as Fe^{2+}. Presumably Cl^-, Na^+, and K^+ would have been important ions, just as they are today. Some of the rarer metals whose concentrations at present are kept low by reactions with CO_3^{2-} or SO_4^{2-}, for example Ba^{2+}, Sr^{2+}, and Mn^{2+}, very likely played a more important role in Precambrian seas.

Changes in this early ocean as the carbon dioxide of the atmosphere decreased and oxygen slowly increased are easy to imagine. The pH would gradually rise; Ca^{2+} and Mg^{2+} would precipitate to form the extensive dolomite beds of the later Precambrian; iron would be oxidized and precipitated as hematite; SO_4^{2-}

would become an important ion except in stagnant basins where primitive organisms flourished. The overall composition would thus gradually shift toward the proportions that have remained nearly constant from the latest Precambrian down to the present day.

Much in this history is speculative, but the general picture of an original slightly acid ocean with abundant dissolved salts of calcium, magnesium, and iron, changing gradually through the Precambrian to the present NaCl- and sulfate-rich ocean, accords both with the sedimentary record and with probable changes in the chemistry of the atmosphere and the solid earth.

One other aspect of the history of seawater is suggested by the numbers in Appendix III. For a few elements, notably chlorine, bromine, sulfur, and boron, the concentrations in seawater seem remarkably high compared with concentrations in common rocks. If the figures are used to estimate the total amounts of these elements in the crust and in the ocean, it appears that a sizable fraction of the earth's supply of each element is dissolved in seawater. The quantity in the sea is far too great to have been supplied by any imaginable process of erosion of ordinary rocks during geologic time. Since the four elements are well known as constituents of volcanic emanations and since their compounds with common cations are all very soluble, the suggestion is natural that compounds of these elements have been added to seawater directly and in large quantity by volcanic activity rather than by erosion, probably all through the geologic past.

22-7 SOME LARGER PROBLEMS

The pH of seawater, touched on briefly in the last section, is an interesting subject for further rumination.

Going back to the discussion of buffers in Chap. 2, we recall that seawater today is protected from major changes in pH by reactions involving carbonic acid and its ions, and to a lesser extent by boric acid and borate ion. This is certainly sufficient to counteract minor additions of acid or alkali, but would it be effective as a long-term control if acid or alkali were added in large quantities? To make the question specific, suppose that the acid gases mentioned in the last section—especially CO_2, HCl, and SO_2—were added to the modern sea continuously and in large amounts from volcanic emanations. The carbonate buffer could take up a great deal of these gases; solid carbonates would be dissolved, CO_3^{2-} would be changed to HCO_3^-, and HCO_3^- eventually to H_2CO_3. The pH would fall slowly as these reactions took place, but even when it dropped to 6.5 or 7, the buffer would still be able to take up H^+.

Eventually, of course, further addition of acid would cause the buffer to break down, and seawater would lose its protection against a drastic lowering of pH. Or would it? A second line of defense can be imagined in the clay minerals. Seawater is in contact with enormous volumes of clay, and ion exchange with the clay minerals could take up huge amounts of H^+. Both the metal ions adsorbed on clay-particle surfaces and interlayer ions like the Na^+ of montmorillonite and the

K^+ of illite would be susceptible to replacement by hydrogen ion. This clay buffer system is slower acting and less well known than the carbonate system, but its capacity is larger, and it may well be the principal regulator of seawater pH—acting, so to speak, behind the scenes while the carbonate system takes care of minor day-to-day fluctuations.

But even ion exchange on clay particles cannot be the ultimate answer, because we can give our imagination free rein and let volcanoes produce enough acid gas to react with clay minerals as well as carbonates. Then what would keep seawater from becoming a solution of hydrochloric, hydrobromic, and sulfuric acids? As a third line of defense we turn to silicate minerals other than clay minerals: feldspars, pyroxenes, amphiboles. These are subject to slow acid attack, breaking down to form silica and kaolinite and contributing their cations to solution. Hence not until our energetic volcanoes had emitted enough acid gas to convert all surface rocks to silica and kaolinite, and to bring all the cations of surface rocks into solution in the ocean, would the acidity of seawater be free to rise.

This series of events has obviously never occurred during geologic time, and a logical next question might be, why not? For an answer we would have to look into the processes by which a volcano generates acid gas. Here we tread on uncertain ground, but in part at least the gases from a volcano represent volatile material released when sedimentary rock is heated sufficiently for melting to start. Reactions of this sort were demonstrated in some of Winkler's experiments described in Sec. 19-5: when shales containing pyrite and added NaCl were heated with water vapor under pressure, HCl and H_2S were noted among the volatile products. Evidently the kinds of sedimentary material formed under near-neutral conditions at the earth's surface are capable of generating acid gases when heated at depth, and the amount of acid gas set free would be limited by the nature of the sediment. In other words we are dealing with a sort of gigantic and complicated equilibrium, in which acid gases and the metal silicates of igneous rocks are the stable forms at high temperature, and acid silicates plus dissolved salts are the ultimate stable forms at low temperature:

$$\text{Low temperature} \rightarrow$$

$$\text{Metal silicates + acid gases} \rightleftarrows \text{hydrogen silicates + dissolved salts}$$

$$\leftarrow \text{High temperature}$$

Reactions in both directions are slow and incomplete.

Another part of the gas emitted from volcanoes must consist of volatile material out of the deep interior, presumably gases trapped when the earth was formed and now on their way slowly to the surface. How important a fraction of volcanic gas this primitive material is at the present time we have no way of knowing. But we can at least be certain, from geologic evidence, that it has not produced major displacements of the above equilibrium in recent geologic time.

Is there a chance that the buffer system of the ocean might break down in the other direction? Could alkali be added in sufficient amount to convert all carbon dioxide from water and air into carbonate, all boric acid to borate, all hydrogen clays to montmorillonite and illite? Here we need not consider the active addition of alkali; it is sufficient to imagine simply that processes of weathering are carried far beyond their usual limits. Metal silicates reacting with water give alkaline solutions (Sec. 4-8), and experimental pH's obtained by grinding common silicates under water have been reported to rise as high as 11. What restraints are there to an indefinitely rising pH, once the capacity of the common buffers is exhausted? For a time the conversion of silica to an ion of silicic acid would act as a brake, but if weathering proceeded long enough, the free silica of the outer crust would be exhausted, and pH's of solutions would thereafter be determined by the hydrolysis of silicate minerals. Specific prediction is difficult, but there seems little to restrain the pH except the solubility of metal hydroxides. Presumably in nature a final control is neutralization of excess alkali by the acid gases from volcanoes.

The essence of such fanciful arguments has been expressed by Urey (1956) in a symbolic equilibrium much simpler than the generalized equation above:

$$CaSiO_3 + CO_2 \rightleftarrows CaCO_3 + SiO_2$$

This equation, in a different context, we have discussed as an example of a simple metamorphic reaction that may occur at granite-limestone contacts (Sec. 19-3). Here we look at the equation from a wider perspective: $CaSiO_3$ stands for any of the common metal silicates, CO_2 for acid gases, either from volcanoes or already present in the atmosphere, $CaCO_3$ is a simple salt, and SiO_2 as the anhydride of silicic acid may stand for the simplest of the acid silicates. The forward reaction, then, symbolizes the reactions of weathering and solution at the earth's surface—the low-temperature processes by which silicates are attacked, cations are set free or fixed in simple compounds, and alkalinity is increased. The reverse reaction shows the formation of complex silicates and acid gases by the heating of simple sedimentary materials to high temperatures. The position of equilibrium in the reaction depends not only on the temperature but on the pressure of CO_2; on a worldwide scale we could read this as the concentration of the constituents of acid gases in the earth's materials. Thus this simple equation summarizes a great deal of geochemistry. It expresses not only the ultimate controls on the pH of seawater, but the principal reactions embodied in the rock cycle.

It is most appropriate that here at the end of our story we find ourselves involved in one final example of equilibrium. The story began with equilibrium, back in Chap. 1, and we have examined equilibria of many kinds and many degrees of complexity. Now in this final discussion, after a long, speculative look at chemical trends down the ages, we hit upon a deceptively simple equation that expresses in the form of an equilibrium much of the subject matter of this and previous chapters. There are many things we do not know about the Urey equilibrium, and about other equilibria as well. But when problems can be formulated in this manner, we have at least achieved a measure of organization of our thinking and a basis for asking pertinent questions.

REFERENCES AND SUGGESTIONS FOR FURTHER READING

Berkner, L. V., and L. C. Marshall, The history of growth of oxygen in the earth's atmosphere, in P. J. Brancazio and A. G. W. Cameron (eds.), "The Origin and Evolution of Atmospheres and Oceans," pp. 102–126, John Wiley & Sons, Inc., New York, 1964. Quantitative estimates of the amounts of oxygen and ozone in the atmosphere at various times in geologic history, based largely on the photochemistry of atmospheric gases.

Birch, F., Speculations on the earth's thermal history, *Geol. Soc. America Bull.*, vol. 76, pp. 133–154, 1965. A review and critique of current data and hypotheses on past temperature distributions within the earth.

Cloud, P., A working model of the primitive earth, *Amer. Jour. Science*, vol. 272, pp. 537–548, 1972. Speculations about early earth history, with emphasis on the chemical consequences of the development of life.

Engel, A. E. J., S. P. Itson, C. G. Engel, and D. M. Stickney, Crustal evolution and global tectonics: a petrogenetic view, *Geol. Soc. America Bull.*, vol. 85, pp. 843–858, 1974. An attempt to relate rock composition to plate tectonics at various times in earth history.

Holland, H. D., The geologic history of sea water—an attempt to solve the problem, *Geochim. et Cosmochim. Acta*, vol. 36, pp. 637–652, 1972. Solubility equilibria and composition of marine evaporites used to set limits on variation in the concentrations of major ions in seawater.

Holland, H. D., "The Chemistry of the Atmosphere and Oceans," John Wiley & Sons, Inc., New York, 1978. Compilation of data and discussion of controls of composition.

Maynard, J. B., The long-term buffering of the oceans, *Geochim. et Cosmochim. Acta*, vol. 40, pp. 1523–1532, 1976. Possible control of major-ion concentrations by reaction of seawater with basalt.

Pytkowicz, R. M., Some trends in marine chemistry and geochemistry, *Earth Science Reviews*, vol. 11, pp. 1–46, 1975. A thoughtful review of recent work, noting the importance of steady-state as well as equilibrium processes and the difficulty in applying thermodynamics to complex systems like seawater.

Rubey, W. W., Geologic history of seawater: an attempt to state the problem, *Geol. Soc. America Bull.*, vol. 62, pp. 1111–1147, 1951. A classical and much quoted paper tracing the origin of atmosphere and oceans to volatile materials from the earth's interior.

Sayles, F. L., and P. C. Mangelsdorf, Equilibration of clay minerals with sea water: exchange reactions, *Geochim. et Cosmochim. Acta*, vol. 41, pp. 951–960, 1977. Control of Na^+ and Ca^{2+} concentrations by exchange reactions with clays brought to the sea by rivers.

Sibley, D. F., and J. T. Wilband, Chemical balance of the earth's crust, *Geochim. et Cosmochim. Acta*, vol. 41, 545–554, 1977. Evidence for no significant changes in the average composition of igneous and sedimentary rocks during the past 1.5×10^9 years.

Urey, H. C., Regarding the early history of the earth's atmosphere, *Geol. Soc. America Bull.*, vol. 67, pp. 1125–1128, 1956. The control of carbon dioxide in the atmosphere by various equilibria, especially by the reaction of calcite and quartz to form wollastonite.

APPENDICES

CONSTANTS AND NUMERICAL VALUES

$\ln_e 10 = 2.3026$

$\ln_e x = 2.3026 \log_{10} x$

$R = 1.987$ cal/deg mole (gas-law constant)

$= 8.314$ joules/deg mole

$R \ln x = 4.576 \log x$ cal/deg mole

$RT \ln x = 1{,}364.3 \log x$ cal/mole at 25°C

$0°C = 273.150$ K

$25°C = 298.15$ K

Avogadro's number $= N_0 = 6.022 \times 10^{23}$ molecules/mole

Volume of 1 mole of a perfect gas

at 0°C and 1 atm pressure $= 22.415$ liters

at 25°C and 1 atm pressure $= 24.47$ liters

Faraday constant $= f = 23.061$ kcal/volt equiv

$= 96{,}487$ coulombs/equiv (or joules/volt equiv)

1 cal $= 4.1840$ joules $= 41.29$ cm^3-atm

1 kcal $= 1$ Cal $= 1000$ cal $= 41.29$ liter-atm $= 0.2390$ kilojoules

1 cm^3-atm $= 0.02422$ cal

1 bar $= 0.987$ atm $= 10^6$ dynes/cm^2 $= 10^5$ pascals

$= 14.504$ pounds per square inch (psi)

$= 1.0197$ kg/cm^2

$= 750.06$ mmHg $= 750.06$ torr

1 atm $= 1.013$ bars

$= 14.70$ psi

$= 1.033$ kg/cm^2

$= 760.0$ mmHg $= 760.0$ torr

1 Å $= 1$ ångström $= 10^{-8}$ cm

SYMBOLS, ATOMIC NUMBERS, AND ATOMIC WEIGHTS OF THE NATURALLY OCCURRING ELEMENTS

Element	Symbol	Atomic number	Atomic weight
Actinium	Ac	89	227.03
Aluminum	Al	13	26.98
Antimony	Sb	51	121.75
Argon	Ar	18	39.95
Arsenic	As	33	74.92
Barium	Ba	56	137.33
Beryllium	Be	4	9.01
Bismuth	Bi	83	208.98
Boron	B	5	10.81
Bromine	Br	35	79.90
Cadmium	Cd	48	112.41
Calcium	Ca	20	40.08
Carbon	C	6	12.01
Cerium	Ce	58	140.12
Cesium	Cs	55	132.91
Chlorine	Cl	17	35.45
Chromium	Cr	24	52.00
Cobalt	Co	27	58.93
Copper	Cu	29	63.55
Dysprosium	Dy	66	162.50
Erbium	Er	68	167.26
Europium	Eu	63	151.96
Fluorine	F	9	19.00
Gadolinium	Gd	64	157.25
Gallium	Ga	31	69.72
Germanium	Ge	32	72.59
Gold	Au	79	196.97
Hafnium	Hf	72	178.49
Helium	He	2	4.003
Holmium	Ho	67	164.93
Hydrogen	H	1	1.008
Indium	In	49	114.82
Iodine	I	53	126.90
Iridium	Ir	77	192.22
Iron	Fe	26	55.85
Krypton	Kr	36	83.80
Lanthanum	La	57	138.91
Lead	Pb	82	207.19

Element	Symbol	Atomic number	Atomic weight
Lithium	Li	3	6.94
Lutetium	Lu	71	174.97
Magnesium	Mg	12	24.31
Manganese	Mn	25	54.94
Mercury	Hg	80	200.59
Molybdenum	Mo	42	95.94
Neodymium	Nd	60	144.24
Neon	Ne	10	20.18
Nickel	Ni	28	58.70
Niobium	Nb	41	92.91
Nitrogen	N	7	14.01
Osmium	Os	76	190.2
Oxygen	O	8	16.00
Palladium	Pd	46	106.4
Phosphorus	P	15	30.97
Platinum	Pt	78	195.09
Polonium	Po	84	209
Potassium	K	19	39.10
Praseodymium	Pr	59	140.91
Protactinium	Pa	91	231.04
Radium	Ra	88	226.03
Radon	Rn	86	222
Rhenium	Re	75	186.21
Rhodium	Rh	45	102.91
Rubidium	Rb	37	85.47
Ruthenium	Ru	44	101.07
Samarium	Sm	62	150.35
Scandium	Sc	21	44.96
Selenium	Se	34	78.96
Silicon	Si	14	28.09
Silver	Ag	47	107.87
Sodium	Na	11	22.99
Strontium	Sr	38	87.62
Sulfur	S	16	32.06
Tantalum	Ta	73	180.95
Tellurium	Te	52	127.60
Terbium	Tb	65	158.93
Thallium	Tl	81	204.37
Thorium	Th	90	232.04
Thulium	Tm	69	168.93
Tin	Sn	50	118.69
Titanium	Ti	22	47.90
Tungsten	W	74	183.85
Uranium	U	92	238.03
Vanadium	V	23	50.94
Xenon	Xe	54	131.30
Ytterbium	Yb	70	173.04
Yttrium	Y	39	88.91
Zinc	Zn	30	65.38
Zirconium	Zr	40	91.22

III

AVERAGE ABUNDANCES OF ELEMENTS IN THE EARTH'S CRUST, IN THREE COMMON ROCKS, AND IN SEAWATER (IN PARTS PER MILLION)

Element	Crust	Granite	Basalt	Shale	Seawater
O	46.4×10^4	48.5×10^4	44.1×10^4	49.5×10^4	880,000
Si	28.2×10^4	32.3×10^4	23.0×10^4	23.8×10^4	2
Al	8.1×10^4	7.7×10^4	8.4×10^4	9.2×10^4	0.002
Fe	5.4×10^4	2.7×10^4	8.6×10^4	4.7×10^4	0.002
Ca	4.1×10^4	1.6×10^4	7.2×10^4	2.5×10^4	412
Na	2.4×10^4	2.8×10^4	1.9×10^4	0.9×10^4	10,770
Mg	2.3×10^4	0.4×10^4	4.5×10^4	1.4×10^4	1,290
K	2.1×10^4	3.2×10^4	0.8×10^4	2.5×10^4	380
Ti	5,000	2,100	9,000	4,500	0.001
H	1,400				110,000
P	1,100	700	1,400	750	0.06
Mn	1,000	500	1,700	850	2×10^{-4}
F	650	800	400	600	1.3
Ba	500	700	300	600	0.02
Sr	375	300	450	400	8.0
S	300	300	300	2,500	905 — from volcanic activity
C	220	320	120	1,000	28
Zr	165	180	140	180	3×10^{-5}
Cl	130	200	60	170	18,800 — from Volcanic act

Element	Crust	Granite	Basalt	Shale	Seawater
V	110	50	250	130	0.0025
Cr	100	20	200	100	3×10^{-4}
Rb	90	150	30	140	0.12
Ni	75	0.8	150	80	0.0017
Zn	70	50	100	90	0.0049
Ce	70	90	30	70	1×10^{-6}
Cu	50	12	100	50	5×10^{-4}
Y	35	40	30	35	1×10^{-6}
La	35	55	10	40	3×10^{-6}
Nd	30	35	20	30	3×10^{-6}
Co	22	3	48	20	5×10^{-5}
Li	20	30	12	60	0.18
N	20	20	20	60	150
Sc	20	8	35	15	6×10^{-7}
Nb	20	20	20	15	1×10^{-5}
Ga	18	18	18	25	3×10^{-5}
Pb	12.5	20	3.5	20	3×10^{-5}
B	10	15	5	100	4.4 — _from volcanic activity_
Th	8.5	20	1.5	12	1×10^{-5}
Pr	8	10	4	9	6×10^{-7}
Sm	7	9	5	7	5×10^{-8}
Gd	7	8	6	6	7×10^{-7}
Dy	6	6.5	4	5	9×10^{-7}
Er	3.5	4.5	3	3.5	8×10^{-7}
Yb	3.5	4	2.5	3.5	8×10^{-7}
Be	3	5	0.5	3	6×10^{-7}
Cs	3	5	1	7	4×10^{-4}
Hf	3	4	1.5	4	7×10^{-6}
U	2.7	5	0.5	3.5	0.0032
Br	2.5	0.5	0.5	5	67 — _from volcanic activity_
Sn	2.5	3	2	6	1×10^{-5}
Ta	2	3.5	1	2	2×10^{-6}
As	1.8	1.5	2	10	0.0037
Ge	1.5	1.5	1.5	1.5	5×10^{-5}
Mo	1.5	1.5	1	2	0.01
Ho	1.5	2	1	1.5	2×10^{-7}
Eu	1.2	1.0	1.5	1.4	1×10^{-8}
W	1.2	1.5	0.8	1.8	1×10^{-4}
Tb	1	1.5	0.8	1	1×10^{-7}
Tl	0.8	1.2	0.2	1	1×10^{-5}
Lu	0.6	0.7	0.5	0.6	2×10^{-7}
Tm	0.5	0.6	0.5	0.6	2×10^{-7}
Sb	0.2	0.2	0.2	1.5	2.4×10^{-4}
I	0.2	0.2	0.1	2	0.06
Cd	0.15	0.1	0.2	0.3	1×10^{-4}
Bi	0.15	0.2	0.1	0.2	2×10^{-5}
In	0.06	0.05	0.07	0.06	1×10^{-7}
Ag	0.07	0.04	0.1	0.1	4×10^{-5}
Se	0.05	0.05	0.05	0.6	2×10^{-4}
Hg	0.02	0.03	0.01	0.3	3×10^{-5}
Au	0.003	0.002	0.004	0.003	4×10^{-6}

Te, Re, and the platinum metals are less than 0.03 ppm in rocks and less than 10^{-5} ppm in seawater. Concentrations of inert gases in seawater: He, 6.8×10^{-6} ppm; Ne, 1.2×10^{-4} ppm; Ar, 4.3×10^{-3} ppm; Kr, 2×10^{-4} ppm; Xe, 5×10^{-5} ppm.

Sources: Values for seawater are from P. G. Brewer, Minor elements in seawater, chap. 7 in "Chemical Oceanography," 2d ed., vol. 1, pp. 417–419, Academic Press Inc., New York, 1975. Values for crust, granite, basalt, and shale are estimated from tables in the following sources:

Turekian, K. K., and K. H. Wedepohl, Distribution of elements in some major units of the earth's crust, *Geol. Soc. America Bull.*, vol. 72, p. 186, 1961.

Vinogradov, A. P., Sredniye soderzhaniya khimicheskikh elementov v glavnykh tipakh izverzhennykh gornykh porod zemnoi kory, *Geokhimiya*, vol. 1962, pp. 560–561.

Taylor, S. R., Abundance of chemical elements in the continental crust, *Geochim. et Cosmochim. Acta*, vol. 28, pp. 1280–1281, 1964.

Mason, B., " Principles of Geochemistry," 3d ed., pp. 45–46, John Wiley & Sons, Inc., New York, 1966.

Parker, R. L., Composition of the earth's crust, *U.S. Geological Survey Professional Paper* 440-D, 1967.

Wedepohl, K. H., ed., "Handbook of Geochemistry," Springer-Verlag, Berlin, 1969–1974.

Notes: The heading " crust " means the continental crust only, and this part of the crust is assumed to be made up of roughly equal parts of basalt and granite. " Granite " includes silica-rich rocks ranging from alkali granite to granodiorite and their volcanic equivalents; " basalt " includes the more common varieties of basaltic lava, diabase, and gabbro; " shale " includes recent clays as well as shales, but not the fine-grained sediments of the deep sea. " Seawater " is normal surface water with a chlorinity of 19‰. No great accuracy can be claimed for any of the values, because they depend on subjective judgments about the kinds of material to be included in each category, because they are subject to change as analytical techniques improve, and because sampling is often inadequate.

IONIC RADII AND ELECTRONEGATIVITIES

Element	Ion†	Radius for 6-coordination ("octahedral"), Å‡	Commonly occurring coordination numbers§	Electro-negativity¶	Approx. ionic character of bond with oxygen,††
Aluminum	Al^{3+}	0.61	4, 6	1.5	60
Antimony	Sb^{3+}	0.88 (5)	6		66
	Sb^{5+}	0.69	4, 6	1.9	48
Arsenic	As^{3+}	0.6 L	4, 6		60
	As^{5+}	0.58	4, 6	2.0	38
Barium	Ba^{2+}	1.44	8–12	0.9	84
Beryllium	Be^{2+}	0.35 (4)	4	1.5	63
Bismuth	Bi^{3+}	1.10	6, 8	1.9	66
Boron	B^{3+}	0.20 (4)	3, 4	2.0	43
Bromine	Br^-	1.88		2.8	
Cadmium	Cd^{2+}	1.03	6, 8	1.7	66
Calcium	Ca^{2+}	1.08	6, 8	1.0	79
Carbon	C^{4+}	0.15 (3) L	3	2.5	23
Cerium	Ce^{3+}	1.09	6, 8	1.1	74
Cesium	Cs^+	1.78	12	0.7	89
Chlorine	Cl^-	1.72		3.0	
Chromium	Cr^{3+}	0.70	6	1.6	53
	Cr^{6+}	0.38 (4)			23
Cobalt	Co^{2+}	0.83	6	1.8	65
Copper	Cu^+	0.96 L	6, 8	1.9	71
	Cu^{2+}	0.81	6	2.0	57

Element	Ion†	Radius for 6-coordination ("octahedral"), Å‡	Commonly occurring coordination numbers§	Electro-negativity¶	Approx. ionic character of bond with oxygen,††
Fluorine	F^-	1.25		4.0	
Gallium	Ga^{3+}	0.70	4, 6	1.6	57
Germanium	Ge^{4+}	0.62	4	1.8	49
Gold	Au^+	1.37 L	8–12	2.4	62
Hafnium	Hf^{4+}	0.79	6	1.3	70
Indium	In^{3+}	0.88	6	1.7	62
Iodine	I^-	2.13		2.5	
	I^{5+}	1.03	6		54
Iron	Fe^{2+}	0.86	6	1.8	69
	Fe^{3+}	0.73	6	1.9	54
Lanthanum	La^{3+}	1.13	8	1.1	77
Lead	Pb^{2+}	1.26	6–10	1.8	72
Lithium	Li^+	0.82	6	1.0	82
Magnesium	Mg^{2+}	0.80	6	1.2	71
Manganese	Mn^{2+}	0.91	6	1.5	72
	Mn^{3+}	0.73	6		51
	Mn^{4+}	0.62	4, 6		38
Mercury	Hg^{2+}	1.10	6, 8	1.9	62
Molybdenum	Mo^{4+}	0.73	6		58
	Mo^{6+}	0.68	4, 6	1.8	47
Nickel	Ni^{2+}	0.77	6	1.8	60
Niobium	Nb^{5+}	0.72	6	1.6	56
Nitrogen	Ni^{5+}	0.12 (3) L	3	3.0	9
Oxygen	O^{2-}	1.32		3.5	
Palladium	Pd^{2+}	0.94	6	2.2	61
Phosphorus	P^{5+}	0.25 (4)	4	2.1	35
Potassium	K^+	1.46	8–12	0.8	87
Radium	Ra^{2+}	1.56 (8)	8–12	0.9	83
Rare-earth metals	$Ce^{3+} - Sm^{3+}$	1.09–1.04	6, 8	1.1–1.2	73–75
	$Eu^{3+} - Lu^{3+}$	1.03–0.94	6	1.2	76
	Eu^{2+}	1.25	8		
Rhenium	Re^{4+}	0.71	6		63
	Re^{7+}	0.65	4, 6		51
Rubidium	Rb^+	1.57	8–12	0.8	87
Scandium	Sc^{3+}	0.83	6	1.3	65
Selenium	Se^{2-}	1.88		2.4	
	Se^{6+}	0.37 (4)	4		26
Silicon	Si^{4+}	0.34 (4)	4	1.8	48
Silver	Ag^+	1.23	8, 10	1.9	71
Sodium	Na^+	1.10	6, 8	0.9	83
Strontium	Sr^{2+}	1.21	8	1.0	82
Sulfur	S^{2-}	1.72		2.5	
	S^{6+}	0.20 (4)	4		20
Tantalum	Ta^{5+}	0.72	6	1.5	63
Tellurium	Te^{2-}	2.2 (estim.)		2.1	
	Te^{6+}	0.56 L	4, 6		36
Thallium	Tl^+	1.58	8–12		79
	Tl^{3+}	0.97	6, 8	1.8	58

Element	Ion†	Radius for 6-coordination ("octahedral"), Å‡	Commonly occurring coordination numbers§	Electronegativity¶	Approx. ionic character of bond with oxygen,††
Thorium	Th^{4+}	1.08	6, 8	1.3	72
Tin	Sn^{2+}	1.30 (8)	6, 8	1.8	73
	Sn^{4+}	0.77	6	1.9	57
Titanium	Ti^{3+}	0.75	6		60
	Ti^{4+}	0.69	6	1.5	51
Tungsten	W^{6+}	0.68	4, 6	1.7	57
Uranium	U^{4+}	1.08 (8)	6, 8		68
	U^{6+}	0.81	6	1.7	62
Vanadium	V^{3+}	0.72	6	1.6	57
	V^{4+}	0.67	6		45
	V^{5+}	0.62	4, 6		36
Yttrium	Y^{3+}	0.98	6	1.2	74
Zinc	Zn^{2+}	0.83	4, 6	1.7	63
Zirconium	Zr^{4+}	0.80	6	1.4	65

† Only ions commonly found in naturally occurring minerals are listed.

‡ *Sources:* All values from E. J. W. Whittaker and R. Muntus, *Geochim. et Cosmochim.Acta*, vol. 34, pp. 945–956, 1970, except for six values (marked L) from L. H. Ahrens, *Geochim. et Cosmochim. Acta*, vol. 2, pp. 155–169, 1952, and one estimated value. All are radii for 6-coordination except a few for which a different coordination is indicated by a number in parentheses. In general, radii for 4-coordination can be estimated from the 6-coordination radii by subtracting 0.13 Å, and radii for 8-coordination by adding 0.13 Å; for most ions these rules give radii within 0.02 Å of the correct values. Accurate radii for many kinds of coordination are given by Whittaker and Muntus.

§ *Source:* F. G. Smith, "Physical Geochemistry," Addison-Wesley Publishing Company, Inc., Reading, Mass., 1963.

¶ *Source:* L. Pauling, "The Nature of the Chemical Bond," Cornell University Press, Ithaca, N.Y., 1960. The numbers are in arbitrary units, ranging from 0.7 for Cs to 4.0 for F.

†† *Source:* Smith, *op. cit.*, calculated by Smith from electronegativity values estimated by A. S. Povarennykh, *Dokl. Akad. Nauk SSSR*, vol. 109, pp. 993–996, 1956.

LARGE-SCALE DATA ABOUT THE EARTH

Mean radius	6,371 km
Volume	1.083×10^{12} km^3 = 1.083×10^{27} cm^3
Mass	5.98×10^{27} g
Radius of core	3471 km
Density	
Mean for entire earth	5.52 g/cm^3
Mean for crust	2.8 g/cm^3
Mean for mantle	4.5 g/cm^3
Mean for core	10.7 g/cm^3
Continental crust	2.7 g/cm^3
Oceanic crust	3.0 g/cm^3
Surface area	5.10×10^8 km^2 = 5.10×10^{18} cm^2
Land area	1.49×10^8 km^2 (about 29.2% of total)
Ocean area	3.61×10^8 km^2 (about 70.8% of total)
Mean height of land above sea level	875 meters
Mean depth of sea	3,800 meters
Seawater	
Volume	1.37×10^9 km^3
Mass	1.41×10^{24} g
Density	1.028 g/cm^3 (for normal seawater at 0°C)
Mass of dissolved salts	4.92×10^{22} g
Mass of freshwater	0.51×10^{21} g
Mass of continental ice	22.83×10^{21} g
Mass of atmosphere	5.12×10^{21} g (dry)
Age of earth	

Time of accretion of planetesimals, approximately 5×10^9 years ago

Time of depletion of iron meteorites in U and Th ("age" of meteorites), 4.55×10^9 years ago

Time of formation of oldest known rocks, approximately 3.8×10^9 years ago.

THE GEOLOGIC TIME SCALE

Era	Period	Epoch	Duration, millions of years	Time since beginning, millions of years
Cenozoic				
	Quaternary	Recent	...	0.01
		Pleistocene	2	2
	Tertiary	Pliocene	5	7
		Miocene	19	26
		Oligocene	11	37
		Eocene	18	55
		Paleocene	10	65
Mesozoic				
	Cretaceous		75	140
	Jurassic		50	190
	Triassic		40	230
Paleozoic				
	Permian		50	280
	Pennsylvanian		40	320
	Mississippian		30	350
	Devonian		55	405
	Silurian		30	435
	Ordovician		65	500
	Cambrian		80	580

Precambrian time extends back to the earliest known rocks, about 3,800 million years old. The Precambrian is often divided into a " Proterozoic era " and an "Archeozoic (or Archean) era," but the boundary between these eras and even the appropriateness of this division are matters of dispute. The Paleozoic, Mesozoic, and Cenozoic eras are often called collectively the "Phanerozoic."

EQUILIBRIUM CONSTANTS

Table VII-1 Solubility products

The numbers are negative logarithms of activity products at 25°C. For example, the number after $PbCl_2$ is 4.8. This means that

$$\text{Solubility product for } \underline{PbCl_2} = a_{Pb^{2+}} \cdot a_{Cl^-}^2 = 10^{-4.8}$$

The physical state of the compounds is not in every case clear from the literature, but for the most part they are probably finely crystalline precipitates. Because of this uncertainty about physical state and because of differences in experimental methods, values in the literature show considerable disagreement. The numbers cannot be expected to agree precisely with solubility products calculated from free energies in Appendix VIII. Most are within 0.8 of values derived from free energies; the four for which the discrepancy is larger are marked with daggers

Chlorides			Sulfides		
CuCl	6.7		Sb_2S_3	90.8	$[Sb^{3+}]^2[S^{2-}]^3$
PbCl$_2$	4.8		Bi_2S_3	100	
Hg$_2$Cl$_2$	17.9	$[Hg_2^{2+}][Cl^-]^2$	CdS	27.0	
AgCl	9.7		CoS	21.3	Alpha
Fluorides			CoS	25.6	Beta
BaF$_2$	5.8		Cu$_2$S	48.5	$[Cu^+]^2[S^{2-}]$
CaF$_2$	10.4	Fluorite	CuS	36.1	
MgF$_2$	8.2†	Sellaite	FeS	18.1	
PbF$_2$	7.5		PbS	27.5	Galena
SrF$_2$	8.5		MnS	10.5	Pink
Sulfates			MnS	13.5	Green
BaSO$_4$	10.0	Barite	HgS	52.7	Metacinnabar
CaSO$_4$	4.5	Anhydrite	HgS	53.3	Cinnabar
CaSO$_4 \cdot 2H_2O$	4.6	Gypsum	NiS	19.4	Alpha

Sulfates				Sulfides		
$PbSO_4$	7.8	Anglesite		NiS	26.6	Gamma
Ag_2SO_4	4.8	$[Ag^+]^2[SO_4^{2-}]$		Ag_2S	50.1	
$SrSO_4$	6.5	Celestite		SnS	25.9†	
Carbonates				ZnS	22.5	Wurtzite
$BaCO_3$	8.3†	Witherite		ZnS	24.7	Sphalerite
$CdCO_3$	13.7			Phosphates		
$CaCO_3$	8.35	Calcite		$AlPO_4 \cdot 2H_2O$	22.1	Variscite
$CaCO_3$	8.22	Aragonite		$Ca_3(PO_4)_2$	28.7	$[Ca^{2+}]^3[PO_4^{3-}]^2$
$CoCO_3$	10.0			$CaHPO_4 \cdot 2H_2O$	6.6	$[Ca^{2+}][HPO_4^{2-}]$
$FeCO_3$	10.7	Siderite		$Cu_3(PO_4)_2$	36.9	
$PbCO_3$	13.1			$FePO_4$	21.6	Amorphous
$MgCO_3$	7.5	Magnesite		$FePO_4 \cdot 2H_2O$	26.4	Strengite
$MgCO_3 \cdot 3H_2O$	5.6	Nesquehonite		$Pb_3(PO_4)_2$	43.5	At 38°
$MnCO_3$	9.3	Rhodochrosite		$PbHPO_4$	11.4	
$NiCO_3$	6.9			$Mg_3(PO_4)_2$	25.2	
$SrCO_3$	9.0†	Strontianite		$(UO_2)_3(PO_4)_2$	49.7	
UO_2CO_3	10.6	$[UO_2^{2+}][CO_3^{2-}]$		UO_2HPO_4	12.2	$[UO_2^{2+}][HPO_4^{2-}]$
$ZnCO_3$	10.0	At 20°		$Zn_3(PO_4)_2$	35.3	
$Cu_2(OH)_2CO_3$	33.8	$[Cu^{2+}]^2[OH^-]^2[CO_3^{2-}]$ malachite		$Ca_5(PO_4)_3OH$	57.8	$[Ca^{2+}]^5$ $[PO_4^{3-}]^3[OH^-]$ hydroxylapatite
				$Ca_5(PO_4)_3F$	60.4	Fluorapatite

† Values calculated from free energies differ by more than 0.8 from the values for these four compounds. From free energies: MgF_2, 10.5; $BaCO_3$, 13.2; $SrCO_3$, 11.8; SnS, 27.5.

Sources of data: R. M. Smith and A. E. Martell, "Critical Stability Constants, vol. 4: Inorganic Complexes," Plenum Press, New York, 1976. L. G. Sillén, Stability constants of metal-ion complexes, Sec. 1: Inorganic ligands, *Chem. Soc. London Spec. Publ.* 17, 1964; and Supplement 1, *Spec. Pub.* 25, 1971.

Table VII-2 Dissociation constants of acids

The numbers are negative logarithms of activity constants at 25°C. For each acid, K_1 shows the dissociation of the first hydrogen ion, K_2 the dissociation of the second, and K_3 the dissociation of the third. For example, for H_3PO_4:

$$H_3PO_4 \rightleftharpoons H^+ + H_2PO_4^- \qquad K_1 = \frac{a_{H^+} \cdot a_{H_2PO_4^-}}{a_{H_3PO_4}} = 10^{-2.1}$$

$$H_2PO_4^- \rightleftharpoons H^+ + HPO_4^{2-} \qquad K_2 = \frac{a_{H^+} \cdot a_{HPO_4^{2-}}}{a_{H_2PO_4^-}} = 10^{-7.2}$$

$$HPO_4^{2-} \rightleftharpoons H^+ + PO_4^{3-} \qquad K_3 = \frac{a_{H^+} \cdot a_{PO_4^{3-}}}{a_{HPO_4^{2-}}} = 10^{-12.4}$$

Acid	Formula	$-\log K_1$	$-\log K_2$	$-\log K_3$
Aluminum hydroxide	H_3AlO_3(amorph)	12.7		
Arsenious	H_3AsO_3(aq)	9.2		
Arsenic	H_3AsO_4(aq)	2.2	7.0	11.5
Boric	H_3BO_3(aq)	9.2		
Carbonic	H_2CO_3(aq)	6.35	10.3	
Hydrofluoric	HF(aq)	3.2		
Water	HOH	14.0		
Phosphoric	H_3PO_4(aq)	2.1	7.2	12.4
Hydrosulfuric	H_2S(aq)	7.0	12.9†	
Sulfuric	H_2SO_4(aq)		2.0	
Hydroselenic	H_2Se(aq)	3.9	15.0	
Selenic	H_2SeO_4(aq)		1.9	
Silicic	H_4SiO_4(aq)	9.9		
Zinc hydroxide	H_2ZnO_2(s, aged)	16.9		

† Determination of this constant is experimentally difficult, and no firm value can be given. Recent estimates range from $10^{-12.9}$ to 10^{-17}.

Sources of data: R. M. Smith and A. E. Martell, "Critical Stability Constants, vol. 4: Inorganic Complexes," Plenum Press, New York, 1976. L. G. Sillén, Stability constants of metal-ion complexes, Sec. 1: Inorganic ligands, *Chem. Soc. London Spec. Publ. 17,* 1964.

Table VII-3 Dissociation constants of hydroxides

The first column in the table below gives the formula of each hydroxide. For metals whose oxides are more stable than the hydroxides, formulas are given as oxide + H_2O; the meaning of the dissociation constants is illustrated by the reaction for ferric oxide:

$$\tfrac{1}{2}Fe_2O_3 + 3/2\ H_2O \rightleftharpoons Fe^{3+} + 3OH^- \qquad K = a_{Fe^{3+}} \cdot a_{OH^-}^3 = 10^{-42.7}$$

The second column indicates the physical state of the hydroxide, if this is specified in the source. Most hydroxides show a range in values of the dissociation constant, the amorphous material first precipitating dissociating somewhat more than an aged precipitate that has partly crystallized, and often considerably more than the equivalent naturally occurring mineral. Hydroxides for which a form is not specified are most likely precipitates that have been allowed to stand, in other words probably very finely crystalline material. Because of the uncertainty about form, these numbers cannot be expected to agree completely with constants calculated from the free energies in Appendix VIII.

Numerical values in the succeeding columns of the table are negative logarithms of the dissociation (activity) constants at 25°C. The third column gives "total" constants, on the assumption of complete dissociation. Numbers in the next three columns are constants for stepwise dissociation; the column headed K_{aq} shows the extent of solution as undissociated molecules; and the last column shows the ability of a hydroxide to dissolve by adding an additional OH^- ion, in other words the extent of its amphoteric character. Two examples will serve as illustrations:

$$Cd(OH)_2 \rightleftharpoons Cd^{2+} + 2OH^- \qquad K_T = a_{Cd^{2+}} \cdot a_{OH^-}^2 = 10^{-14.4}$$

$$Cd(OH)_2 \rightleftharpoons CdOH^+ + OH^- \qquad K_1 = a_{CdOH^+} \cdot a_{OH^-} = 10^{-10.5}$$

$$CdOH^+ \rightleftharpoons Cd^{2+} + OH^- \qquad K_2 = \frac{a_{Cd^{2+}} \cdot a_{OH^-}}{a_{CdOH^+}} = 10^{-3.9}$$

$$Cd(OH)_2(s) \rightleftharpoons Cd(OH)_2(aq) \qquad K_{aq} = a_{Cd(OH)_2(aq)} = 10^{-6.7}$$

$$Cd(OH)_2 + OH^- \rightleftharpoons Cd(OH)_3^- \qquad K_A = \frac{a_{Cd(OH)_3^-}}{a_{OH^-}} = 10^{-4.1}$$

$$Al(OH)_3(amorph) \rightleftharpoons Al^{3+} + 3OH^- \qquad K_T = a_{Al^{3+}} \cdot a_{OH}^3 = 10^{-31.6}$$

$$Al(OH)_3(amorph) \rightleftharpoons Al(OH)_2^+ + OH^- \qquad K_1 = a_{Al(OH)_2^+} \cdot a_{OH^-} = 10^{-12.3}$$

$$Al(OH)_2^+ \rightleftharpoons AlOH^{2+} + OH^- \qquad K_2 = \frac{a_{AlOH^{2+}} \cdot a_{OH^-}}{a_{Al(OH)_2^+}} = 10^{-10.3}$$

$$AlOH^{2+} \rightleftharpoons Al^{3+} + OH^- \qquad K_3 = \frac{a_{Al^{3+}} \cdot a_{OH^-}}{a_{AlOH^{2+}}} = 10^{-9.0}$$

$$Al(OH)_3(amorph) + OH^- \rightleftharpoons Al(OH)_4^- \qquad K_A = \frac{a_{Al(OH)_4^-}}{a_{OH^-}} = 10^{+1.1}$$

Hydroxide	Form	$-\log K_T$	$-\log K_1$	$-\log K_2$	$-\log K_3$	$-\log K_4$	$-\log K_{aq}$	$-\log K_A$
NH_4OH	Dissolved	4.7						
$Al(OH)_3$	Amorphous	31.6	12.3	10.3	9.0			−1.1
$Al(OH)_3$	Gibbsite	34.1	14.8	10.3	9.0			1.4
$AlOOH + H_2O$	Boehmite	34.2	14.9	10.3	9.0			1.5
$Be(OH)_2$	Amorphous	21.0						2.2
$Cd(OH)_2$		14.4	10.5	3.9			6.7	4.1(?)
$Cr(OH)_3$		29.8			10.1			
$Co(OH)_2$		14.9	10.6	4.3			6.5	5.2
$Co(OH)_3$		44.5						

Table VII-3 (continued)

Hydroxide	Form	$-\log K_T$	$-\log K_1$	$-\log K_2$	$-\log K_3$	$-\log K_4$	$-\log K_{aq}$	$-\log K_A$
$\frac{1}{2}Cu_2O + \frac{1}{2}H_2O$		14.7						
$Cu(OH)_2$		19.3	13.0	6.3				2.9
$CuO + H_2O$	Tenorite	20.3						
$Fe(OH)_2$		15.1	10.6	4.5			8.4(?)	5.1
$Fe(OH)_3$	Amorphous	38.8	16.5	10.5	11.8			4.4
$\frac{1}{2}Fe_2O_3 + \frac{3}{2}H_2O$	Hematite	42.7						
$FeOOH + H_2O$	Goethite	41.5						
$PbO + H_2O$	Red	15.3	9.0	6.3			4.4	1.4
$Mg(OH)_2$		11.2	8.6	2.6				
$Mn(OH)_2$		12.8	9.4	3.4				5.1
$HgO + H_2O$	Red	25.4	14.8	10.6			3.6	4.5
$Ni(OH)_2$		15.2	11.1	4.1			7	4
$\frac{1}{2}Ag_2O + \frac{1}{2}H_2O$		7.7					5.7	3.4
$Th(OH)_4$	Amorphous	44.7			10.3	10.8		5.8
$ThO_2 + 2H_2O$		49.7						
$SnO + H_2O$		26.2	15.8(?)	10.4(?)				
$UO_2 + 2H_2O$		56.2				13.3		
$UO_2(OH)_2$		22.4	14.2	8.2				3.6
$V(OH)_3$		34.4			11.7			
$VO(OH)_2$		23.5	15.2	8.3				
$Zn(OH)_2$	Amorphous	15.5	10.5	5.0			4.4(?)	1.9
$ZnO + H_2O$		16.7						

Sources of data: R. M. Smith and A. E. Martell, "Critical Stability Constants, vol. 4: Inorganic complexes," Plenum Press, New York, 1976. L. G. Sillén, Stability constants of Metal-ion Complexes, Sec. 1: Inorganic ligands, *Chem. Soc. London Spec. Publ.* 17, 1964; and Supplement 1, *Spec. Publ.* 25, 1971.

Table VII-4 Equilibrium constants for complex ions and molecules

The numbers are negative logarithms of activity constants at 25°C. For example:

$$AgCl_2^- \rightleftharpoons AgCl(s) + Cl^- \qquad K = \frac{a_{Cl^-}}{a_{AgCl_2^-}} = 10^{+4.4}$$

$$CuCO_3(aq) \rightleftharpoons Cu^{2+} + CO_3^{2-} \qquad K = \frac{a_{Cu^{2+}} \cdot a_{CO_3^{2-}}}{a_{CuCO_3(aq)}} = 10^{-6.8}$$

For many of the numbers the original data show wide disagreement. The values in the table should be considered only approximate

$AlF^{2+} \rightleftharpoons Al^{3+} + F^-$	7.0
$CuCl_2^- \rightleftharpoons CuCl(s) + Cl^-$	-1.2
$CuCl_3^{2-} \rightleftharpoons CuCl_2^- + Cl^-$	0.2
$CuCl^+ \rightleftharpoons Cu^{2+} + Cl^-$	0.4
$CuCO_3(aq) \rightleftharpoons Cu^{2+} + CO_3^{2-}$	6.8
$Cu(CO_3)_2^{2-} \rightleftharpoons CuCO_3(aq) + CO_3^{2-}$	3.2
$FeCl^{2+} \rightleftharpoons Fe^{3+} + Cl^-$	1.5
$FeF^{2+} \rightleftharpoons Fe^{3+} + F^-$	6.0
$PbCl_2(s) \rightleftharpoons PbCl_2(aq)$	3.2
$PbCl_2(aq) \rightleftharpoons PbCl^+ + Cl^-$	0.2
$PbCl^+ \rightleftharpoons Pb^{2+} + Cl^-$	1.6
$HgS_2^{2-} \rightleftharpoons HgS(s) + S^{2-}$	0.6
$AgCl_2^- \rightleftharpoons AgCl(s) + Cl^-$	-4.4
$AgCl(s) \rightleftharpoons AgCl(aq)$	6.4
$SnCl^+ \rightleftharpoons Sn^{2+} + Cl^-$	1.8
$SnF^+ \rightleftharpoons Sn^{2+} + F^-$	Approx. 4.1
$SnF_6^{2-} \rightleftharpoons Sn^{4+} + 6F^-$	Approx. 25
$SnS_3^{2-} \rightleftharpoons SnS_2(s) + S^{2-}$	5.0
$UO_2(CO_3)_2^{2-} \rightleftharpoons UO_2^{2+} + 2CO_3^{2-}$	14.6
$UO_2(CO_3)_2^{2-} \rightleftharpoons UO_2CO_3(s) + CO_3^{2-}$	4.0
$UO_2(CO_3)_3^{4-} \rightleftharpoons UO_2(CO_3)_2^{2-} + CO_3^{2-}$	3.8
$ZnCl^+ \rightleftharpoons Zn^{2+} + Cl^-$	0.4
$ZnF^+ \rightleftharpoons Zn^{2+} + F^-$	1.2

Sources of data: R. M. Smith and A. E. Martell, "Critical Stability Constants, vol. 4: Inorganic Complexes," Plenum Press, New York, 1976. L. G. Sillén, Stability constants of metal-ion complexes, Sec. 1: Inorganic ligands, *Chem. Soc. London Spec. Publ.* 17, 1964; and Supplement 1, *Spec. Publ.* 25, 1971.

VIII

STANDARD FREE ENERGIES, ENTHALPIES, AND ENTROPIES

The second column gives the physical state of each substance, insofar as it is known. Abbreviations: s, solid, form not specified in source; 1, liquid; g, gas; aq, dissolved in water at an activity of $1M$. The columns headed $\Delta G°$ and $\Delta H°$ give standard free energies and enthalpies of formation from the elements at 25°C and 1 atm pressure, in kilocalories per mole. The column headed $S°$ gives entropies in standard entropy units, calories per mole per degree. The last column refers to the sources of data listed at the end of the table.

To find the standard free-energy change for a reaction, subtract the sum of $\Delta G°$ values for the reactants from the sum of $\Delta G°$ values for the products. To find the equilibrium constant for a reaction at 25°, use the relation

$$\log K_a = -\frac{\Delta G°}{1.364}$$

By using this equation in reverse, free energies of formation for many compounds and ions not given in the table may be calculated from the equilibrium constants in Appendix VII.

To find approximate free-energy changes for a reaction at temperatures other than 25°C, use the relation

$$\Delta G° = \Delta H° - 0.001 T \, \Delta S°$$

(*Caution:* *Do not* use this equation for an individual substance, substituting $S°$ for $\Delta S°$. The $\Delta S°$ term is a *difference* in entropies of products and reactants for a complete reaction.)

Numbers in the table are of widely varying accuracy, and are subject to continual revision as new data are reported in the literature.

Formula	Form	$\Delta G°$	$\Delta H°$	$S°$	Source
		Aluminum			
Al	s	0	0	6.77	NBS
Al_2O_3	Corundum	-378.2	-400.5	12.2	NBS
AlOOH	Boehmite	-218.2	-236.0	11.6	NBS
$Al(OH)_3$	Gibbsite	-273.4	-306.3	16.8	NBS
$Al(OH)_3$	Amorphous	-271.9	-304.9	17	Latimer
$Al_2Si_2O_5(OH)_4$	Kaolinite	-903.0	-979.6	48.5	NBS
Al^{3+}	aq	-116.0	-127.0	-76.9	NBS
$Al(OH)_4^-$	aq	-310.2	-356.2	28.0	NBS
		Arsenic			
As	Metallic	0	0	8.4	NBS
As	g	$+62.4$	$+72.3$	41.6	NBS
As_4O_6	Claudetite	-275.8	-313.0	56	NBS
As_2O_5	s	-187.0	-221.1	25.2	NBS
AsH_3	g	$+16.5$	$+15.9$	53.2	NBS
H_3AsO_3	aq	-152.9	-177.4	46.6	NBS
H_3AsO_4	aq	-183.1	-215.7	44	NBS
As_2S_3	Orpiment	-40.3	-40.4	39.1	NBS
$H_2AsO_3^-$	aq	-140.4	-170.8	26.4	NBS
AsO_4^{3-}	aq	-155.0	-212.3	-38.9	NBS
		Barium			
Ba	s	0	0	16.0	RW
BaO	s	-132.0	-139.1	16.8	RW
BaF_2	s	-273.6	-286.0	23.0	KE
BaS	s	-104.5	-106.0	18.7	KE, KK
$BaSO_4$	Barite	-325.3	-352.1	31.6	RW
$BaCO_3$	Witherite	-278.4	-297.5	26.8	RW
$BaSiO_3$	s	-368.1	-388.7	26.8	KE, K62
Ba^{2+}	aq	-134.0	-128.7	3	Latimer
		Boron			
B	s	0	0	1.40	NBS
B_2O_3	s	-285.3	-304.2	12.9	NBS
H_3BO_3	s	-231.6	-261.6	21.2	NBS
$H_4BO_4^-$	aq	-275.7	-321.2	24.5	NBS
		Calcium			
Ca	s	0	0	9.95	RW
CaO	s	-144.4	-151.8	9.5	RW
$Ca(OH)_2$	Portlandite	-214.7	-235.6	19.9	RW
CaF_2	Fluorite	-280.1	-292.6	16.4	Nordstrom
CaS	s	-113.1	-114.3	13.5	RW
$CaCO_3$	Calcite	-269.9	-288.6	22.2	RW
$CaCO_3$	Aragonite	-269.7	-288.7	21.2	RW
$CaMg(CO_3)_2$	Dolomite	-518.7	-557.6	37.1	RW
$CaSO_4$	Anhydrite	-316.5	-343.3	25.5	RW
$CaSO_4 \cdot 2H_2O$	Gypsum	-430.1	-484.0	46.4	RW
$Ca_3(PO_4)_2$	Whitlockite	-932.8	-986.2	57.6	RW
$CaSiO_3$	Wollastonite	-370.3	-390.6	19.6	RW
$CaAl_2Si_2O_8$	Anorthite	-955.6	-1009.3	48.5	RW
$CaMgSi_2O_6$	Diopside	-725.8	-767.4	34.2	RW
Ca^{2+}	aq	-132.3	-129.7	-12.7	Nordstrom

Formula	Form	$\Delta G°$	$\Delta H°$	$S°$	Source
		Carbon			
C	Graphite	0	0	1.37	RW
C	Diamond	0.69	0.45	0.57	RW
CH_4	g	-12.13	-17.88	44.49	NBS
C_2H_6	g	-7.86	-20.24	54.85	NBS
C_3H_8	g	-5.61	-24.82	64.51	NBS
C_4H_{10}	g	-4.10	-30.15	74.12	NBS
C_2H_4	g	$+16.28$	$+12.49$	52.45	NBS
C_6H_6	l	$+30.99$	$+19.82$	64.34	NBS
CO	g	-32.78	-26.42	47.22	NBS
CO_2	g	-94.25	-94.05	51.06	NBS
H_2CO_3	aq	-148.94	-167.22	44.8	NBS
HCO_3^-	aq	-140.26	-165.39	21.8	NBS
CO_3^{2-}	aq	-126.17	-161.84	-13.6	NBS
		Chlorine			
Cl_2	g	0	0	53.29	NBS
HCl	g	-22.8	-22.1	44.65	NBS
Cl^-	aq	-31.4	-39.95	13.5	NBS
		Copper			
Cu	s	0	0	7.92	NBS
Cu_2O	Cuprite	-34.9	-40.3	22.3	NBS
CuO	Tenorite	-31.0	-37.6	10.2	NBS
$Cu(OH)_2$	s	-85.3	-106.1	19	Latimer
CuCl	s	-28.7	-32.8	20.6	NBS
Cu_2S	Chalcocite	-20.6	-19.0	28.9	NBS
CuS	Covellite	-12.8	-12.7	15.9	NBS
$Cu_2(OH)_2CO_3$	Malachite	-213.6	-251.3	44.5	NBS
Cu^+	aq	$+12.0$	$+17.1$	9.7	NBS
Cu^{2+}	aq	$+15.7$	$+15.5$	-23.8	NBS
$CuCl_2^-$	aq	-58.1	-66.3		Rose
		Fluorine			
F_2	g	0	0	48.44	NBS
HF	g	-65.3	-64.8	41.5	NBS
HF	aq	-71.0	-76.5	21.2	NBS
F^-	aq	-66.4	-79.1	-2.7	Nordstrom
		Gold			
Au	s	0	0	11.33	NBS
AuCl	s	-3.6	-8.3	22.2	KEA
$AuCl_3$	s	-10.8	-27.5	35.4	KEA
Au^+	aq	$+39.0$			Latimer
Au^{3+}	aq	$+103.6$			Latimer
$AuCl_2^-$	aq	-36.1			NBS
$AuCl_4^-$	aq	-56.2	-77.0	63.8	NBS
		Hydrogen			
H_2	g	0	0	31.21	NBS
H^+	aq	0	0	0	

Formula	Form	$\Delta G°$	$\Delta H°$	$S°$	Source
		Iron			
Fe	s	0	0	6.52	NBS
$Fe_{0.947}O$	Wüstite	-58.6	-63.6	13.7	NBS
Fe_3O_4	Magnetite	-242.7	-267.3	35.0	NBS
Fe_2O_3	Hematite	-177.4	-197.0	20.9	NBS
$Fe(OH)_2$	s	-116.3	-136.0	21	NBS
$Fe(OH)_3$	Amorphous	-166.5	-196.7	25.5	NBS
FeOOH	Goethite	-116.4	-133.6	14.4	Langmuir(2), NBS
FeS	Troilite	-24.0	-23.9	14.4	NBS
FeS_2	Pyrite	-39.9	-42.6	12.7	NBS
$FeCO_3$	Siderite	-159.4	-177.0	22.2	NBS
Fe_2SiO_4	Fayalite	-329.6	-353.7	34.7	NBS
Fe^{2+}	aq	-18.9	-21.3	-32.9	NBS
Fe^{3+}	aq	-1.1	-11.6	-75.5	NBS
		Lead			
Pb	s	0	0	15.49	NBS
Pb	g	$+38.7$	$+46.6$	41.9	NBS
PbO	Red	-45.2	-52.3	15.9	NBS
PbO	g	$+5.0$	$+10.1$	57.3	C, KK
PbO_2	s	-52.0	-66.3	16.4	NBS
$Pb(OH)_2$	s	-100.6			NBS
$PbCl_2$	Cotunnite	-75.1	-85.9	32.5	NBS
PbS	Galena	-23.6	-24.0	21.8	NBS
$PbSO_4$	Anglesite	-194.4	-219.9	35.5	NBS
$PbCO_3$	Cerussite	-149.5	-167.1	31.3	NBS
$PbSiO_3$	s	-253.9	-273.8	26.2	NBS
Pb^{2+}	aq	-5.8	-0.4	2.5	NBS
$Pb(OH)_3^-$	aq	-137.6			NBS
		Magnesium			
Mg	s	0	0	7.81	RW
MgO	Periclase	-136.1	-143.8	6.4	RW
$Mg(OH)_2$	Brucite	-199.5	-221.2	15.1	RW
MgF_2	Sellaite	-256.0	-268.7	13.7	RW
MgS	s	-82.6	-83.0	10.2	KEA
$MgCO_3$	Magnesite	-246.1	-266.1	15.7	RW
$MgCO_3 \cdot 3H_2O$	Nesquehonite	-412.7			Langmuir(1)
$MgSiO_3$	Clinoenstatite	-349.4	-370.1	16.2	RW
Mg_2SiO_4	Forsterite	-491.9	-520.4	22.8	RW
Mg^{2+}	aq	-108.8	-111.5	-32.7	Langmuir(1)
		Manganese			
Mn	s	0	0	7.65	NBS
MnO	Manganosite	-86.7	-92.1	14.3	NBS
Mn_3O_4	Hausmannite	-306.7	-331.7	37.2	NBS
Mn_2O_3	s	-210.6	-229.2	26.4	NBS
MnO_2	Pyrolusite	-111.2	-124.3	12.7	NBS
$Mn(OH)_2$	Precipitate	-147.0	-166.2	23.7	NBS

Formula	Form	$\Delta G°$	$\Delta H°$	$S°$	Source
		Manganese			
MnS	Alabandite	-52.2	-51.2	18.7	NBS
$MnCO_3$	Rhodochrosite	-195.2	-213.7	20.5	NBS
$MnSiO_3$	Rhodonite	-296.5	-315.7	21.3	NBS
Mn_2SiO_4	Tephroite	-390.1	-413.6	39.0	NBS
Mn^{2+}	aq	-54.5	-52.8	-17.6	NBS
MnO_4^-	aq	-106.9	-129.4	45.7	NBS
		Mercury			
Hg	l	0	0	18.17	NBS
Hg	g	$+7.6$	$+14.7$	41.8	NBS
HgO	Red	-14.0	-21.7	16.8	NBS
Hg_2Cl_2	Calomel	-50.4	-63.4	46.0	NBS
HgS	Cinnabar	-12.1	-13.9	19.7	NBS
HgS	Metacinnabar	-11.4	-12.8	21.1	NBS
Hg_2^{2+}	aq	$+36.7$	$+41.2$	20.2	NBS
Hg^{2+}	aq	$+39.3$	$+40.9$	-7.7	NBS
$HgCl_4^{2-}$	aq	-106.8	-132.4	70	NBS
HgS_2^{2-}	aq	$+10.0$			NBS
		Molybdenum			
Mo	s	0	0	6.85	NBS
MoO_3	s	-159.7	-178.1	18.6	NBS
MoS_2	Molybdenite	-54.0	-56.2	15.0	NBS
$CaMoO_4$	Powellite	-344.0	-369.5	29.3	RW
		Nickel			
Ni	s	0	0	7.14	NBS
NiO	s	-50.6	-57.3	9.1	NBS
$Ni(OH)_2$	s	-106.9	-126.6	21	NBS
NiS	s	-19.0	-19.6	12.7	NBS
$NiCO_3$	s	-146.4			NBS
Ni^{2+}	aq	-10.9	-12.9	-30.8	NBS
		Nitrogen			
N_2	g	0	0	45.77	NBS
N_2O	g	$+24.9$	$+19.6$	52.5	NBS
NO	g	$+20.7$	$+21.6$	50.3	NBS
NH_3	g	-3.9	-11.0	46.0	NBS
NH_4OH	aq	-63.0	-87.5	43.3	NBS
NO_3^-	aq	-26.6	-49.6	35.0	NBS
NH_4^+	aq	-19.0	-31.7	27.1	NBS
		Oxygen			
O_2	g	0	0	49.00	NBS
H_2O	l	-56.69	-68.32	16.71	NBS
H_2O	g	-54.63	-57.80	45.10	NBS
OH^-	aq	-37.59	-54.97	-2.57	NBS
		Potassium			
K	s	0	0	15.48	RW
KCl	Sylvite	-97.7	-104.4	19.7	RW
K_2SiO_3	s	-343.3	-365.9	33.0	KEA

Formula	Form	$\Delta G°$	$\Delta H°$	$S°$	Source
		Potassium			
$KAlSi_3O_8$	Microcline	−892.8	−946.3	52.5	RW
$KAlSi_2O_6$	Leucite	−681.6	−721.7	44.1	RW
$KAl_3Si_3O_{10}(OH)_2$	Muscovite	−1330.1	−1421.2	69.0	RW
K^+	aq	−67.3	−60.0	24.2	Latimer, K62
		Silicon			
Si	s	0	0	4.50	RW
SiO_2	α-Quartz	−204.6	−217.6	9.88	RW
SiO_2	α-Cristobalite	−204.1	−216.9	10.38	RW
SiO_2	α-Tridymite	−204.1	−216.9	10.50	RW
SiO_2	Glass	−203.3	−215.9	11.33	RW
$SiCl_4$	g	−147.5	−157.0	79.0	NBS
SiF_4	g	−375.9	−386.0	67.5	NBS
SiH_4	g	+13.6	+8.2	48.9	NBS
H_4SiO_4	aq	−312.5	−349.1	43	Siever, NBS
		Silver			
Ag	s	0	0	10.17	NBS
Ag_2O	s	−2.7	−7.4	29.0	NBS
AgCl	Cerargyrite	−26.2	−30.4	23.0	NBS
AgF	s	−44.2	−48.5	20.0	KEA
Ag_2S	Acanthite	−9.7	−7.8	34.4	NBS
Ag^+	aq	+18.4	+25.2	17.4	NBS
$AgCl_2^-$	aq	−51.1	−58.2	55.3	Rose, NBS
		Sodium			
Na	s	0	0	12.24	RW
NaCl	Halite	−91.8	−98.3	17.2	RW
Na_2SiO_3	s	−349.8	−372.2	27.2	K62, KK
$NaAlSi_3O_8$	Low albite	−884.0	−937.1	50.2	RW
$NaAlSiO_4$	Nepheline	−469.7	−497.0	29.7	RW
Na^+	aq	−62.5	−57.3	14.0	Latimer, KK
		Strontium			
Sr	s	0	0	12.5	RW
SrO	s	−137.3	−144.4	13.0	RW
$SrSO_4$	Celestite	−319.8	−346.6	28.2	RW
$SrCO_3$	Strontianite	−275.5	−294.6	23.2	RW
$SrSiO_3$	s	−369.7	−389.8	28.5	K62
Sr^{2+}	aq	−133.2	−130.4	−9.4	Latimer
		Sulfur			
S	Orthorhombic	0	0	7.60	NBS
S_2	g	+18.96	+30.68	54.51	NBS
H_2S	g	−8.02	−4.93	49.16	NBS
H_2S	aq	−6.66	−9.5	29	NBS
SO_2	g	−71.75	−70.94	59.30	NBS
SO_3	g	−88.69	−94.58	61.34	NBS
S^{2-}	aq	+20.5	+7.9	−3.5	NBS
HS^-	aq	+2.88	−4.2	15.0	NBS
SO_4^{2-}	aq	−177.97	−217.32	4.8	NBS
HSO_4^-	aq	−180.69	−212.08	31.5	NBS

Formula	Form	$\Delta G°$	$\Delta H°$	$S°$	Source
		Tin			
Sn	s	0	0	12.32	NBS
SnO	s	−61.4	−68.3	13.5	NBS
SnO_2	Cassiterite	−124.2	−138.8	12.5	NBS
$Sn(OH)_2$	Precipitated	−117.5	−134.1	37	NBS
$SnCl_4$	g	−103.3	−112.7	87.4	NBS
SnS	s	−23.5	−24	18.4	NBS
SnS_2	s	−38.0	−40.0	20.9	KEA
Sn^{2+}	aq	−6.5	−2.1	−4	NBS
Sn^{4+}	aq	+0.6	+7.3	−28	NBS
SnF_6^{2-}	aq	−420	−474.7	0	Latimer
$Sn(OH)_6^{2-}$	aq	−310.5			Latimer
		Titanium			
Ti	s	0	0	7.32	RW
TiO	s	−116.9	−123.9	8.3	KE
TiO_2	Rutile	−212.6	−225.8	12.0	RW
TiS_2	s	−78.9	−80.0	18.7	KEA
TiO^{2+}	aq	−138			Latimer
		Uranium			
U	s	0	0	12.00	RW
UO_2	Uraninite	−246.6	−259.2	18.6	RW
UO_3	s	−275.5	−294.0	23.6	KEA
UF_6	g	−484.8	−505.0	89.8	Latimer, KK
U^{4+}	aq	−138.4	−146.7	−78	Latimer
UO_2^{2+}	aq	−236.4	−250.4	−17	Latimer
		Zinc			
Zn	s	0	0	9.95	NBS
Zn	g	+22.75	+31.25	38.45	NBS
ZnO	Zincite	−76.1	−83.2	10.4	NBS
$Zn(OH)_2$	s	−132.3	−153.4	19.4	NBS
ZnS	Sphalerite	−48.1	−49.2	13.8	NBS
$ZnCO_3$	Smithsonite	−174.9	−194.3	19.7	NBS
Zn_2SiO_4	Willemite	−364.1	−391.2	31.4	NBS
Zn^{2+}	aq	−35.1	−36.8	−26.8	NBS
$Zn(OH)_4^{2-}$	aq	−205.2			NBS

Sources of data: Letters and names preceding the references are the abbreviations used in the last column of the table.

C — Coughlin, J. P., Contributions to the data on theoretical metallurgy. XII. Heats and free energies of formation of inorganic oxides, *U.S. Bur. Mines Bull.* 542, 1954.

K62 — Kelley, K. K., Heats and free energies of formation of anhydrous silicates, *U.S. Bur. Mines Rept. Inv.*, 5901, 1962.

KK — Kelley, K. K., and E. G. King, Contributions to the data on theoretical metallurgy. XIV. Entropies of the elements and inorganic compounds, *U.S. Bureau of Mines Bull.* 592, 1961.

KEA — Kubaschewski, O., E. L. Evans, and C. B. Alcock, "Metallurgical Thermochemistry," 4th ed., Pergamon Press, New York, 1967.

Langmuir (1) Langmuir, D., Stability of carbonates in the system $MgO-CO_2-H_2O$, *Jour. Geology*, vol. 73, pp. 730–754, 1965.

Langmuir (2) Langmuir, D., The Gibbs free energies of substances in the system $Fe-O_2-H_2O-CO_2$ at 25°C, *U.S. Geol. Survey Prof. Paper* 650-B, pp. 180–184, 1969.

NBS Wagman, D. D., W. H. Evans, V. B. Parker, I. Halow, S. M. Bailey, and R. H. Schumm, Selected values of chemical thermodynamic properties, *National Bureau of Standards Technical Notes* 270-3 and 270-4, 1968 and 1969.

Nordstrom Nordstrom, K. N., and E. A. Jenne, Fluorite solubility equilibria in selected geothermal waters, *Geochim. et Cosmochim. Acta*, vol. 41, pp. 175–188, 1977.

†RW Robie, R. A., and D. R. Waldbaum, Thermodynamic properties of minerals and related substances at 298.15°K and one atmosphere pressure and at higher temperatures, *U.S. Geol. Survey Bull.* 1259, 1968.

Rose Rose, A. W., Effect of cuprous chloride complexes in the origin of red-bed copper and related deposits, *Econ. Geology*, vol. 71, pp. 1036–1048, 1976.

Siever Siever, R., The silica budget in the sedimentary cycle, *Am. Mineralogist*, vol. 42, pp. 821–841, 1957.

† An updated version of this reference was published while this book was in press: Robie, R. A., B. S. Hemingway, and J. R. Fisher, Thermodynamic properties of minerals and related substances at 298.15 K and 1 bar pressure and at higher temperatures, *U.S. Geol. Survey Bull.* 1452, 1978. In the new version values for $\Delta G°$ and $\Delta H°$ are given in kilojoules (kJ) rather than kilocalories, and values for $S°$ in joules (J). Conversion factors: 1 kcal = 4.184 kJ and 1 cal = 4.184 J.

STANDARD ELECTRODE POTENTIALS

The value of $E°$ for each half-reaction is its potential in volts referred to the H_2-H^+ half-reaction, which is assigned the arbitrary value zero. The values are given for 25°C and 1 atm pressure, with all substances at unit activity. All pure substances whose state is not specified in the equations are assumed to be in their standard states at 25°C and 1 atm.

The equation for each couple is written so that the reducing agent is at the left. Potential differences for complete reactions may be obtained by subtracting potentials for the appropriate half-reactions, provided that formulas of oxidizing and reducing agents are identical in the half-reactions and the complete reaction. $E°$ values for half-reactions not shown in the table may be calculated from the free energies in Appendix VIII, by using the equation $E° = \Delta G°/23.1n$, where n is the coefficient of e^- in the half-reaction

$$E = E° + \frac{.059}{n} \log K$$

$$-\log \gamma = A z^2 I^{1/2}$$

$$A = .51$$

Potentials in acid solutions		
$K \rightleftharpoons K^+ + e^-$	STRONG	-2.93
$Ca \rightleftharpoons Ca^{2+} + 2e^-$	REDUCING	-2.87
$Na \rightleftharpoons Na^+ + e^-$	AGENTS	-2.71
$Mg \rightleftharpoons Mg^{2+} + 2e^-$		-2.37
$Th \rightleftharpoons Th^{4+} + 4e^-$		-1.90
$U \rightleftharpoons U^{3+} + 3e^-$		-1.80
$Al \rightleftharpoons Al^{3+} + 3e^-$		-1.66
$Mn \rightleftharpoons Mn^{2+} + 2e^-$		-1.18
$V \rightleftharpoons V^{2+} + 2e^-$		-1.18
$Si + 2H_2O \rightleftharpoons SiO_2 + 4H^+ + 4e^-$		-0.99
$Zn \rightleftharpoons Zn^{2+} + 2e^-$		-0.76
$Cr \rightleftharpoons Cr^{3+} + 3e^-$		-0.74
$H_2Te(aq) \rightleftharpoons Te + 2H^+ + 2e^-$		-0.74
$U^{3+} \rightleftharpoons U^{4+} + e^-$		-0.61
$Fe \rightleftharpoons Fe^{2+} + 2e^-$		-0.41
$Cr^{2+} \rightleftharpoons Cr^{3+} + e^-$		-0.41

The reduced form of a couple will react with the oxidized form of a couple below it, but not with the oxidized form of a couple above it.

Potentials in acid solutions

$H_2Se(aq) \rightleftharpoons Se + 2H^+ + 2e^-$	-0.40
$Co \rightleftharpoons Co^{2+} + 2e^-$	-0.28
$V^{2+} \rightleftharpoons V^{3+} + e^-$	-0.26
$Ni \rightleftharpoons Ni^{2+} + 2e^-$	-0.24
$Sn \rightleftharpoons Sn^{2+} + 2e^-$	-0.14
$Pb \rightleftharpoons Pb^{2+} + 2e^-$	-0.13
$H_2 \rightleftharpoons 2H^+ + 2e^-$	0.00
$H_2S(aq) \rightleftharpoons S + 2H^+ + 2e^-$	$+0.14$
$Sn^{2+} \rightleftharpoons Sn^{4+} + 2e^-$	$+0.15$
$Cu^+ \rightleftharpoons Cu^{2+} + e^-$	$+0.16$
$S^{2-} + 4H_2O \rightleftharpoons SO_4^{2-} + 8H^+ + 8e^-$	$+0.16$
$H_2SO_3(aq) + H_2O \rightleftharpoons SO_4^{2-} + 4H^+ + 2e^-$	$+0.17$
$Ag + Cl^- \rightleftharpoons AgCl + e^-$	$+0.22$
$As + 2H_2O \rightleftharpoons HAsO_2(aq) + 3H^+ + 3e^-$	$+0.25$
$U^{4+} + 2H_2O \rightleftharpoons UO_2^{2+} + 4H^+ + 2e^-$	$+0.33$
$Cu \rightleftharpoons Cu^{2+} + 2e^-$	$+0.34$
$V^{3+} + H_2O \rightleftharpoons VO^{2+} + 2H^+ + e^-$	$+0.34$
$S + 3H_2O \rightleftharpoons H_2SO_3(aq) + 4H^+ + 4e^-$	$+0.45$
$Cu \rightleftharpoons Cu^+ + e^-$	$+0.52$
$2I^- \rightleftharpoons I_2(s) + 2e^-$	$+0.54$
$3I^- \rightleftharpoons I_3^- + 2e^-$	$+0.54$
$HAsO_2(aq) + 2H_2O \rightleftharpoons H_3AsO_4(aq) + 2H^+ + 2e^-$	$+0.56$
$Pd + 4Cl^- \rightleftharpoons PdCl_4^{2-} + 2e^-$	$+0.62$
$Pt + 4Cl^- \rightleftharpoons PtCl_4^{2-} + 2e^-$	$+0.73$
$Se + 3H_2O \rightleftharpoons H_2SeO_3(aq) + 4H^+ + 4e^-$	$+0.74$
$Fe^{2+} \rightleftharpoons Fe^{3+} + e^-$	$+0.77$
$2Hg \rightleftharpoons Hg_2^{2+} + 2e^-$	$+0.79$
$Ag \rightleftharpoons Ag^+ + e^-$	$+0.80$
$Hg \rightleftharpoons Hg^{2+} + 2e^-$	$+0.85$
$Pd \rightleftharpoons Pd^{2+} + 2e^-$	$+0.92$
$NO(g) + 2H_2O \rightleftharpoons NO_3^- + 4H^+ + 3e^-$	$+0.96$
$Fe^{2+} + 3H_2O \rightleftharpoons Fe(OH)_3 + 3H^+ + e^-$	$+0.98$
$Au + 4Cl^- \rightleftharpoons AuCl_4^- + 3e^-$	$+1.00$
$VO^{2+} + H_2O \rightleftharpoons VO_2^+ + 2H^+ + e^-$	$+1.00$
$2Br^- \rightleftharpoons Br_2(l) + 2e^-$	$+1.07$
$2Br^- \rightleftharpoons Br_2(aq) + 2e^-$	$+1.09$
$HgS \rightleftharpoons S + Hg^{2+} + 2e^-$	$+1.11$
$H_2SeO_3(aq) + H_2O \rightleftharpoons SeO_4^{2-} + 4H^+ + 2e^-$	$+1.15$
$\frac{1}{2}I_2(s) + 3H_2O \rightleftharpoons IO_3^- + 6H^+ + 5e^-$	$+1.20$
$2H_2O \rightleftharpoons O_2(g) + 4H^+ + 4e^-$	$+1.23$
$Mn^{2+} + 2H_2O \rightleftharpoons MnO_2(s) + 4H^+ + 2e^-$	$+1.23$
$2Cr^{3+} + 7H_2O \rightleftharpoons Cr_2O_7^{2-} + 14H^+ + 6e^-$	$+1.33$
$2Cl^- \rightleftharpoons Cl_2(g) + 2e^-$	$+1.36$
$Pb^{2+} + 2H_2O \rightleftharpoons PbO_2(s) + 4H^+ + 2e^-$	$+1.46$
$Au \rightleftharpoons Au^{3+} + 3e^-$	$+1.50$
$Mn^{2+} \rightleftharpoons Mn^{3+} + e^-$	$+1.51$
$Mn^{2+} + 4H_2O \rightleftharpoons MnO_4^- + 8H^+ + 5e^-$	$+1.51$
$IO_3^- + 3H_2O \rightleftharpoons H_5IO_6(aq) + H^+ + 2e^-$	$+1.6$
$Au \rightleftharpoons Au^+ + e^-$	Approx. $+1.68$
$Co^{2+} \rightleftharpoons Co^{3+} + e^-$	$+1.82$
$2F^- \rightleftharpoons F_2(g) + 2e^-$	$+2.87$

STRONG OXIDIZING AGENTS

Potentials in basic solutions

$Mg + 2OH^- \rightleftharpoons Mg(OH)_2 + 2e^-$	-2.69
$U + 4OH^- \rightleftharpoons UO_2 + 2H_2O + 4e^-$	-2.39
$Al + 4OH^- \rightleftharpoons Al(OH)_4^- + 3e^-$	-2.32
$Mn + 2OH^- \rightleftharpoons Mn(OH)_2 + 2e^-$	-1.55
$Zn + 2OH^- \rightleftharpoons Zn(OH)_2 + 2e^-$	-1.25
$SO_3^{2-} + 2OH^- \rightleftharpoons SO_4^{2-} + H_2O + 2e^-$	-0.93
$Se^{2-} \rightleftharpoons Se + 2e^-$	-0.92
$Sn + 3OH^- \rightleftharpoons Sn(OH)_3^- + 2e^-$	-0.91
$Sn(OH)_3^- + 3OH^- \rightleftharpoons Sn(OH)_6^{2-} + 2e^-$	-0.90
$Fe + 2OH^- \rightleftharpoons Fe(OH)_2 + 2e^-$	-0.89
$H_2 + 2OH^- \rightleftharpoons 2H_2O + 2e^-$	-0.83
$V(OH)_3 + OH^- \rightleftharpoons VO(OH)_2 + H_2O + e^-$	-0.64
$Fe(OH)_2 + OH^- \rightleftharpoons Fe(OH)_3 + e^-$	-0.55
$Pb + 3OH^- \rightleftharpoons Pb(OH)_3^- + 2e^-$	-0.54
$S^{2-} \rightleftharpoons S + 2e^-$	-0.44
$2Cu + 2OH^- \rightleftharpoons Cu_2O + H_2O + 2e^-$	-0.36
$Cr(OH)_3 + 5OH^- \rightleftharpoons CrO_4^{2-} + 4H_2O + 3e^-$	-0.13
$Cu_2O + 2OH^- + H_2O \rightleftharpoons 2Cu(OH)_2 + 2e^-$	-0.08
$Mn(OH)_2 + 2OH^- \rightleftharpoons MnO_2 + 2H_2O + 2e^-$	-0.05
$SeO_3^{2-} + 2OH^- \rightleftharpoons SeO_4^{2-} + H_2O + 2e^-$	$+0.05$
$Pd + 2OH^- \rightleftharpoons Pd(OH)_2 + 2e^-$	$+0.07$
$Hg + 2OH^- \rightleftharpoons HgO(red) + H_2O + 2e^-$	$+0.10$
$Mn(OH)_2 + OH^- \rightleftharpoons Mn(OH)_3 + e^-$	$+0.1$
$Co(OH)_2 + OH^- \rightleftharpoons Co(OH)_3 + e^-$	$+0.17$
$PbO(red) + 2OH^- \rightleftharpoons PbO_2 + H_2O + 2e^-$	$+0.25$
$I^- + 6OH^- \rightleftharpoons IO_3^- + 3H_2O + 6e^-$	$+0.26$
$4OH^- \rightleftharpoons O_2 + 2H_2O + 4e^-$	$+0.40$

Sources of data: L. G. Sillén, Stability constants of metal-ion complexes. Sec. I: Inorganic ligands, *Chem. Soc. London Spec. Pub.* 17, 1964; and Supplement 1, *Spec. Pub.* 25, 1971. W. M. Latimer, "Oxidation Potentials," 2d ed., Prentice-Hall, Inc., Englewood Cliffs, N.J., 1952.

BALANCING OXIDATION AND
REDUCTION EQUATIONS

Most oxidation-reduction reactions of geologic interest can be balanced by inspection or by simple trial-and-error methods. For the occasional more complicated equation, a set of formal rules is often helpful. With a little practice one soon learns to judge which of the rules may be safely bypassed in a given situation.

1. *Note the pH range in which the reaction occurs.* Usually it suffices to know whether the process takes place in a strongly acid, weakly acid, weakly basic, or strongly basic solution.
2. *Write down the formulas of the substances which are oxidized, the substances which are reduced, and the substances formed from each.* Be sure that the formulas are accurate and that they are appropriate for the pH range. For example, ferric iron would be written $Fe(OH)_3$ in basic and weakly acid solutions, Fe^{3+} in strong acid; divalent sulfur would be H_2S in acid, HS^- in weakly basic solutions, S^{2-} in strong base; sexivalent chromium would be $Cr_2O_7^{2-}$ in acid and CrO_4^{2-} in base.
3. *If any one of the elements oxidized or reduced does not balance at this stage, put in the appropriate coefficients by inspection.* These coefficients are only tentative. (This step is often not necessary.)
4. *Balance the oxidation process against the reduction process, making sure that the total number of electrons removed from one (or more) kind of atom is balanced by the number added to another kind (or kinds) of atom.* It is often helpful to write the oxidation number of each atom oxidized or reduced above the formula containing it on each side of the equation; then the necessary coefficients must be inserted to make the differences between these numbers identical.

5. *Make sure that all atoms except H and O are balanced.*
6. *Balance the number of + and − charges shown by the formulas so far established by adding H^+ or OH^- to either side of the equation. Use H^+ if the reaction takes place in an acid solution, OH^- if it takes place in a basic solution.*
7. *Balance the number of H atoms by adding H_2O to the side of the equation where H is deficient.*
8. *The equation should now be balanced. Check for balance by counting O atoms on each side.*

For a simple example, consider the oxidation of Cl^- to Cl_2 gas by MnO_2 in acid solution. The only stable form of manganese with a lower oxidation state in acid is Mn^{2+}, so that for step 2 we have

$$Cl^- + MnO_2 \rightleftharpoons Cl_2 + Mn^{2+}$$

Since one atom of Cl appears on the left and two on the right, we add a coefficient in accordance with step 3:

$$2Cl^- + MnO_2 \rightleftharpoons Cl_2 + Mn^{2+}$$

These formulas show a gain of 2 electrons by one Mn atom (changing its oxidation state from +4 to +2) and a loss of 2 electrons by two Cl^- ions (changing the oxidation state from −1 to 0). Hence the oxidation and reduction already balance each other, and nothing further is needed to complete step 4. Nor is any change necessary to complete step 5. Now on the right side of the equation we have represented two + charges and on the left side two − charges; to bring these into balance (step 6) we need four H^+ on the left side:

$$4H^+ + 2Cl^- + MnO_2 \rightleftharpoons Cl_2 + Mn^{2+}$$

Now there are four H atoms on one side and none on the other, so we add the necessary H_2O (step 7):

$$4H^+ + 2Cl^- + MnO_2 \rightleftharpoons Cl_2 + Mn^{2+} + 2H_2O$$

Inspection shows two O atoms on each side; therefore the equation is balanced (step 8).

For a more complicated example, consider the oxidation of native gold by MnO_2 in an acid solution in the presence of Cl^-. The manganese is again reduced to Mn^{2+}, and the gold is dissolved as the complex ion $AuCl_4^-$. Hence for step 2:

$$Au + MnO_2 \rightleftharpoons AuCl_4^- + Mn^{2+}$$

Only one atom of Au and one of Mn are shown on each side, so that step 3 is unnecessary. For the oxidation and reduction, we write oxidation numbers above Au and Mn:

$$\overset{0}{Au} + \overset{+4}{MnO_2} \rightleftharpoons \overset{+3}{AuCl_4^-} + \overset{+2}{Mn^{2+}}$$

Three electrons are lost by each Au atom, and two are gained by each Mn. To balance this transfer, we need two Au atoms and three Mn (step 4):

$$2Au + 3MnO_2 \rightleftharpoons 2AuCl_4^- + 3Mn^{2+}$$

Now all atoms (besides O and H) are balanced except Cl. It was originally specified that Cl^- must be present; hence we complete step 5 by adding this ion on the left:

$$8Cl^- + 2Au + 3MnO_2 \rightleftharpoons 2AuCl_4^- + 3Mn^{2+}$$

To balance charges, we evidently need $12H^+$ on the left (step 6), and then to balance H atoms we must add $6H_2O$ on the right (step 7):

$$12H^+ + 8Cl^- + 2Au + 3MnO_2 \rightleftharpoons 2AuCl_4^- + 3Mn^{2+} + 6H_2O$$

A count of O atoms shows six on each side, and hence the equation is balanced.

For a final example, take the oxidation of chromite by atmospheric oxygen in the presence of an alkaline solution. Here two elements are oxidized: chromium to chromate ion and iron to ferric hydroxide. Oxygen is reduced to its usual state of -2, but we need not represent this by a separate formula since the reduced oxygen may appear in any of the oxygen-containing formulas among the products. Hence for step 2:

$$FeCr_2O_4 + O_2 \rightleftharpoons Fe(OH)_3 + CrO_4^{2-}$$

Since only one Cr atom appears on the right, we double the CrO_4^{2-} (step 3). At the same time we may write down the oxidation numbers of the elements oxidized and reduced:

$$\overset{+2\ +3}{FeCr_2O_4} + \overset{0}{O_2} \rightleftharpoons \overset{+3}{Fe(OH)_3} + \overset{+6\ -2}{2CrO_4^{2-}}$$

[The -2 for O could equally well be written over the O in $Fe(OH)_3$.] For every mole of $FeCr_2O_4$ oxidized, an Fe atom loses 1 electron and two Cr atoms lose 6 electrons, giving a total loss of 7 electrons. Now in the reduction of oxygen each atom must gain 2 electrons; hence it is necessary to double every formula containing Fe or Cr and to show a total of 7 oxygen atoms (step 4):

$$2FeCr_2O_4 + \tfrac{7}{2}O_2 \rightleftharpoons 2Fe(OH)_3 + 4CrO_4^{2-}$$

(Alternatively one can avoid fractional coefficients by taking $7O_2$ and multiplying the other coefficients by 4.) All atoms except H and O are now accounted for, and step 5 can be skipped. Since an alkaline solution is specified, we balance charges this time with OH^- (step 6):

$$8OH^- + 2FeCr_2O_4 + \tfrac{7}{2}O_2 \rightleftharpoons 2Fe(OH)_3 + 4CrO_4^{2-}$$

Eight H atoms appear on the left and only 6 on the right; hence a molecule of water must be added (step 7):

$$8OH^- + 2FeCr_2O_4 + \tfrac{7}{2}O_2 \rightleftharpoons 2Fe(OH)_3 + 4CrO_4^{2-} + H_2O$$

The total of 23 oxygen atoms on each side shows that the equation is balanced.

THE THERMODYNAMIC
BACKGROUND OF GEOCHEMISTRY

This appendix is a somewhat more formal approach to thermodynamics, designed to supplement Chap. 8 for students who would like to see how the elementary thermodynamic equations are derived. The discussion is by no means complete. It gives only the bare framework of the subject, without the details and the illustrative examples that should flesh it out, and it is limited to parts of thermodynamics that have a direct application to geologic problems. Additional material can be found in any good textbook of physical chemistry or thermodynamics. A particularly clear and interesting account, on which the present discussion leans heavily, is given in Lewis and Randall's " Thermodynamics," as revised by Pitzer and Brewer (McGraw-Hill Book Company, New York, 2d ed., 1961). A useful briefer treatment, more directly slanted toward geochemical applications, is included by Barth in "Theoretical Petrology" (John Wiley & Sons, Inc., New York, 2d ed., 1962).

THE FIRST LAW

Let E stand for the internal energy of a system, i.e., the sum of the kinetic and potential energies of its constituent atoms. The absolute value of E cannot be easily determined, but for most purposes we are interested only in its changes. If the energy of a system changes from E_1 to E_2, we write

$$\Delta E = E_2 - E_1$$

Suppose that the change is brought about by adding heat to the system and that the system does mechanical work as a result. By the law of conservation of energy, which is also called the first law of thermodynamics,

$$\Delta E = q - w \qquad \text{(XI-1)}$$

where q is the heat added and w is the work done. For an infinitesimal addition of heat,

$$dE = dq - dw \qquad \text{(XI-2)}$$

To measure the work, suppose the system is arranged so that work is done by expansion against an external pressure. Specifically, suppose that the system is confined in a vertical cylinder and that expansion results in raising a mass m at the top of the cylinder against the force of gravity. If the vertical distance moved is h, the mechanical energy expended is mgh, where g is the acceleration of gravity. This energy can also be expressed as $P \, \Delta V$, where P is pressure and ΔV is volume change, since

$$P = \frac{\text{force}}{\text{area}} = \frac{mg}{\text{area}} \qquad \text{and} \qquad \Delta V = \text{area} \times h$$

If the expansion is performed slowly, so that internal and external pressures never differ by more than infinitesimal amounts (in other words, if expansion occurs reversibly), P may be regarded not as the effect of an added mass, but as the prevailing external pressure acting on the system. Then we write in general

$$\Delta E = q - P \, \Delta V \qquad \text{(XI-3)}$$

or for infinitesimal expansion

$$dE = dq - P \, dV \qquad \text{(XI-4)}$$

The energy E is a *property* of the system, determined by the nature and arrangements of its constituent particles. By contrast, q and w are not properties of the system, but quantities that can be varied at will, within limits, of course. By different experimental arrangements, the same change in energy, ΔE, can be produced with different combinations of q and w.

ENTHALPY

Define another property of the system called *enthalpy* as

$$H = E + PV \qquad \text{(XI-5)}$$

or

$$dH = dE + P \, dV + V \, dP \qquad \text{(XI-6)}$$

If P is constant, $dP = 0$, and

$$dH = dE + P \, dV = dq \qquad \text{(XI-7)}$$

We may write Eq. (XI-5) for each substance taking part in a reaction and add together enthalpies of products and enthalpies of reactants. If pressure is constant, the sums will be

$$H_{pr} = E_{pr} + PV_{pr}$$
$$H_{re} = E_{re} + PV_{re}$$

where E_{pr} is the sum of internal energies of the products, and so on. Subtraction gives

$$H_{pr} - H_{re} = \Delta H = \Delta E + P \, \Delta V = q \qquad \text{(XI-8)}$$

where $\Delta E = E_{pr} - E_{re}$, or the change in internal energy during the reaction, $\Delta V = V_{pr} - V_{re}$, or the total volume change, and q is the heat absorbed. Hence, for a reaction taking place at constant pressure, ΔH is identical with the heat of reaction (except for an opposite sign, since q is heat absorbed and heat of reaction is generally regarded as heat evolved). For this reason H is often called the "heat content" of a system. As defined above, however, H has a broader meaning, not limited to isobaric processes, so that the name "enthalpy" is preferable.

If a reaction takes place with both pressure and volume constant,

$$\Delta H = q = \Delta E \qquad \text{(XI-9)}$$

so that the entire energy change is represented by the enthalpy change.

HEAT CAPACITY

The heat capacity of a substance is the amount of heat required to raise the temperature of one mole by 1°C. More rigorously, it may be defined as the limit of the ratio of heat added to temperature change produced, as the latter approaches zero:

$$c = \frac{dq}{dT} \qquad \text{(XI-10)}$$

If V is constant during the addition of heat, the heat absorbed equals the change in internal energy [Eq. (XI-3)]:

$$c_V = \left(\frac{\partial E}{\partial T}\right)_V \dagger \qquad \text{(XI-11)}$$

---

† A "partial derivative" like $(\partial E/\partial T)_V$ means "the rate of change of E with temperature when V is constant." It may be treated as an ordinary derivative if the restriction of constant volume is kept in mind.

where c_V is heat capacity at constant volume. If P is constant rather than V, heat absorbed goes into both temperature rise and expansion; hence

$$c_P = \left(\frac{\partial E}{\partial T}\right)_P + P\left(\frac{\partial V}{\partial T}\right)_P = \left(\frac{\partial H}{\partial T}\right)_P \qquad \text{(XI-12)}$$

For the change in heat capacity during a chemical reaction,

$$\Delta c_V = \left(\frac{\partial \Delta E}{\partial T}\right)_V \qquad \text{and} \qquad \Delta c_P = \left(\frac{\partial \Delta H}{\partial T}\right)_P \qquad \text{(XI-13)}$$

The second equation may be integrated to give

$$\Delta H = \int \Delta c_P \, dT \qquad \text{(XI-14)}$$

Hence, if Δc_P is known as a function of temperature, the change of ΔH with temperature may be calculated (Sec. 8-3).

ENTROPY

Suppose that a sample of gas is confined under pressure in a cylinder by a piston that moves without friction, and suppose that an amount of heat q is added to the gas. The gas responds by expanding against the pressure P exerted by the piston and thereby does an amount of work $P \, \Delta V$. Suppose further that the addition of heat and the resulting expansion take place slowly, so that the pressure of the gas on the piston is never more than infinitesimally higher than the external pressure, and so that the temperature of the gas remains constant. At any time during the expansion, it would be possible, by an infinitesimal increase in the external pressure, to reverse the operation—to do work on the gas by moving the piston inward, the work being converted into heat which would escape to the surroundings. Under these conditions we say that the expansion takes place *reversibly*, the expression meaning that infinitesimal changes can make the process reverse itself at any time. Let us then define a new property of the gas, the *entropy*, by the equation

$$\Delta S = \frac{q}{T} \qquad \text{(XI-15)}$$

Expressed in words, this says that the *increase in entropy* of the gas is equal to the heat absorbed in a reversible process divided by the absolute temperature. We treat the entropy, then, as a quantity like E or H, whose changes can be measured but whose absolute value remains unspecified. Later we shall find that entropy differs from E and H in that absolute values *can* be assigned, but for the moment we are concerned only with entropy differences.

If the expansion does not take place reversibly, Eq. (XI-15) no longer holds. Suppose, for example, that the piston is not frictionless, so that an excess of internal pressure over external is necessary to make it move. Then more q will be required to perform the same amount of work, or to make the gas expand by a given amount. In other words, the relation between q and w is arbitrary, depending on the arrangement of the apparatus. The q for reversible expansion is a particular value of the heat absorbed, and it is only this value which defines the entropy difference. By this restriction we make the entropy a *property* of the gas, meaning that the entropy difference is the same no matter how the expansion is accomplished, provided only that the initial and final states of the gas are the same. The q, by contrast, is not a property, because differing amounts of heat can be used to produce the same expansion. In this respect also the quantity S resembles the functions E and H.

If we idealize the experiment one step further and require that the gas be an ideal gas (meaning that its molecules behave as if they have no finite volume and that they exert no appreciable forces on each other), then expansion at constant temperature involves no change in internal energy. In other words, the heat supplied has been expended entirely in forcing the piston to move outward, and the gas molecules at the end of the expansion have the same average speed as before. From Eq. (XI-1),

$$\Delta E = q - w = q - P \, \Delta V = 0$$

so that

$$q = P \, \Delta V = T \, \Delta S \tag{XI-16}$$

Suppose, still using an ideal gas, we perform the experiment differently: we insert a shutter at the original position of the piston, move the piston back to its position at the end of the preceding experiment, evacuate the space between shutter and piston, then withdraw the shutter and let the gas expand freely to fill the enlarged space. The end result of the experiment is identical with that of the first experiment, for an ideal gas undergoes no temperature change during free expansion. This time we have added no heat and obtained no work, so that both q and w are zero. The entropy change, however, must be the same as before, since the initial and final states of the gas are identical in the two experiments. Equation (XI-16) no longer holds, because the process this time is highly *irreversible;* there is no possible way, while the gas is expanding, by an infinitesimal change in external conditions to make the molecules move back to their former restricted volume.

These two experiments illustrate the dual role that entropy plays. In a reversible process, the increase in entropy of a system is measured by the heat absorbed divided by the absolute temperature. In an irreversible process the entropy change may have nothing to do with heat, but is defined by some change in the configuration of the system, in this case by the expansion of the gas to a greater volume. The *measure* of the entropy change is still the heat that *would* be absorbed *if* the process were carried out reversibly, and the entropy change is the same whether the process is reversible or not.

Let us look again at the reversible process, in which heat absorbed by the gas is converted into work. After the gas has expanded, could more work be obtained from it by adding more heat? The answer is yes, but in smaller amount, because the pressure is now lower. In a hypothetical experiment, with a cylinder infinitely long and a device for making the external pressure indefinitely small, the conversion of heat to work could go on forever. But in any practical sense the conversion soon loses interest, because to maintain a lower and lower external pressure would require more work than the piston produces. By reason of the first expansion, then, we could say that the gas has lost some of its capacity for doing work. How could this capacity be restored? Obviously, by recompressing the gas. But this would mean doing work on the gas, and the work required would be at least as great as the work originally produced. Can we find some other way to restore the gas to its earlier condition that would require less work? One possibility is to cool the expanded gas, since at lower temperatures the work for a given amount of compression is smaller. So we let the gas cool itself by pushing the piston further, this time supplying no heat from the outside; at the lower temperature we compress it, keeping the temperature low by letting the heat produced escape into the surroundings; and finally, in the last stages of compression, we prevent the further escape of heat and let compression warm the gas up to its original temperature. Now we have completed the cycle, and from the original heat supplied we have obtained a net amount of work equal to the difference between work done during expansion at the high temperature and work supplied during compression at the low temperature. Notice, however, that an essential step in the cycle is *to let heat escape* at the lower temperature. In other words, only a part of the original heat can be converted into work:

$$w = w_{exp} - w_{comp} = q_1 - q_2 \qquad \text{(XI-17)}$$

where q_1 is the heat supplied and q_2 is the heat that must necessarily escape.

The net entropy change of the gas during this cycle of changes must be zero, since the gas returns to its original condition. If all steps in the cycle are carried out reversibly, entropy changes can occur only when heat from the outside is being added to the gas (during expansion) and when heat is escaping from the gas to the outside (during compression). In parts of the cycle when the gas is cooling itself by expansion against the piston, and when it is being heated by compression, no entropy change occurs because heat is not being added or subtracted. During the time when heat is entering the gas, the entropy increases by an amount q_1/T_1, as we have seen previously [Eq. (XI-15)]. When heat is escaping at the lower temperature, the entropy of the gas diminishes by the amount q_2/T_2. Since the total change is zero, these two quantities must be equal:

$$\frac{q_1}{T_1} = \frac{q_2}{T_2} \qquad \text{(XI-18)}$$

Combination of Eqs. (XI-17) and (XI-18) gives

$$\frac{w}{q_1} = \frac{T_1 - T_2}{T_1}$$

(XI-19)

which expresses the fraction of the total heat energy supplied to the system (q_1) which can be turned into work. This is the famous relation obtained by Sadi Carnot a century and a half ago, and the hypothetical cycle of changes which leads to this result is called the *Carnot cycle* in his honor.

The cycle refers to a piston-and-cylinder machine operated by compressed gas and therefore serves immediately as a sort of idealized model of a steam engine or hot-air engine. The expression for w/q_1, however, turns out to be of much more general application, though it would take us too far afield to demonstrate this rigorously. The fraction $(T_1 - T_2)/T_1$ is the proportion of any given quantity of heat that can be turned into work by *any* machine which operates reversibly between the two temperatures T_1 and T_2. Note that some heat must *always* escape from the system at the lower temperature, however the machine is designed; in other words, heat flows through the machine from a high temperature T_1 to a low temperature T_2, and during the flow a fraction of the heat is turned into work. No matter how much heat is available, no work can be obtained from it unless a region of lower temperature is also available into which some of the heat can flow. The fraction of work obtainable becomes larger as the temperature difference is made greater but can never become 100% unless T_2 is reduced to absolute zero. This is a basic limitation on all processes for obtaining work from heat energy.

It is important to keep in mind the entropy relationships. Reversibility of all steps in the cycle, and the return of the system to its starting condition at the end of each cycle, ensure that the net entropy change will remain zero. An increase in entropy accompanies the feeding in of heat and the doing of work; the system attains its maximum entropy when the gas has reached the limit of its expansion, hence when it has lost most of its capacity for doing work. The entropy becomes a minimum when the gas is compressed and hence most able to do work. One definition of entropy can be framed from these relationships: *entropy is a measure of the extent to which a system has lost its capacity for doing work;* or in more colloquial language, entropy measures the extent to which a system has run down. Note also that entropy changes within the gas of our hypothetical experiment, or more generally within the *system* which is doing the work, must be accompanied by entropy changes in the surroundings from which the heat comes and to which the heat goes. As heat is added to the system during expansion of the gas, entropy must be lost by the surroundings, and as heat is subtracted during the cool part of the cycle, entropy must be gained by the surroundings. Furthermore, the amount gained or lost by the system must be equal to the amount lost or gained by the surroundings. For a reversible process, then, entropy is conserved, just as energy is conserved.

Reversibility is an ideal which can never be attained in practice. Always friction and uncontrolled escape of heat to the surroundings must cut down the efficiency of any real engine. Hence in real engines the fraction w/q_1 can *never* reach the amount $(T_1 - T_2)/T_1$. This is an ideal or maximum efficiency, which may be approached as an engine is made more nearly reversible but which never is actually attainable.

Let us return to the first step in the operation of the ideal engine we have been discussing, in which a perfect gas does work by reason of heat absorbed at constant temperature, and for which the relations in a reversible process are

$$w_{max} = q = P \, \Delta V = T \, \Delta S$$

Now if the process is made to some extent irreversible (but retaining the ideal gas so that there is no complication from change in internal energy), the work done will be less than this maximum. The extreme of irreversibility is represented by the experiment described on page 576, in which the gas expands into a vacuum and therefore does no work at all. Here $P = 0$, $w = 0$, and $q = 0$; but $T \, \Delta S$, *representing the heat absorbed if the process were carried out reversibly*, retains its value. To indicate the irreversibility of the change, we might relabel the entropy change ΔS_{irrev}. If irreversibility is not quite so extreme (say, if the space behind the piston is evacuated partly but not completely, so that the external pressure P is small but not zero), $T \, \Delta S$ will be made up of two parts: $T \, \Delta S_{rev}$, which arises from the q that is being absorbed and that is converted into work, and $T \, \Delta S_{irrev}$ which arises from the fact that the gas is expanding in part freely, without doing the maximum work of which it is capable. The amount of entropy that comes from the surroundings is now less than the original ΔS, since less work is being done and less heat is being absorbed; but the total entropy acquired by the gas is of course the same as before, since it depends only on the state of the gas and not on how the expansion is carried out. When the gas is recompressed, the entropy given by the gas to its surroundings must be the original ΔS, since again the initial and final states are identical. The surroundings, therefore, acquire ΔS units of entropy but give up only ΔS_{rev} units; the remainder of the entropy acquired by the gas, ΔS_{irrev}, is therefore *new* entropy which was not present originally. In an irreversible process, total entropy is not conserved but increases, and the net increase is symbolized by ΔS_{irrev}. Since all actual processes are irreversible, this means that, *in any energy change, the total entropy of all systems involved increases. The amount of the net entropy increase measures the extent of irreversibility.* This is an alternative way of defining the concept of entropy.

Difficulty in understanding entropy often arises because of confusion regarding the systems under consideration. A "system" is merely a region of space, a part of the universe temporarily marked off for discussion. We have been talking at length about a sample of gas in a cylinder; this gas is a system. Occasionally we have mentioned the surroundings of the cylinder, from which heat can flow and to which heat can be given; the surroundings are a second system. Or we can think of the cylinder plus its surroundings as making up a third, larger system. As the gas expands and contracts, the entropy changes depend on which system we are considering. The important statement at the end of the last paragraph refers to the overall, composite system: in this system, representing all parts of the universe involved in our hypothetical experiments, total entropy remains constant when reversible changes occur and increases when irreversible changes occur. In the individual systems, on the other hand, entropy may increase or decrease or remain constant during various parts of a cycle of changes. In reversible processes the entropy changes in one system, or one part of a system, must be balanced by opposite changes in another system; in irreversible processes the entropy increase in some system or systems is not completely balanced by entropy decreases elsewhere.

THE SECOND LAW

This long discussion of entropy centers attention on some fundamental characteristics of heat energy which differentiate it from all other kinds of energy. These characteristics we can now summarize in the generalization called the second law of thermodynamics.

A very general statement of the second law can be formulated as follows: in any energy transformation, some of the original energy always appears in the form of heat energy which is no longer available for conversion into other forms of energy. If the original energy is heat energy, some of it must flow to a cold region, where it is no longer available for further transformations. If the original energy is of some other kind, part of it always goes into friction or into heating the surroundings, and the energy so lost is no longer available. Heat energy is "available" only when a difference in temperature exists, and even in the best of circumstances only a part of it can be recovered.

More restricted statements of the second law are often useful: (1) Heat always flows from hot objects to cold objects, never spontaneously from cold to hot. (2) It is impossible for a self-acting machine to transfer heat from a cold object to a hot object. (3) Operation of a refrigerator to keep one side of a heat engine cold requires more mechanical energy than is produced by operation of the heat engine.

Entropy provides a means of making such statements quantitative. One not-very-accurate statement of the second law is simply that the total entropy of the universe is increasing. The same idea can be worded more precisely by saying that all natural processes are to some extent irreversible, and the increase in entropy is a measure of the irreversibility. Alternatively we can focus attention on reversible processes and say that the increase in entropy of a system or a part system during a reversible change is a measure of the decrease in its ability to do work. In symbols, the second law may be stated by repeating our original definition of entropy change,

$$\Delta S = \frac{q}{T} \qquad \text{(XI-15)}$$

But this statement is so succinct and requires so much qualifying explanation that it fails to convey the meaning of the second law except to those who have had long practice in its use.

A combination of the first and second laws [Eqs. (XI-1) and (XI-15)] gives

$$\Delta E = T \, \Delta S - w \tag{XI-20}$$

or in differential form

$$dE = T \, dS - dw \tag{XI-21}$$

and if work is restricted to pressure-volume work

$$dE = T \, dS - P \, dV \tag{XI-22}$$

This equation applies to reversible processes if dS is defined as dq_{rev}/T, or it may be treated as a general expression for a closed system if dS is understood to include also the amount of new entropy produced in an irreversible process. From this equation it follows that

$$dH = dE + P \, dV + V \, dP = T \, dS + V \, dP \tag{XI-23}$$

MAXIMUM WORK FROM CHEMICAL REACTIONS

If a chemical reaction occurs in or adjacent to the confined gas we have been discussing, we can use the ideas of preceding paragraphs to derive expressions for the maximum work obtainable from the chemical energy released during the reaction.

The energy of the reaction serves to increase the energy of the gas, so that we may begin with the usual expression for the first law:

$$\Delta E = q - w \tag{XI-1}$$

where ΔE is the energy supplied by the reaction, q is the heat added to the gas, and w the work done by the gas on its surroundings. By analogy with the preceding discussion, we might guess that the work will be a maximum if the reaction is carried out reversibly.

Now what does it mean to carry out a chemical reaction reversibly? This is a much more difficult question than for the simple expansion of a gas. For a mixture near equilibrium, say a mixture of hydrogen, oxygen, and water vapor at high temperatures, one can indeed imagine that the reaction could be carried out so that it could always be reversed by a slight change in external conditions. Or the reaction in an electrolytic cell could be permitted to take place against a potential difference kept nearly as large as that produced by the cell, so that the reaction could be reversed at any time by a slight increase in the external voltage. But for the great majority of chemical reactions a setup permitting an approach to reversibility would be difficult to imagine. Take the gasoline and air in the cylinder of an automobile, for example: to make this reversible, one would have to suppose that some arrangement could be devised so that gasoline and air could be generated from water and carbon oxides by a slight change in external conditions. Despite the difficulty of visualization, the idea of a reaction occurring reversibly is a convenient basis for setting a limit to the possible work obtainable from it.

If a reaction occurs reversibly, the heat produced is measured by the difference in entropy between products and reactants; hence we may write

$$\Delta E = T \, \Delta S - w$$

and the w in this expression is the maximum work that the reaction can perform:

$$w_{max} = T \, \Delta S - \Delta E \tag{XI-24}$$

In any actual process the work obtained would be less than this figure, since more of the energy available would be dissipated in the form of heat.

If a chemical reaction is carried out at constant pressure, some change in volume may be produced by the reaction itself, quite apart from temporary changes that may result from the heat liberated. This change is small in reactions involving only solids and liquids but may be large if gases are produced or consumed. The complete combustion of gasoline, for example, may be symbolized:

$$C_7H_{16} + 11O_2 \rightarrow 7CO_2 + 8H_2O$$

Fifteen gas molecules are produced at the expense of 12, so that, if the products of reaction are brought back to the original pressure, they must occupy a larger volume than the reactants; this has nothing to do with the violent expansion of the heated gases during the explosion, from which the major part of the work is obtained. In considering reactions at constant pressure, it is often useful to separate out this work of necessary overall expansion, measured as $P \, \Delta V$, from the rest of the work produced. For a reversible process,

$$w'_{max} = T \, \Delta S - \Delta E - P \, \Delta V = T \, \Delta S - \Delta H \tag{XI-25}$$

FREE ENERGY

Let two new quantities, G and F, be defined as

$$G = H - TS = E + PV - TS \tag{XI-26}$$

$$F = E - TS \tag{XI-27}$$

G is called the *Gibbs free energy* and F the *Helmholtz free energy*. (Some American writers use F in place of G and A in place of F.) The Gibbs free energy is sometimes called "free enthalpy," or "Gibbs function," and the Helmholtz free energy is sometimes called "work content." For our purposes G is the more important of the two, and it will hereafter be designated simply as "free energy."

Since G and F are defined as differences of terms in E, S, and H, they must, like these quantities, be properties of a system and must vary in amount with the amount of material present. Like E and H (but not like S), G and F do not have readily measurable absolute values, and we are generally concerned only with their changes. To express changes in these quantities, we differentiate the defining equations and make substitutions from Eqs. (XI-23) and (XI-22):

$$dG = dH - T\,dS - S\,dT = V\,dP - S\,dT \qquad \text{(XI-28)}$$

$$dF = dE - T\,dS - S\,dT = -P\,dV - S\,dT \qquad \text{(XI-29)}$$

For a chemical reaction we may write ΔG as the sum of G's for all the products minus the sum of G's for the reactants, and ΔF as a similar summation of individual F's, just as we have previously used ΔE, ΔH, and ΔS to express overall changes in these variables during a reaction. So for the megascopic changes during a reaction that takes place at constant temperature $(dT = 0)$,

$$\Delta G = \Delta H - T\,\Delta S \qquad \text{(XI-30)}$$

$$\Delta F = \Delta E - T\,\Delta S \qquad \text{(XI-31)}$$

Comparison of these equations with Eqs. (XI-24) and (XI-25) shows that

$$w_{max} = T\,\Delta S - \Delta E = -\Delta F \qquad \text{(XI-32)}$$

$$w'_{max} = T\,\Delta S - \Delta H = -\Delta G \qquad \text{(XI-33)}$$

so that ΔF is a measure of the maximum work obtainable from a reaction carried out reversibly and isothermally, and ΔG is a measure of this same maximum work with expansion work subtracted. The signs are reversed because ΔF and ΔG are defined as energy gained by the reacting system, while w_{max} and w'_{max} refer to work performed by the reaction on its surroundings.

If a reaction mixture has reached a state of equilibrium, it is no longer capable of doing work, so that, for infinitesimal changes, $dG = 0$ and $dF = 0$. For an isothermal process $(dT = 0)$, the condition that $dG = 0$ implies that dP also be zero [Eq. (XI-28)], and the condition that $dF = 0$ implies that dV be zero [Eq. (XI-29)]. In other words, for a reaction taking place at constant pressure, a condition of equilibrium is that $dG = 0$ for any infinitesimal change; for a reaction occurring at constant volume, equilibrium requires that $dF = 0$. Since most chemical reactions of geologic interest take place at constant pressure rather than constant volume, G is generally a more useful function than F in defining conditions of equilibrium. We shall henceforth limit the discussion to equations involving G; this means that reactions will be considered isobaric unless specified otherwise.

If a reaction is far from equilibrium and is capable of occurring spontaneously (for example, a mixture of H_2 and O_2 at ordinary temperatures), the maximum work it can do is given by $-\Delta G$. In other words, ΔG serves as a criterion for the occurrence of a spontaneous process: if $\Delta G = 0$, the mixture is at equilibrium and no reaction can occur, but if $\Delta G < 0$, a reaction can take place. If ΔG is a positive quantity, the work the reaction can do is negative work, which means that energy must be supplied in order for the reaction to occur at all. The great importance of ΔG in geochemistry is this characteristic; its sign indicates whether or not a given reaction can occur spontaneously, and its magnitude indicates to what extent the reaction can go before equilibrium is attained. We have seen many examples of this use of ΔG in preceding chapters.

MEASUREMENT OF ΔG

The direct measurement of ΔG for most reactions is a difficult operation. One method, applicable to many oxidation-reduction processes, depends on measurements of electromotive force and use of the relation

$$\Delta G^\circ = -w' = nf E^\circ \tag{XI-34}$$

This equation expresses the fact that electromotive force (E°) multiplied by the amount of electric charge moving through a cell (the charge carried by a mole of electrons, f, times the number of moles, n) gives the maximum electrical work that the cell reaction can accomplish. Another possible method involves measurement of activities of reactants and products when a reaction has reached equilibrium, and use of the equation

$$\Delta G^\circ = -RT \ln K_a$$

where K_a is the activity constant (see page 593). But obviously it would be desirable to have a general method for determining free-energy changes from heat measurements alone. Such a method might be based on the relation

$$\Delta G = \Delta H - T \Delta S \tag{XI-30}$$

if ΔH and ΔS can be determined independently. The ΔH term is easily found by measuring the heat of reaction in a calorimeter. Now if we had a way of obtaining ΔS also, the problem of measuring ΔG would be solved.

THE THIRD LAW

It turns out that values of ΔS can be found by determining absolute entropies from heat-capacity measurements.

From the defining equations for heat capacity and entropy change,

$$c = \frac{dq}{dT} \quad \text{and} \quad dS = \frac{dq}{T}$$

it follows that

$$S = \int \frac{c\,dT}{T} \qquad \text{(XI-35)}$$

If we knew an absolute value for S at one temperature, its value at any other temperature could be found by carrying out this integration. Arbitrarily let us assign a value of zero to the entropy of some crystal form of each element at the absolute zero of temperature. Then the entropy of this crystal form at a temperature T can be determined by measuring its heat capacity over the temperature interval 0 to T (involving slight extrapolation at the lower limit, since absolute zero is experimentally unattainable) and integrating between these limits. Now a striking experimental fact emerges if we then determine the entropy differences between various crystal forms of the same element: the differences become smaller and smaller as the temperature drops toward absolute zero. In other words, if we make the assumption of zero entropy at $T = 0$ for one crystal form, then the entropy of other crystal forms is also zero at this temperature. Even more striking is the experimental fact that entropy differences in reactions involving combinations of crystalline elements to form pure crystalline compounds also fall toward zero as T approaches zero. We can generalize from such experimental results: *if the entropy of each element in some crystalline state is assumed to be zero at $T = 0$, then the entropy of other pure crystalline solids is also zero at this temperature.* This statement is called the third law of thermodynamics. Although formulated much later than the first and second laws and for a time seriously questioned, its validity has now been established by many different kinds of experiments.

The third law makes possible the determination of absolute entropies for most pure crystalline solids. (It should be noted that purity and perfection of crystal form are essential in a careful statement of the third law; glassy solids, solid solutions, and imperfect crystals would all have finite positive entropies at absolute zero.) Entropies of liquids and gases are obtainable from those of the corresponding solids by adding the entropies of melting or vaporization (heats of fusion or vaporization divided by absolute temperature at the transition points; see page 588). Absolute entropies of gases may be calculated also from spectroscopic data by the methods of statistical mechanics. Entropies of ions and of undissociated molecules in solution may be found, with a few additional assumptions, from heats of solution. Thus absolute entropies are obtainable, at least in principle, for most substances of geochemical interest. Some of these are listed in Appendix VIII.

The ΔS's for reactions are simply differences between the combined absolute entropies of reactants and products. These, multiplied by the absolute temperature and subtracted from ΔH's, give values for ΔG. This is the most generally useful method of finding free energies of reaction. Furthermore, if ΔG for a reaction has been found at one temperature, the same equation ($\Delta G = \Delta H - T\,\Delta S$) may be used to get approximate values at other temperatures, since ΔH and ΔS for many reactions are nearly constant over moderate temperature ranges. (For an example, see Sec. 8-11.)

RELATIONS AMONG THERMODYNAMIC FUNCTIONS

Several useful equations may be derived by applying Eqs. (XI-26), (XI-28), and (XI-30) to special situations.

Change of ΔG with Temperature

If pressure is constant, the term $V\,dP$ drops out of Eq. (XI-28), and we may write

$$\left(\frac{\partial G}{\partial T}\right)_P = -S \tag{XI-36}$$

From Eq. (XI-26), this is equivalent to

$$\left(\frac{\partial G}{\partial T}\right)_P = \frac{G - H}{T} \tag{XI-37}$$

In a reaction, since the same expression can be set up for reactants and products and since the pressure restriction applies to both,

$$\left(\frac{\partial \Delta G}{\partial T}\right)_P = -\Delta S = \frac{\Delta G - \Delta H}{T} \tag{XI-38}$$

If the entropy change in a reaction, for example, is 5 cal/mole deg, the free-energy change must decrease by 5 cal for each degree rise in temperature. To put Eq. (XI-38) in another useful form, note that, by the elementary rules of differentiation,

$$\frac{d(\Delta G/T)}{dT} = \frac{1}{T}\frac{d\Delta G}{dT} + \Delta G\frac{d(1/T)}{dT} = \frac{1}{T}\frac{d\Delta G}{dT} - \frac{\Delta G}{T^2} \tag{XI-39}$$

Multiply Eq. (XI-38) by $1/T$:

$$\frac{1}{T}\frac{d\Delta G}{dT} = \frac{\Delta G}{T^2} - \frac{\Delta H}{T^2} \tag{XI-40}$$

and substitute in Eq. (XI-39):

$$\frac{d(\Delta G/T)}{dT} = -\frac{\Delta H}{T^2} \tag{XI-41}$$

or

$$\frac{\Delta G}{T} = -\int \frac{\Delta H}{T^2}\,dT \tag{XI-42}$$

This is the integral we have used before to find an expression for the change of ΔG with temperature in cases where ΔH cannot be considered constant (Sec. 8-11).

Change of ΔG with Pressure

By a similar argument, at constant temperature the term $S \, dT$ in Eq. (XI-28) becomes zero and we may write

$$\left(\frac{\partial G}{\partial P}\right)_T = V \quad \text{and} \quad \left(\frac{\partial \Delta G}{\partial P}\right)_T = \Delta V \tag{XI-43}$$

If the products of a reaction have a molal volume 2 cm^3 greater than the reactants, for example, the free-energy change increases by 2 cm^3-atm, or 0.048 cal, as the total pressure increases by 1 atm. For a reaction involving gases the change in ΔV is relatively large, so large that changes in molal volumes of solids and liquids may be neglected in comparison. If perfect-gas behavior is assumed, the gas law gives

$$\Delta V = \frac{RT}{P} \tag{XI-44}$$

for the net evolution of one mole of gas. Therefore

$$\Delta G = \int \Delta V \, dP = \int \frac{RT}{P} \, dP = RT \int d \ln P \tag{XI-45}$$

The change in ΔG when the pressure is increased from P_1 to P_2 is given by the definite integral

$$\Delta G_2 - \Delta G_1 = RT \int_{P_1}^{P_2} d \ln P = RT \ln \frac{P_2}{P_1} \tag{XI-46}$$

Consider now this last equation as applied to the special case of vaporization of a liquid. If we define ΔG for $P_1 = 1$ atm as a standard free-energy change, $\Delta G°$, the equation becomes

$$\Delta G_2 - \Delta G° = RT \ln P_2 \tag{XI-47}$$

At some one value of P_2, defined as the vapor pressure at the given temperature, the liquid and its vapor are in equilibrium and $\Delta G_2 = 0$. Hence, if the vapor pressure is P,

$$-\Delta G° = RT \ln P \tag{XI-48}$$

For this simple case the vapor pressure is identical with the equilibrium constant of the reaction liquid $\rightleftharpoons$ vapor, so that

$$\Delta G° = -RT \ln K \tag{XI-49}$$

This is an equation we have used many times. To show that it is true for any equilibrium constant, not only for the vapor pressure of a liquid, requires a longer argument (page 593).

General Equation for Absolute Entropy

Now consider Eq. (XI-30) as applied to a reaction mixture at equilibrium. Equilibrium requires that $\Delta G = 0$, so Eq. (XI-30) becomes

$$\Delta H - T\Delta S = 0$$

and therefore

$$\Delta S = \frac{\Delta H}{T} \tag{XI-50}$$

When two phases are in equilibrium at their transition point (for example, liquid and vapor at the boiling point, solid and liquid at the freezing point, two polymorphous solids at the transition point), this means that the entropy change is equal to the latent heat of transition divided by the absolute temperature.

Using this relation, we may refine the calculation of entropy for a crystalline solid at a given temperature: the total entropy is the sum of entropy changes at all transition points as given by Eq. (XI-50), plus entropy changes between transition points as given by Eq. (XI-35). Thus, if a substance has transition points at T_1, $T_2, \ldots$, its absolute entropy at a temperature T_n is

$$S = \int_0^{T_1} \frac{C}{T} dT + \frac{\Delta H_1}{T_1} + \int_{T_1}^{T_2} \frac{C}{T} dT + \frac{\Delta H_2}{T_2} + \cdots + \int_{T_{n-1}}^{T_n} \frac{C}{T} dT \tag{XI-51}$$

The Clapeyron, Clausius-Clapeyron, and Van't Hoff Equations

From Eq. (XI-28) for a reaction at equilibrium, we may write for free-energy changes of reactants and products resulting from infinitesimal changes in temperature and pressure

$$dG_{pr} = V_{pr}\, dP - S_{pr}\, dT$$

$$dG_{re} = V_{re}\, dP - S_{re}\, dT$$

Subtracting gives

$$dG_{pr} - dG_{re} = d\Delta G = \Delta V\, dP - \Delta S\, dT \tag{XI-52}$$

Now if equilibrium is maintained as P and T change, ΔG will remain at all times 0 and $d\Delta G$ must also be 0. Hence the condition for maintenance of equilibrium may be expressed

$$\Delta V\, dP - \Delta S\, dT = 0 \tag{XI-53}$$

or

$$\frac{dP}{dT} = \frac{\Delta S}{\Delta V} \tag{XI-54}$$

Substitution of Eq. (XI-50) gives

$$\frac{dP}{dT} = \frac{\Delta H}{T\, \Delta V} \tag{XI-55}$$

This is the Clapeyron equation, derived originally to show the dependence of vapor pressure on temperature, enthalpy change, and volume change. The above derivation is perfectly general, however, so the equation holds as well for solid-solid and liquid-solid transitions as for vaporization.

For the special case of vaporization, this equation may be modified by substituting for ΔV its value as given by the perfect-gas law for one mole [Eq. (XI-44)]:

$$\frac{dP}{dT} = \frac{\Delta H}{T}\frac{P}{RT} \qquad \text{or} \qquad \frac{d\ln P}{dT} = \frac{\Delta H}{RT^2} \qquad \text{(XI-56)}$$

Equation (XI-56) is the Clausius-Clapeyron equation. In this simple case we note once more that P is identical with the equilibrium constant, so that

$$\frac{d\ln K}{dT} = \frac{\Delta H}{RT^2} \qquad \text{(XI-57)}$$

This is another familiar relation, often called the van't Hoff equation, which we have used previously to calculate the change of equilibrium constant with temperature (Sec. 8-2). Like Eq. (XI-49), it can be shown to hold for equilibrium constants in general (page 593).

SOLUTIONS

The discussion so far has concerned pure substances exclusively, but the relations we have developed require only slight modification to be applicable to solutions also.

Consider first a solution whose constituents can be mixed without appreciable energy change; two unreactive gases at low pressure, or two isotopes of the same element, would be good examples. The free energy of the mixture will be simply the sum of the free energies of its constituents. If G_1, G_2, and so on, represent the free energies of one mole of each constituent in the pure state, and if x_1, x_2, and so on, represent the mole fractions of each constituent in the solution, the free energy per mole of the solution may be symbolized

$$G = G_1 x_1 + G_2 x_2 + \cdots \qquad \text{(XI-58)}$$

Clearly any of these free energies—the molal free energies G, G_1, G_2, and so on, or the constituent free energies $G_1 x_1, G_2 x_2$, and so on—can be used in any of the free-energy formulas we have developed in preceding paragraphs. Solutions are seldom as simple as this, but if no other information about a solution is available, the assumption of additivity of free energies often gives useful approximate results.

Now suppose, still in our simple ideal solution, that the number of moles of each constituent is changed by an infinitesimal amount dn_1, dn_2, and so on. Some of the dn's will be positive, some negative, if the amount of the whole solution is to remain constant. The resulting change in the overall free energy is

$$dG = G_1\ dn_1 + G_2\ dn_2 + \cdots \tag{XI-59}$$

If only a single constituent changes, all other dn's become 0, and the equation reduces to

$$dG = G_1\ dn_1 \quad \text{or} \quad G_1 = \left(\frac{\partial G}{\partial n_1}\right)_{n_2, n_3, \ldots} \tag{XI-60}$$

(The partial derivative merely emphasizes the constancy of the other constituents.) Equation (XI-60) tells us that the rate of change of total free energy of the solution per mole of any constituent is equal to the molal free energy of that constituent—a statement so obvious as to be silly, for the simple solution we have been considering.

In most solutions the situation is complicated by the energy change on mixing; recall, for example, the heat evolved when sulfuric acid dissolves in water. Now the total free energy can no longer be a simple sum of free energies of pure constituents. We can still set up an equation like (XI-58), however, provided that we *define* G_1, G_2, and so on, by Eq. (XI-60). In other words, we *make* Eq. (XI-58) true by changing the definition of individual free energies: G_1 is now the rate of change of total free energy per mole of constituent 1, which is no longer the same as the molal free energy of the pure constituent. We call G_1 so defined the *partial molal free energy* (Sec. 8-14) of constituent 1, and write it with a barred symbol:

$$\bar{G}_1 = \left(\frac{\partial G}{\partial n_1}\right)_{T, P, n_2, n_3, \ldots} \tag{XI-61}$$

In a similar manner we define partial molal quantities for any of the extensive thermodynamic properties we have discussed:

$$\left(\frac{\partial V}{\partial n_1}\right) = \bar{V}_1 \qquad \left(\frac{\partial E}{\partial n_1}\right) = \bar{E}_1 \qquad \left(\frac{\partial H}{\partial n_1}\right) = \bar{H}_1$$
$$\left(\frac{\partial S}{\partial n_1}\right) = \bar{S}_1 \qquad \left(\frac{\partial C}{\partial n_1}\right) = \bar{C}_1 \tag{XI-62}$$

P, T, and all n's except n_1 are assumed to be held constant.

Both the molal free energy of a pure substance and the partial molal free energy of a substance in solution are often given another name, the *chemical potential*, symbolized by the Greek letter μ.

Note that partial molal free energies (and other partial molal quantities), unlike the simple molal quantities, change in a complex manner as the strength of the solution changes. The measurement of partial molal quantities at various dilutions is often a complicated operation, which we cannot discuss here. But once these quantities are known, they may be used in place of the corresponding properties of pure substances in any of the equations we have derived; for example,

$$\left(\frac{\partial \bar{G}_1}{\partial T}\right)_P = \left(\frac{\partial \mu_1}{\partial T}\right)_P = -\bar{S}_1 \qquad \left(\frac{\partial \bar{G}_1}{\partial P}\right)_T = \bar{V}_1$$

$$\bar{G}_1 = \mu_1 = \bar{H}_1 - T\bar{S}_1$$

$$\left(\frac{\partial \bar{E}_1}{\partial T}\right)_V = \bar{C}_V \qquad \left(\frac{\partial \bar{H}_1}{\partial T}\right)_P = \bar{C}_P$$

The partial molal free energy provides a convenient means of defining equilibrium for reactions involving several phases, some or all of which may be solutions. Equilibrium demands that no substance have a tendency to move from one phase to another, in other words that the "escaping tendency" (Sec. 8-14) of each substance be the same in all phases. This means that the molal free energy in all phases must be the same; for if a substance had a higher free energy in one phase than in another, the free energy of the system would be lowered by movement of the substance into the second phase, and the system by definition would not be in equilibrium. Hence for any hypothetical movement from one phase to another, the differentials dG_1, dG_2, and so on, must equal zero. If some of the phases are solutions, we need only use $d\bar{G}_1$, $d\bar{G}_2$, and so on, in the same statement. In other words, a general criterion of equilibrium is simply that the chemical potential of each substance must be the same in every phase.

FUGACITY AND ACTIVITY

In developing Eqs. (XI-44) through (XI-48), and again in Eq. (XI-56), we have made the assumption of perfect-gas behavior. This assumption is never completely accurate but leads to little error for gases at low pressures and well above the boiling points of the corresponding liquids. At high pressures and near the boiling points, however, the assumption is far from accurate. To make the equations valid, pressure should be replaced by *fugacity*, as defined previously [Eq. (8-25)]:

$$d \ln f = \frac{dG}{RT} \qquad \text{(at constant temperature)} \qquad \text{(XI-63)}$$

where f and G are the fugacity and molal free energy of the gas. In effect, this means that we define fugacity so that Eq. (XI-45) will be true if f is substituted for P, and then devise means to measure f for various values of P. Clearly the two become equal for any substance that behaves as a perfect gas.

Fugacities may be used for solids and liquids as well as gases. We may speak of the fugacity of a vapor above a liquid, or the fugacity of the liquid itself, just as the vapor pressure of a liquid may be considered a property of either the vapor or the liquid. Vapor pressure, in fact, is equal to fugacity as long as the vapor behaves as a perfect gas. Measurement of vapor pressures is the most common method of obtaining approximate values of fugacity for solids and liquids; for greater accuracy the usual gas-law corrections must be applied. Fugacities of constituents of solutions may be defined by

$$d \ln f_1 = \frac{d\bar{G}_1}{RT} \tag{XI-64}$$

and for volatile solutes may be measured (approximately) by determining partial vapor pressures.

Just as Eqs. (XI-48) and (XI-56) contain the assumption of perfect-gas behavior, so do the more general equations (XI-49) and (XI-57) imply that solutions taking part in an equilibrium behave as ideal solutions. This assumption is often very far from true, even when the solutions are fairly dilute.

To make the equations accurate in the general case, we use in place of concentration the *activity* (Secs. 3-2 and 8-14). This is defined as a ratio of fugacities,

$$a = \frac{f}{f^\circ} \tag{XI-65}$$

where f° is fugacity in some standard state, often taken as a hypothetical solution of unit concentration that obeys Henry's law. This definition is perfectly general, but for nonvolatile solutes some other method than vapor-pressure measurements must be used to determine fugacities.

The meaning of Eq. (XI-65) may be clarified by applying it first to an ideal solution. Suppose that a constituent of such a solution has unit concentration, say $1M$. Its fugacity, by definition, would be f°. Now if the concentration of this constituent is increased to $2M$, Henry's law for an ideal solution tells us that the fugacity will also be doubled, so that $a = f/f^\circ = \frac{2}{1} = 2$. Thus for an ideal solution the activity is numerically equal to concentration. In real solutions, however, fugacity is not proportional to concentration except at extreme dilutions, so that in concentrated solutions activities and concentrations may deviate widely.

THE EQUILIBRIUM CONSTANT

Equation (XI-63) may be integrated to give

$$G - G' = RT \ln \frac{f}{f'} \tag{XI-66}$$

where f and f' are fugacities, and G and G' free energies, of a substance in two different states or at two different pressures (but at the same temperature). If f' is taken as fugacity in a standard state, f°, and if G° is the corresponding free energy, the equation becomes

$$G - G^\circ = RT \ln \frac{f}{f^\circ} = RT \ln a \tag{XI-67}$$

Or, if the substance is a constituent of a solution,

$$\bar{G} - \bar{G}° = RT \ln a \qquad \text{(XI-68)}$$

Now for a chemical reaction

$$bB + dD = yY + zZ$$

Eq. (XI-67) or (XI-68) can be set up for each substance. The overall change in free energy during the reaction would then be

$$\Delta G = yG_Y + zG_Z - bG_B - dG_D \qquad \text{(XI-69)}$$

$$= \Delta G° + RT \ln \Delta a \qquad \text{(XI-70)}$$

$$= yG_Y° + zG_Z° - bG_B° - dG_D° + RT \ln a_Y^y$$

$$+ RT \ln a_Z^z - RT \ln a_B^b - RT \ln a_D^d \qquad \text{(XI-71)}$$

The symbols in these equations are ones we have used many times before: B, D, Y, Z are chemical formulas, b, d, y, z are coefficients, G_Y is the molal free energy of substance Y, a_Y is the activity of Y; each free energy and each activity is taken as many times as the coefficient indicates, since the G's refer to free energies per mole. We now combine the $G°$'s into $\Delta G°$ and the logarithms into a quotient:

$$\Delta G = \Delta G° + RT \ln \frac{a_Y^y a_Z^z}{a_B^b a_D^d} \qquad \text{(XI-72)}$$

At equilibrium, $\Delta G = 0$, and

$$\Delta G° = -RT \ln \frac{a_Y^y a_Z^z}{a_B^b a_D^d} \qquad \text{(XI-73)}$$

where the activities now are the particular values for equilibrium between products and reactants. This quotient must be a constant if ΔG is to remain zero. We call it the equilibrium constant, and write

$$\Delta G° = -RT \ln K_a \qquad \text{(XI-74)}$$

where the subscript on the K specifies that this is a quotient of activities. Equations (XI-72) and (XI-74) are identical with Eqs. (8-13) and (8-15).

An expression for the variation of the equilibrium constant with temperature can be derived by writing Eq. (XI-74) in the form

$$\ln K = -\frac{1}{R}\left(\frac{\Delta G°}{T}\right)$$

and differentiating:

$$\frac{d \ln K}{dT} = -\frac{1}{R}\frac{d(\Delta G°/T)}{dT}$$

From Eq. (XI-41) this is equivalent to

$$\frac{d \ln K}{dT} = \frac{\Delta H^\circ}{RT^2} \tag{XI-75}$$

This is the van't Hoff equation, which we have used many times before [Eq. (8-2)], and which we derived for the special case of vaporization as Eq. (XI-57).

DERIVATION OF THE PHASE RULE

In Chap. 13 the phase rule was derived inductively, by showing that it would express observed relations among a number of phases, a number of components, and a number of degrees of freedom for several simple systems. The rule may be derived more generally in several ways, of which the following is an example.

We consider a general system made up of c components (c_1, c_2, c_3, etc.) distributed among p phases (p_1, p_2, p_3, ...). Each component has a certain chemical potential (molal free energy, or partial molal free energy) in each phase; we designate the chemical potential of component c_1 in phase p_1 by $\mu_{c_1}^{p_1}$, etc. The condition for equilibrium to exist is that the chemical potential of c_1 must be the same in all phases, of c_2 must be the same in all phases, and so on:

$$\mu_{c_1}^{p_1} = \mu_{c_1}^{p_2} = \mu_{c_1}^{p_3} \cdots$$
$$\mu_{c_2}^{p_1} = \mu_{c_2}^{p_2} = \mu_{c_2}^{p_3} \cdots$$
$$\cdots\cdots\cdots\cdots\cdots$$

For each component the number of these equations connecting the μ's is $(p - 1)$, and the total number of equations is therefore $c(p - 1)$.

The chemical potential of each component is a function of temperature, pressure, and the concentration of the component in each phase. To express concentration we use mole fractions; other concentration units might be used, but mole fractions have the advantage that for each phase they add up to unity. The mole fraction of component c_1 in phase p_1 is written $x_{c_1}^{p_1}$, etc., and the μ's may be expressed:

$$\mu_{c_1} = \text{function } (T, P, x_{c_1}^{p_1}, x_{c_1}^{p_2}, x_{c_1}^{p_3} \cdots)$$
$$\mu_{c_2} = \text{function } (T, P, x_{c_2}^{p_1}, x_{c_2}^{p_2}, x_{c_2}^{p_3} \cdots)$$
$$\cdots\cdots\cdots\cdots\cdots\cdots$$

In these equations the total number of variables is $2 + cp$. (The "2" refers to T and P, which are uniform throughout the whole system, and cp is the sum of all the mole fractions in all the phases.)

Because the sum of mole fractions in each phase is equal to 1, a set of equations connecting the x's can be written:

$$x_{c_1}^{p_1} + x_{c_2}^{p_1} + x_{c_3}^{p_1} \cdots = 1$$
$$x_{c_1}^{p_2} + x_{c_2}^{p_2} + x_{c_3}^{p_2} \cdots = 1$$
$$\cdots\cdots\cdots\cdots\cdots\cdots$$

The number of these equations is p, and this number added to the number of equations for the μ's gives a total of $c(p-1)+p$ equations expressing relationships among the $2+cp$ variables.

Now if the number of equations is equal to the number of variables, the system is completely determined; the set of equations can be solved to give specific values for all the variables. Thus if $c(p-1)+p=2+cp$, or if $p=c+2$, equilibrium can exist only at particular values of P, T, and concentrations of the various components in each phase. In other words, if the number of phases is 2 more than the number of components, the system is invariant: any change in pressure, temperature, or concentration will require the disappearance of at least one phase. On the other hand, if the number of phases is less than $c+2$, meaning that the number of variables is greater than the number of equations, the system is partly indeterminate. One or more variables may be given arbitrary values without destroying equilibrium. The number of variables that may be changed is called the "variance," or the "number of degrees of freedom," and is found by subtracting the number of equations from the number of variables:

$$\text{Variance} = f = 2 + cp - [c(p-1)+p] = c - p + 2$$

This is Gibbs' phase rule, derived now for any system.

As an illustration, consider a system in which solid NaCl is in equilibrium with a solution containing both NaCl and KCl. The number of components is 3, the number of phases is 2. We write 5 equations, 3 connecting μ's and 2 connecting x's:

$$\mu_{KCl}^{solution} = \mu_{KCl}^{solid} \qquad \mu_{NaCl}^{solution} = \mu_{NaCl}^{solid} \qquad \mu_{H_2O}^{solution} = \mu_{H_2O}^{solid}$$

$$x_{KCl}^{solution} + x_{NaCl}^{solution} + x_{H_2O}^{solution} = 1 \qquad x_{KCl}^{solid} + x_{NaCl}^{solid} + x_{H_2O}^{solid} = 1$$

Some of these terms, of course, are vanishingly small (for example, the mole fraction of KCl in the solid NaCl). The total number of variables is 8:

$$P,\ T,\ x_{NaCl}^{solid},\ x_{KCl}^{solid},\ x_{H_2O}^{solid},\ x_{NaCl}^{solution},\ x_{KCl}^{solution},\ x_{H_2O}^{solution}$$

and the variance is therefore $8-5=3$. The same number is obtained from the phase rule:

$$f = c - p + 2 = 3 - 2 + 2 = 3$$

This means that changes may be made in pressure, temperature, and concentration of KCl without destroying equilibrium between solid salt and solution.

ANSWERS TO END-OF-CHAPTER PROBLEMS

CHAPTER 1

3 (a) $6.7 \times 10^{-5} M$; 6.7×10^{-4} g/100 ml; 2.7 ppm Ca. (b) $9.0 \times 10^{-8} M$. (c) 7,500.

4 $[Ag^+]^2[SO_4^{2-}] = 6.7 \times 10^{-5}$.

5 $[O_2] = 4 \times 10^{-88}$ atm $= 1 \times 10^{-65}$ molecule/liter (equivalent to a single molecule in a space equal to the size of the galaxy).

7 $[Pb^{2+}]/[Zn^{2+}] = 10^{-2.8}$. In a solution where $[Zn^{2+}]/[Pb^{2+}] = 100$, sphalerite would be replaced by galena.

9 $[Ca^{2+}] = 5.8 \times 10^{-3} M$, $[Ba^{2+}] = 1.7 \times 10^{-8} M$, $[SO_4^{2-}] = 5.8 \times 10^{-3} M$.

CHAPTER 2

1 Large concentrations: Na^+, S^{2-}, HS^-, OH^-. Small concentrations: H_2S, H^+.

3 Equivalents of + charge: $1.95 + 0.33 = 2.28$ equiv/liter. Equivalents of − charge: $2.31 + 0.02 = 2.33$ equiv/liter. Hence solution is acid. Concentration of $H^+ = 0.05M$, and pH $= 1.3$.

4 $(H^+) = 10^{-5.2} M$, pH $= 5.2$, fraction ionized $= 10^{-1.2} = 0.06 =$ about 6%. In $0.01M$ H_2CO_3, fraction ionized is 0.006, or 0.6%.

5 120 ppm $SiO_2 = 0.12$ g/liter $= 0.0020M$ $H_4SiO_4 = 10^{-2.7} M$ H_4SiO_4; pH $=$ approx. 6.3.

6 $[PbCl_2]/[PbCl^+] = 10^{0.2}$; $[PbCl^+]/[Pb^{2+}] = 10^{1.6}$.

7 pH = 11.7.

8 pH = approx. 12.0.

9 At pH 4: $Al(OH)_2^+ = 10^{-2.3}M$, $AlOH^{2+} = 10^{-2.6}M$, $Al^{3+} = 10^{-1.6}M$, $Al(OH)_4^- = 10^{-8.9}M$. Dominant ion: Al^{3+}. Total concentration of $Al = 10^{-1.5}M = 0.03M$.

10 At pH 8: $FeOH^+ = 10^{-4.6}M$, $Fe^{2+} = 10^{-3.1}M$, $Fe(OH)_2(aq) = 10^{-4.3}M$, $Fe(OH)_3^- = 10^{-11.1}M$. A hydroxide is appreciably amphoteric only if $-\log K_A$ is less than about 2.

11 pH = 6.7 for (a) and pH = 5.0 for (b). The latter is more realistic. Precipitation of $Zn(OH)_2$ from $0.001M$ Zn^{2+} begins at pH 7.7, for either (a) or (b).

16 $[H_2CO_3] = 0.07M$; $[HCO_3^-] = 0.03M$; $[CO_3^{2-}] = 10^{-5.8}M$. Solution is unsaturated with $CaCO_3$.

17 $HCO_3^- + H^+ \rightarrow H_2CO_3$; pH = 5.9.

CHAPTER 3

5 $I = 0.8$, $a_{Ca^{2+}} = 0.0023M$.

6 Solubility of calcite = $0.003M$; pH = 6.7.

7 (a) $10^{-5.4}$. (b) No effect. (c) Decreased.

8 Molar concentrations: $[Ca^{2+}] = 0.01M$, $[F^-] = 6.9 \times 10^{-5}M$. Activities: $a_{Ca^{2+}} = 2.3 \times 10^{-3}M$, $a_{F^-} = 4.8 \times 10^{-5}M$. Product of ion activities = $a_{Ca^{2+}} \cdot a_{F^-}^2 = 5.3 \times 10^{-12} = 10^{-11.3}$; this is less than the solubility product. Authigenic fluorine-containing sedimentary minerals: fluorapatite, clays.

10 $K = a_{CO_3^{2-}}/a_{F^-}^2 = 10^{2.15}$. Calcite would replace fluorite.

12 Celestite.

15 Ion product $[Ba^{2+}][SO_4^{2-}] = 10^{-8.38}$ and product of activities is $10^{-9.66}$. This is close to the standard activity product, $10^{-10.0}$, showing that seawater is approximately saturated with $BaSO_4$.

CHAPTER 4

4 (a) $CaCO_3 + H_2CO_3 \rightarrow Ca^{2+} + 2HCO_3^-$

(b) $2Ca_3Al_2Si_3O_{12} + 12H_2CO_3 + 2H_2O \rightarrow 6Ca^{2+} + Al_4Si_4O_{10}(OH)_8 + 2H_4SiO_4 + 12HCO_3^-$

(c) $ZnS + 2H_2CO_3 \rightarrow H_2S + Zn^{2+} + 2HCO_3^-$

(d) $4NaAlSiO_4 + 6H_2O \rightarrow 4Na^+ + Al_4Si_4O_{10}(OH)_8 + 4OH^-$

CHAPTER 7

8 See Ans. Fig. 7-1.

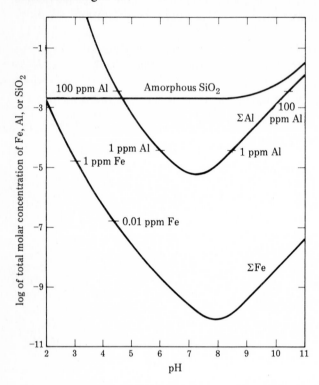

CHAPTER 8

1 $\Delta H°$: -138.8, $+4.4$, -40.8 kcal; $\Delta G°$: -124.2, -1.2, -43.1 kcal. More stable: tin oxide, hematite.

2 $K = 10^{-11.4}$ at $100°$ and $10^{-7.5}$ at $200°$; $(H_2)/(H_2O) = 10^{-10.4}$ at $100°$ and $10^{-6.5}$ at $200°$.

3 $K = 10^{5.7}$ at $25°$ and $10^{-1.2}$ at $250°$.

4 Solubility product $= 10^{-28.0}$; calculated solubility $= 10^{-14.0}M$. Hydrolysis of Pb^{2+} and S^{2-} would increase solubility.

5 At $100°C$: $\Delta G = -22{,}770$ cal, $\log K = +13.3$, $[H_2] = [Cl_2] = 10^{-13.3}$ atm. At $1000°C$: $\Delta G = -24{,}500$ cal, $\log K = +4.2$, $[H_2] = [Cl_2] = 10^{-4.2}$ atm.

6 $K_1 = 10^{-6.2}$ and $K_2 = 10^{-11.1}$ at $100°C$, so $[S^{2-}]$ is higher.

7 For reaction $PbS + 2OH^- + H_2O \rightleftharpoons Pb(OH)_3^- + HS^-$ $\Delta G = +20.8$ kcal and $\log K = -15.3$. Even at $pH = 14$, $Pb(OH)_3^-$ is only $10^{-7.7}M$, so solubility is not appreciable. For reaction $PbS + 3OH^- \rightleftharpoons Pb(OH)_3^- + S^{2-}$, $\Delta G = 19.3$ kcal and $\log K = -14.1$. So at $pH = 14$, $Pb(OH)_3^- = 10^{-7.1}M$, again too small.

CHAPTER 9

1 Decreasing pH: C, A, D, E, G, B, F (positions of A and D, and of E and G, may be reversed). Decreasing Eh: F, C, A, E, B, D, G (positions of C, A, and E uncertain).

4 $2Fe^{2+} + UO_2^{2+} + 4H^+ \rightleftarrows 2Fe^{3+} + U^{4+} + 2H_2O$
$6Cl^- + 2MnO_4^- + 8H^+ \rightleftarrows 3Cl_2 + 2MnO_2 + 4H_2O$
$2V(OH)_3 + O_2 + 6OH^- \rightleftarrows 2VO_4^{3-} + 6H_2O$
$3CH_4 + 4SO_4^{2-} + 8H^+ \rightleftarrows 3CO_2 + 4S + 10H_2O$

5 (a) $E° = -1.52$ volts, $\Delta G° = -70.1$ kcal. (b) $E° = +0.30$ volt, $\Delta G° = +13.8$ kcal. (c) $E° = -0.62$ volt, $\Delta G° = -85.8$ kcal. (d) $E° = +0.15$ volt, $\Delta G° = +20.8$ kcal. (e) $E° = -1.01$ volts, $\Delta G° = -46.6$ kcal.

6 $0.044M$.

7 $+0.77$ volt.

CHAPTER 10

2 At pH 6.5 and Eh $+0.30$ volt, $a_{Fe^{2+}} = 10^{-8.0}M$. At pH 8.4 and Eh -0.30 volt, $a_{Fe^{2+}} = 10^{-3.6}M$.

3 $MnO_2 + 2Fe(OH)_2 + 2H_2O \rightleftarrows Mn(OH)_2 + 2Fe(OH)_3$ $\Delta G° = -22.8$ kcal
$MnO_2 + 2Fe^{2+} + 4H_2O \rightleftarrows Mn^{2+} + 2Fe(OH)_3 + 2H^+$ $\Delta G° = -11.7$ kcal
For the second reaction, $\log K = \log a_{Mn^{2+}} + 2 \log a_{H^+} - 2 \log a_{Fe^{2+}} = 8.6$. If pH $= 4$, $\log a_{Mn^{2+}} - 2 \log a_{Fe^{2+}} = 16.6$. Hence $a_{Mn^{2+}} \gg a_{Fe^{2+}}$ for $a_{Fe^{2+}} >$ approximately $10^{-14}M$.

7 For $a_{Fe^{2+}} = a_{Ca^{2+}} = 10^{-4}M$ and total carbonate $= 1M$, limits are approximately Eh < 0.0 volt and pH between 5.0 and 6.5. The range is greater if $a_{Fe^{2+}} > a_{Ca^{2+}}$.

8 At Eh 0.4 volt and pH 8.2, $a_{Mn^{2+}}$ in equilibrium with MnO_2 is $10^{-4.7}M$, or 1.1 ppm. In seawater the concentration required for precipitation is four or five times greater than this.

9 For line B: $H_2S + 4H_2O \rightleftarrows SO_4^{2-} + 10H^+ + 8e^-$
Eh $= E° + 0.0074 \log a_{SO_4^{2-}} - 0.0074 \log a_{H_2S} - 0.074$pH
If $a_{SO_4^{2-}} = a_{H_2S}$, this becomes Eh $= E° - 0.074$pH. Lines A, B, C are not changed by increasing total dissolved sulfur; lines D and E would shift so as to make the field of native sulfur larger.

CHAPTER 15

2 (a) 2,700 atm (assuming 2.7 g/cm³ as density of average rock); (b) 5,400 atm; (c) 100 to 500°C; (d) approximately 700°C; (e) 2.7×10^{18} g; (f) 11.0×10^{18} cm³; (g) 9.3×10^{18} cm³.

3 0.8 km if pressure assumed lithostatic; 2.2 km if pressure assumed hydrostatic.

CHAPTER 16

1 For $2H_2 + O_2 \rightleftarrows 2H_2O$, ΔG at $1000°C$ is -88.6 kcal and $\log K = 15.2$. For $H_2 + Cl_2 \rightleftarrows 2HCl$, $\log K = 8.6$. In a mixture of 1 atm H_2O and 0.1 atm HCl, $[H_2] =$ approximately $10^{-5.0}$ atm, $[O_2] = 10^{-5.2}$ atm, $[Cl_2] = 10^{-5.6}$ atm.

3 $\Delta H° = -33.4$ kcal; $\Delta G° = -39.0$ kcal at $25°C$ and -52.7 kcal at $727°C$. In the reaction, four gas molecules are produced from three. Hence high temperature and low pressure favor precipitation of SiO_2; low temperature, high pressure, and a high $[HCl]^2/[H_2O]$ ratio favor volatilization.

4 $\log K = -2.1$ at $627°$ and -1.5 at $827°C$. Hence deposition of galena is favored by falling temperature and a high $[H_2S]/[HCl]^2$ ratio. Total pressure has little effect.

5 For reaction $N_2 + 3H_2 \rightleftarrows 2NH_3$, $\Delta G = -22.0 + 0.0474T$. Hence if all gases are at 1 atm partial pressure, the equilibrium is displaced to the left at temperatures over about $200°C$. Therefore the amount of NH_3 is appreciable only at high pressures, fairly low temperatures, and with abundant H_2.

CHAPTER 17

2 $Cu_2S + 4Cl^- + H_2O \rightleftarrows 2CuCl_2^- + HS^- + OH^-$; $\Delta G° = +52.0$ kcal
$\log K = -38.1 = 2 \log a_{CuCl_2-} + \log a_{HS-} + \log a_{OH-} - 4 \log a_{Cl-}$
At $1M$ NaCl, $a_{CuCl_2-} = 10^{-10.3}M$; at $10M$ NaCl, $a_{CuCl_2-} = 10^{-8.9}M$. Hence Cu_2S is not appreciably soluble in Cl^- at $25°$. Solubility increases as pH decreases.

3 To give $10^{-5}M$ $Pb(OH)_3^-$ would require $[OH^-] = 400M$ at $25°$; hence galena does not dissolve appreciably in alkaline solutions found in geologic environments.

4 For $MoS_2 + \frac{3}{2}O_2 \rightleftarrows MoO_3 + S_2$, $\Delta G° = -77.3$ kcal at $627°C$ and $\log K = +18.9 = \log [S_2] - \frac{3}{2} \log [O_2]$. If $\log [S_2] = -3.4$ and $\log [O_2] = -17$, $\log [S_2] - \frac{3}{2} \log [O_2] = 22.1$. This is greater than the equilibrium value, so the reaction goes to the left.

5 $SO_2 + 3H_2 \rightleftarrows 2H_2O + H_2S$, $\Delta G = -49.6 + 13.5T$
If $[H_2O] = 10^3$ atm and $[H_2] = 10^{-3}$ atm, the ratio $[H_2S]/[SO_2] = 10^{+5.0}$ at $200°$ and $10^{-1.9}$ at $400°C$.

6 For $MgCO_3$ and MgF_2, ratio $[CO_3^{2-}]/[F^-]^2$ must be greater than $10^{+2.4}$, or 250.

8 pH $= 6.3$; pressure $= 1,760$ atm.

CHAPTER 18

6 $a_{UO_2^{2+}} = 10^{-7.4}M$; $a_{U^{4+}} = 10^{-24.2}M$; $a_{UO_2OH^+} = 10^{-7.2}M$;
$a_{UO_2(CO_3)_2^{2-}} = 10^{-8.2}M$; $a_{UO_2(CO_3)_3^{4-}} = 10^{-12.1}M$.

INDEX